Audio and Video Signal Processing

Audio and Video Signal Processing

Edited by
Celia Lopez

Willford Press
www.willfordpress.com

Published by Willford Press,
118-35 Queens Blvd., Suite 400,
Forest Hills, NY 11375, USA

ISBN: 978-1-68285-336-8

Cataloging-in-publication Data

Audio and video signal processing / edited by Celia Lopez.
p. cm.
Includes bibliographical references and index.
ISBN 978-1-68285-336-8
1. Signal processing--Digital techniques. 2. Sound--Recording and reproducing--Digital techniques. 3. Streaming technology. 4. Streaming audio. 5. Streaming video. I. Lopez, Celia.
TK5102.9 .A83 2017
621.382 2--dc23

For information on all Willford Press publications
visit our website at www.willfordpress.com

Willford Press

Printed in China.

Contents

Preface

Audio and Video Signal Processing is the rapidly expanding field of study. It delves into the manipulation as well as alteration of audio and video signals, primarily through filters, to achieve desired results. The ever growing need of advanced technology is a reason that has fueled the research in this field in recent times. This book is a compilation of chapters that discusses the most vital concepts and emerging trends in the field of audio and video processing. It glances at various advancements in this area along with their applications as well as ramifications. This book will serve as a reference to a broad spectrum of readers including researchers, experts and students engaged in this field.

After months of intensive research and writing, this book is the end result of all who devoted their time and efforts in the initiation and progress of this book. It will surely be a source of reference in enhancing the required knowledge of the new developments in the area. During the course of developing this book, certain measures such as accuracy, authenticity and research focused analytical studies were given preference in order to produce a comprehensive book in the area of study.

This book would not have been possible without the efforts of the authors and the publisher. I extend my sincere thanks to them. Secondly, I express my gratitude to my family and well-wishers. And most importantly, I thank my students for constantly expressing their willingness and curiosity in enhancing their knowledge in the field, which encourages me to take up further research projects for the advancement of the area.

Editor

1

Localization based stereo speech source separation using probabilistic time-frequency masking and deep neural networks

Yang Yu[1*†], Wenwu Wang[2†] and Peng Han[1]

Abstract

Time-frequency (T-F) masking is an effective method for stereo speech source separation. However, reliable estimation of the T-F mask from sound mixtures is a challenging task, especially when room reverberations are present in the mixtures. In this paper, we propose a new stereo speech separation system where deep neural networks are used to generate soft T-F mask for separation. More specifically, the deep neural network, which is composed of two sparse autoencoders and a softmax regression, is used to estimate the orientations of the dominant source at each T-F unit, based on low-level features, such as mixing vector (MV), interaural level, and phase difference (IPD/ILD). The dataset for training the networks was generated by the convolution of binaural room impulse responses (RIRs) and clean speech signals positioned in different angles with respect to the sensors. With the training dataset, we use unsupervised learning to extract high-level features from low-level features and use supervised learning to find the nonlinear functions between high-level features and the orientations of dominant source. By using the trained networks, the probability that each T-F unit belongs to different sources (target and interferers) can be estimated based on the localization cues which is further used to generate the soft mask for source separation. Experiments based on real binaural RIRs and TIMIT dataset are provided to show the performance of the proposed system for reverberant speech mixtures, as compared with a model-based T-F masking technique proposed recently.

Keywords: Deep learning, Deep neural networks, Source separation, Soft mask

1 Introduction

Robust speech separation is an attractive research field and provides a useful front-end for many applications, e.g., hearing aids, mobile communication device, and automatic speech recognition system. Many methods have been applied to this problem, such as independent component analysis (ICA) [1–3], beamforming [4], and computation auditory scene analysis (CASA) [5, 6]. The performance of these algorithms, however, is still limited in complex acoustic environment, especially when room reverberation is present in the mixtures. This is in contrast to human auditory system which is skillful in listenting into a particular conversation in a cocktail party environment with the presence of background noise and interfering sound. There is a big performance gap between the human auditory system and machine-based listening system. An influential view in auditory scene analysis is that the human auditory system splits the sound mixtures to fragments (e.g., regions in the time-frequency plane), and the fragments which belong to the same acoustic source will be assigned to a same cluster. Based on this idea, time-frequency (T-F) masking technique has been proposed for speech source separation where the mask can be derived from various cues based on the analysis of temporal, spectral, or spatial features of the sources. Recently, a time-frequency masking technique has been proposed in [7] where the mixing vector (MV) [8] and interaural phase and level difference (IPD/ILD) [9] have been integrated using a Gaussian mixture model (GMM) whose

*Correspondence: nwpuyuy@nwpu.edu.cn
†Equal contributors.
[1]School of Marine Science and Technology, Northwestern Polytechnical University, Xi'an, China
Full list of author information is available at the end of the article

parameters are estimated iteratively using an expectation maximization (EM) algorithm. These methods provide a nice probabilistic framework for incorporating complementary information to deal with the uncertainties in T-F assignment. However, the performance of these algorithms is also limited by the accuracy of model fitting especially when room reverberation is present.

The GMM is essentially a shallow architecture of neural network which contains at most one layer of nonlinear feature transformation and is shown to offer good performance in source separation for anechoic mixtures [10] or mixtures with a relatively low level of reverberation. The shallow architecture, however, has a limited representation ability, which may cause performance degradation when applied in the complex real-world problems, such as speech separation in highly reverberant environments. Recent studies in speech recognition have shown that a deep architecture with more hidden layers can increase the representation abilities of a neural network, and it can be used to build internal representation for rich sensory data [11–14]. The deep architecture is regarded as being similar to the hierarchical structures within human visual and auditory systems, where the raw image or speech waveforms are transformed to a high-order linguistic level by these hierarchical structures [15–18]. The deep structure has the potential to reduce the performance gap between the human auditory system and machine listening system, as shown in recent works in the area of natural language processing and speech recognition systems [19–22]. The success of deep neural networks (DNNs) in these applications inspires us to investigate its potential for improving the performance of stereo speech source separation algorithms.

In this paper, we focus on the multiuser stereo speech source separation in reverberation environments and present a new approach for T-F assignment and mask estimation based on DNNs [23]. The network is trained with the low-level of features (i.e., MV and ILD/IPD) extracted from a training dataset of observed speech signals. In the separation stage, the trained network is used to estimate the orientations (i.e., directions of arrivals) of the target and interferers which is further exploited to derive the source occupation probability (and thereby the mask) at each T-F unit of the mixture. Our experimental results show that the proposed method performs significantly better than the GMM/EM-based baseline method [7] in terms of both signal to distortion ratio (SDR) and perceptual evaluation of speech quality (PESQ).

The remainder of the paper is organized as follows. Section 2 briefly discusses the related works. Section 3 outlines the proposed system. Section 4 discusses the low-level features to be used as inputs to the network. Section 5 presents the details about the deep network, including its structure, the training method, and how it is used for separation. Section 6 shows the experiments using real RIRs and TIMIT data before the conclusion is drawn in Section 7.

2 Relation to prior work

Several recent works have explored the potential of using DNNs for monaural/stereo speech separation. In [24], Wang et al. explored the use of monaural features for classification-based speech segregation. To deal with noise in the mixtures, a group Lasso approach and SVM classifier have been applied for generating the ideal binary mask for noise cancellation by combining different features. The experimental results show that (1) the complementary feature set is shown to give stable performance in experiments and outperforms each of its components significantly and (2) the unit-level features give better performance than frame-level features in unmatched test condition. In [25], Xu et al. presented a regression-based speech enhancement framework using DNNs, and the restricted Boltzmann machines (RBMs) have been used to learn a deep generative model for pre-training. They found that (1) using the large training dataset could result in a good generalization capability in mismatched testing conditions and (2) the two and three hidden layer DNNs have the similar performance. In [26], Narayanan and Wang proposed a feature enhancement algorithm for improving noise robustness of automatic speech recognition systems. The algorithm estimates a smoothed ideal ratio mask in the Mel spectrogram domain using DNNs, which is then used to filter out noise before cepstral transformation. In [27, 28], Huang et al. proposed to jointly optimize the deep learning models (deep neural networks and recurrent neural networks) with an extra masking layer to enforce a reconstruction constraint. They used a discriminative training criterion for the neural networks to further enhance the monaural source separation performance. In [29], Jin and Wang proposed a supervised learning approach to monaural segregation of reverberant speech using the multiresolution cochleagram (MRCG) features [30].

In [31], Jiang et al. first introduced DNNs to stereo speech separation. Similar to the work in [25] and [26], the RBMs were used to get the initial parameters of the DNNs and the output of the DNNs are the estimated ideal binary mask (IBM). They found that the DNN-based algorithm with joint binaural and monaural features could achieve better results than the representative binaural separation algorithms, especially when reverberation is present in the environment, and the target and interfering sources are either collocated or close to each other.

Compared with the monaural segregation of reverberant speech in [29], the stereo speech separation in [31] tends to be more robust due to the use of spatial information. In [7], GMM is used to model the MV and IPD/ILD cues that contain spatial information and the EM algorithm is used to estimate the model parameters and to derive the T-F mask. The combination of IPD/ILD and MV improves the separation quality as compared with the use of either only IPD/ILD or only MV and is achieved by using a coarse search to find the optimum set of weighting parameters which adjust the contribution of these cues. However, the optimum set of weighting parameters varies with different acoustic environment, i.e., the level of reverberation. In addition, the GMM used in [7] is a classical shallow architecture, and its representation ability is limited and can cause the performance degradation when the reverberation is present in the mixtures.

In this paper, similar to [7] and [31], we also consider multiuser stereo source separation problem. Instead of using GMM and EM or the RBMs, however, we use DNNs (composed of sparse autoencoders and softmax classifier) to estimate the source occupation likelihood at each T-F point. More specifically, the sparse autoencoders are used to learn the general model and the combined features—IPD/ILD and MV were used as the input of the DNNs. In other words, the low level features, i.e., IPD/ILD and MV, are now modelled with DNNs composed of sparse autoencoders and softmax classifier, and the output of the DNNs is an estimated soft mask (ratio mask). The network parameters are obtained through training by a greedy layer-wise training method [32] based on a training dataset containing observed speech signals (with one source speech signal placed at a different direction with respect to the microphones). With the trained sparse autoencoder and softmax classifer, we extract high-level features (i.e., spatial information of the sources) from the low-level features of the mixtures and generate the soft mask based on the softmax regression. The weighting parameters which are used to adjust the contribution of different cues will be learned automatically by the deep neural networks. Hence, different from [7, 31], we improve the separation quality by using the deep neural networks to find the optimum set of parameters and weighting the contributions of the cues (IPD/ILD, MV) automatically.

3 System overview

Our proposed system consists of the following four stages: (1) extraction of the low-level features (i.e. MV and ILD/IPD) (details in Section 4), (2) training of the deep networks (details in Section 5.1), (3) estimation of the probabilities that each T-F unit of the mixtures belongs to different sources and generation of the soft mask (details in Section 5.2), and (4) reconstruction of the target signal from the soft mask and the mixture signal. The system architecture is shown in Fig. 1. It should be noted that the neural nets are trained using isolated utterances (utterances originating from a single direction, i.e., clean speech utterances convolved with the binaural room impulse responses (BRIRs) corresponding to that direction) rather than the mixtures in stage (2).

The inputs to the system are the stereo (left and right) channel mixtures. We perform short-time Fourier transform (STFT) to both channels and obtain the T-F representation of the input signals, $X_L(m,f)$ and $X_R(m,f)$ where $m = 1, \cdots, M$ and $f = 1, \cdots, F$ are the time frame and frequency bin indices, respectively. The low-level features, i.e., MV and IPD/ILD, are then estimated at each T-F unit (details in Section 4). Next, we group the low-level features into N blocks (only along the frequency bins f). Each block includes K frequency bins, for example, the n-th block contains the bins $((n-1)K + 1, \cdots, nK)$, where $K = \frac{F}{N}$. We build N deep networks with each corresponding to one block and use them to estimate the direction of arrivals (DOAs) of the sources. Through unsupervised learning and the sparse autoencoder [11] in deep networks, high-level features (coded positional information of the sources) are extracted and used as inputs for the output layer (i.e., the softmax regression) of the networks. The output of softmax regression is a source occupation probability (i.e., the soft mask) of each block (through the ungroup operation, T-F units in the same block are assigned with the same source occupation probability) of the mixtures. Then, the sources can be recovered by the inverse STFT (ISTFT).

The key point of our proposed system is the training of deep networks and the generation of soft mask. From the view point of practical applications, we create a dataset of sensor signals with each containing only a single source (i.e., source speech convolved with RIRs) from different directions with respect to the sensors for training and use the orientations of single source as the ground truth (described in Section 5.2). With the training dataset, the deep networks are trained by using a greedy layer-wise training method [32]. With the trained deep networks, we can get the probability of each T-F block of the input mixtures associated with different DOAs. Using a predefined threshold, we can estimate the number of sources, the DOAs of the sources, and a matrix of probabilities which we call "Probability Mask" in Fig. 1. Through the ungroup operation, we assign the same probability to T-F units that are belonging to the same block. Then, we obtain the soft mask for speech separation from the Probability Mask.

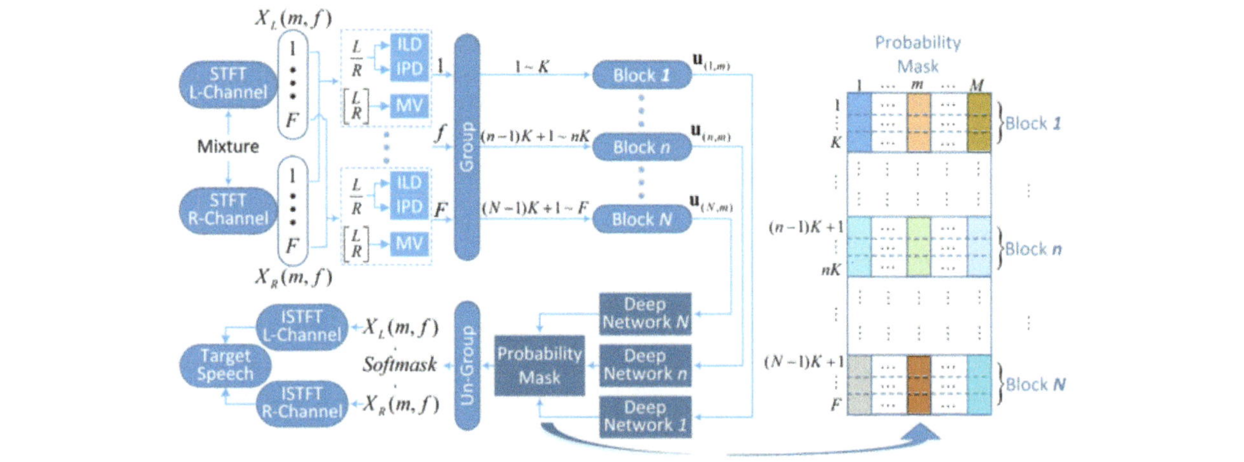

Fig. 1 System architecture for deep neural network-based time-frequency masking for stereo source separation

The N deep networks in our proposed system have the same architecture, and the details about the architecture and training method can be found in Section 5. Next, we discuss the low-level features used in our proposed system.

4 The low-level features for localization based separation

Many features can be used for stereo speech separation, such as IPD or ITD [33], ILD or interaural intensity differences (IID) [33], and the MV cue. It is widely acknowledged that ITD or IPD tends to be more robust in the low frequency range, whereas ILD or IID is more robust in the high-frequency range [34]. In [7], Alinaghi et al. found that the MV cues are more distinct compared to binaural cues (IPD/ILD) for the sources placed close to each other, whereas binaural cues IPD/ILD offer better separation results when the sources are distant from each other. These observations motivated Alinaghi et al. to combine these cues, introducing a new robust algorithm to improve the speech separation quality. We follow the work in [7] and use IPD/ILD and MV as the low-level features and the inputs to the neural networks. The nonlinear relationship between the source occupation probabilities and the input low-level features can be found by the deep networks and thus has the potential to further improve the speech separation quality. These low-level features are used to derive high-level features to be classified by sparse autoencoders.

The MV and the IPD/ILD cues can be calculated from the mixtures.

The MV [8] can be derived as

$$\mathbf{z}(m,f) = \frac{\mathbf{W}(f)\widetilde{\mathbf{x}}(m,f)}{\left\|\mathbf{W}(f)\widetilde{\mathbf{x}}(m,f)\right\|} \tag{1}$$

with $\widetilde{\mathbf{x}}(m,f) = \frac{[X_L(m,f),X_R(m,f)]^T}{\left\|[X_L(m,f),X_R(m,f)]^T\right\|}$, where $\mathbf{W}(f)$ is a whitening matrix, with each row being one eigenvector of $E(\widetilde{\mathbf{x}}(m,f)\widetilde{\mathbf{x}}^H(m,f))$, the superscript H is Hermitian transpose, and $\|\bullet\|$ is Frobenius norm.

ILD and IPD are the phase and amplitude difference between the left and right channel and calculated as follows [9]:

$$\alpha(m,f) = 20\log_{10}\left(\left|\frac{X_L(m,f)}{X_R(m,f)}\right|\right) \tag{2}$$

$$\phi(m,f) = \angle\left(\frac{X_L(m,f)}{X_R(m,f)}\right) \tag{3}$$

where $|\bullet|$ takes the absolute value of its argument, and $\angle(\bullet)$ finds the phase angle.

Concatenating the MV and ILD/IPD features, a feature vector can be obtained at each T-F unit, which is $\widetilde{\mathbf{u}}(m,f) = \left[\widehat{\mathbf{z}}^T(m,f), \alpha(m,f), \phi(m,f)\right]^T \in \mathbb{R}^6$. Since the inputs to the DNNs are real numbers, we use the real part and imaginary part of $\mathbf{z}$ as the features, i.e., $\widehat{\mathbf{z}}(m,f) = \left[\mathrm{Re}(z_L(m,f)), \mathrm{Im}(z_L(m,f)), \mathrm{Re}(z_R(m,f)), \mathrm{Im}(z_R(m,f))\right]$. Then, we group all the feature vectors $\widetilde{\mathbf{u}}(m,f)$ into N blocks (only along the frequency bins). For each block, we get a $6K$-dimensional feature vector $\mathbf{u}_{(n,m)} = \left[\widetilde{\mathbf{u}}^T(m,(n-1)K+1), \cdots, \widetilde{\mathbf{u}}^T(m,nK)\right]^T \in \mathbb{R}^{6K}$, where K is the number of the frequency bins in each block, as the input to the deep networks.

5 The deep networks

As described in Section 3, we group the low-level features into N blocks and build N individual deep networks which have the same architecture to classify the DOAs of the current input mixture in each block. The architecture of the deep network is shown in Fig. 2 and composed of deep autoencoder [11] and softmax classifier. More specifically,

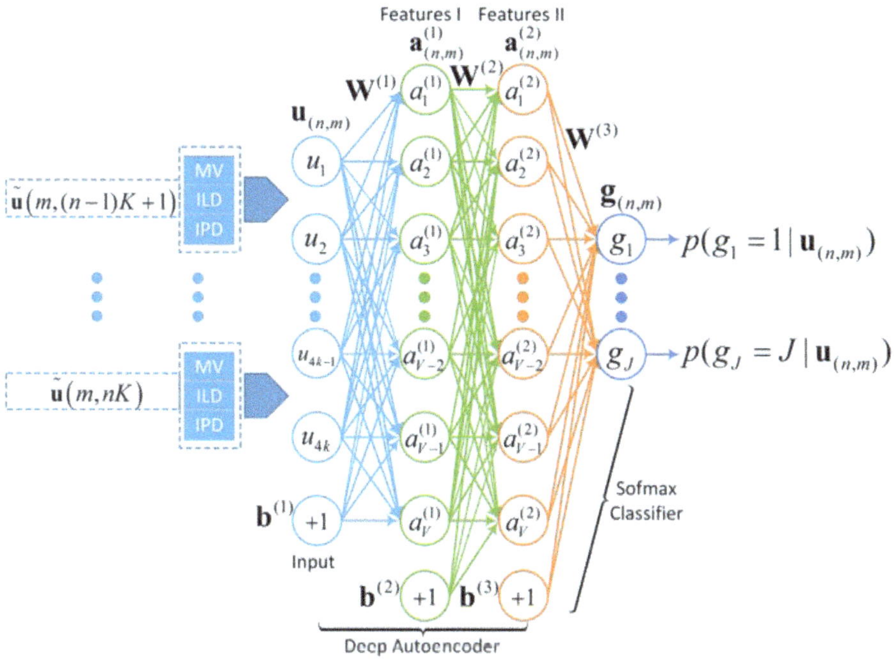

Fig. 2 The architecture for the deep neural networks in our proposed system. The deep neural networks are composed by the deep autoencoder for high-level feature extraction and the softmax classifier for soft mask generation and was trained by the greedy layer-wise training method

for stereo speech separation task, the target location is a natural choice of the output of the network. As shown in Fig. 3, we split the whole space to J ranges with respect to the sensors and separate the target and interferers based on different orientation ranges (DOAs with respect to the listener) where they are located. We apply the softmax classifier (to be discussed in Section 5.2) to perform the classification task and the inputs to the classifier, i.e., the high-level features: $\mathbf{a}^{(2)}$ which are extracted from the low-level features (ILD/IPD and MVs), are produced by the deep autoencoder. Assuming that the position of the target in the current input sample remains unchanged, the deep network estimates the probability $p\left(g_j = j|\mathbf{u}_{(n,m)}\right)$ of the orientation of the current input sample belonging to the orientation index j, where g_j is the j-th output unit of the network and the $\mathbf{u}_{(n,m)}$ is the m-th input sample of the n-th block (group). With the estimated orientation (obtained by selecting the maximum probability index) of each input sample, we cluster the samples which have the same orientation index to get the probability mask and obtain the soft mask from the probability mask through the ungroup operation. Note that each T-F unit in the same block is assigned the same probability. The number of sources can also be estimated from the probability mask by using a predefined probability threshold, typically chosen as 0.1 in our experiments (we only considered two or three sources and found empirically this value to be suitable).

5.1 Deep autoencoder

An autoencoder (shown in Fig. 4) is an unsupervised learning algorithm based on backpropagation. It aims to learn an approximation $\widehat{\mathbf{u}}$ of the input $\mathbf{u}$. It appears to be learning a trivial identity function; but by using some constraints on the learning process, such as limiting the number of neurons activated (sparsity constraint), it discloses some interesting structures about the data [11, 35, 36]. As shown in Fig. 4, the output of the autoencoders can be defined as $\widehat{\mathbf{u}} = \text{sigm}\left(\widehat{\mathbf{W}}^{(2)}\mathbf{a}^{(1)} + \widehat{\mathbf{b}}^{(2)}\right)$ with $\mathbf{a}^{(1)} = \text{sigm}\left(\mathbf{W}^{(1)}\mathbf{u} + \mathbf{b}^{(1)}\right)$, where the function $\text{sigm}(\mathbf{u}) = \frac{1}{1+\exp(-\mathbf{u})}$ is the logistic function, $\mathbf{W}^{(1)} \in \mathbb{R}^{V\times Y}$, $\mathbf{b}^{(1)} \in \mathbb{R}^{V}$, $\widehat{\mathbf{W}}^{(2)} \in \mathbb{R}^{Y\times V}$, and $\widehat{\mathbf{b}}^{(2)} \in \mathbb{R}^{Y}$, V is the number of hidden layer neurons, and Y is the number of input layer neurons, which is the same as that of the output layer neurons [37]. In our proposed system, we set $Y = 6K$, where K is the number of frequency bin in each block (group).

With the sparsity constraint, most of the neurons in the autoencoder are assumed to be inactive. More specifically, $\widehat{\rho}_v = \frac{1}{M}\sum_{m=1}^{M} a_v\left(\mathbf{u}_{(m)}\right)$ is the average activation of unit v with the m-th input sample $\mathbf{u}_{(m)}$, where $a_v = \text{sigm}\left(\mathbf{w}_v^{(1)\prime}\mathbf{u} + b_v^{(1)}\right)$ denotes the activation value of the hidden layer unit v in the autoencoder and M is the number of training samples. With the sparsity constraint $\widehat{\rho}_v =$

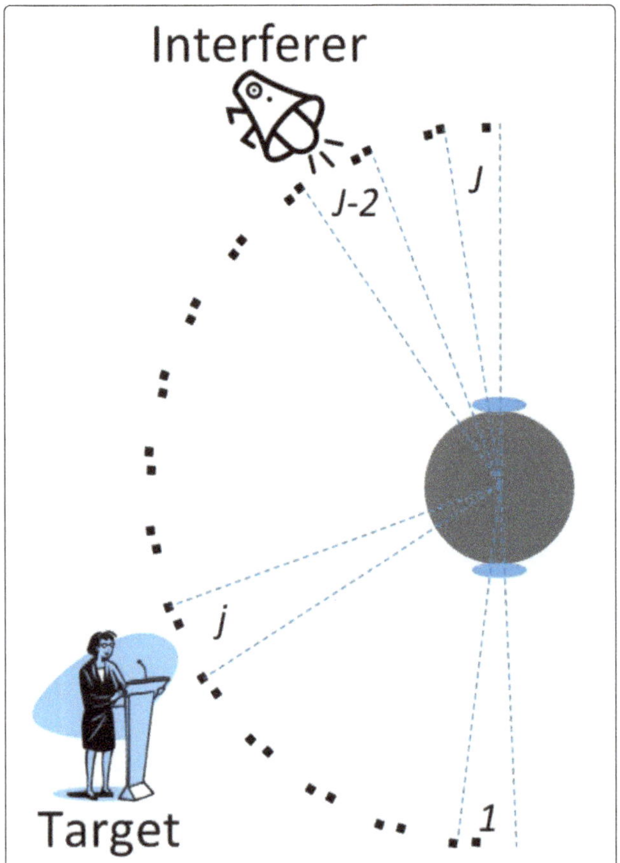

Fig. 3 Split the space to J ranges. We split the whole space to J ranges with respect to the sensors, and separate the target and interferers based on different orientation ranges (DOAs with respect to the listener) where they are located

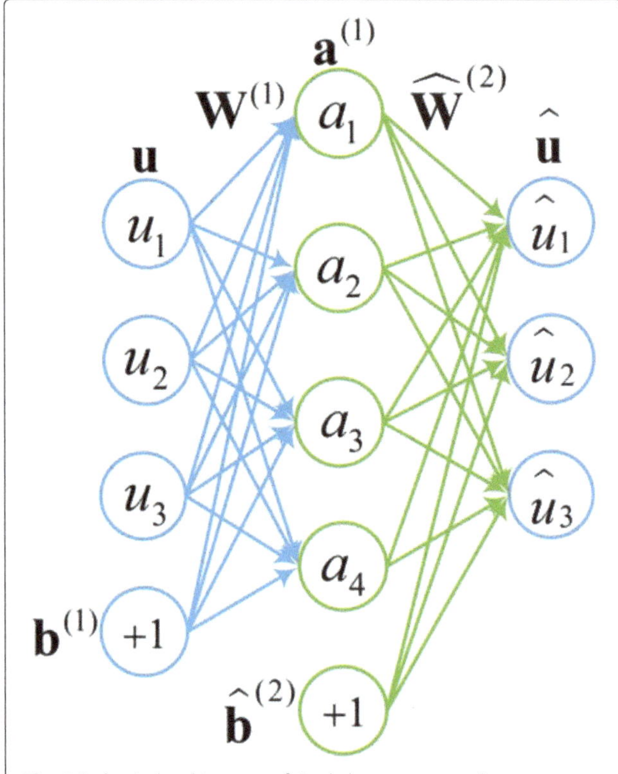

Fig. 4 A classical architecture of single layer autoencoder

ρ, the cost function $J_{\text{sparse}}(\mathbf{W}, \mathbf{b})$ of the sparse autoencoder can be written as follows

$$J_{\text{sparse}}(\mathbf{W}, \mathbf{b}) = \frac{1}{2}\|\mathbf{u} - \widehat{\mathbf{u}}\|^2 + \beta \sum_{v=1}^{V} KL(\rho \parallel \widehat{\rho}_v)$$
$$\sum_{v=1}^{V} KL(\rho \parallel \widehat{\rho}_v) = \sum_{v=1}^{V} \rho \log \frac{\rho}{\widehat{\rho}_v} + (1-\rho) \log \frac{1-\rho}{1-\widehat{\rho}_v} \tag{4}$$

where β controls the weight of the penalty term and ρ is a parameter preset before training, typically very small [37]. More details about the sparse autoencoder can be found in online lecture notes [38, 39].

In our proposed system, the cost function $J_{\text{sparse}}(\mathbf{W}, \mathbf{b})$ is minimized using the limited memory BFGS (L-BFGS) optimization algorithm [40, 41] and the single-layer sparse autoencoder is trained by using the backpropagation algorithm.

After the finishing of the training of single-layer sparse autoencoder, we discard the output layer neurons, the relative weights $\widehat{\mathbf{W}}^{(2)}$, bias $\widehat{\mathbf{b}}^{(2)}$, and only save the input layer neurons $\mathbf{W}^{(1)}$ and $\mathbf{b}^{(1)}$. The output of the hidden layer—$\mathbf{a}^{(1)}$ are used as the input samples of the next single-layer sparse autoencoder. We can build a deep autoencoder by repeating these steps and stacking two or more layers of independently trained sparse autoencoders. The stacking procedure is shown on the right side of Fig. 5. The features II shown on figure are the high-level features and can be used as the training dataset for the softmax regression discussed next.

Many studies on deep autoencoders have shown that with the deep architecture (more than one hidden layer), more complex representation can be obtained from the simple low-level features. As a result, the underlying regularities of the data can be captured, leading to better performance, e.g., in recognition [23]. This motivates us to use deep autoencoder (two hidden layers) in our proposed system.

5.2 Softmax classifier

In our proposed system, the softmax classifier [37], based on softmax regression, was used to estimate the probabilities of the current input, i.e., the m-th sample $\mathbf{u}_{(n,m)}$ in the n-th block, belonging to the orientation index j, by

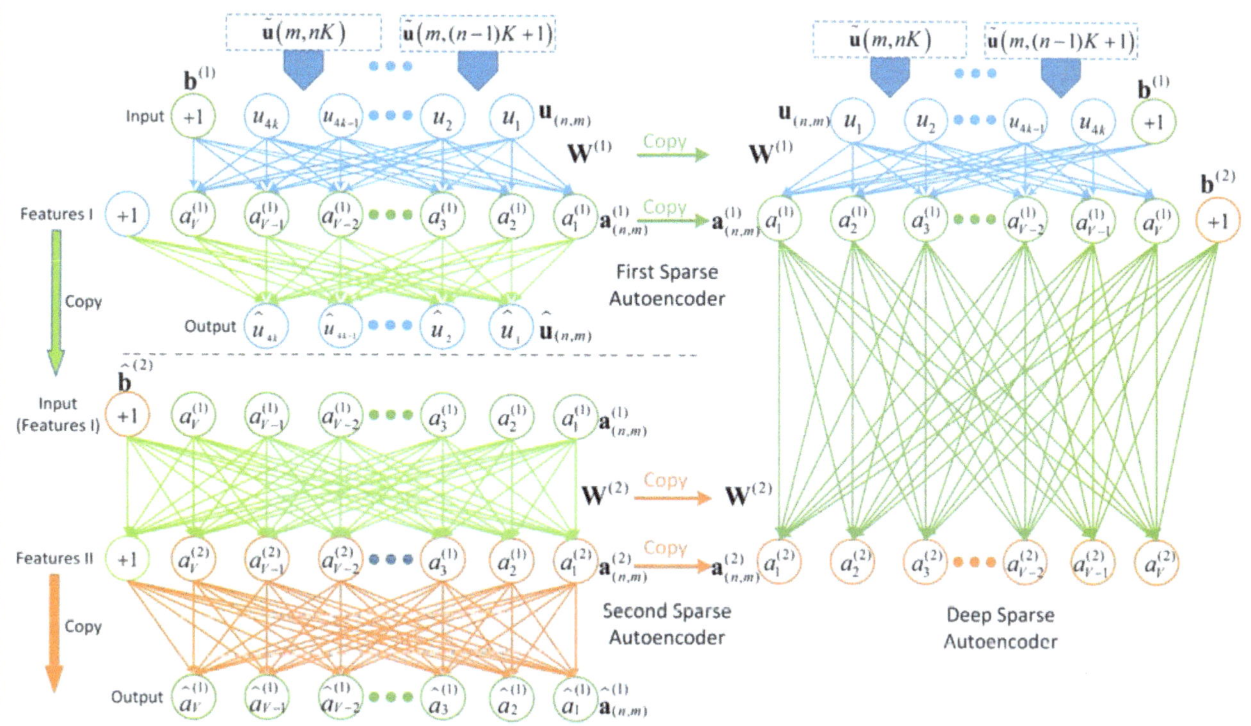

Fig. 5 The illustration of greedy layer-wise training and stacking. The *left part* of the figure shows the procedure of greedy layer-wise training, and the *right part* of the figure shows the procedure of stacking of sparse autoencoders

the deep autoencoder with the extracted high-level features $\mathbf{a}^{(2)}_{(n,m)}$ as inputs of the classifier. The architecture of the softmax classifier we used is shown in Fig. 6. In our proposed system, we represent the label of the training dataset as a one-hot vector (with 1 for the target class and 0 for others): $\mathbf{g}_{(n,m)} \in \mathbb{R}^J$. Then, the cross-entropy loss (cost function) of the softmax classifier can be written as follows:

$$J_{\text{softmax}}(\mathbf{W}^{(3)}) = -\frac{1}{M}\left[\sum_{m=1}^{M}(\mathbf{g}_{(n,m)})^T \mathbf{h_W}\left(\mathbf{a}^{(2)}_{(n,m)}+\mathbf{b}\right)\right] + \frac{\lambda}{2}\sum_{j=1}^{J}\sum_{i=1}^{I}(w_{j,i})^2$$

$$\mathbf{h_W}\left(\mathbf{a}^{(2)}_{(n,m)}\right) = \frac{1}{\sum_{j=1}^{J} e^{\mathbf{w}_j\left(\mathbf{a}^{(2)}_{(n,m)}+\mathbf{b}\right)}}\left[e^{\mathbf{w}_1^T\left(\mathbf{a}^{(2)}_{(n,m)}+\mathbf{b}\right)} \cdots e^{\mathbf{w}_J^T\left(\mathbf{a}^{(2)}_{(n,m)}+\mathbf{b}\right)}\right]^T \quad (5)$$

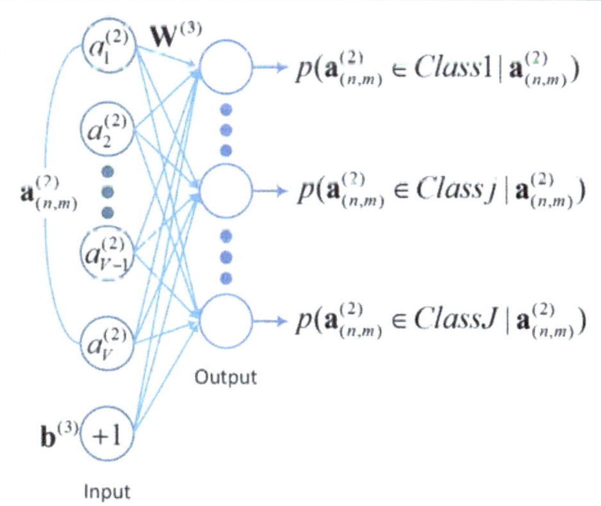

Fig. 6 A classical architecture of the softmax classifier

The softmax classifier can be trained by using the L-BFGS algorithm based on a dataset, in order to find an optimal parameter set $\mathbf{W}^{(3)}$ for minimizing the cost function $J_{\text{softmax}}\left(\mathbf{W}^{(3)}\right)$. In our proposed system, the dataset for softmax classifier training is composed by two parts. The first part is the input sample—$\mathbf{a}^{(2)}_{(n,m)}$ (features II), calculated from the last hidden layer of the deep autoencoder. The second part is the data label—$\mathbf{g}_{(n,m)} \in \mathbb{R}^J$, where the j-th element—g_j of $\mathbf{g}_{(n,m)}$ will be set to 1 when the input sample belongs to the source located in the range of DOAs of index j.

5.3 Stacking deep autoencoder and softmax classifier

We stack the softmax classifier and deep autoencoder together after the training is completed, as shown on the left part of Fig. 7. Finally, we use the training dataset and the L-BFGS algorithm to fine-tune the deep network with the initialized parameters $\mathbf{W}^{(1)}$, $\mathbf{b}^{(1)}$, $\mathbf{W}^{(2)}$, $\mathbf{b}^{(2)}$, $\mathbf{W}^{(3)}$, $\mathbf{b}^{(3)}$ obtained from the sparse autoencoders and softmax classifier training. The training phase of the sparse autoencoders and softmax classifier are called pre-training phase, and the stacking/training of the overall network, i.e., deep network, is called fine-tuning phase. In the pre-training phase, the shallow neural networks, i.e., sparse autoencoders and softmax classifier, are training individually, using the output of current layer as the input for the next layer. In the fine-tuning phase, we use the L-BFGS algorithm (i.e., a gradient descent method) to minimize the difference between the output of the deep network and the label of the training dataset. The gradient descent works well because the initialized parameters obtained from the pre-training phase include a significant amount of "prior" information about the input data through unsupervised learning [11].

6 Experiments

In this section, we first describe the generation of the datasets for training and testing and the setup of the training parameters of the deep networks. Similar to [7], different sentences from different speakers were convolved with real BRIRs to generate the stereo mixtures with room effects. The algorithms in [7] and [42] are used as baselines. We then apply both our proposed system and the basline algorithms to these mixtures to separate the target source. The separation quality is evaluated in terms of both signal distortion and perceptual speech quality.

6.1 Dataset generation

Similar to [7], the datasets that we used for training and testing are generated by the convolution of the original speech signal with real BRIRs. The original speech sources (target and interferer) were randomly selected from the TIMIT dataset which is a continuous speech corpus containing 6300 sentences: 10 sentences spoken by each of 630 speakers from 8 major dialect regions of the USA [43]. 10 sentences spoken by 2 female speakers, who were randomly selected from the training usage set of the TIMIT, as the training dataset; another 30 sentences spoken by 4 male and 2 female speakers, who were randomly selected from the test usage set of the TIMIT, as the test dataset, where 10 sentences spoken by 2 males as the target source, 5 sentences spoken by 1 male, and 5 sentences spoken by 1 female as the interferer 1; and the remaining sentences which were spoken by 1 male and 1 female as the interferer 2. Details about the sentences and speaker IDs can be found in Table 1. All the sentences were normalized to have equal root mean square magnitude and cut to a same length (about 2.6 s for each sentence) for consistency.

The BRIR datasets used in our experiments were recorded using a dummy head and torso in five different types of room, named as X, A, B, C, and D at the University of Surrey, measured by Hummersone [44] and can be download from the website [45]. Room X is a very large room, and the reflections were truncated in the recordings

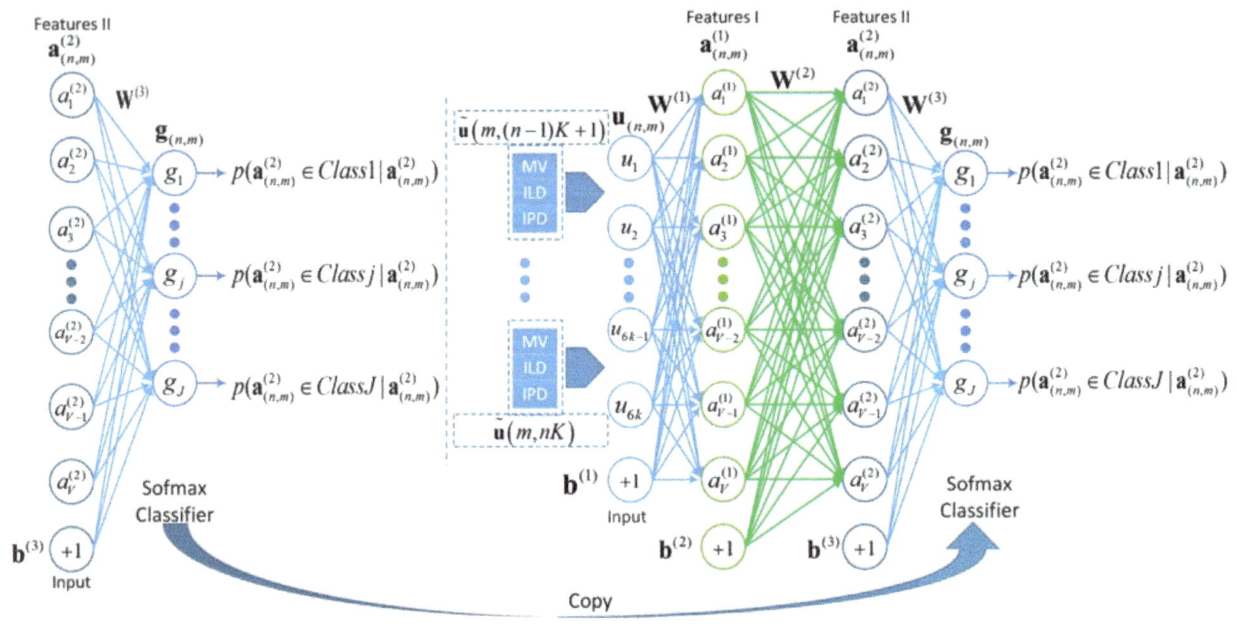

Fig. 7 The illustration of stacking deep autoencoder and softmax classifier

Table 1 Details about the speakers and the sentences, including speakers and sentences ID, their genders, dialect regions (DR), used in training or test dataset and used as the target or interferers

ID	Sex	DR	Sentence ID					Dataset	Source type
TBRO	F	1	21	201	381	111	291	Training	/
MEMO	F	1	297	333	207	387	117	Training	/
BJKO	M	2	95	275	5	185	365	Testing	Target
RMS1	M	7	407	137	317	47	227	Testing	Target
CTTO	M	5	28	208	388	118	298	Testing	Interferer 1
PKTO	F	3	8	188	368	98	278	Testing	Interferer 1
BDGO	M	3	383	113	293	23	203	Testing	Interferer 2
UTBO	F	5	124	304	34	214	394	Testing	Interferer 2

to produce anechoic recordings. We aim to evaluate the speech separation quality in reverberant environments. For this reason, in our experiments, we only used the BRIRs recorded in rooms A, B, C, and D. Different from other similar datasets, such as [46], this dataset has higher angular resolution and many different acoustic properties, which enabled us to evaluate the performance of the system over different acoustic environments with finer resolution. Table 2 shows the different acoustical properties of the rooms used in our evaluation. In each room, acoustic sources were placed 1.5 m away from the dummy head and had the same height as the dummy head, and the head related transfer function (HRTF) is applied in the BRIRs to mimic sound sources that would have been heard by human ears.

In our experiments, the training dataset, used to train the deep networks, is generated by the convolution of real BRIRs with the clean speech signals of two randomly selected speakers from the TIMIT dataset. More specifically, we use the speech signals observed at the microphones with a single source placed in different orientations with respect to the microphones, rather than the mixtures, to train the DNNs, and the orientations of the source are used as the ground truth. We consider speakers of different genders in training and test dataset for the evaluation of the generalization ability of the proposed system. More specifically, the training data set is generated by the convolution of clean New England female speech signals with all the real BRIRs (from $-90°$ to $+90°$ with a step of $5°$), and the sentences spoken by the male speakers from a different dialect region are (DR in Table 1) are used as the target source in the test set.

Table 2 Room acoustic properties in initial time delay gap (ITDG), direct-to-reverberant ratio (DRR), and reverberation time (T60)

Room	Type	ITDG (ms)	DRR (dB)	T60 (s)
A	Medium office	8.73	6.09	0.32
B	Small class room	9.66	5.31	0.47
C	Large lecture theatre	11.9	8.82	0.68
D	Large seminar room	21.6	6.12	0.89

Different from the training dataset, the test set is composed by mixtures and used for the evaluation of speech separation quality, including determined and underdetermined cases, i.e., for two sources (target and interferer 1) and three sources (target, interferer 1 and interferer 2) with just two microphones as receivers. More specifically, similar to [7], the mixtures in the test dataset were generated by adding the reverberant target and interfering signals together which is equivalent to assuming superposition of their respective sound fields. The target and interfering signals are the randomly selected sentences from different male and female speakers, each convolved with the real BRIRs. For the determined case, the target source was located at $0°$ azimuth, and the azimuth of interferer 1 is varied from $-90°$ to $+90°$ with the step of $5°$. For the underdetermined case, we add the speech signals from interferer 2 which was located at $30°$ to the mixtures of the determined case.

6.2 Experimental setting

Even though all the sources (including both the target and interferers) at different azimuths are recovered in our proposed system, the performance of the system is reported based on the quality of the recovered target located at $0°$ azimuth, with the azimuths of the interferers varied from $-90°$ to $+90°$ with step of $5°$, similar to [7]. The sampling rate f_s used in signal sampling, STFT and ISTFT operation was 16 kHz (f_s = 16 kHz). We used a Hanning window of 2048 (128 ms) samples with 75 % overlap between the neighboring windows for the STFT. The frequency grouping parameters K and N are set to 16 and 128, respectively. Hence, we use 128 deep networks to generate the soft mask, with each deep network corresponding to a block. For each deep network, the input layer includes 96 units and $V = 256$ neurons for each of the hidden layers. $J = 37$ neurons were used in the output layer, corresponding to azimuths from $-90°$ to $+90°$ with a step of $5°$.

The learning parameters are set as follows, the weight decay parameter $\lambda = 1 \times 10^{-4}$, the weight of the penalty term $\beta = 3$, and the sparsity parameter $\rho = 4 \times 10^{-3}$. The maximum number of iterations is set to 300. The parameters for the training of the softmax classifier are set as follows. The weight decay parameter $\lambda = 1 \times 10^{-4}$ and the maximum number of iterations was set to 200. In the fine-tuning phase, the weight decay parameter was changed to $\lambda = 3 \times 10^{-3}$.

For speech separation performance evaluation, we consider SDR [47] and PESQ [48] and the algorithms in [7, 42] as the baseline. In the evaluation, we consider both determined and undetermined cases and test the performance of our proposed system in different reverberation conditions, spatial diffuse noises, training dataset conditions, unseen rooms, block size K, and network types. The separation results including the comparison with the baseline methods are shown in Section 6.3.

6.3 Experimental results

In this section, we first test the performance of the proposed system under different training dataset configurations (the training set with full or half of the azimuths as discussed earlier) and different levels of reverberation for determined and underdetermined cases. Finally, we present the separation results for the mixtures corrupted by different levels of spatially diffuse noises (SNR = 5 and 10 dB).

6.3.1 Reverberation effect

We use the four reverberant rooms, i.e., rooms A, B, C and D, to evaluate the performance. The acoustical properties of the four rooms can be found in Table 2 in Section 6.1.

Figure 8 presents the SDRs of the separated signals with different DOAs of the interferer 1 and different rooms, for the determined case, where the deep networks used for soft mask generation were trained by the training set. Compared with the two baseline methods, we obtain at least 2-dB improvement in rooms B and D, when the interfering speech is placed far away from the target source. However, we obtain similar performance to the baseline in rooms A and C. It can be seen that with different reverberation times (T60s) and direction to reverberation ratios (DRRs), the proposed system performs generally more robust than the baseline methods and the performance of the proposed system does not decrease as much as the baseline methods when the level of room reverberation increases. Similar to [7], it can be seen that the separation

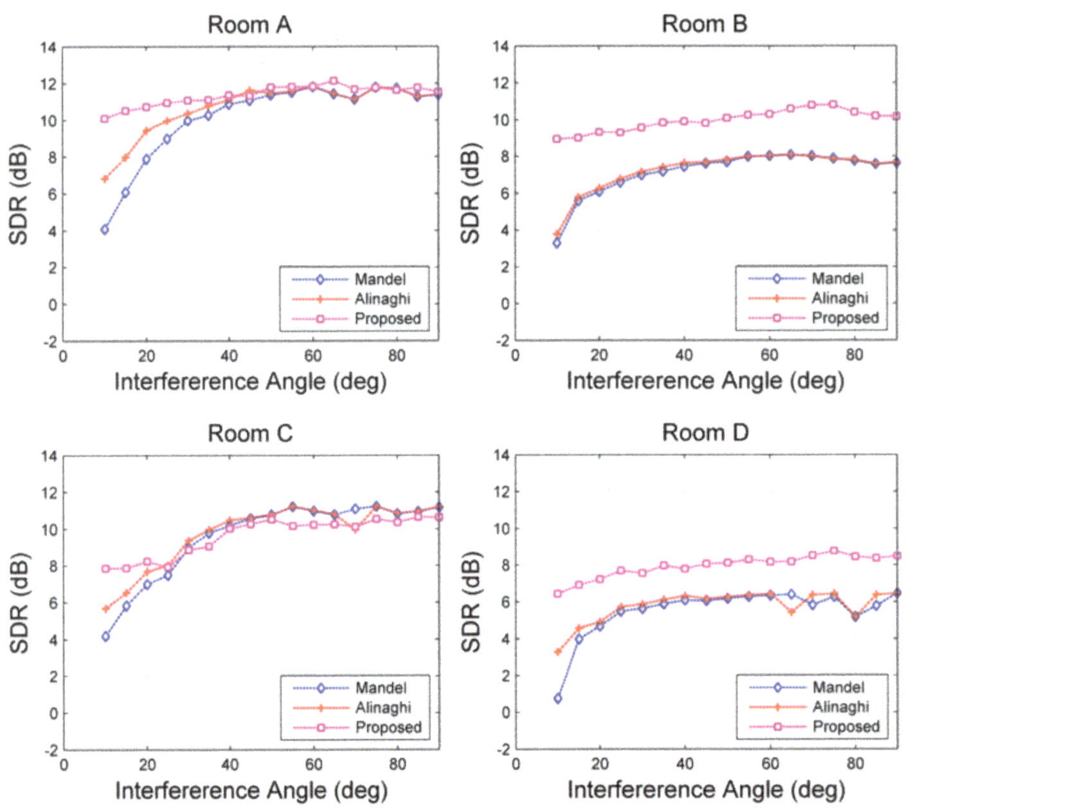

Fig. 8 SDRs performance comparison between the proposed system and the baseline methods among the rooms A, B, C, and D, in determined case, without noise

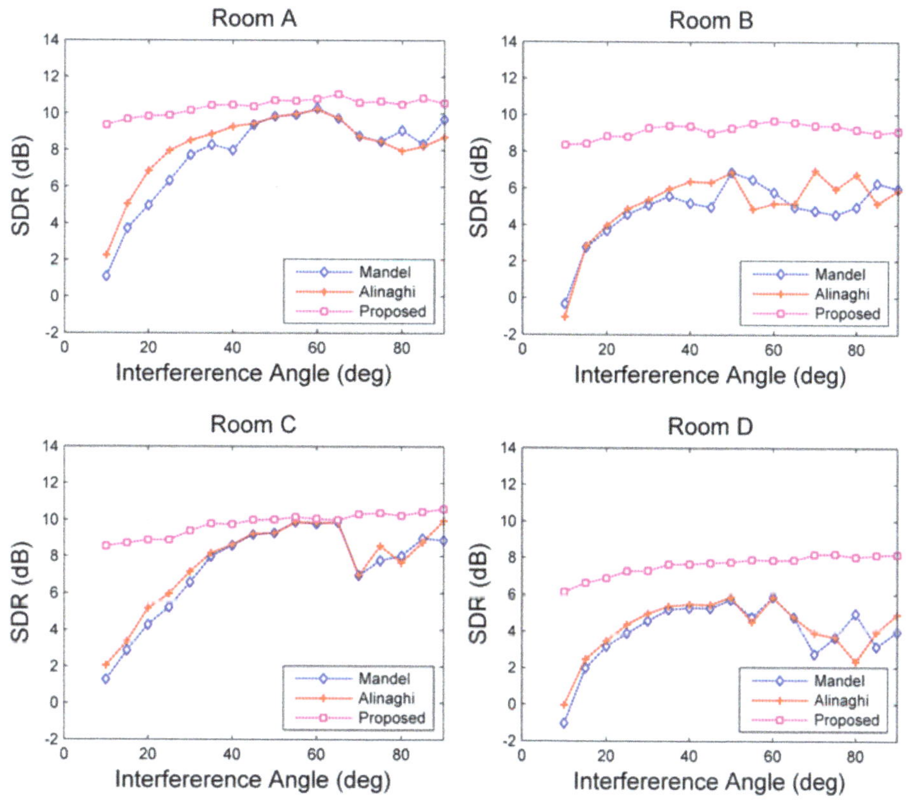

Fig. 9 SDRs performance comparison between the proposed system and the baseline methods among the rooms A, B, C, and D, in underdertermined case, without noise

quality of our proposed system depends on the acoustic parameters T_{60} and DRR.

The separation result for the underdetermined case is presented in Fig. 9. It can be seen that, compared with the two baseline methods, we obtain about 1-dB improvement for rooms B and D, and similar performance for rooms A and C, except for the situation that the target and interferer are close to each other. Compared with Fig. 8, it can be seen that the SDRs of the proposed system decrease for about 1 dB and the performance of our proposed system decreases with the increase in the number of the sources within the mixtures. Compared with Fig. 8, it can be seen that the SDRs of the proposed system decrease for about 2 dB and the PESQs of the proposed system decrease about 0.5. It can be seen that, similar to the baseline methods, the performance of our proposed system decreases with the increase in the number of sources within the mixtures.

From Figs. 8 and 9, we see that the proposed system is more robust to the acoustic parameters, i.e. the DRRs and T60s than the baseline methods, with at least 1 dB improvement in SDR. A summary of the PESQ results is represented in Table 3.

The comparison between the proposed system and the baseline methods suggests that the deep networks are able to provide more robust estimation results for the time-frequency mask even though blocking was used in our system.

6.3.2 *Spatially diffuse noise*

Similar to [7], we also evaluated the performance of the proposed system in the case of the mixtures corrupted by spatially diffuse noise. Same as Section 6.3.1, we repeat the experiments, but adding two different levels of noise

Table 3 Results of the baseline methods and the proposed method, for reverberant mixtures with the average over rooms A, B, C, and D in perceptual evaluation of speech quality (PESQ)

Case	Methods	Room A	Room B	Room C	Room D	Mean
Determined	Mandel	2.34	2.07	2.34	1.96	2.18
	Alinaghi	2.39	2.11	2.38	2.01	2.22
	Propsoed	2.34	2.22	2.30	2.14	2.25
Underdetermined	Mandel	2.1	1.85	2.14	1.81	1.98
	Alinaghi	2.15	1.87	2.18	1.84	2.01
	Propsoed	2.06	1.97	2.19	1.85	2.02

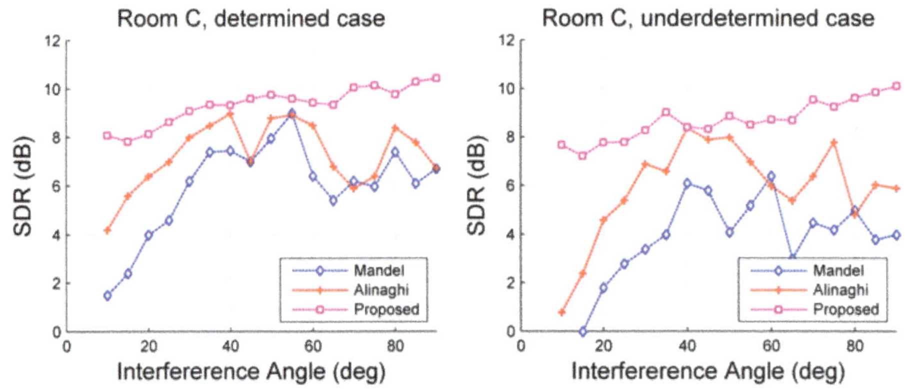

Fig. 10 SDR performance comparison between the proposed system and baseline methods in room C, with the SNR = 10 dB, for the determined and underdetermined cases

in the mixtures, i.e., the signal-to-noise-ratios (SNRs) were set to 5 and 10 dB (with respect to the mixture), respectively.

Figure 10 presents the SDRs comparison between the proposed system and the baseline methods in room C, with the SNR = 10 dB, for the determined and underdetermined cases. It can be seen that, for the determined case, the proposed system gives about 1 dB improvement in all of the azimuths, and for the underdetermined case, it also gives about 1 dB improvement in most of the azimuths. Similar to the results without noise, the performance of the deep network-based time-frequency masking technique also decreases with the increase in the number of sources presented in mixtures.

Figure 11 presents the SDRs for determined and underdetermined cases with SNR = 5 or 10 dB of the mixtures. We see that the SDRs of the proposed system decrease

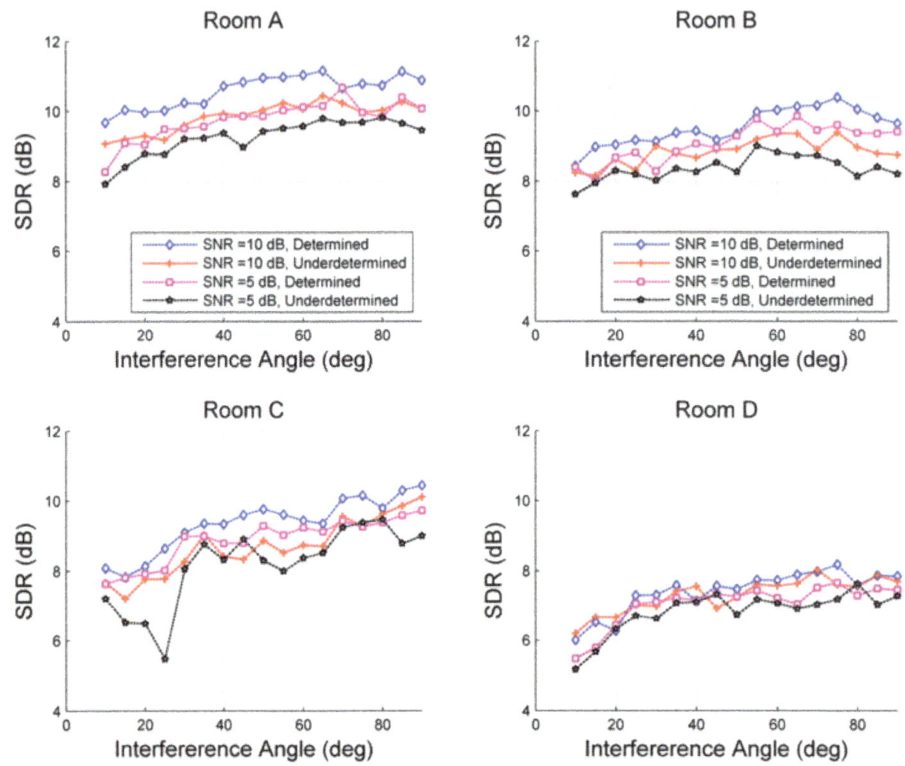

Fig. 11 SDRs for determined and underdetermined cases with SNR = 5 dB or 10 dB of the mixtures

about 1.5 dB in determined case and about 1 dB in underdetermined case when the SNR of mixtures is varied from 10 to 5 dB. Furthermore, as compared with the SDRs in Section 6.3.1 without noise, we can see that there is only about 1 dB performance drop when the SNR = 10 dB.

The Fig. 12 shows a separation example for room D, including the spectrogram of the mixture signals (Fig. 13a), original target source signals (Fig. 13b), separated target source signals (Fig. 13d), and the soft mask for separation (Fig. 13c), with the deep networks trained using the full training set, for the determined case (the interferer 1 was located at +15°).

6.3.3 *Generalization to different rooms*

In this subsection, we consider the generalization performance of our proposed system to unseen rooms in the determined and underdetermined cases. To this end, we selected each of the BRIRs recording from the four rooms in turn to generate the training set and use all the BRIRs to generate the test set. For instance, as shown in the top left plot of Fig. 13, we choose the BRIR of the room A to generate the training dataset and use all the BRIRs to generate the test dataset. The interferer is varied from −90° to +90° with a step of 5°. As shown in Fig. 13 (determined case) and Fig. 14 (underdetermined case), the system that was trained by the BRIRs of room D got the best generalization performance and the system trained by the BRIR of room A got the worst. Consider the different acoustic properties of these four rooms, we could find that the generalization performance of the proposed system increases with the complexity of the acoustic properties of the room. Compared with Figs. 8 and 9, the SDR performance of the proposed system trained by room D decreases about 4 dB in rooms A and C and increases about 1 dB in rooms B and D, both in determined and underdetermined cases.

6.3.4 *Evaluation in different block size K*

As mentioned in Section 3, we group K frequency bins to a block and use a corresponding DNN to generate a probability from the input features for each block. In this subsection, we evaluate the effect of different block size K for the determined case. As shown in Fig. 15, the system gives the best performance when the block size $K = 4$ and the SDR performance decreases with the increase of the K. However, we chose $K = 8$ in our proposed system,

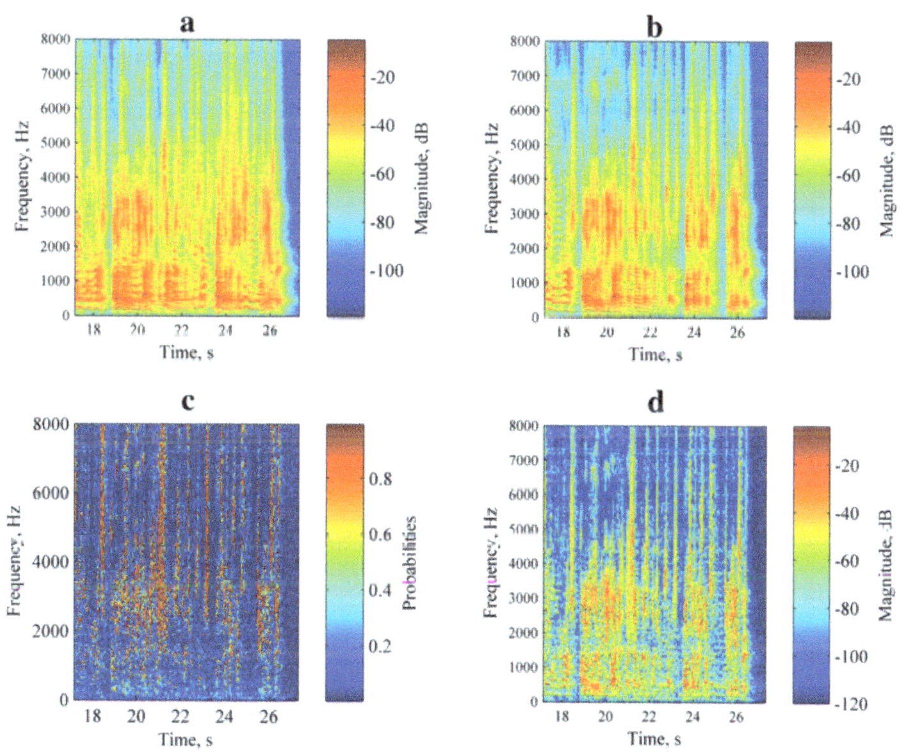

Fig. 12 A separation example for room D, including the spectrogram of the mixture signals, original target source signals, separated target source signals, and soft mask for separation. The mixtures include two sources (interferer 1 was located at +15°), without the noise. **a** Magnitude spectrogram of the mixture signals. **b** Magnitude spectrogram of the original target source signals. **c** The soft mask for separation. **d** Magnitude spectrogram of separated target source signals

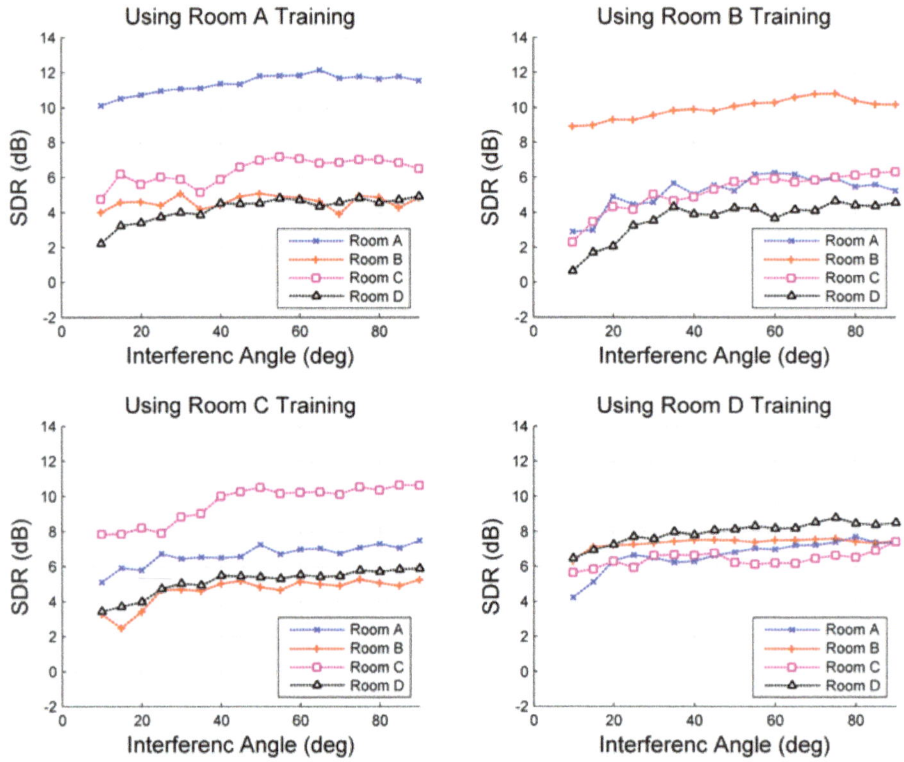

Fig. 13 SDRs to unseen rooms, in determined case. We select each of the four rooms to generate the training dataset and use the BRIRs from the four rooms (one by one respectively) to generate the test dataset. The interferer is varied from −90° to +90° with the step of 5°

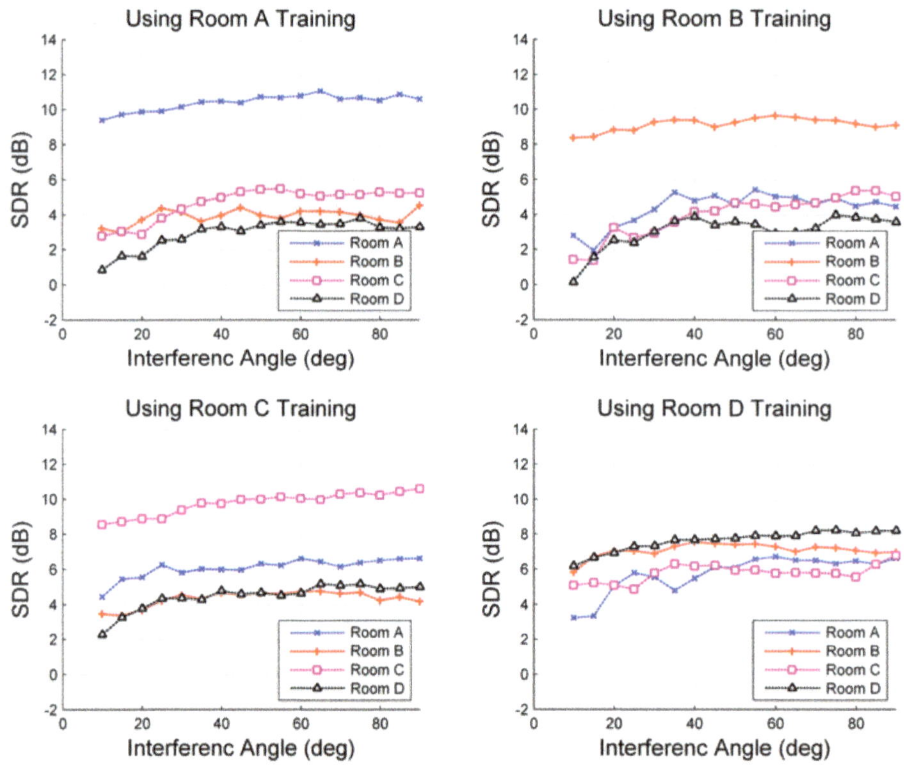

Fig. 14 SDR to unseen rooms, in underdetermined case. We select each of the four rooms to generate the training dataset and use the BRIRs from the four rooms (one by one respectively) to generate the test dataset. The interferer is varied from −90° to +90° with the step of 5°

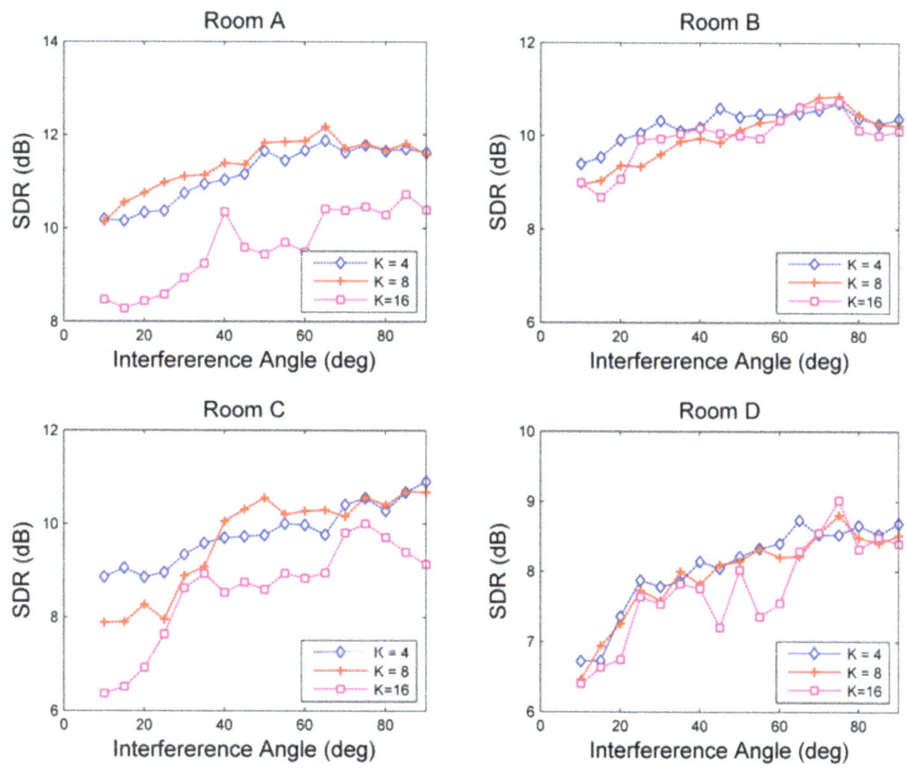

Fig. 15 SDR performance vs. different *K*, in determined case

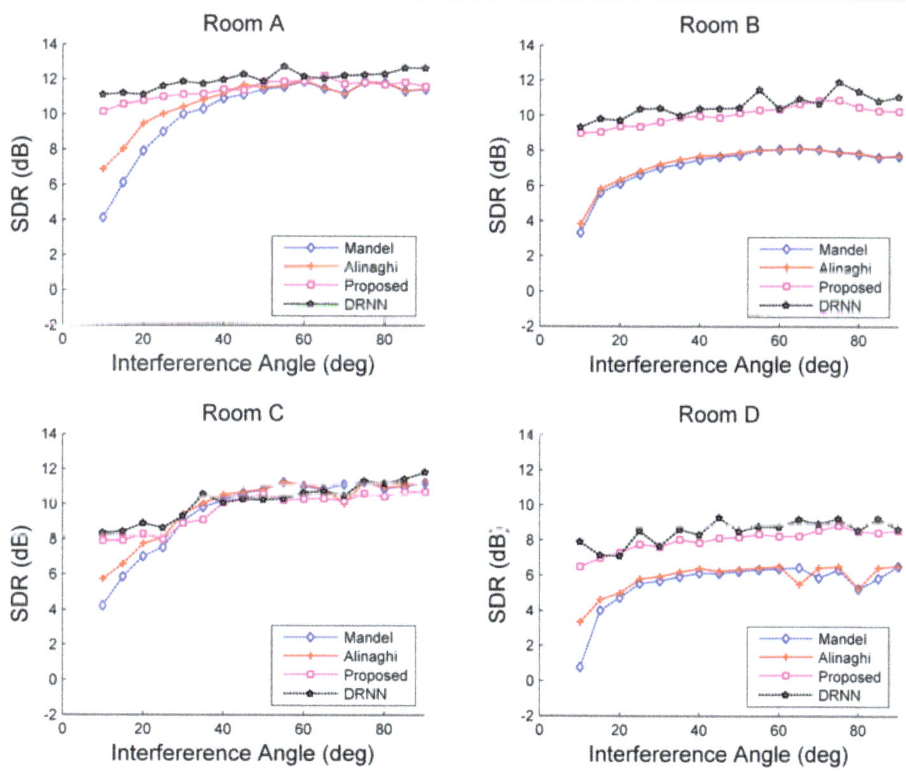

Fig. 16 SDRs comparisons among the proposed system, the RNNs method, Mandel method, and Alinaghi method, in underdetermined case

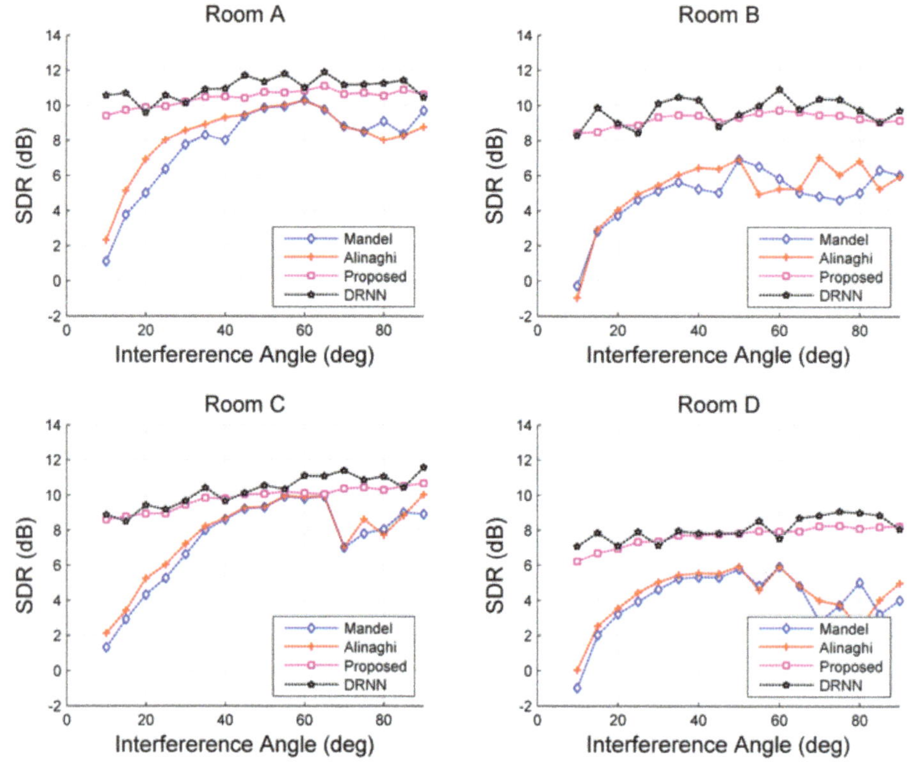

Fig. 17 SDRs comparisons among the proposed system, Mandel method, Alinaghi method, and the RNNs method, in underdetermined case

for the similar SDR performance and less computational complexity.

6.3.5 Evaluation in different neural network type

Room reverberation effect on speech signal can be regarded as signal extension in time. From this viewpoint, the recurrent neural networks (RNNs) may perform better than the DNNs in dealing with reverberation. In this subsection, we evaluate the use of the deep recurrent neural networks (DRNNs) in our system, instead of using DNNs. The DRNNs which were originally used by Huang et al. for monaural speech separation [27, 28] are employed here. The differences between the proposed system and the method in [27, 28] reside in the training dataset and ground truth. More specifically, we use the orientations of the sources as the ground truth and the isolated observed speech signals as training dataset, instead of using the separated source speech and mixture as the ground truth and training dataset. As shown in Figs. 16 and 17, we compare the SDR performance among the proposed system (DNNs method), the RNNs method, the Mandel method, and the Alinaghi method in four rooms, for the determined and underdetermind cases. It can be seen that the RNNs method get the best performance in all rooms, with about 1 dB improvement over the DNNs method. It is worth noting that the computational complexity of the RNNs method appears to be high and deserves further study in our future work.

7 Conclusions

We have presented a new localization-based stereo speech separation system using deep networks. Compared with GMM/EM-based algorithm in [7, 42], the deep network-based techniques provide better results in SDR and PESQ when room reverberation is presented in the mixtures. It is also shown that they are robust to spatially diffuse noise. In our future work, it would be interesting to compare the proposed method with other existing deep network-based separation algorithms such as [30, 31].

Competing interests

The authors declare that they have no competing interests.

Acknowledgements

This study was performed when the first author was an academic visitor in the Center for Video Speech and Signal Processing, University of Surrey, and he wishes to thank Qingju Liu, Philip JB Jackson, and Atiyeh Alinaghi for providing help for issues related to the algorithm in [7]. This research was supported partially by the Natural Science Basis Research Plan in Shaanxi Province of China (Program No.2014JQ8355). The authors wish to thank the anonymous reviewers for their helpful comments in improving this paper.

Author details

[1] School of Marine Science and Technology, Northwestern Polytechnical University, Xi'an, China. [2] Centre for Vision Speech and Signal Processing, University of Surrey, Guildford, UK.

References

1. P Comon, C Jutten, *Handbook of blind source separation: independent component analysis and applications*. (Academic Press, New York, 2010)
2. A Hyvärinen, J Karhunen, E Oja, *Independent component analysis*, vol. 46. (Wiley, New York, 2004)
3. A Hyvärinen, E Oja, Independent component analysis: algorithms and applications. Neural Netw. **13**(4), 411–430 (2000)
4. BD Van Veen, KM Buckley, Beamforming: a versatile approach to spatial filtering. IEEE ASSP Mag. **5**(2), 4–24 (1988)
5. D Wang, GJ Brown, *Computational auditory scene analysis: principles, algorithms, and applications*. (Wiley-IEEE Press, New York, 2006)
6. GJ Brown, D Wang, in *Speech Enhancement*. Separation of speech by computational auditory scene analysis (Springer, Berlin Heidelberg, 2005), pp. 371–402
7. A Alinaghi, PJ Jackson, Q Liu, W Wang, Joint mixing vector and binaural model based stereo source separation. IEEE/ACM Trans. Audio Speech Lang. Process. **22**(9), 1434–1448 (2014). doi:10.1109/TASLP.2014.2320637
8. H Sawada, S Araki, S Makino, Underdetermined convolutive blind source separation via frequency bin-wise clustering and permutation alignment. IEEE Trans. Audio Speech Lang. Process. **19**(3), 516–527 (2011)
9. MI Mandel, RJ Weiss, DP Ellis, Model-based expectation-maximization source separation and localization. IEEE Trans. Audio Speech Lang. Process. **18**(2), 382–394 (2010)
10. G Kim, Y Lu, Y Hu, PC Loizou, An algorithm that improves speech intelligibility in noise for normal-hearing listeners. J. Acoust. Soc. Am. **126**(3), 1486–1494 (2009)
11. Y Bengio, A Courville, P Vincent, Representation learning: a review and new perspectives. IEEE Trans. Pattern Anal. Mach. Intel. **35**(8), 1798–1828 (2013). doi:10.1109/TPAMI.2013.50
12. J Schmidhuber, Deep learning in neural networks: an overview. ArXiv e-prints (2014). arxiv: url1404.7828
13. D Yu, L Deng, Deep learning and its applications to signal and information processing [exploratory dsp]. IEEE Signal Process. Mag. **28**(1), 145–154 (2011). doi:10.1109/MSP.2010.939038
14. Y Bengio, Learning deep architectures for ai. Found. Trends® Mach. Learn. **2**(1), 1–127 (2009)
15. J Baker, L Deng, J Glass, S Khudanpur, C-H Lee, N Morgan, D O'Shaughnessy, Developments and directions in speech recognition and understanding, part 1 [dsp education]. IEEE Signal Process. Mag. **26**(3), 75–80 (2009)
16. J Baker, L Deng, S Khudanpur, C-H Lee, JR Glass, N Morgan, D O'Shaughnessy, Updated minds report on speech recognition and understanding, part 2 [dsp education]. IEEE Signal Process. Mag. **26**(4), 78–85 (2009). doi:10.1109/MSP.2009.932707
17. L Deng, in *Computational Models of Speech Pattern Processing*. Computational models for speech production (Springer, Berlin Heidelberg, 1999), pp. 199–213
18. L Deng, in *Mathematical Foundations of Speech and Language Processing*. Switching dynamic system models for speech articulation and acoustics (Springer, New York, 2004), pp. 115–133
19. L Deng, J Li, J-T Huang, K Yao, D Yu, F Seide, M Seltzer, G Zweig, X He, J Williams, Y Gong, A Acero, in *Proceedings of the IEEE International Conference on Acoustics, Speech and Signal Processing*. Recent advances in deep learning for speech research at microsoft, (2013), pp. 8604–8608. doi:10.1109/ICASSP.2013.6639345
20. J-T Huang, J Li, D Yu, L Deng, Y Gong, in *Proceedings of the IEEE International Conference on Acoustics, Speech and Signal Processing*. Cross-language knowledge transfer using multilingual deep neural network with shared hidden layers, (2013), pp. 7304–7308. doi:10.1109/ICASSP.2013.6639081
21. SM Siniscalchi, J Li, C-H Lee, Hermitian polynomial for speaker adaptation of connectionist speech recognition systems. IEEE Trans. Audio Speech Lang. Process. **21**(10), 2152–2161 (2013). doi:10.1109/TASL.2013.2270370
22. SM Siniscalchi, D Yu, L Deng, C-H Lee, Exploiting deep neural networks for detection-based speech recognition. Neurocomputing. **106**, 148–157 (2013)
23. G Hinton, L Deng, D Yu, GE Dahl, A Mohamed, N Jaitly, A Senior, V Vanhoucke, P Nguyen, TN Sainath, B Kingsbury, Deep neural networks for acoustic modeling in speech recognition: the shared views of four research groups. IEEE Signal Process. Mag. **29**(6), 82–97 (2012). doi:10.1109/MSP.2012.2205597
24. Y Wang, K Han, D Wang, Exploring monaural features for classification-based speech segregation. IEEE Trans. Audio Speech Lang. Process. **21**(2), 270–279 (2013)
25. Y Xu, J Du, L-R Dai, C-H Lee, An experimental study on speech enhancement based on deep neural networks. Signal Process. Letters, IEEE. **21**(1), 65–68 (2014)
26. A Narayanan, D Wang, in *2013 IEEE International Conference on Acoustics, Speech and Signal Processing (ICASSP)*. Ideal ratio mask estimation using deep neural networks for robust speech recognition (IEEE, 2013), pp. 7092–7096
27. P-S Huang, M Kim, M Hasegawa-Johnson, P Smaragdis, in *2014 IEEE International Conference on Acoustics, Speech and Signal Processing (ICASSP)*. Deep learning for monaural speech separation (IEEE, 2014), pp. 1562–1566
28. P Huang, M Kim, M Hasegawa-Johnson, P Smaragdis, Joint optimization of masks and deep recurrent neural networks for monaural source separation. IEEE/ACM Trans. Audio Speech Lang. Process. **PP**(99), 1–1 (2015). doi:10.1109/TASLP.2015.2468583
29. Z Jin, D Wang, A supervised learning approach to monaural segregation of reverberant speech. IEEE Trans. Audio Speech Lang. Process. **17**(4), 625–638 (2009). doi:10.1109/TASL.2008.2010633
30. J Chen, Y Wang, D Wang, in *Proceedings of the IEEE International Conference on Acoustics, Speech and Signal Processing*. A feature study for classification-based speech separation at very low signal-to-noise ratio, (2014), pp. 7039–7043. doi:10.1109/ICASSP.2014.6854965
31. Y Jiang, D Wang, R Liu, Z Feng, Binaural classification for reverberant speech segregation using deep neural networks. IEEE/ACM Trans. Audio Speech Lang. Process. **22**(12), 2112–2121 (2014). doi:10.1109/TASLP.2014.2361023
32. Y Bengio, P Lamblin, D Popovici, H Larochelle, Greedy layer-wise training of deep networks. Adv. Neural Inf. Process. Syst. **19**, 153 (2007)
33. J Blauert, *Spatial hearing: the psychophysics of human sound localization*, (1997)
34. FL Wightman, DJ Kistler, The dominant role of low-frequency interaural time differences in sound localization. J. Acoust. Soc. Am. **91**(3), 1648–1661 (1992)
35. M Ranzato, FJ Huang, Y-L Boureau, Y LeCun, in *Proceedings of the IEEE International Conference on Computer Vision and Pattern Recognition*. Unsupervised learning of invariant feature hierarchies with applications to object recognition, (2007), pp. 1–8. doi:10.1109/CVPR.2007.383157
36. M Ranzato, Unsupervised learning of feature hierarchies. PhD thesis, NEW YORK UNIVERSITY (2009)
37. CM Bishop, et al., *Pattern recognition and machine learning*, vol. 1. (Springer, New York, 2006)
38. A Ng, Lecture notes on sparse autoencoders. http://web.stanford.edu/class/cs294a/sparseAutoencoder.pdf
39. A Ng, Sparse autoencoder. CS294A Lecture notes. **72**, 1–19 (2011)
40. J Dean, G Corrado, R Monga, K Chen, M Devin, M Mao, A Senior, P Tucker, K Yang, QV Le, et al., in *Advances in Neural Information Processing Systems 25 (NIPS 2012)*. Large scale distributed deep networks, (NIPS, 2012), pp. 1223–1231
41. Y Chauvin, DE Rumelhart, *Backpropagation: theory, architectures, and applications*. (Psychology Press, London, 1995)
42. MI Mandel, RJ Weiss, DPW Ellis, Model-based expectation-maximization source separation and localization. Audio Speech Lang. Process. IEEE Trans. **18**(2), 382–394 (2010). doi:10.1109/TASL.2009.2029711
43. JS Garofolo, LF Lamel, WM Fisher, JG Fiscus, DS Pallett, DARPA TIMIT acoustic-phonetic continous speech corpus CD-ROM. NIST speech disc n.1. NASA STI/Recon Technical Report N. **93**, 27403 (1993)
44. C Hummersone, A psychoacoustic engineering approach to machine sound source separation in reverberant environments. PhD thesis, University of Surrey (2011)
45. Binaural room impulse responses captured in real rooms. http://iosr.surrey.ac.uk/software/index.php
46. BG Shinn-Cunningham, N Kopco, TJ Martin, Localizing nearby sound sources in a classroom: binaural room impulse responses. J. Acoust. Soc. Am. **117**(5), 3100–3115 (2005)

47. E Vincent, R Gribonval, C Fevotte, Performance measurement in blind audio source separation. IEEE Trans. Audio Speech Lang. Process. **14**(4), 1462–1469 (2006). doi:10.1109/TSA.2005.858005
48. L Di Persia, D Milone, HL Rufiner, M Yanagida, Perceptual evaluation of blind source separation for robust speech recognition. Signal Process. **88**(10), 2578–2583 (2008)

2

JND-based spatial parameter quantization of multichannel audio signals

Li Gao[1,2,3], Ruimin Hu[1,2,3*], Xiaochen Wang[1,2,3], Gang Li[1,2,3], Yuhong Yang[1,2,3] and Weiping Tu[1,2,3]

Abstract

In multichannel spatial audio coding (SAC), the accurate representations of virtual sounds and the efficient compressions of spatial parameters are the key to perfect reproduction of spatial sound effects in 3D space. Just noticeable difference (JND) characteristics of human auditory system can be used to efficiently remove spatial perceptual redundancy in the quantization of spatial parameters. However, the quantization step sizes of spatial parameters in current SAC methods are not well correlated with the JND characteristics. It results in either spatial perceptual distortion or inefficient compression. A JND-based spatial parameter quantization (JSPQ) method is proposed in this paper. The quantization step sizes of spatial parameters are assigned according to JND values of azimuths in a full circle. The quantization codebook size of JSPQ was 56.7 % lower than one of the quantization codebooks of MPEG surround. Average bit rate reduction on spatial parameters for standard 5.1-channel signals reached up to approximately 13 % compared with MPEG surround, while preserving comparable subjective spatial quality.

Keywords: Spatial audio coding, Multichannel signals, Spatial parameters, Just noticeable difference

1 Introduction

Along with the trend towards high-quality audio, audio systems have evolved through mono, stereo, to multichannel audio systems. The multichannel audio systems developed from 5.1 to 22.2 (NHK) [1], 64 (Dolby Atmos [2]) channels loudspeaker systems and other multichannel audio systems with even more loudspeakers (e.g., loudspeaker systems with Ambisonics [3] or WFS [4]). With more loudspeakers configured in three-dimensional (3D) space, current multichannel audio systems can freely reproduce virtual sound image in 3D space. However, with the increasing number of loudspeakers, it will bring challenges to the efficient storage and transmission of large amounts of channel data in current multichannel audio systems. Take the 22.2-channel audio system for example, the data rate will reach 28 Mbps before data compression. Even with general efficient audio codec such as MP3 with data rate 128 kbps used to compress all the channel data, the data amounts will still reach up to 10 G bits for an hour's 22.2-channel signals, which are often not affordable for the storage device and transmission bandwidth in real application. So high-efficient compression schemes for multichannel audio signals play an important role on popularization of current 3D multichannel audio systems.

Spatial audio coding (SAC) [5] has been an ongoing research topic in recent years for the high-efficient compression performance. SAC represents two or more channel signals as one or two downmix signals, accompanied with spatial parameters extracted from channels to model spatial attributes of the auditory scene generated by original channel signals. With the increasing channel number, the total amount of coded information does not notably increase in SAC compared with conventional stereo or multichannel audio coding schemes.

In multichannel audio systems, the virtual sounds are widely distributed in space and a large amount of complex spatial information is contained. It is not that easy to find a few finite spatial parameters to comprehensively represent all spatial attributes of the auditory scene in multichannel system. Many aspects might be very important, such as the basic audio quality, auditory spatial image, auditory

*Correspondence: hurm1964@gmail.com
[1] State Key Laboratory of Software Engineering, Wuhan University, Wuhan, China
[2] National Engineering Research Center for Multimedia Software, the Key Laboratory of Multimedia and Network Communication Engineering, Computer School, Wuhan University, Wuhan, China
Full list of author information is available at the end of the article

object location/width, ambience, listener envelopment, etc. [6]. Many processes in multichannel system might add artifacts to these aspects and need to be optimized, such as multichannel downmixing, quantization of the downmix, and 3D reproduction. For example, the downmix signals in SAC are closely related to both basic and spatial audio quality and often encoded by an outstanding perceptual encoder such as AAC [7] to ensure basic quality.

Although there may be many complex spatial parameters associated with spatial quality, and the exact spatial location of perceived sound may be not the most important aspect in a multichannel system, the accurate representation and efficient compression of spatial information of virtual sound is still very important to the perfect and vivid reconstruction of spatial sound effects in SAC. In the general SAC methods, spatial parameters mainly include spatial parameters between two channel signals such as inter-channel level difference (ICLD) and inter-channel time difference (ICTD) [5], as well as spatial parameters among more than two channel signals such as azimuths of virtual sound images, which are closely related to the spatial location of virtual sound.

In recent years, lots of research achievements in spatial psychological acoustics model have contributed a lot to the development of SAC. The spatial perceptual features of human auditory system, for example, the perceptual sensitivity limits of human auditory system, have played important roles in the perceptual redundancy removal of spatial parameter quantization in SAC. The change of sound property (for instance, the direction or location of sound) has to reach a certain amount to be detectable for the human auditory system, and such required minimum detectable change is often referred to as just noticeable difference (JND) [8–11]. Although not all the spatial perceptual aspects of the human auditory system have the JND characteristics, and for most spatial perceptual aspects the JND characteristics are still not very clear at present, many researches have been done on the directional perceptual JND characteristics of the human auditory system. And the relevant research results have been used in the perceptual redundancy removal of spatial parameter quantization in general SAC methods. Thus, with the perceptual characteristics of JND, if the spatial distortion caused by quantization errors of spatial parameters can be limited below the JND thresholds, then the quantization is almost perceptual lossless.

2 Related work

The spatial perceptual features of human auditory system have played important roles in the perceptual redundancy removal of spatial parameters in SAC. Early in 1877, Strutt has stated that binaural cues take the leading role in the human auditory system in the judgment of sound direction in horizontal plane [12]. Binaural cues are mainly due to the comprehensive filter action of reflection, absorption, and attenuation in the transmission process of sound; thus, the sounds that arrive at humans' two ears are different from each other and also different with the original sound. The major binaural cues include interaural level difference (ILD), and iInteraural time difference (ITD). In the case of playback with headphones, the signals played out of the headphones are almost the same with the signals that arrive at the human ears. Thus, the inter-channel cues (also called spatial parameters, such as ICLD) are almost the same with inter-aural cues (also called binaural cues).

As for the auditory scene reproduced by stereo loudspeaker system, without considering the transmitting procedures of sound from loudspeakers to ears, inter-channel cues can be regarded as approximations of binaural cues. For example, one of the most important spatial parameters ICLD can be regarded as an approximation of interaural level difference (ILD) [11]. Then the spatial perceptual features of binaural cues (such as JND of ILD) can be used as a reference in the quantization of spatial parameters (such as ICLD) to remove perceptual redundancy. These spatial hearing characteristics of human auditory system were used in the first SAC framework Binaural Cue Coding (BCC) to reconstruct spatial effect in stereo signal with a bit rate of only 2 kbps for spatial parameters [13, 14].

However, the spatial parameters are uniformly quantized in BCC, which means that the quantization step sizes between different quantization codewords are the same. But in fact, the perceptual sensitivities for different cues in human auditory system are not the same. For example, the JNDs of ILDs with different values are not the same [15]. When ILD is 0 dB, this ILD has the smallest JND that is about 0.5 dB. As absolute ILD increases, JND increases too. For ILD of 15 dB, the JND amounts to about 2 dB. So as not to introduce audible spatial distortion, the quantization errors of spatial parameters should be controlled below JNDs. Different spatial parameters with different JNDs should be assigned with different quantization step sizes.

Ignoring the transmitting procedures from sound source to ears with earphone or stereo audio system, ICLD can be regarded as an approximation of ILD. In this case, the JND characteristic of ILD can be referred in the quantization codebook design of ICLD. Thus, in parametric stereo (PS), the uniform quantization of spatial parameters are replaced by nonuniform quantization according to the perceptual sensitivity attributes of interaural cues. Different quantization step sizes are set between quantization codewords. For example, small quantization steps are assigned to ICLDs with small JNDs. For ICLD value of 0 dB, the quantization step size is the smallest. As for bigger ICLD values, the quantization step sizes are bigger.

With the nonuniform quantization of spatial parameters, PS was known as one of the most efficient stereo coding schemes and was standardized as MPEG-4 AAC Plus v2 or 3GPP Enhanced AAC+ (E-AAC+) [16, 17].

SAC was extended from stereo to multichannel signals and standardized as ISO/MPEG multichannel coding standard MPEG surround [15]. With two kinds of elementary coding blocks one-to-two (OTT) or two-to-three (TTT), multichannel signals are combined pairwise or per three channel signals to form coding tree structures. OTT/TTT referring to two/three channel signals are represented as one/two downmix accompanied with extracted spatial parameters. Each OTT block has the same coding procedure as in PS, as well as the same quantization methods of spatial parameters as in PS. All OTT blocks for different channel pair signals share the same quantization codebooks of spatial parameters.

Although ICLD can be regarded as an approximation of ILD ignoring the transmission filter procedures in earphone or stereo audio system, the transmission procedures are not ignorable in multichannel audio systems. Owing to the filter function of transmission process from multiple loudspeakers to ears, significant differences exist between inter-channel cues and interaural cues in the case of playback with loudspeakers. For example, with two loudspeakers with azimuths −110° and 30°, the difference of ICLD and corresponding ILD could be up to 55 dB [18].

The correlations of inter-channel parameters and interaural cues (such as ICLD and ILD) closely depend on the location of loudspeaker pairs. In the case of loudspeaker pairs with arbitrary locations in multichannel systems, there are significant differences between ICLD and ILD, which cannot be treated as the approximation correlation. It is obviously inappropriate to design quantization codebook for ICLDs of arbitrary channel pairs uniformly according to JND of ILD. The same quantization codebooks for spatial parameters of channel pairs in different directions will definitely degrade the spatial quality.

Besides JND of binaural cues often used to analyze the spatial perceptual sensitivity of human auditory system, the JND of sound angle change (also referred to as minimum audible angle, MAA) has also been investigated by lots of previous researchers. Early in 1958, Mills presented a classic method to measure the JND of human auditory system when sound angle changed [10]. The experimental measurements showed that for the sounds with frequencies from 500–3700 Hz located in front of subjects, when the sound azimuths changed about 1–3.1°, the subjects can just notice the change.

Supposing the JND values for all azimuths in a circle of 360° are a fixed value 3°, Choi designed a quantization codebook of spatial parameters. Owing to the different configurations of different loudspeaker pairs (such as the different interval angel between each two loudspeakers), Choi designed a specific quantization codebook for each specific loudspeaker pair in 5.1 loudspeaker system. However, the quantization codebooks had no essential differences with MPEG surround [19].

It is known that the ICLD value between two channel signals may be any value from 0 to infinite. When ICLD reaches some threshold, even if ICLD continues to increase, the spatial sound image will limitlessly approach one of the two loudspeakers but never go outside of the region between two loudspeakers. No matter how ICLD changes, the spatial sound image will always locate between two loudspeakers. Although ICLD may be infinite values without threshold, but the azimuth of virtual sound image has threshold and can only be value between −180° and 180° in a circle.

Thus, instead of spatial parameter ICLD between channel pair to represent spatial direction of virtual sound, Cheng proposed to extract the azimuth of virtual sound as spatial parameter. Cheng designed three different quantization codebooks for azimuths in three regions: front, side, and rear regions [20]. The quantization resolutions of azimuths in three regions were different. But in each region, the azimuth quantization resolution was uniform. The front region (from −30° to 30°) had the smallest quantization step sizes with two kinds of quantization resolution 2° or 3°. The quantization resolution for side regions (from 30° to 110°, or from −30° to −110°) were 6.6°/20°. As for the rear region from −110° to 110°, the azimuth resolutions were fixed 17.5°/35°. Since the azimuth quantization errors of Cheng's method were much bigger than the azimuthal JNDs, obvious perceptual distortion of spatial quality was introduced.

To improve the spatial quality of multichannel spatial audio coding, Elfitri presented a closed-loop encoding system applied on MPEG surround [21]. The closed-loop procedure inversely quantized downmix and spatial parameters at the encoder to get decoded signals and then obtained new residual signals between original and decoded signals. The improved spatial quality mainly benefited from the accurate extraction of residual signals, however, had nothing to do with the quantization method of spatial parameters. Meanwhile, the closed-loop procedure will significantly increase the computational complexity at the encoder.

The abovementioned spatial audio coding methods are mainly based on the elementary mechanism of parametric stereo and focus on the quantization of spatial parameters between two channel signals. The inter-channel spatial parameters such as ICLD can be extracted from only two channel signals. Cheng proposed a method to extract the azimuth of virtual sound from an arbitrary number of loudspeaker signals [22]. In Cheng's method, downmix signal of multichannel signals were obtained as the signal of virtual sound source. Then, the signal and the azimuth of virtual sound source were coded. It seemed similar with what spatial audio object coding (SAOC) schemes do [23, 24]. But they are different in fact. The spatial parameters in SAC represent spatial information of virtual sound. However, the parameters in SAOC are correlation parameters between audio object signals, which have nothing to do with the spatial location information of sound. The quantization of parameters in SAOC is beyond the scope of this paper. Cheng designed uniform quantization codebooks with fixed step sizes of 2° or 3° for azimuths in horizontal plane. Since high quantization resolution of azimuths in Cheng's method, the quantization codebook sizes for azimuths in a full circle were almost twice or triple of that in MPEG surround and resulted in high bit rates of spatial parameters [22].

Since Mills' experiments conducted to get azimuthal JND in 1958, lots of experiments have been conducted by researchers to study the directional perceptual sensitivity of human auditory system. The results showed that when sound was located in front of subjects in a horizontal plane (where sound azimuth was 0°), azimuthal JND was the smallest; when sound azimuth increased, JND increased too; when sound azimuth was 90° or −90° (sound was located at the sides of subjects), JND was the biggest; azimuthal JND of the rear sound was almost twice that of the front sound [25–27].

With the different JNDs for different azimuths, different azimuths in a full circle should be assigned with different quantization step sizes according to azimuthal JND data. The azimuthal JND characteristics were used in the quantization method of spatial parameters as a first attempt in [28]. An adaptive quantization scheme of spatial parameter ICLD according to arbitrary loudspeaker configurations was proposed. However, the spatial parameter ICLD was still extracted from channel pair signals. The quantization method for spatial parameters extracted from multichannel signals was not discussed in detail. The objective and subjective experiments about quantization performance was limitedly discussed and mainly compared with MPEG surround.

In this paper, the quantization of spatial parameters between two or more channel signals are both discussed. The main contributions and works include the following: to accurately represent the virtual sound, a method to estimate spatial parameter azimuth and the signal of virtual sound from an arbitrary number of loudspeakers was proposed; an azimuthal JND based spatial parameters quantization method (JSPQ) was proposed; and the generation procedure of azimuth quantization codebook was elaborated in details. The quantization of spatial parameter is more consistent with the JND values of azimuths in a full circle, and thus, perceptual redundancy of spatial parameter can be efficiently removed with coding bit rate as low as possible. Objective experiments and subjective evaluation were conducted to confirm that the proposed JSPQ outperformed reference quantization methods of spatial parameters in the respects of codebook sizes, quantization errors, coding bit rates, and spatial qualities.

3 Spatial audio coding

3.1 Codec structures of SAC

According to the extraction method of spatial parameter, there are mainly two kinds of coding structures for high-efficient multichannel SAC. One typical representative is the tree-structured layer-by-layer coding schemes such as in MPEG surround [15] and MPEG 3D audio standard [24]. Take MPEG surround for example; after T/F transformation, multichannel signals are combined pairwise or per three channel signals using elementary coding blocks (OTT or TTT). As illustrated in Fig. 1, spatial parameters (ICLD, ICC, etc.) and downmix signals are acquired in each OTT blocks before quantization in the encoder. In the decoder, decoded downmix signals and spatial parameters are processed in inverse OTT blocks to get recovered multichannel T/F signals. In each OTT or TTT block in the cascaded tree structure, a set of spatial parameters are extracted from different channel signals or downmix signals. However, for each kind of spatial parameter (such as ICLD), the same quantizer is used as applied in parametric stereo coders.

The other type of spatial audio coding structure is based on virtual sound source information representation as in Fig. 2 [22]. Given that only one virtual sound source generated by loudspeakers exists for each time-frequency bin, multichannel signals can be represented with a virtual sound source with its spatial location information. The signal and location information of the virtual sound source can be extracted from an arbitrary number of loudspeaker signals in encoder. With inverse-quantized information of virtual sound in decoder, virtual sound can be reproduced by general panning techniques such as vector-based amplitude panning (VBAP) [29], or Ambisonics techniques such as higher order Ambisonics (HOA) [30] to get recovered multichannel T/F signals.

The proposed JND-based spatial parameter quantization methods (JSPQ) in this paper are independent of the SAC structure. Thus, the modified quantization methods for spatial parameters ICLD and azimuth can be used as substitutions respectively in these two types of SAC structures.

3.2 Quantization of spatial parameters

ICLD is one of the most important spatial parameters to represent the direction information of virtual sound image and is commonly used in SAC such as E-AAC+ [17] and MPEG surround [15]. In MPEG surround for coding multichannel signals, ICLD values are extracted from pairwise channel signals pair by pair among all channels.

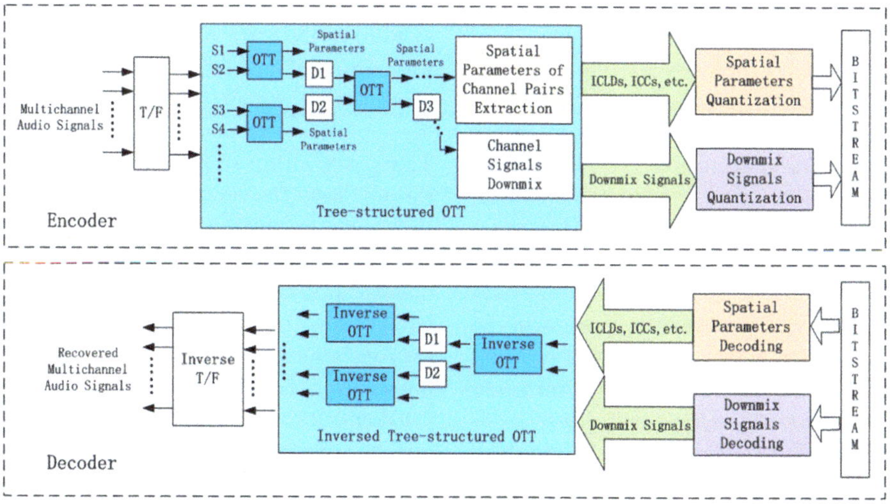

Fig. 1 Spatial audio coding structure in MPEG surround

Given the frequency amplitude of two loudspeakers, $X_1(k)$ and $X_2(k)$, respectively, k is the index of frequency bin and ICLD can be calculated in each frequency Bark band b ($b \in \{0, 1, 2, .., 23\}$) as:

$$ICLD(b) = 10 \log_{10} \left(\frac{E_1(b)}{E_2(b)} \right), \tag{1}$$

$$E_1(b) = \sum_{k=k_b}^{k_{b+1}-1} X_1^2(k), \tag{2}$$

$$E_2(b) = \sum_{k=k_b}^{k_{b+1}-1} X_2^2(k) \tag{3}$$

Except ICLD, the azimuth of virtual sound is also commonly used as one of the spatial parameters as in [22], especially used for more than two loudspeakers. The azimuth of virtual sound extracted from two loudspeaker signals by:

$$\theta_0 = \arctan \left(\frac{g_L - g_R}{g_L + g_R} \tan \theta \right), \tag{4}$$

in which g_L and g_R are the gain parameters of two loudspeakers signals, and θ is half the intersection angle between two loudspeakers. Given there is only one sound source that can be perceived by human auditory system in each frequency bin, ICLD and azimuth can be converted to each other under the condition of two loudspeakers.

The quantization codewords distribution of spatial parameters ICLDs and azimuths in different methods

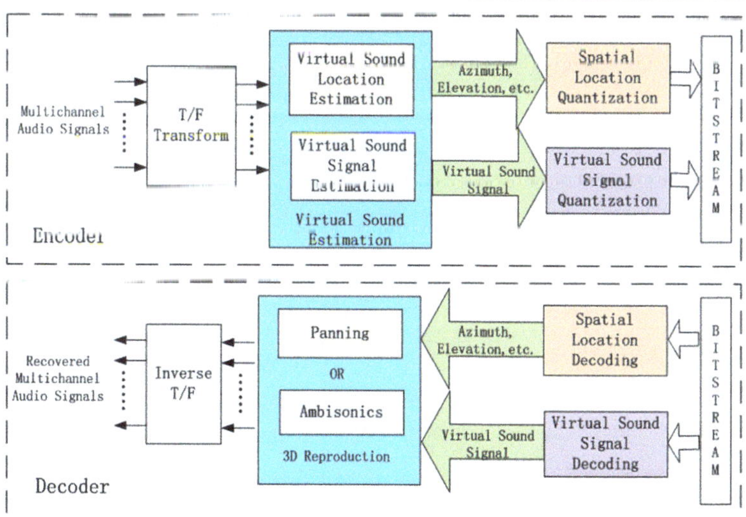

Fig. 2 Spatial audio coding structure based on virtual sound source information representation

(MPEG surround [15], Cheng's [20], Choi's [19] methods and SLQP [22]) are intuitively illustrated in Fig. 3.

MPEG surround and Choi's methods provide quantization of spatial parameters ICLDs between channel pairs. ICLDs extracted from four kinds of channel pairs under 5.1 loudspeaker configuration are converted to azimuths in Fig. 3a–c. Some methods have two kinds of quantization resolutions; thus, H and L, respectively, represent high and low resolution quantization methods. Each red point on the circles represents one quantization codeword of azimuth. Every two adjacent solid lines form the quantization interval region for the inside quantization codeword. That means the azimuths in each quantization interval region will be quantized as the quantization codeword inside the region. Then, are these distribution of azimuth codewords in Fig. 3a–f reasonable for azimuthal perceptual quantization or not?

Researches about the directional perceptual sensitivity of human auditory system showed that: humans have different azimuthal JNDs for sound in different directions. The smallest azimuthal JND corresponds to front directions, bigger JND for rear directions and biggest azimuthal JND for side directions [8–11]. According to the azimuthal JND features of human auditory system, the azimuths with small JNDs allow small quantization errors, the azimuths with big JNDs allow big quantization errors. The quantization step sizes of front azimuths should be the smallest. The quantization step sizes of side azimuths should be the biggest.

But it can be observed from Fig. 3 that the previous quantization methods of spatial parameters are not correlated well with the azimuthal JND. It is evident that there are four regions that are obviously filled with dense solid lines both in Fig. 3a, c, which correspond to the four loudspeakers's locations (L, R, Ls, Rs) in 5.1 audio system. The quantization codewords around the four loudspeakers are the densest in Fig. 3a, c. In Fig. 3a, b, d and e, the most sparse distribution of codewords all lie in rear regions. In Fig. 3f, all the azimuth codewords are equally spaced for the uniform quantization with fixed step sizes. Thus, the quantization step sizes of spatial parameters in current SAC methods are not in coincidence with the JND

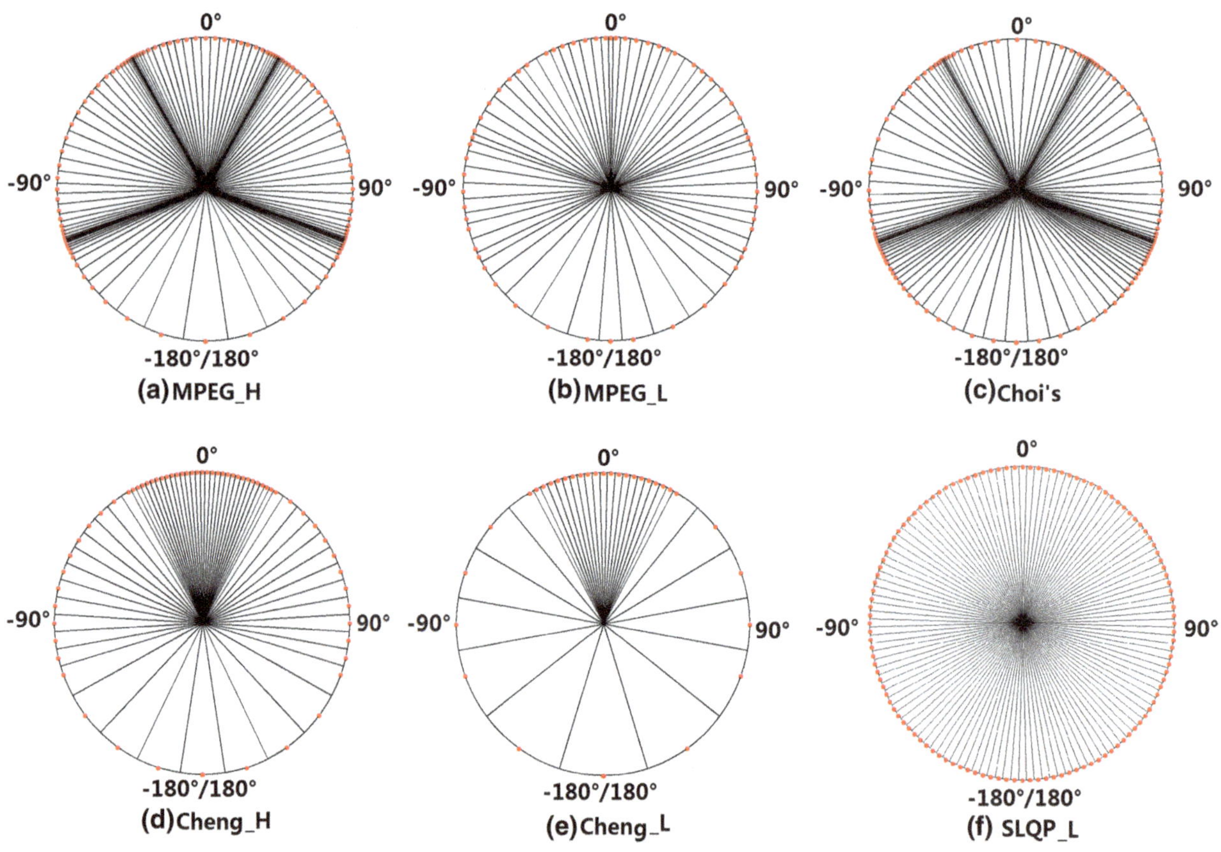

Fig. 3 Quantization codewords distributions in different methods. **a** High resolution quantization of spatial parameters in MPEG Surround [15], denoted as MPEG_H; **b** Low resolution quantization of spatial parameters in MPEG Surround [15], denoted as MPEG_L; **c** Choi's method [19]; **d** High resolution quantization in Cheng's method [20], denoted as Cheng_H; **e** Low resolution quantization in Cheng's method [20], denoted as Cheng_L; **f** Low resolution quantization in SLQP [22], denoted as SLQP_L

characteristics of different azimuths in a full circle. These quantization methods of spatial parameters will result in either spatial perceptual distortions of virtual sound images, or inefficient compressions of spatial parameters.

The intention of this paper is to effectively use the characteristic of azimuthal JND to design the quantization of spatial parameters. Thus, in the proposed quantization codebook, front azimuths with small JND will have small quantization steps, side azimuths with big JND will have big quantization steps, and rear azimuths will have medium quantization steps.

4 Proposed quantization methods of spatial parameters

In SAC, spatial parameters are extracted from channel signals to represent the spatial information of virtual sound source generated by loudspeakers. The most important spatial parameters extracted in SAC schemes often include azimuth of virtual sound generated by multiple loudspeakers and ICLD between channel pairs. The perceptual lossless coding of spatial parameters is the important guarantee of spatial quality for the spatial sound reproduction in decoder.

4.1 Derivation and quantization of azimuth

4.1.1 Estimation of virtual sound source

With multiple correlated sound signals simultaneously rendered by two or multiple loudspeakers, humans can only perceive a summing virtual sound. Regarding the human head as a rigid sphere and without considering the effect of the head on the sound transmission, the sound field around listener generated by loudspeakers can be approximatively equivalent to the sound field generated by a real sound at the location of virtual sound. Given a fixed frequency, if the sound field in the head region can be approximatively reproduced with multiple loudspeakers, then the sounds at a human's two ears in reproduced sound field will be approximatively equivalent to those in the original sound field. In sound reproduction, two essential properties, sound pressure and particle velocity [1], are often used to describe the properties of sound field. The total sound pressure and particle velocity at the listening point (center of listener's head) generated by multiple loudspeakers are approximatively equivalent to those by the single sound (or loudspeaker) at the location of virtual sound. With the peer properties of reproduced sound field by multiple loudspeakers and the sound field by original single sound source, the virtual sound generated by multiple loudspeakers can be taken as the original single sound and estimated from multichannel signals. Since particle velocity is highly correlated with the direction of arrival of sound, the azimuth of virtual sound source can be estimated from loudspeaker signals based on the equations of particle velocity in two sound fields [1].

After 1024-point short time Fourier transform (STFT) with 50 % overlapped window is applied in each frame of each channel signal, the virtual sound source is estimated in each frequency bin. Given loudspeakers with azimuths and distances to the coordinate system's origin in Fig. 4, the total sound particle velocity vector at the coordinate system's origin as the listening point generated by M loudspeakers of frequency bin k can be written as:

$$u = A\begin{pmatrix} \sum\limits_{m=1}^{M} \frac{e^{-ik_{wn}r_m}}{r_m} X_m \cos\theta_m \\ \sum\limits_{m=1}^{M} \frac{e^{-ik_{wn}r_m}}{r_m} X_m \sin\theta_m \end{pmatrix}, \tag{5}$$

where A is the proportionality coefficient relevant to sound transmission, X_m is the amplitude of loudspeaker S_m of frequency k, m is loudspeaker index,$m \in \{1, 2, .., M\}, M \geq 2$, θ_m is the azimuth of loudspeaker S_m, r_m is the distance from loudspeaker S_m to the receiving point, i is imaginary unit, k_{wn} is the wave number, $k_{wn} = \frac{2\pi k}{c}$, k is frequency, and c is sound speed.

Given a sound source S_0 with azimuth θ_0 and distance r_0 to the coordinate system's origin in Fig. 4, the particle velocity vector generated at the coordinate system's origin as the receiving point is written as:

$$u_0 = A\begin{pmatrix} \frac{e^{-ik_{wn}r_0}}{r_0} \cos\theta_0 \\ \frac{e^{-ik_{wn}r_0}}{r_0} \sin\theta_0 \end{pmatrix} X_0. \tag{6}$$

If the virtual sound source generated by loudspeakers is located with azimuth θ_0 and distance r_0 to the coordinate system's origin, then

$$u_0 = u \tag{7}$$

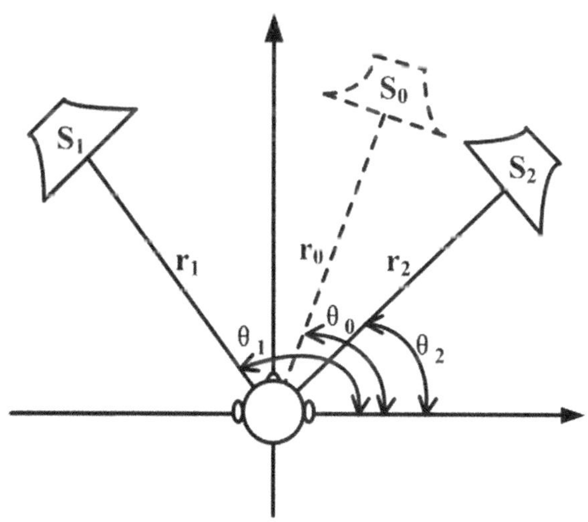

Fig. 4 Virtual sound image generated by loudspeaker pair in the coordinate system

holds. Given the loudspeakers arranged on a sphere surface with the same distance to the coordinate system's origin as the receiving point, the azimuth of the virtual sound can be obtained as [31]:

$$\theta_0 = \arctan\left(\frac{\sum_{m=1}^{M} X_m \sin\theta_m}{\sum_{m=1}^{M} X_m \cos\theta_m}\right). \quad (8)$$

The estimation of virtual sound includes azimuth, as well as the signal estimation of virtual sound. To keep the same energy of virtual sound signal with total energy of original loudspeaker signals, the signal of virtual sound can be estimated as:

$$X_0 = \sum_{m=1}^{M} X_m \frac{\sqrt{\sum_{m=1}^{M} X_m^2}}{\left|\sum_{m=1}^{M} X_m\right|}. \quad (9)$$

Figure 5 illustrated the sound fields (expressed with sound pressure) around listener's head region generated respectively by two loudspeakers and single loudspeaker at the location of estimated virtual sound. It can be deduced from Fig. 5 that the sounds at the listener's two ears in the two sound fields are approximate.

Thus, the multichannel signals are expressed with the virtual sound signal as well as spatial information. The virtual sound signal can be coded with conventional mono perceptual codec, such as AAC, and then transmitted to the decoder. The spatial parameter azimuth will be coded as side information and transmitted to the decoder. The coding method of azimuth is mainly concerned and analyzed in the following subsections.

4.1.2 Quantization of azimuth

For any azimuth in a circle of 360°, if the azimuth is quantized with quantization error below the corresponding azimuthal JND (or MAA), then the azimuth is perceptually losslessly quantized. This is the original intention of the proposed quantization method.

Given the quantization value of some azimuth is *C*, and the JND of *C*, then the quantized interval region of *C* can be computed. All azimuths in the quantized interval region should be quantized as *C* and the biggest quantization error of all azimuths would not be bigger than their JNDs. If the whole circle of 360° is divided to these quantization interval regions adjacent to each other, then all azimuths can be perceptually losslessly quantized.

The azimuthal JND data can only be obtained by subjective listening tests. The JND data obtained by different listeners are different. The listening environments also influence the test results for the same listener. Thus, the final JND data is often an average among multiple listeners. Meanwhile, owing to the operation limitation, it is impractical to get JND data of all azimuths by practical listening tests. A few of typical azimuths with limited resolution are often used for listening tests. Thus, the JND data obtained from the listening tests often contain a finite number of sparsely distributed values and need to be interpolated to get higher resolution JND data. JND data in our previous research [32] are used in the following experiments. Since the JND data obtained from the tests are angles with one decimal place, JND data are interpolated at a resolution of 0.1° for azimuths from 0°

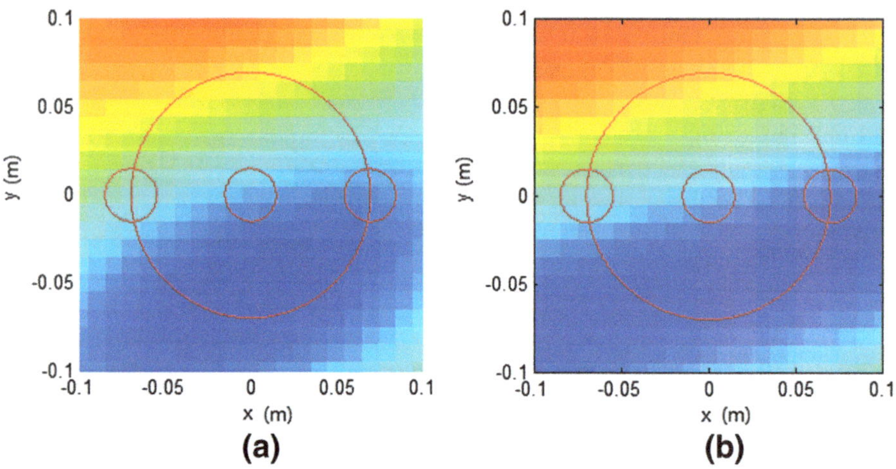

Fig. 5 Sound fields (expressed with sound pressure) in the zone of 0.1 m*0.1 m around mimic human heads (*two big red circles*) in the horizontal plane, different color levels represent different sound pressure levels, and *small red circles* correspond to the center of head and two ears. **a** Sound field generated with two loudspeakers (azimuth 60°, −60°; elevation 0°, 0°; radius 2.5 m, 2.5 m; 1000 Hz signal with gains 0.518 and 1.316, respectively, allocated to two loudspeakers). **b** Sound field generated with one loudspeaker at the estimated location of virtual sound generated by two loudspeakers in (**a**)

(front) to 180° (rear) in the horizontal plane before the quantization codebook design.

JND data of azimuths were often incrementally obtained in subjective audiometries from 0° to 180°. Given an azimuth A_n between 0° and 180°, with the increase of A_n, if the directional difference between the sound with azimuth A_n and $A_{n+1} = A_n + a$ was just noticeable by the listener in the subjective audiometry, JND of azimuth A_n is $JND\_A_n = a$. This can be expressed as $A_{n+1} = A_n + JND\_A_n$ or $A_n = A_{n+1} - JND\_A_n$. With these, we can get the quantization codebook of azimuths with the procedures as follows.

At first, the first quantization value is selected as a start point value. One of the azimuths whose JND is the biggest is selected as one quantization value $C_0 = angle(\max(JND\_A_n))$ and set as the start point. JND_A_n is the JND value of azimuth A_n, $n \in \{ 0°,0.1°,0.2°,..,180°\}$. In general, $C_0 = 90°$. There are two different procedures from C_0 to get all other quantization values. One procedure begins from C_0 to get the quantization values between C_0 and 180°, named backward procedure. Another procedure begins from C_0 to get the quantization values between 0° and C_0, named forward procedure. In the two procedures, all quantization values between C_0 and 180° and the interval region of each quantization value are obtained.

In the backward procedure, quantization values of azimuths are obtained from C_0 to 180°. For quantization value C_i (C_i=C_0 at the beginning), one of the endpoints I_i ($i \geq 0$) of quantization interval region is calculated as

$$I_i = C_i + JND\_C_i. \tag{10}$$

If $C_i + JND\_C_i > 180°$, then set $I_i = 180°$ and end up the procedure. Afterwards, the quantization value C_{i+1} next to C_i is obtained by

$$C_{i+1} = I_i + JND\_I_i. \tag{11}$$

If $I_i + JND\_I_i > 180°$, then set $C_{i+1} = 180°$ and end up the procedure. In the same way, the other endpoint of quantization interval region for C_{i+1} is calculated as

$$I_{i+1} = C_{i+1} + JND\_C_{i+1}. \tag{12}$$

Then, the interval region of C_{i+1} is $[I_i, I_{i+1})$. All the azimuths in this interval region will be quantized as quantization value C_{i+1}. The backward procedure ends up when I_i or C_i reach 180°.

In the forward procedure, quantization values of azimuths are obtained from C_0 to 0°. For quantization value C_i' ($C_i' = C_0$ at the beginning), the other endpoint I_i' ($i \geq 0$) of quantization interval region can be obtained to meet

$$I_i' = C_i' - JND\_I_i'. \tag{13}$$

If $C_i' - JND\_I_i' < 0°$, then set $I_i' = 0°$ and end up the procedure. Afterwards, the quantization value C_{i+1}' next to C_i' can be obtained to meet

$$C_{i+1}' = I_i' - JND\_C_{i+1}'. \tag{14}$$

If $I_i' - JND\_C_{i+1}' < 0°$, then set $C_{i+1}' = 0°$ and end up the procedure. The forward procedure ends up when I_i' or C_i' reach 0°.

Together with all C_i and C_i' in the above two procedures, we can get the whole quantization codebook for azimuths from 0° to 180°. For the approximate bilateral symmetry, the obtained codebook from one side region (from 0° to 180°) can be duplicated for the other side region (from 0° to −180°) and finally get the azimuthal codebook of the full circle of 360°.

Such procedure can be illustrated in Fig. 6, for example. Generally, $C_0 = 90°$ corresponding to the side of human head. The right endpoint of quantized interval for C_0 is calculated as $I_0 = C_0 + JND\_C_0$. The next quantization value C_1 is calculated as $C_1 = I_0 + JND\_I_0$. Similarly, I_1 is calculated and the quantized interval of C_1 is $[I_0, I_1)$.

4.1.3 *Azimuth distribution in quantization codebook*

In the proposed JND-based spatial parameters quantization method (JSPQ), the quantization codebooks of spatial parameters are relevant to the direction information of virtual sound source. The quantization codewords densities are dependent on azimuthal JNDs. Small quantization step sizes are assigned for azimuths with small JNDs. Big quantization step sizes are assigned for azimuths with big JNDs. This is the main difference between proposed JSPQ and previous quantization methods of spatial parameters.

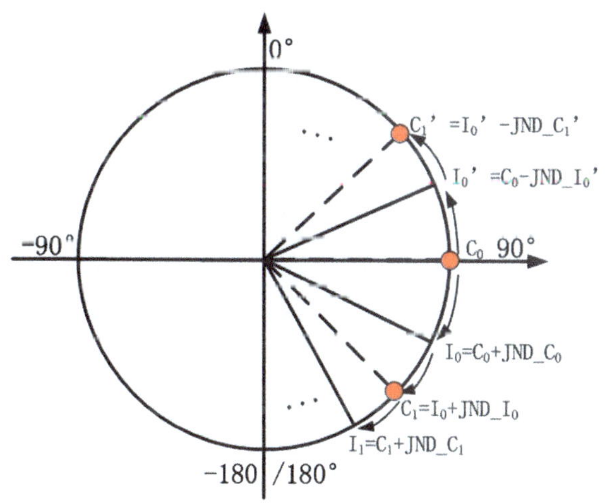

Fig. 6 Demonstration of quantization codebook generation of azimuth. All quantization codewords of azimuths are obtained with JND, trying to control the quantization error of each azimuth in accordance with the JND of the quantized azimuth

Since spatial parameters are to represent the direction information of virtual sound source, the direction information are expressed as azimuths on a circle as illustrated in Fig. 7. With the same symbols as in Fig. 3, each red point on the circle represents one quantization codeword of azimuth, and every two adjacent solid lines form the quantization interval region for the inside quantization codeword. The azimuths on the circle in the quantization interval region will be quantized as the inside quantization codeword.

With essential differences with previous methods in Fig. 3, the illustration of JSPQ in Fig. 7 shows that relative denser quantization codewords are assigned to the front region, fewer quantization codewords are for rear region, and the distribution of quantization codewords for side region are the most rare.

4.1.4 Coding of quantization index

After azimuths are quantized as quantization codewords according to the quantization codebook, the quantization indices of quantization values will be further coded. Owing to the short-time stationary of most sound signals, the sound location information in the same frequency bin in two adjacent time frames will not change rapidly. The difference between the virtual sound azimuths in each two adjacent time frames will fluctuate in a tiny range. Thus, differential Huffman coding will benefit the quantization indices compression. Both natural binary coding and differential Huffman coding are used in JSPQ to code the quantization indices of azimuths. The quantization index of azimuth of each time-frequency bin for the first time frame, *index_q*, is coded by natural binary coding with 6 bits. From the second to the last time frame, the difference of quantization index in each two adjacent time frames is calculated:

$$diff\_index(tf,k) = index\_q(tf,k) - index\_q(tf-1,k), \quad (15)$$

in which tf is the index of time frame, $tf > 1$, k is the index of frequency bins, $k = 0, \ldots 23$, and $index\_q(tf,k)$ is the quantization index for azimuth of frequency bin k in time frame tf. Then, $diff\_index(tf,k)$ is coded with Huffman coding.

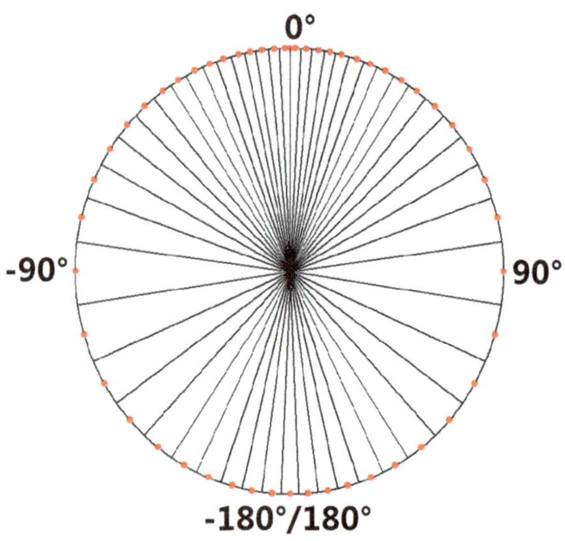

Fig. 7 Quantization codewords distribution of azimuths in proposed JSPQ. Azimuths in the front region have the smallest quantization step sizes. Azimuths in side regions have the biggest quantization step sizes. Azimuths in the rear region have the medium quantization steps

4.2 Quantization of spatial parameter ICLD

In the case of two loudspeakers, both azimuth and ICLD can be extracted from two channel signals as spatial parameters to represent direction information of virtual sound images. Since ICLD can be converted to azimuth, the quantization of ICLD can be easily achieved based on the quantization codebook of azimuth as the process procedure illustrated in Fig. 8.

In the encoder, ICLD can be calculated with two loudspeaker signals by Eq. (1). In each frequency subband, azimuth from Eq. (8) can also be expressed with the signals and azimuths of two loudspeakers as:

$$\theta_0 = \arctan\left(\frac{\sqrt{E_1(b)}\sin\theta_1 + \sqrt{E_2(b)}\sin\theta_2}{\sqrt{E_1(b)}\cos\theta_1 + \sqrt{E_2(b)}\cos\theta_2}\right). \quad (16)$$

With Eqs. (1) and (16), the azimuth of virtual sound θ_0 and ICLD can be converted to each other:

$$\theta_0 = \arctan\left(\frac{10^{\frac{ICLD(b)}{20}}\sin\theta_1 + \sin\theta_2}{10^{\frac{ICLD(b)}{20}}\cos\theta_1 + \cos\theta_2}\right) \quad (17)$$

$$ICLD(b) = 10\log_{10}\left(\frac{\sin\theta_2 - \tan\theta_0\cos\theta_2}{\tan\theta_0\cos\theta_1 - \sin\theta_1}\right)^2. \quad (18)$$

Azimuth θ_0 is quantized according to the quantization codebook of azimuths to get the quantization index. Then, natural binary coding or differential Huffman coding is used to code the quantization index. In the decoder, the azimuth is inversely quantized and converted to retrieve ICLD. Further details are not discussed in this paper (for details, see [28]).

5 Experiments

Objective and subjective experiments were conducted to verify the effectiveness and performance of proposed JND-based spatial parameters quantization method (JSPQ). Reference methods include quantization methods of spatial parameters in MPEG surround [15], Choi's [19], SLQP [22], and Cheng's methods [20].

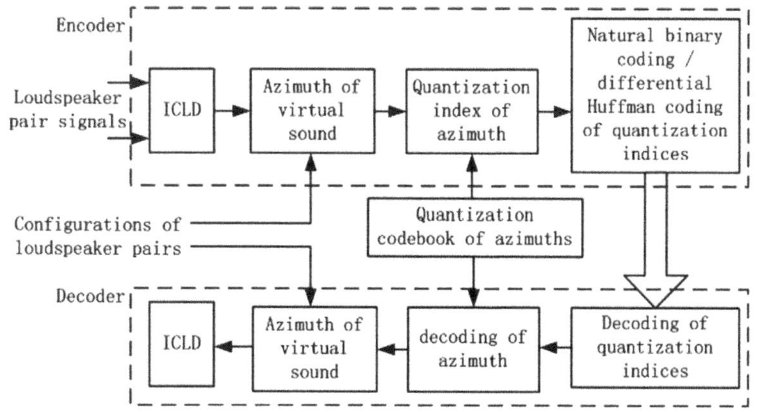

Fig. 8 Framework of ICLD quantization

There are three objective experiments conducted to compare the quantization codebook sizes, spatial distortions, and coding bit rates of spatial parameters in different methods. Subjective experiments aimed to evaluate the spatial quality of different quantization methods for spatial parameters, in which listeners need to rank the perceived spatial quality of audio excerpts encoded and decoded with different quantization methods against the original unprocessed audio excerpts. The specific experimental methods and audio materials are described in detail in the following subsections.

5.1 Comparison of quantization codebook sizes

The codebook sizes of spatial parameters directly determine the bit rates required for transmission of spatial parameters. In the current experiment, the codebook sizes required for quantization of spatial direction information in a circle of 360° at horizontal plane were compared between proposed JSPQ and reference quantization methods.

In MPEG surround and Choi's quantization methods, the main spatial parameters used to represent the spatial direction information are ICLDs. In Cheng's method and SLQP, the spatial parameters used to represent the spatial direction information are azimuths. Thus, reference quantization codebooks include quantization codebooks of ICLDs in MPEG surround and Choi's method, as well as quantization codebooks of azimuths in Cheng's method and SLQP.

Since ICLD is computed from channel pair signals, four channel signals (L, R, Ls, Rs) in standard 5.1 loudspeaker system are pairwise used to extract ICLD in this experiment, as illustrated in Fig. 9. ICLD values are extracted, respectively, with four channel pair signals (L&R, L&Ls, Ls&Rs, R&Rs) to represent spatial direction information in the whole circle of 360° at horizontal plane. The quantization codebooks of ICLD for these four channel pair signals are compared among different methods.

The extraction of azimuth from channel signals does not depend on channel pair signals. The number of channel signals used to extract azimuth of virtual sound can be two or more. In this experiment, the same channel pair signals are used for the extraction of azimuth as for ICLD. The quantization codebooks of azimuths for channel pair signals are compared among different methods.

To represent spatial direction information in a circle of 360° at horizontal plane, the total number of quantization codewords (codebook sizes) for spatial parameters ICLD and azimuth in different methods are illustrated in Fig. 3. Meanwhile, the respective codebook sizes in different regions (front, side, and rear) are also illustrated. As for MPEG surround, SLQP, and Cheng's method, each of them offers two kinds of quantization codebooks: one is coarse quantization with low precision and another is fine quantization with high precision. The quantization

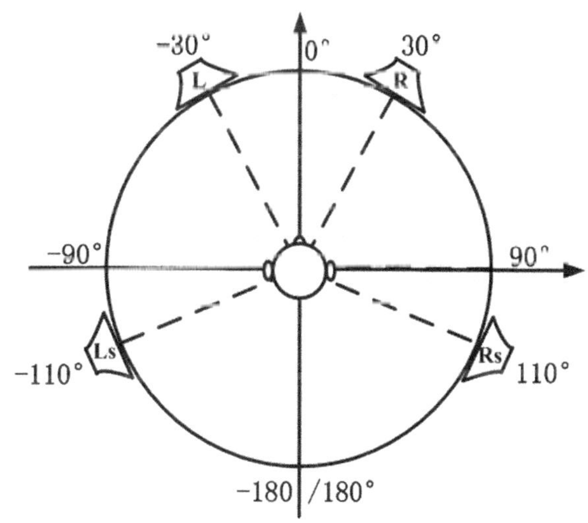

Fig. 9 Configuration of loudspeakers *L*, *R*, *Ls*, and *Rs* in the 5.1 loudspeaker system

codebooks with low and high precision are distinguished with L and H, respectively, in Fig. 10.

The quantization method of spatial parameters in MPEG surround comes from that in E-AAC+ for stereo signals. Thus, the quantization codebooks for spatial parameter ICLD of arbitrary channel pairs are the same, no matter where the loudspeaker pairs are, and no matter how big the intersection angle between two loudspeakers is. Each channel pair signal needs 15 and 30 ICLD values to quantize spatial direction information between two loudspeakers for coarse quantization and fine quantization, respectively. Given four loudspeakers located in a circle, the total numbers of ICLD values to quantize spatial direction information in circle are 60 and 120 for coarse quantization and fine quantization, respectively.

In SLQP, uniform quantization codebooks are used for spatial parameters azimuths. There are two quantization step sizes 3° and 2° for coarse quantization and fine quantization, respectively. Thus, the quantization errors of azimuths will not exceed 1.5° and 1°, which are much smaller than JND values of most azimuths. Significant perceptual redundancies in the quantization for azimuths surely exist, which leads to inefficient compressions of spatial parameters.

In Cheng's two quantization codebooks, the total numbers of azimuths are 32 and 64 for coarse and fine quantization, respectively. The quantization step sizes are nonuniform for azimuths in different directions. It is well known that the JND values of azimuths at the rear are smaller than that at side regions. But in Cheng's two quantization codebooks, contrary to what one might suppose, the total number of quantization azimuths at the rear region is much less than that at side regions. There are only three or seven quantization values for azimuths in the range of 140 degrees from −110° to 110°. These coarse quantization for spatial direction information will definitely result in obvious spatial perceptual distortions.

In the quantization codebook of proposed JSPQ for azimuths, small step sizes are set for front and rear regions, and big step sizes are set for side region. Fewer quantization values are set for azimuths at side regions compared with most of the other methods. The codebook size of JSPQ is smaller than most of other methods, which decreases by 13.3 and 56.7 % compared with two codebooks of MPEG surround and decreases by 56.7 and 71.1 % compared with SLQP's two codebooks.

5.2 Comparison of quantized azimuthal errors

The quantization error of directional information will directly influence the perceptual spatial quality. In this experiment, the absolute quantized errors of azimuths in a half circle of 360° were calculated with different quantization methods. The azimuths from 0° to 180° in horizontal plane (0° corresponds to the front) were chosen with intervals of 0.1° in the calculations. There are 1801 azimuths in total. The absolute quantized error of azimuth is calculated by

$$E_q = |A - C_q| , \tag{19}$$

in which C_q is the quantized value of azimuth A, $A \in \{0°, 0.1°, 0.2°, \ldots, 180°\}$, $q \in \{0, 1, 2, \ldots, 180\}$.

The local maximums of quantized azimuth error with different methods are computed and compared with azimuthal JND data in Fig. 11. The data of JND curve

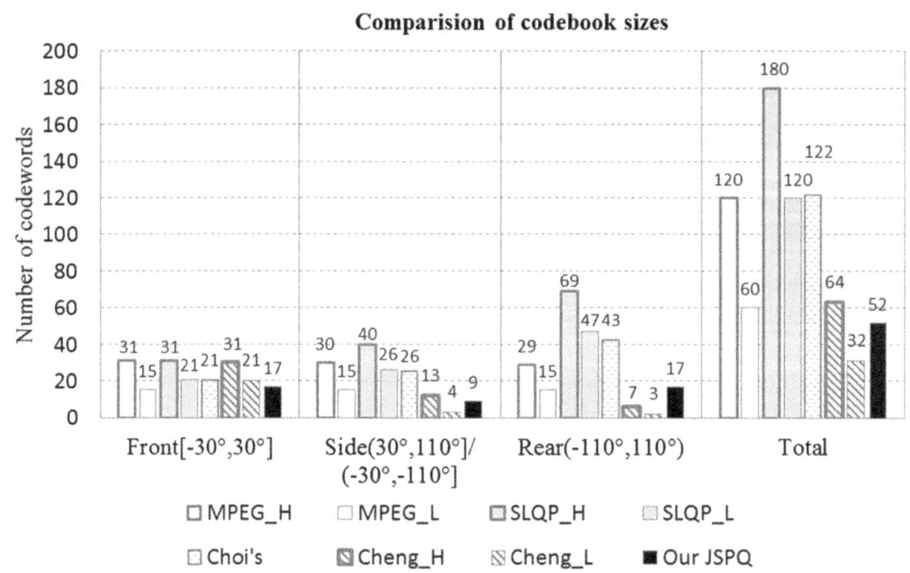

Fig. 10 Codebook sizes comparison of spatial parameters in different methods. The total numbers of codewords for spatial parameters in different regions (*front, side,* and *rear regions*), and in a full circle in horizontal plane are shown, respectively

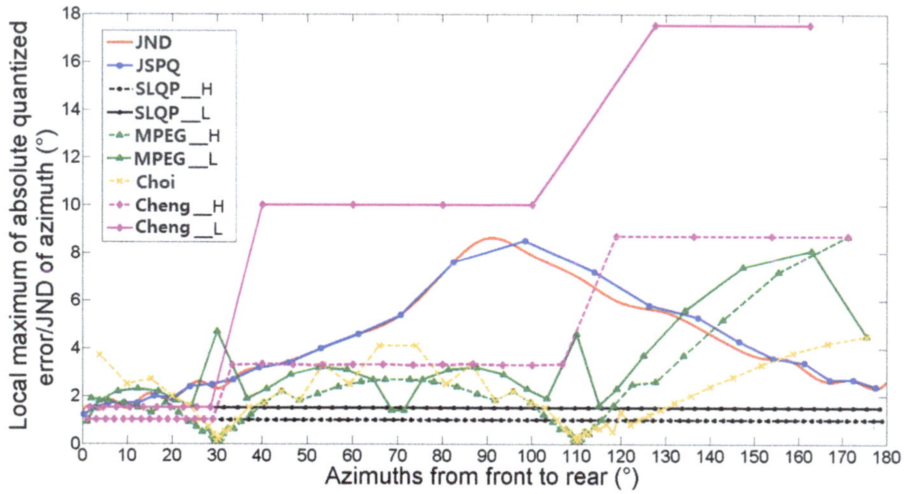

Fig. 11 Quantization errors of virtual sound azimuths in different methods compared with azimuthal JND data. Smaller quantization error than JND means that obvious perceptual redundancy exists; in contrast, bigger quantization error than JND means that sensible spatial distortion exists. Quantization errors of azimuths coincided with azimuthal JND curve means nearly perceptual lossless quantization

come from [32] and are used as references to evaluate the perceptual distortion of azimuthal quantization. Quantized error that is bigger than JND will induce perceptual distortion, and smaller quantized error than JND means perceptual redundancy exists.

Owing to the compact quantization step sizes in SLQP, the quantization errors are the smallest. The quantization errors are much smaller than JND, and it means that obvious perceptual redundancies exist and will lead to inefficient compressions of spatial parameters.

The biggest quantization errors were caused by Cheng's quantization method with lower precision. Most of the quantization errors of azimuths are much bigger than JND. Especially in the rear region, the biggest azimuthal error even exceeds 17°. It will lead to significant spatial distortions with Cheng's methods.

As for other reference methods, most quantization errors in front region are smaller than JND. However, in side regions, most errors are significantly smaller than JND, and errors in rear region are significantly larger than JND.

The proposed JSPQ has small quantization errors in front and rear regions and bigger quantization errors in side regions. The error curve of JSPQ is almost in accordance with the JND curve. Compared with reference methods, the improvement of spatial quality with JSPQ is mainly manifest in the rear region.

5.3 Comparison of bit rates

This experiment aims to compare the coding bit rates of spatial parameters with different quantization methods. The theoretical bit rates calculated according to the codebook sizes of different methods in Fig. 10 and the actual bit rates required to code spatial parameters from standard 5.1-channel signals are both presented.

According to the codebook sizes of different methods in Fig. 10, the theoretical bit rates required to code spatial parameters are calculated and shown in Table 1, which are calculated with simple binary coding without entropy coding methods such as differential Huffman coding. Based on four specific loudspeaker pairs in the front, side (left and right), and rear regions, the theoretical coding bit rates for spatial parameters between several specific channel pairs with different methods are illustrated in Table 1, as well as the bit rates for spatial parameters in a full circle in a horizontal plane with 24 frequency bands and 20 ms time frame.

As for MPEG, the uniform codebook is designed for all channel pairs; thus, the bit rates of MPEG in Table 1 for different channel pairs are the same. With the biggest codebook size, SLQP has the biggest bit rate with nearly 9 kbps for spatial parameters in a full circle in a horizontal plane. Although Cheng_L has the lowest bit rate for the full circle, the bit rates for side and rear channel pairs are so low that obvious spatial distortion will be caused by quantization, as the quantization error presented previously in Fig. 11. Except Cheng_L, the proposed JSPQ has the lowest theoretical bit rate.

Except theoretical bit rates, actual bit rates of spatial parameters for standard 5.1-channel signals are also calculated and presented. Four standard 5.1-channel signals (shown in Table 2) from MPEG and AVS (the advanced audio and video coding standard working group of China) were employed for comparison. The comparison quantization methods include the following: MPEG surround, Choi's method [19], Cheng's method [20], and proposed

Table 1 According to the codebook sizes of different methods in Fig. 10, the theoretical bit rates (kbps) required to code spatial parameters in a full circle in a horizontal plane with 24 frequency bands and 20 ms time frame

Methods	MPEG_H	MPEG_L	SLQP_H	SLQP_L	Choi's	Cheng_H	Cheng_L	Our JSPQ
L&R pair	5.89	4.69	5.95	5.27	5.27	5.95	5.27	4.90
L&Ls/R&Rs	5.89	4.69	6.39	5.64	5.64	4.44	2.40	3.80
Ls&Rs	5.89	4.69	7.33	6.67	6.51	3.37	1.90	4.90
Full circle	8.29	7.09	8.99	8.29	8.32	7.20	6.00	6.84

JSPQ. Since SLQP is mainly applied to multichannel signals with 3D loudspeakers configurations, the alternative method of Cheng's was employed in this experiment. And since Cheng's low resolution quantization method introduces obvious spatial distortion for side and rear regions and has non-comparability with other methods, it is not presented in the experiment.

To encode and decode the 5.1-channel signals, the 5.1-to-mono MPEG surround codec with OTT mode was used in the comparison. Spatial parameter ICLDs are hierarchically extracted and quantized from the six channels signals in MPEG surround codec. In order to focus on the spatial quality and not to be affected by the coding quality of downmix signals, all the downmix signals in comparison were specially coded with 96 kbps AAC codec for high fidelity of basic audio quality. The primary concern of following experiments is the compression performance and spatial quality of spatial parameters. In this comparison experiment, the core quantization of spatial parameters ICLDs were replaced with different quantization methods. For fair comparison, the quantization indices of spatial parameters have been coded with the same inter-frame difference Huffman coding in all methods to further improve compression performance. The total coding bit rates of spatial parameters ICLDs for six channel signals were averaged through time frames for each method and presented in Table 3.

Since the virtual sound distributions in different test audio signals differ greatly, the coding performances of different methods for different test audio signals are quite different as we can see in Table 3. The highest bit rate is provided by MPEG_H which reached 12.21 kbps. The lowest bit rate for all four test signals among different methods is provided by the proposed JSPQ, which is the only one that gets an average bit rate less than 8 kbps. And JSPQ is also the only one of which the highest bit rate for all four test items did not exceed 10 kbps. Compared with MPEG_H, MPEG_L, Choi's, and Cheng's_H, the average bit rate of JSPQ has decreased by 12.77, 9.44, 9.99, and 5.17 %, respectively.

5.4 Subjective evaluation

Subjective experiments were performed to evaluate the spatial quality of proposed quantization method for spatial parameters. Listeners need to rank the perceived spatial quality of audio excerpts encoded and decoded with different quantization methods against the original unprocessed audio excerpts. Spatial parameters ICLD and azimuth are to represent the spatial direction information of virtual sound image between loudspeakers. Thus, the spatial direction and location of virtual sound image are mainly considered in the subjective experiment to evaluate the spatial perceptual distortion caused by quantization of spatial parameters with different methods.

In some experiments on sound source localization, laser pointer-based methods [33–35] are often used to locate the exact location of sound source perceived and pointed by listeners. Since we mainly focus on the difference of perceived virtual sound location between processed and unprocessed multichannel audio excerpts, the listener just needs to tell how about the differences, obvious or not. Considering it is easier for listener to point out if there are differences between perceived locations than exactly pointing to the perceived locations by hands, the standard subjective evaluation scheme MUSHRA of ITU-R BS.1534-1 [36, 37] was employed in this experiment, instead of laser pointer-based methods.

There were seven audio signals used in each MUSHRA test: original audio signal from Table 2 as a reference; original audio signal as a hidden reference; spatial quality degraded original audio signal as a hidden anchor; and test signals processed from original audio signals with

Table 2 Standard 5.1-channel test audio signals

Item names	Source	Content descriptions	Duration(s)
pcm_mps_44khz_HQ.wav	MPEG	Complex electronic music	10.84
AVS_LL6001_48_24_6.wav	AVS	Complex electronic music	8.98
AVS_LL6003_48_24_6.wav	AVS	Complex electronic music	11.00
AVS_LL6004_48_24_6.wav	AVS	Complex electronic music	10.00

Table 3 Coding bit rates (kbps) for spatial parameters with standard 5.1-channel test signals. The average bit rate of JSPQ decreased by 12.77, 9.44, 9.99, and 5.47 % compared with MPEG_H, MPEG_L, Choi's , and Cheng_H, respectively

Item names	MPEG_H	MPEG_L	Choi's	Cheng_H	Our JSPQ
pcm_mps_44khz_HQ.wav	12.21	11.68	11.48	10.87	9.92
AVS_LL6001_48_24_6.wav	9.17	8.98	8.92	8.57	8.14
AVS_LL6003_48_24_6.wav	5.88	5.68	5.92	5.69	5.68
AVS_LL6004_48_24_6.wav	9.16	8.74	8.97	8.47	8.02
Average bit rates	9.10	8.77	8.82	8.40	7.94
Bit rates reduction	12.77 %	9.44 %	9.99 %	5.47 %	—

different spatial parameter quantization methods (four reference methods and proposed JSPQ).

When original audio signals are rendered with six loudspeakers in 5.1 audio system, it may reproduce virtual sound in arbitrary directions in a full circle. Since the intention of our evaluation is about spatial quality, the hidden anchor is a processed signal with degraded spatial quality, instead of 3.5 kHz low pass filtered signal in traditional evaluations for basic audio quality. Downmix of original 5.1-channel signals is a mono signal without spatial information. The channel signals of anchor are the same signal derived from the downmix signal, which can ensure that the virtual sound reproduced with anchor signal will always stay in the middle of active loudspeakers. Since the virtual sound of original signal may be located in arbitrary location among loudspeakers, the spatial quality of anchor signal is severely degraded compared with the original signal. Energy compensation is used for the anchor signal to ensure that the energies of anchor signal and original signal are equal. Thus, the basic audio qualities of anchor signal and original signal are the same, but the spatial information of them are totally different.

To eliminate the effect of basic audio quality on subjective evaluation of spatial quality, the quantization of downmix signals were nearly perceptual lossless for test signals in four reference methods and proposed JSPQ. The basic audio qualities of test signals are almost comparable with the original signal. Owing to the quantization of spatial parameters, the spatial qualities of test signals are different with that of the original signals. The spatial quality about spatial location of the virtual sound is the key concern in our subjective evaluation.

There were a total of 15 male and female graduate students chosen as test listeners in the tests, whose research areas are audio signal processing. They were aged between 22 and 35 and were trained before attending the listening tests. They were trained to mainly focus on evaluating the virtual sound images among different test audio signals. With the original audio signal as a reference, listeners should give scores to evaluate the spatial quality about accuracy of virtual sound location reproduced with different test audio signals coded and processed by different methods. The scoring criteria for subjective evaluation is shown in Table 4. Listeners need to compare the reproduced virtual sound images by processed multichannel audio signals by different methods with that of the original reference signals. The more similar the reproduced virtual sound image is with the original virtual sound image, the higher score should be given to the corresponding method. The scores were filtered to exclude some extreme scores before averaged to get the final results with standard deviations.

Average scores are calculated as:

$$\bar{u}_{jk} = \frac{1}{N}\sum_{i=1}^{N} u_{ijk}, \tag{20}$$

in which u_i is the score of listener i for audio sequence k with method j, and N is the total number of contributing listeners. The confidence interval with confidence coefficient 95 % is:

$$\left[\bar{u}_{jk} - \delta_{jk}, \bar{u}_{jk} + \delta_{jk}\right], \tag{21}$$

in which $\delta_{jk} = t_{0.05}\frac{S_{jk}}{\sqrt{N}}$, $S_{jk} = \sqrt{\sum_{i=1}^{N} \frac{(\bar{u}_{jk}-u_{ijk})^2}{N-1}}$.

The mean scores with 95 % confidence interval of MUSHRA for spatial audio quality of different quantization methods are illustrated in Fig. 12.

The feedback from testers revealed that when virtual sounds were reproduced in the front region, there were little differences among the spatial qualities of virtual sounds

Table 4 Scoring criteria for subjective evaluation of MUSHRA

Score interval	Spatial quality about accuracy of reproduced virtual sound
100–90	Precise
90–80	Tiny bias
80–60	Small deviation
60–40	Big distortion
40–20	Serious distortion
20–0	Error

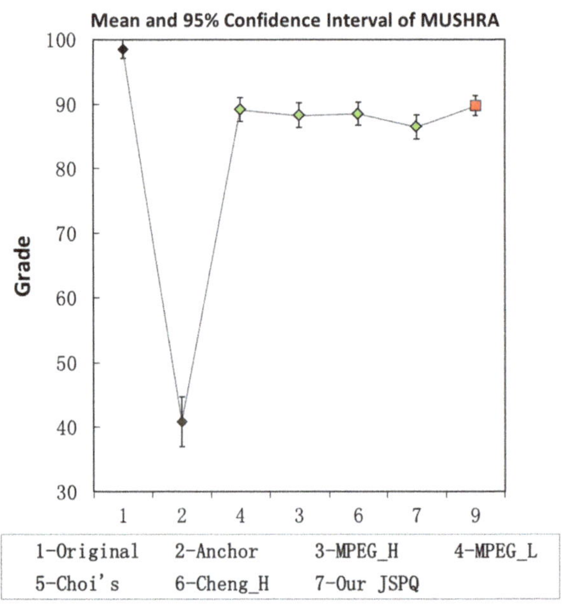

Fig. 12 Subjective evaluation MUSHRA scores for original, anchor references and different quantization methods. The average scores of most reference methods (3, 4, 5) and proposed JSPQ (7) are nearly at equivalent levels

by different quantization methods. This is probably due to that for the front region from −30° to 30°, the quantization codebooks of spatial parameters in most methods are fine enough to avoid obvious spatial perceptual distortion.

In addition, it was found that when virtual sounds were reproduced in the side regions, most reference methods did not introduce obvious spatial perceptual distortion for their enough fine quantization codebooks. Although the quantization precision of JSPQ in the side region was lower than in the front region, since the quantization errors were controlled to below perceptual JND, obvious spatial perceptual distortions were not perceived in the subjective listening tests.

As for the virtual sounds reproduced in the rear region in subjective listing tests, listeners reflected that the perceptual distortions of Cheng's methods were the most serious. This is probably attributed to the big quantization error of 35° for azimuths behind the listeners in Cheng's methods.

JSPQ assigned quantization values for spatial parameters in all directions according to the azimuthal perceptual JND. Thus, although with the lowest coding bit rate for spatial parameters, insignificant spatial distortion existed for JSPQ in all regions. As illustrated in Fig. 12, JSPQ has a comparative score with most of reference methods.

5.5 Summary of experiments

There are four respects of coding performance are analyzed by above objective and subjective experiments: quantization codebook sizes, quantization errors and coding bit rates of spatial parameters, as well as subjective MUSHRA evaluations of coding performance.

In the respect of quantization codebook sizes, it requires only 52 quantization codewords in JSPQ for the the quantization of azimuths in a full circle in horizontal plane. The quantization codebook size of JSPQ is obviously smaller than most of reference methods, for example, it has decreased by 13.3 and 56.7 % than two quantization codebooks of MPEG surround.

Meanwhile, the quantization errors of spatial parameters in JSPQ are more consistent with the JNDs of azimuths in a full circle than all of the reference methods. It means that the advantage of JSPQ is to balance between quantization error and quantization redundancy.

In respect of coding bit rates for standard test signals, the average bit rate of JSPQ to quantize spatial parameters in a full circle is lower than most of the reference methods.

In addition, subjective MUSHRA results testified that the quantization of JSPQ did not introduce obvious spatial perceptual distortion. The average MUSHRA score of JSPQ is almost comparable to the reference methods such as MPEG surround.

To sum up, a good trade-off between bit rates of spatial parameters and spatial quality is obtained with JSPQ.

5.6 Limitation and future work

According to the above objective and subjective experiments, the overall performance of JSPQ was better than the reference methods. Nevertheless, there were still some limitations in the proposed JSPQ and conducted experiments.

Although the proposed JSPQ also applies the spatial parameters quantization in different horizontal planes with different elevation angles, the implementation process and related experiments were not discussed in this paper. In 3D audio, the virtual sounds distribute widely in 3D sound field. In 3D audio, it will contain a large amount of spatial information, which is much more than in 2D horizontal plane. Thus, the application of JSPQ will achieve more significant gain in 3D audio multichannel signals coding and will be implemented in future.

In this paper, we only focused on the most important and generally used spatial parameters ICLDs and azimuths. Besides, other stereo parameters such as ICTD are also mainly related to the direction perception of virtual sound. ICTDs can be mapped to azimuth of virtual sound too. Thus, the JND characteristics of directional perception of human auditory system can be also used in the perceptual quantization of ICTDs, the perceptual redundancy removals of these spatial parameters can be the emphases of next research work.

6 Conclusions

In spatial audio coding, the spatial information of virtual sounds generated by loudspeakers are extracted as spatial parameters, as well as downmix signals are obtained from loudspeaker signals. In the case of multichannel loudspeaker systems, the virtual sounds may be widely distributed in 3D space. It contains a large amount of spatial information to achieve vivid spatial sound effects. Thus, the accurate representation of virtual sound and the efficient compression of spatial parameters are the key to the perfect reconstruction of spatial sound effects.

Human auditory system has the perceptual limitations of just noticeable difference (JND) characteristics for spatial sound location estimation. If the spatial distortion caused by quantization errors of spatial parameters can be limited below, the correlated JND threshold, then the quantization can be regarded as nearly perceptual lossless.

However, the quantization step sizes of spatial parameters in current SAC methods are not consistent well with the JND characteristics of different azimuths in a full circle. It will result in either spatial perceptual distortions of virtual sound images, or inefficient compressions of spatial parameters.

Therefore, in this paper, the characteristic of azimuthal JND was effectively used to design the quantization of spatial parameters in multichannel signals. In the proposed quantization codebook of JSPQ, the quantization step sizes of azimuths are assigned according to the JND values of azimuths in a full circle. Different azimuths from front region to rear region have different quantization step sizes.

Objective experiments and subjective evaluations confirmed the coding performance of proposed JSPQ compared with reference quantization methods of spatial parameters in respect of codebook sizes, quantization errors, coding bit rates, and spatial qualities.

The quantization codebook size of JSPQ was 13.3 and 56.7 % lower than two quantization codebooks of MPEG surround. Average bit rate reduction on spatial parameters for standard 5.1-channel signals reached up to approximately 9 and 13 % compared with MPEG Surround, while preserving comparable subjective spatial quality.

For future work, we plan to implement the framework on top of state-of-the-art 3D audio processing platforms. In particular, we are interested in advanced technologies for perceptual sensitivity analysis of complex multivariate spatial parameters [38, 39] as well as promising computing schemes for optimal 3D audio reproduction in sophisticated applications in practice [40–42].

Competing interests

The authors declare that they have no competing interests.

Acknowledgements

This work has been supported by the National Nature Science Foundation of China (61231015, 61471271), National High Technology Research, and Development Program of China (863 Program, 2015AA016306).

Author details

[1] State Key Laboratory of Software Engineering, Wuhan University, Wuhan, China. [2] National Engineering Research Center for Multimedia Software, the Key Laboratory of Multimedia and Network Communication Engineering, Computer School, Wuhan University, Wuhan, China. [3] Collaborative Innovation Center of Geospatial Technology, Wuhan, China.

References

1. A Ando, Conversion of multichannel sound signal maintaining physical properties of sound in reproduced sound field. IEEE Trans. Audio Speech Lang. Process. **19**(6), 1467–1475 (2011)
2. G Sergi, Knocking at the door of cinematic artifice: Dolby Atmos, challenges and opportunities. New Soundtrack. **50**(10), 107–121 (2013)
3. DB Ward, TD Abhayapala, Reproduction of a plane-wave sound field using an array of loudspeakers. IEEE Trans. Speech Audio Process. **9**(6), 697–707 (2001)
4. S Spors, R Rabenstein, J Ahrens, in *Proceedings of the 124th Convention of the Audio Engineering Society (AES)*. The theory of wave field synthesis revisited (Audio Engineering Society, Amsterdam, The Netherlands, 2008)
5. S Disch, C Ertel, C Faller, et al, in *Proceedings of the 117th Convention of the Audio Engineering Society (AES)*. Spatial audio coding: next-generation efficient and compatible coding of multi-channel audio (Audio Engineering Society, San Francisco, CA, USA, 2004)
6. F Rumsey, Spatial quality evaluation for reproduced sound: terminology, meaning, and a scene-based paradigm. J. Audio Eng. Soc. **50**(9), 651–666 (2002)
7. M Bosi, RE Goldberg, *Introduction to digital audio coding and standards*. (Springer Science & Business Media, Dordrecht, Netherlands, 2012)
8. RM Hershkowitz, NI Durlach, Interaural time and amplitude jnds for a 500-Hz Tone. J. Acoust. Soc. Am. **46**(6), 1464–1465 (1969)
9. JC Blauert, *Spatial Hearing: The psychophysics of human sound localisation*. (MIT Press, Cambridge, MA, US, 1997)
10. AW Mills, On the minimum audible angle. J. Acoust. Soc. Am. **30**(4), 237–246 (1958)
11. J Breebaart, S van de Par, A Kohlrausch Schuijers, Parametric coding of stereo audio. EURASIP J. Appl. Signal Process. **9**, 1305–1322 (2005)
12. JW Strutt, *The theory of sound*. (Dover Publications, Mineola, NY, US, 1877)
13. F Baumgarte, C Faller, Binaural cue coding—part I. Psychoacoustic fundamentals and design principles. IEEE Trans. Speech Audio Process. **11**(6), 509–519 (2003)
14. C Faller, F Baumgarte, Binaural cue coding—part II: schemes and applications. IEEE Trans. Speech Audio Process. **11**(6), 520–531 (2003)
15. J Breebaart, C Faller, *Spatial audio processing: MPEG surround and other applications*. (Wiley, Hoboken, NJ, US, 2008)
16. ISO/IEC Standard 14496-3, Technical description of parametric coding for high quality audio (2005)
17. 3GPP TS 26.401, General audio codec audio processing functions; enhanced aacPlus general audio codec; General Description (2004)
18. T Wu, R Hu, L Gao, et al, *Analysis and comparison of inter channel level difference and interaural level difference, multimedia modeling*. (Springer International Publishing, Miami, 2016), pp. 586–595
19. SJ Choi, Y Jung, HJ Kim, H Oh, in *Paper presented in Proceedings of the 120th Convention of the Audio Engineering Society (AES)*. New CLD quantisation method for spatial audio coding, (Paris, France, 2006)
20. B Cheng, C Ritz Burnett, Psychoacoustic-based quantisation of spatial audio cues. Electron. Lett. **44**(18), 1098–1099 (2008)
21. I Elfitri, B Guenel, AM Kondoz, Multichannel audio coding based on analysis by synthesis. Proc. IEEE. **99**(4), 657–670 (2011)
22. B Cheng, C Ritz, I Burnett, X Zheng, A general compression approach to multi-channel three-dimensional audio. IEEE Trans. Audio Speech Lang. Process. **21**(8), 1676–1688 (2013)
23. J Engdegard, B Resch, C Falch, et al, in *Proceedings of the 124th Convention of the Audio Engineering Society (AES'08)*. Spatial audio object coding

(SAOC)-the upcoming MPEG standard on parametric object based audio coding (Audio Engineering Society, Amsterdam, The Netherlands, 2008)
24. ISO/IEC JTC1/SC29/WG11 International Standard ISO/MPEG 23008-3, 3D Audio, Geneva (2015)
25. DR Perrott, Role of signal onset in sound localization. J. Acoust. Soc. Am. **45**(2), 436–445 (1969)
26. DR Perrott, K Saberi, Minimum audible angle thresholds for sources varying in both elevation and azimuth. J. Acoust. Soc. Am. **87**(4), 1728–1731 (1990)
27. DW Grantham, BWY Hornsby, EA Erpenbeck, Auditory spatial resolution in horizontal, vertical, and diagonal planes. J. Acoust. Soc. Am. **114**(2), 1009–1022 (2003)
28. L Gao, R Hu, Y Yang, X Wang, W Tu, T Wu, *Azimuthal perceptual resolution model based adaptive 3D spatial parameter coding, multimedia modeling*. (Springer International Publishing, Sydney, 2015)
29. V Pulkki, M Karjalainen, Multichannel audio rendering using amplitude panning [DSP applications]. Signal Process. Mag. IEEE. **25**(3), 118–122 (2008)
30. J Daniel, S Moreau, R Nicol, in *Proceedings of the 114th Convention of the Audio Engineering Society (AES)*. Further investigations of high-order ambisonics and wavefield synthesis for holophonic sound imaging (Audio Engineering Society, Amsterdam, The Netherlands, 2003)
31. S Ke, X Wang, L Gao, T Wu, Y Yang, in *16th Pacific-Rim Conference on Multimedia (PCM), Gwangju, Korea*. Physical properties of sound field based estimation of phantom source in 3D, (2015)
32. H Wang, C Zhang, R Hu, W Tu, X Wang, *The perceptual characteristics of 3D orientation, multimedia modeling*. (Springer International Publishing, Dublin, 2014)
33. P Majdak, R Baumgartner, B Laback, Acoustic and non-acoustic factors in modeling listener-specific performance of sagittal-plane sound localization. Front. Psychol. **5**(319), 1–10 (2014)
34. R Baumgartner, P Majdak, Modeling localization of amplitude-panned virtual sources in sagittal Planes. J. Audio Eng. Soc. **63**(7/8), 562–569 (2015)
35. P Majdak, MJ Goupell, B Laback, 3-D localization of virtual sound sources: effects of visual environment, pointing method, and training. Atten. Percept. Psychophys. **72**(2), 454–469 (2010)
36. ITU-R Rec. BS.1534-2, *Method for the subjective assessment of intermediate quality level of coding systems(MUSHRA)*. (International Telecommunications Union, Geneva, 2014)
37. M Schoeffler, F Stoter, B Edler, J Herry, in *1534 (MUSHRA),1st web audio conference*. Towards the next generation of web-based experiments: a case study assessing basic audio quality following the itu-r recommendation bs (Audio Engineering Society, Paris, France, 2015)
38. D Chen, X Li, D Cui, L Wang, D Lu, Global synchronization measurement of multivariate neural signals with massively parallel nonlinear interdependence analysis. IEEE Trans. Neural Syst. Rehabil. Eng. **22**(1), 33–43 (2014)
39. D Chen, X Li, D Cui, L Wang, S Khan, J Wang, K Zeng, C Cai, Fast and scalable multi-way analysis of massive neural data. Comput. IEEE Trans. **64**(3), 707–719 (2015)
40. D Li, R Hu, X Wang, S Yang, W tu, in *16th Pacific-Rim Conference on Multimedia (PCM)*. Multichannel simplification based on deviation of loudspeaker positions (Audio Engineering Society, Gwangju, Korea, 2015), pp. 544–553
41. D Li, R Hu, X Wang, W Tu, S Yang, in *15th Pacific-Rim Conference on Multimedia (PCM)*. Automatic multichannel simplification with low impacts on sound pressure at ears (Audio Engineering Society, Kuching, Malaysia, 2014)
42. D Chen, L Wang, A Zomaya, M Dou, J Chen, Z Deng, S Hariri, Parallel simulation of complex evacuation scenarios with adaptive agent models. Parallel Distributed Syst. IEEE Trans. **26**(3), 847–857 (2015)

3

Geometric bounding box interpolation: an alternative for efficient video annotation

Pedro Gil-Jiménez*, Hilario Gómez-Moreno, Roberto López-Sastre and Saturnino Maldonado-Bascón

Abstract

In video annotation, instead of annotating every frame of a trajectory, usually only a sparse set of annotations is provided by the user: typically its endpoints plus some key intermediate frames, interpolating the remaining annotations between these key frames in order to reduce the cost of the video labeling. While a number of video annotation tools have been proposed, some of which are freely available, and bounding box interpolation is mainly based on image processing techniques whose performance is highly dependent on image quality, occlusions, etc. We propose an alternative method to interpolate bounding box annotations, based on cubic splines and the geometric properties of the elements involved, rather than image processing techniques.

The algorithm proposed is compared with other bounding box interpolation methods described in the literature, using a set of selected videos modeling different types of object and camera motion. Experiments show that the accuracy of the interpolated bounding boxes is higher than the accuracy of the other evaluated methods, especially when considering rigid objects. The main goal of this paper is related with the bounding box interpolation step, and we believe that our design can be integrated seamlessly with any annotation tool already developed.

Keywords: Video annotation, Bounding box interpolation, Projective geometry, Cubic splines

1 Introduction

The growth in image and video processing demands larger quantities of annotated training data. A still image can be annotated at different levels. At the image level, annotations consist of simply indicating the objects present in the image, without any spatial information. At the pixel level, each pixel must be assigned to a label. Intermediate solutions may consist of drawing polygons surrounding the object or simple geometric shapes, such as rectangles or ellipses. The complexity, and thus, the cost or time required to annotate an image, depends on the level at which we need to annotate it.

Large-scale image databases are useful for supervised learning of visual object appearance. There are many publicly available databases. For instance, the PASCAL VOC Challenge [1] provides a standard dataset of annotated images at the bounding box level, for object categorization and detection. The images are collected from the flickr photo-sharing web site, and annotations are done at a single annotation "party", to ensure consistency. The ImageNet database [2] offers a collection of tens of millions of annotated images hierarchically organized, in this case, at the image level.

Many applications has been proposed in the literature for the task of image annotation, both for personal use and for massive annotations. In [3], Russell et al. proposed *LabelMe*, a web-based open tool that allows polygon-based image annotations. This tool can be accessed freely by any researcher, either to annotate their own images or to download the already annotated image collections. In [4], von Ahn and Dabbish took advantage of the *Games With A Purpose* (GWAP) philosophy to develop an online game which allows image annotation while the users are enjoying the game. In this case, images can only be annotated at the image level. The *Human Intelligence Task* (HIT) service provided by *Amazon Mechanical Turk* was explored by Sorokin and Forsyth [5] to outsource human annotation services at a low cost. This approach enables the annotation of large image databases at the bounding box level in a short time.

Video annotation can also be done at different levels. For instance, ANVIL [6] is a free tool that allows video

*Correspondence: pedro.gil@uah.es
Signal Theory and Communication Department, Universidad de Alcala, Crta Madrid-Barcelona km 33.600, Alcala de Henares, Spain

annotations at the frame level, that is, without any spatial information. However, when object location is required, the temporal evolution of the annotated objects must also be provided. Manually annotating the object location for all the frames in the video sequence is a tedious task. This is typically solved by allowing the user to provide only a limited set of annotations for a given object, then propagating this to the rest of the frames. The frames where the annotations are supplied are called *key frames*, and this set must usually include the first and last frames where the object is visible, in order for the system to know the endpoints of the object's trajectory.

The number of key frames needed depends on many factors, such as the object type (rigid or non-rigid), trajectory length (number of frames), type of object motion (straight, curved, or chaotic), and the accuracy required for the propagated annotations. The number and position of these key frames can be fixed or defined by the user. Vondrick et al. [7] discovered that 'a fixed key frame schedule is significantly faster than a user-defined key frame schedule.' However, the key frame frequency needs to be adjusted depending on the type of motion to annotate. When an object moves in a chaotic trajectory, more annotations are needed in order to accurately handle the chaotic motion. In this paper, however, we are not concerned with this issue and will use both schedules indiscriminately.

Propagation across frames has typically been performed by means of image and video processing techniques. In [8], Badrinarayanan et al. proposed a propagation scheme, given the pixel-level annotation of the first and last frames of a video sequence. The algorithm is based on a hidden Markov model and labels every pixel in all the frames between the endpoints. In [9], Liu et al. proposed a contour tracker to track the contour provided by the user in the first frame. The user can specify a time-varying depth value for each annotated object, intended to handle occlusions and 3D linear interpolation. The user may also be requested for an inspection to correct tracker mistakes. Contour tracking is also used in [10] to track annotated curves in a video sequence for *rotoscoping*, i.e., the process of tracking shapes throughout a video sequence. In [11], Kalal et al. proposed a long-term tracker of a bounding box provided by the user in the first frame. The system is able to automatically re-detect the object after a period when it is not in the field of view or after tracking failure, by learning the object appearance from the previous frames.

Propagation can also be done by means of the geometric properties of the elements involved. In this case, the user provides a set of simple (rectangle, ellipse, etc.) or complex shapes, and the system computes the coordinates of the polygon for the remaining frames. For this case, the main techniques described in the literature will be analyzed in Section 2, where different publicly available annotation tools will be introduced. Section 3 describes the alternative procedure proposed in this paper. Section 4 describes the experiments performed to compare the existing software applications. Section 5 concludes the paper.

2 Related works

In this paper, we propose an alternative method to propagate bounding box annotations between key frames. For that reason, we will analyze here the most popular software packages for video annotation at the bounding box level, focusing on the propagation algorithms proposed in each.

2.1 Available software

VIdeo Performance Evaluation Resource (ViPER) is one of the earliest video annotation tools provided by the research community [12]. The initial goal was to create a flexible ground truth format and a tool for ground truth generation and sharing. It is a JAVA application that allows the user to define object categories (people, cars, animals, etc.) and different attributes, including spatial ones (bounding box, bounding ellipse, points, etc.). The software includes a default linear interpolation utility. For this interpolator, if we define $\mathbf{b}_i$ to be the coordinates of a bounding box at $t = t_i$, then

$$\mathbf{b}_i = \begin{bmatrix} x_i & y_i & w_i & h_i \end{bmatrix}, \tag{1}$$

where (x_i, y_i) are the coordinates of the center of the bounding box and (w_i, h_i) are its width and height. Given two consecutive annotations $\mathbf{b}_i$ and $\mathbf{b}_{i+1}$, the coordinates of the bounding box for a given time $t = t_j$ are

$$\mathbf{b}_j = \left(\frac{t_{i+1} - t_j}{t_{i+1} - t_i}\right)\mathbf{b}_i + \left(\frac{t_j - t_i}{t_{i+1} - t_i}\right)\mathbf{b}_{i+1}. \tag{2}$$

Linear interpolation can be defined as the simplest approach [7] for label propagation. This method has two main drawbacks. On the one hand, linear interpolation can only handle straight trajectories, since interpolation between adjacent segments are computed independently. On the other hand, it can not accurately capture scale transformations due to perspective effects. A recent update can be found in [13], which allows semiautomatic video annotation. Nevertheless, the system is based on a foreground/background motion detection, which limits its use to videos recorded with static cameras.

Vondrick et al. proposed in [7] a user interface to be deployed on Amazon Mechanical Turk (MT) to outsource bounding box annotations using its HIT services. In their work, they analyzed several aspects related to this HIT service, such as interface design, the evaluation of a worker's talents, and the optimal key frame frequency. Focusing on the bounding box propagation technique, they proposed two different algorithms, the simplest one being the linear interpolation discussed above. The other method is based on dynamic programming, using a visual model of the annotated object. They extract the model from the N bounding boxes provided, compute a feature descriptor composed of HOG and color features, and train a linear SVM classifier from positive and negative (background) samples. The main drawback is that the algorithm is highly dependent on the image quality, and the system can even fail in the case of object occlusions. The system includes a constant to allow the visual interpolator to gracefully degrade to linear interpolation when the former becomes inaccurate.

Instead of relying on visual properties, the interpolation can be done using the geometric properties of the elements involved. In this case, when interpolating coordinates, we can use image or spatial coordinates.

Consider a point $\mathbf{Q}$ moving in front of a camera, as in Fig. 1. At $t = t_i$, the point will be located at $\mathbf{Q}_i$ in spatial coordinates, and its image will be at $\mathbf{q}_i$, and similarly for $t = t_{i+1}$. If we suppose constant motion for the point, at $t_j = (t_i + t_{i+1})/2$, the point will be located at $\mathbf{Q}_j$, at the intermediate position between the points in space and its image at $\mathbf{q}_j$. However, if we interpolate using image coordinates, the point $\hat{\mathbf{q}}_j$ will be located at the intermediate position between the image points, and as can be seen in Fig. 1, the coordinates $\mathbf{q}_j$ and $\hat{\mathbf{q}}_j$ do not coincide, due to the effects of perspective.

Perspective effects get canceled only when the object moves parallel to the image plane. These effects increase as the component of the motion normal to the image plane increases. For instance, in Fig. 2, we can see two annotated trajectories. Object 1 moves almost parallel to the image plane, so that the perspective effects are minimal and interpolation using image coordinates can be an accurate solution.

However, parallel motion is not the general situation. For object 2 in Fig. 2, the variation in depth is large. In this situation, the last assumption does not hold, and the accuracy in the bounding box interpolation decreases, as we will see later, in Section 4.

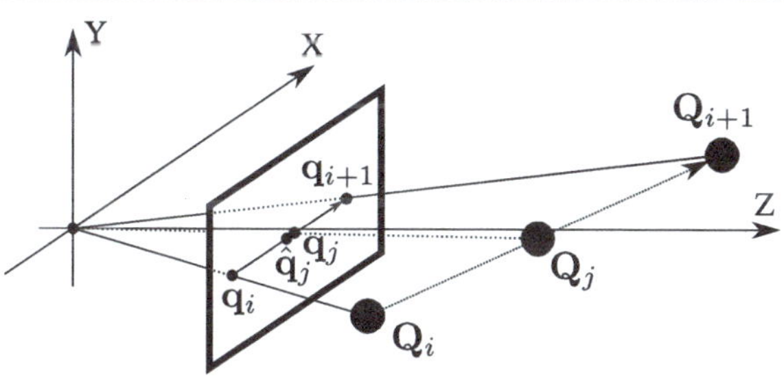

Fig. 1 Interpolation of the intermediate position between $\mathbf{Q}_i$ and $\mathbf{Q}_{i+1}$ using image ($\hat{\mathbf{q}}_j$) and spatial ($\mathbf{q}_j$) coordinates

Fig. 2 Perspective effects for an object which moves parallel (*1*) and for an object which does not move parallel (*2*) to the image plane

To take into account this effect, Yuen et al. [14] pointed out that 'a constant velocity in space does not project to a constant velocity in the image plane, due to perspective effects.' In their work, they proposed an alternative method to project the coordinates of the annotated object to the image plane for any time between two key frames.

Assuming that an object in space is moving with constant motion between two key frames, any point of the object will therefore move in a straight line. The line in the image plane joining the same point of the object in both key frames will be the image of the line followed by this point in space. If all the points of the object move in parallel trajectories, all points in space will meet at infinity, and so the cross point in the image plane of all lines will be the vanishing point for this trajectory. The coordinates of any point of the object for a given time t is computed using

$$(x(t), y(t)) = \left(\frac{x_0 + \lambda(t) x_v}{\lambda(t) + 1}, \frac{y_0 + \lambda(t) y_v}{\lambda(t) + 1} \right) \tag{3}$$

where (x_0, y_0) are the coordinates of the point in the frame at $t = 0$, and $p_v = (x_v, y_v)$ are the coordinates of the vanishing point computed as described above. $\lambda(t)$ is the displacement of the point along the direction of the motion in space, and since constant motion is assumed, we have $\lambda(t) = vt$, where v is the velocity of the point in space and can be computed, given a second key frame at $t = t_1$, using

$$v = \frac{x_1 - x_0}{t_1 \cdot (x_v - x_1)}. \tag{4}$$

This approach has two main drawbacks. Suppose that we have two bounding boxes of an object at $t = t_i$ and $t = t_{i+1}$, as in Fig. 3, and we trace the lines joining the four pair of corners. As stated before, the cross point would correspond to the vanishing point of the trajectory. However, in the general situation, annotation errors and changes in object aspect ratio will make the lines fail

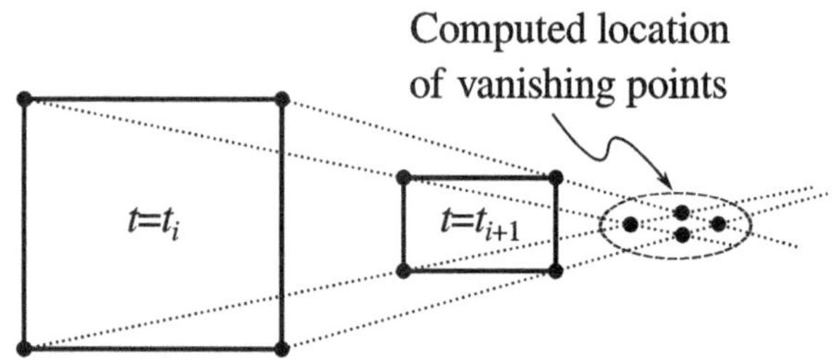

Fig. 3 Errors in computation of the vanishing point from a pair of bounding boxes

to meet at the same point. For some points in the last figure, some of the cross points are even located outside the image plane. Furthermore, for motion parallel to the image plane, the vanishing point will lie at infinity, so that in (3), there is an indetermination that must be resolved some other way. The second drawback is that this schema can only be employed between a pair of key frames, while assuming straight paths in between, which is not a correct assumption for object motion in general.

The second drawback is solved by Lee et al. in [15, 16] by interpolation using cubic splines [17]. In their work, they proposed a tool for collaborative video annotation, mainly intended for interactive services (i.e., smart TV) in the multimedia industry. They also proposed a dynamic sampling-based cubic spline interpolation (DSCSI) algorithm to reduce the size of the trajectory data on a "clickable video". Focusing on the interpolation part, the object position is computed using cubic spline interpolation (CSI) from a sparse set of annotations. Nevertheless, the interpolation is performed using image coordinates instead of spatial coordinates, so that, as explained above, it is not able to accurately model object trajectories with large depth variations.

To address these problems, in this paper, a new interpolation algorithm that combines a 3D reconstruction schema is proposed, similar to the one proposed in [14], with the cubic spline interpolation used in [16]. However, the main difference with respect to [14] is that the algorithm proposed in our paper performs a continuous reconstruction in the spatial coordinates, as will be described in Section 3.3. This improvement allows us to use cubic splines to interpolate the spatial bounding box coordinates, instead of using linear interpolation, to model object trajectories. The main difference with respect to [16] is that the proposed 3D reconstruction allows us to model more accurately the trajectory of objects when they exhibit large depth variations.

3 Geometric interpolation

In this section, we will describe the process designed to compute the bounding box coordinates for any frame from the known bounding box annotations. The overall schema is depicted in Fig. 4. Suppose we have two annotations $\mathbf{b}_i$ and $\mathbf{b}_{i+1}$ of an object, the key idea is finding the coordinates $\mathbf{B}_i$ and $\mathbf{B}_{i+1}$ of the bounding box in space. Having that, computing the coordinates $\mathbf{b}_j$ of the bounding box in the image plane for any time $t = t_j$ is straightforward, assuming we know the motion model of the object in space.

We will demonstrate that using only bounding box coordinates, we can perform a 3D reconstruction of the bounding box positions. If the camera's internal parameters are not known, i.e., we are working with an uncalibrated camera, the 3D positions can be computed only up to a scale transformation. Nevertheless, we will see that actual 3D coordinates are irrelevant for bounding box interpolation, and so, no camera calibration is needed.

Prior to defining the interpolation schema proposed in this paper, we introduce briefly the basic concepts related with central projection and the camera model needed. The notation has been taken from [18] where possible. We will first describe the interpolation technique for a simpler two-dimensional case and then extend the concepts to three dimensions.

3.1 Two-dimensional camera model

In the two-dimensional case, central projection is based on the *line camera* shown in Fig. 5. This camera projects points on the plane XZ onto an *image line*. Let the center of projection be the point $\mathbf{C} = (X_C, Z_C, 1)^T$, expressed in homogeneous coordinates, and let $\mathbf{l} = (a, b, c)^T$ be the homogeneous vector representing the image line. The distance from the line $\mathbf{l}$ to the point $\mathbf{C}$ is f, which is called the *focal length* of the camera. The line from the center of projection, normal to the image line, is called the *principal axis* of the camera, and the cross point between the image line and the principal axis is called the *principal point*.

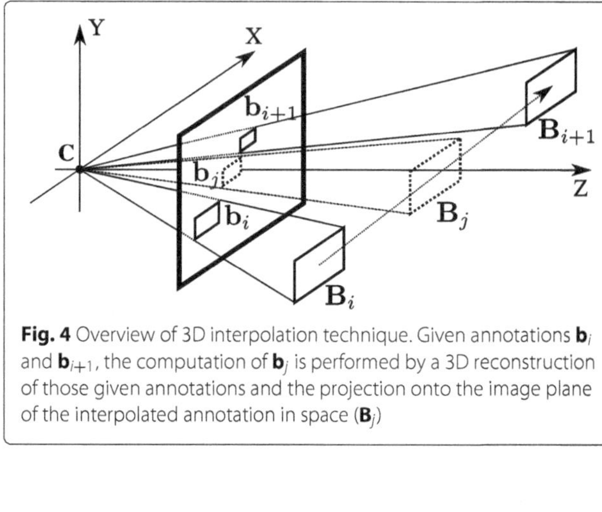

Fig. 4 Overview of 3D interpolation technique. Given annotations $\mathbf{b}_i$ and $\mathbf{b}_{i+1}$, the computation of $\mathbf{b}_j$ is performed by a 3D reconstruction of those given annotations and the projection onto the image plane of the interpolated annotation in space ($\mathbf{B}_j$)

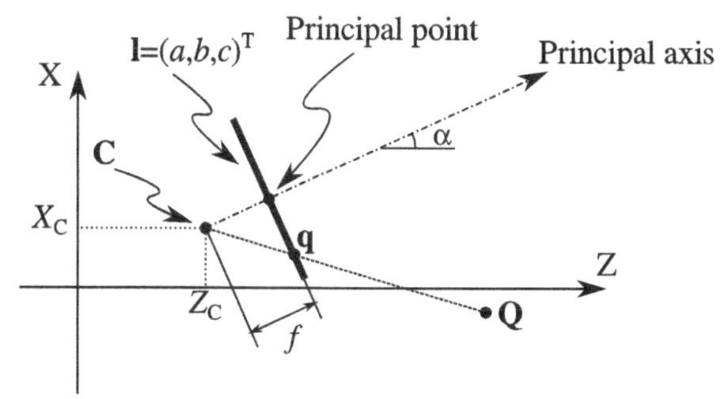

Fig. 5 Model for a two-dimensional projective camera

Under this model, a point $\mathbf{Q} = (X_Q, Z_Q, 1)^T$ in the plane is mapped to the point $\mathbf{q} = (x_q, w_q)^T$ in homogeneous coordinates in the image line. This point will be located where a line joining the point $\mathbf{Q}$ with the center of projection $\mathbf{C}$ meets the image line. The homogeneous coordinates of $\mathbf{q}$ can be computed from the *projection matrix* P:

$$\mathrm{P} = \mathrm{K} \cdot \mathrm{R} \cdot [\mathrm{I} \mid -\widetilde{\mathbf{C}}] \quad (5)$$

where K is the 2×2 camera *internal parameter* matrix, R is the 2×2 camera *rotation* matrix, I is the 2×2 identity matrix, and $\widetilde{\mathbf{C}} = \{X_C, Z_C\}$ is the center of projection expressed in inhomogeneous coordinates. In order to simplify the problem, suppose that the camera is placed at a canonical position, that is, suppose that the center of projection is the origin of the coordinate system, $\mathbf{C} = (0, 0, 1)^T$, and that the principal axis is the Z axis ($\alpha = 0$), as in Fig. 6. Furthermore, we can have the principal point be the origin of the image coordinate system. In this situation, P reduces to

$$\mathrm{P} = \begin{pmatrix} f & 0 \\ 0 & 1 \end{pmatrix} \cdot \begin{pmatrix} 1 & 0 \\ 0 & 1 \end{pmatrix} \cdot \begin{pmatrix} 1 & 0 & 0 \\ 0 & 1 & 0 \end{pmatrix} = \begin{pmatrix} f & 0 & 0 \\ 0 & 1 & 0 \end{pmatrix}. \quad (6)$$

A point $\mathbf{Q} = \left(X_Q, Z_Q, 1\right)^T$ in the plane is projected onto the image line at

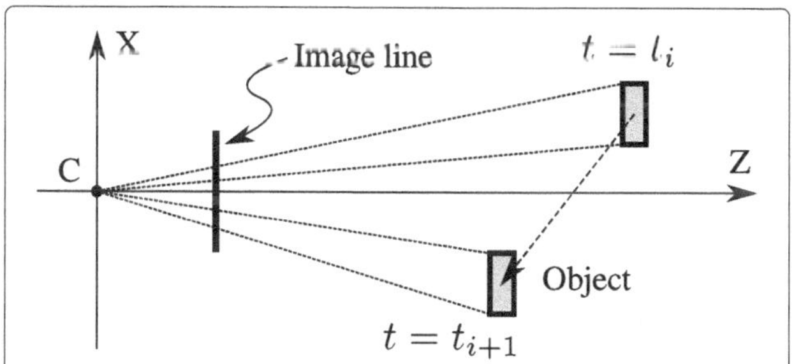

Fig. 6 An object moving in the plane *XZ*, with the camera placed at a canonical position

$$\mathbf{q} = \begin{pmatrix} x_q \\ w_q \end{pmatrix} = \mathrm{P} \cdot \mathbf{Q} = \begin{pmatrix} f \cdot X_Q \\ Z_Q \end{pmatrix} \quad (7)$$

which can be expressed in inhomogeneous coordinates as

$$\widetilde{q} = \frac{f \cdot X_Q}{Z_Q}. \quad (8)$$

3.2 Reconstruction for a single segment

Now, consider an object moving in front of a line camera from $t = t_i$ to $t = t_{i+1}$, as shown in Fig. 6. In this section, we will consider a trajectory defined by only one segment, that is, only two key frames, at the start and endpoints of the trajectory, without intermediate key frames. In the following sections, we will extend the method to any number of segments.

If we call a the lower edge of the object, and b the upper one, as can be seen in Fig. 7, the edges of the object will have inhomogeneous coordinates $\widetilde{\mathbf{Q}}_i^a = \{X_i^a, Z_i\}$ and $\widetilde{\mathbf{Q}}_i^b = \{X_i^b, Z_i\}$, respectively. Note that we have assumed the object to be parallel to the image line, thus the coordinate Z for both points is the same. Under a camera matrix P, the point a is imaged onto the image line at

$$\widetilde{q}_i^a = \frac{f \cdot X_i^a}{Z_i} \quad (9)$$

as shown in Fig. 8. If the shift of the object from t_i to t_{i+1} is $\Delta_i = \{\Delta X_i, \Delta Z_i\}$, then the coordinates for point a at $t = t_{i+1}$ will be

$$\widetilde{\mathbf{Q}}_{i+1}^a = \widetilde{\mathbf{Q}}_i^a + \Delta_i \quad (10)$$

and will be imaged by the camera at

$$\widetilde{q}_{i+1}^a = \frac{f \cdot (X_i^a + \Delta X_i)}{Z_i + \Delta Z_i} \quad (11)$$

and similarly for point b. Gathering all the equations, we have

$$\begin{cases} \widetilde{q}_i^a = \frac{f \cdot X_i^a}{Z_i} \\ \widetilde{q}_i^b = \frac{f \cdot X_i^b}{Z_i} \\ \widetilde{q}_{i+1}^a = \frac{f \cdot (X_i^a + \Delta X_i)}{Z_i + \Delta Z_i} \\ \widetilde{q}_{i+1}^b = \frac{f \left(X_i^b + \Delta X_i\right)}{Z_i + \Delta Z_i}. \end{cases} \quad (12)$$

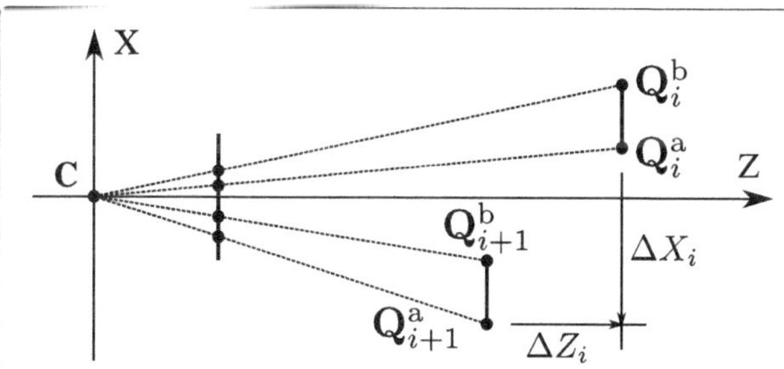

Fig. 7 Point coordinates as the object moves from $t = t_i$ to $t = t_{i+1}$

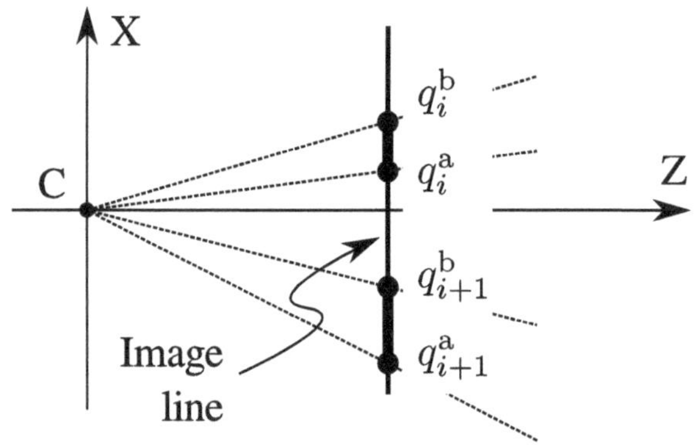

Fig. 8 Projection of the object points onto the camera image line

Before going on, we will analyze the effect of the focal length f. As can be seen in Fig. 9, for an image point $\mathbf{q}_i$ at $t = t_i$, if we assume the point depth to be $Z = Z_i$ and use a camera with focal length f, the point is back projected to $\widetilde{\mathbf{Q}}_i = \{X_i, Z_i\}$. Now, changing the camera focal length to $f' = k_E \cdot f$ while keeping the same depth Z_i, the point coordinates will change to $\widetilde{\mathbf{Q}}'_i = \{X'_i, Z_i\}$. By similar triangles, we have

$$\frac{\widetilde{q}_i}{f} = \frac{X_i}{Z_i} \tag{13}$$

and

$$\frac{\widetilde{q}_i}{f'} = \frac{X'_i}{Z_i} \tag{14}$$

which gives $X_i = k_E \cdot X'_i$, and the same for $t = t_{i+1}$. Since our final interest is not the reconstruction itself, but the computation of the images of intermediate positions, it can easily be seen that the interpolation is independent of the particular focal length chosen. For instance, if we interpolate the image coordinates for the point at the middle position between t_i and t_{i+1}, assuming constant motion, the coordinates for the point when focal length is f will be

$$\widetilde{\mathbf{Q}}_j = \left\{\frac{X_i + X_{i+1}}{2}, \frac{Z_i + Z_{i+1}}{2}\right\} \tag{15}$$

which will be imaged at

$$\widetilde{q}_j = f\left(\frac{X_i + X_{i+1}}{Z_i + Z_{i+1}}\right). \tag{16}$$

In the same way, if the focal length is f', the interpolated position will be

$$\widetilde{q}'_j = f'\left(\frac{X'_i + X'_{i+1}}{Z_i + Z_{i+1}}\right) = f'\left(\frac{X_i + X_{i+1}}{k_E \cdot (Z_i + Z_{i+1})}\right) = \widetilde{q}_j, \tag{17}$$

that is, both points are imaged onto the same image coordinate. Figure 9 shows a graphical example.

The last proof shows that the coordinate of the interpolated points is independent of the particular focal length chosen, and therefore, the internal camera parameters are not needed. Thus, to simplify the problem, we choose $f = 1$. Reorganizing the initial equations,

$$\begin{cases} \widetilde{q}^a_i Z_i - X^a_i = 0 \\ \widetilde{q}^b_i Z_i - X^b_i = 0 \\ \widetilde{q}^a_{i+1} Z_i + \widetilde{q}^a_{i+1}\Delta Z_i - X^a_i - \Delta X_i = 0 \\ \widetilde{q}^b_{i+1} Z_i + \widetilde{q}^b_{i+1}\Delta Z_i - X^b_i - \Delta X_i = 0 \end{cases}. \tag{18}$$

The last result forms a system of linear equations in the unknowns $\mathbf{B}_i = [X^a_i, X^b_i, Z_i, \Delta X_i, \Delta Z_i]$, which can be computed from its null space. Since (18) is a homogeneous system of equations, $\mathbf{B}_i$ can be found up to a constant factor k_S. The meaning from the geometric point of view is that the space is reconstructed up a scale transformation. However, under central projection, two points related by a scale transformation are imaged onto the same image coordinates. This means that the particular constant factor k_S chosen for $\mathbf{B}_i$ is not important, since only the relation between all variables is important. We will come back to this concept when analyzing trajectories composed of more than one segment.

With a little computation from (18), we get

$$\begin{cases} \left(\widetilde{q}^a_{i+1} - \widetilde{q}^a_i\right) Z_i + \widetilde{q}^a_{i+1}\Delta Z_i - \Delta X_i = 0 \\ \left(\widetilde{q}^b_{i+1} - \widetilde{q}^b_i\right) Z_i + \widetilde{q}^b_{i+1}\Delta Z_i - \Delta X_i = 0 \end{cases} \tag{19}$$

and

$$\begin{cases} X^a_i = \widetilde{q}^a_i Z_i \\ X^b_i = \widetilde{q}^b_i Z_i \end{cases}. \tag{20}$$

Solving (19) using the cross product, we obtain

$$\begin{cases} Z_i = \widetilde{q}^b_{i+1} - \widetilde{q}^a_{i+1} = s_{i+1} \\ \Delta Z_i = \left(\widetilde{q}^b_i - \widetilde{q}^a_i\right) - \left(\widetilde{q}^b_{i+1} - \widetilde{q}^a_{i+1}\right) = s_i - s_{i+1} \\ \Delta X_i = \widetilde{q}^b_i \widetilde{q}^a_{i+1} - \widetilde{q}^a_i \widetilde{q}^b_{i+1} \end{cases} \tag{21}$$

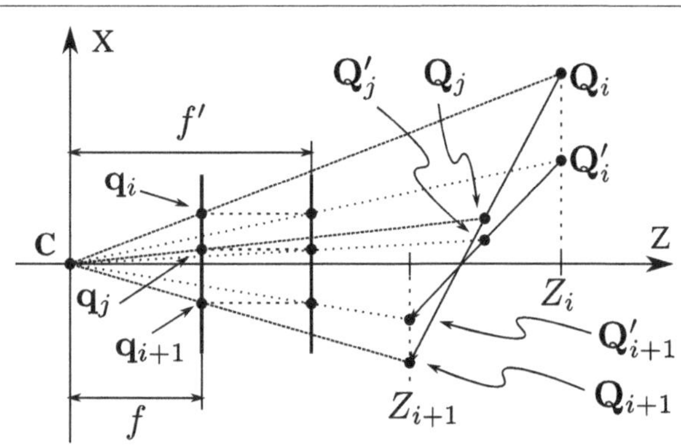

Fig. 9 Effect of the camera focal length on the interpolated points. The inhomogeneous coordinate $\widetilde{q}_j$ of the interpolated point is independent of the particular focal length f chosen

where $s = \widetilde{q}^{\text{b}} - \widetilde{q}^{\text{a}}$ is the size of the object in the image line. Finally, (20) can be solved using these last results:

$$\begin{cases} X_i^{\text{a}} = \widetilde{q}_i^{\text{a}} \cdot s_{i+1} \\ X_i^{\text{b}} = \widetilde{q}_i^{\text{b}} \cdot s_{i+1} \end{cases} . \tag{22}$$

3.3 Reconstruction from multiple segments

The results obtained in the previous section allow us to interpolate the coordinates for any frame between two key frames. For a trajectory composed of N key frames, we could apply the same process independently for the $N - 1$ segments. However, since the computations are independent, the results would give disjoint reconstructed segments. To see this, suppose we have three different annotations for an object, $\mathbf{b}_i$, $\mathbf{b}_{i+1}$, and $\mathbf{b}_{i+2}$, as in Fig. 10.

The reconstruction for the segment defined by $\mathbf{b}_i$ and $\mathbf{b}_{i+1}$ gives $\mathbf{B}_i$ and $\mathbf{B}_{i+1}$. Likewise, for points $\mathbf{b}_{i+1}$ and $\mathbf{b}_{i+2}$ we have $\mathbf{B}'_{i+1}$ and $\mathbf{B}'_{i+2}$. The reconstructed segments will be, in general, disjoint. Although both reconstructions are correct, they are only useful if further interpolations are performed independently for each segment, i.e. assuming straight paths between key frames. However, typical object motion is better modeled with interpolation schemes which need more samples. As we will see later, we will use cubic spline interpolation, a method which uses more than two samples to compute any interpolation. To this end, we need the reconstructed trajectory to be continuous along the whole path.

Continuity can be achieved easily taking into account that reconstruction can only be computed up to a scale transformation. We can adjust the scale factor k_s, introduced before, for the second segment to have $\hat{\mathbf{B}}'_{i+1} = \mathbf{B}_{i+1}$:

$$\hat{\mathbf{B}}'_{i+1} = k_i^{\text{s}} \cdot \mathbf{B}'_{i+1} = \mathbf{B}_{i+1} \tag{23}$$

where $k_i^{\text{s}} = Z_{i+1}/Z'_{i+1}$ is the scale factor to be applied to the second segment, that is, $\hat{\mathbf{B}}'_{i+1} = k_i^{\text{s}} \cdot \mathbf{B}'_{i+1}$ and $\hat{\mathbf{B}}'_{i+2} = k_i^{\text{s}} \cdot \mathbf{B}'_{i+2}$. Figure 11 shows how, applying this scale correction, we can make the trajectory continuous. Furthermore, for the next segment, a new scale factor $k_{i+1}^{\text{s}} = Z'_{i+2}/Z''_{i+2}$

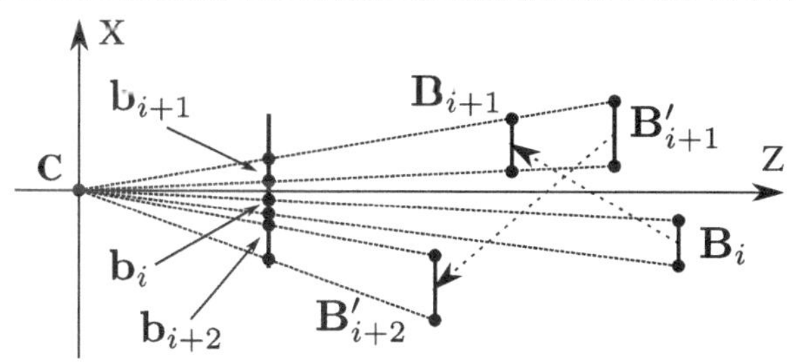

Fig. 10 Independent reconstruction for two segments of an object trajectory

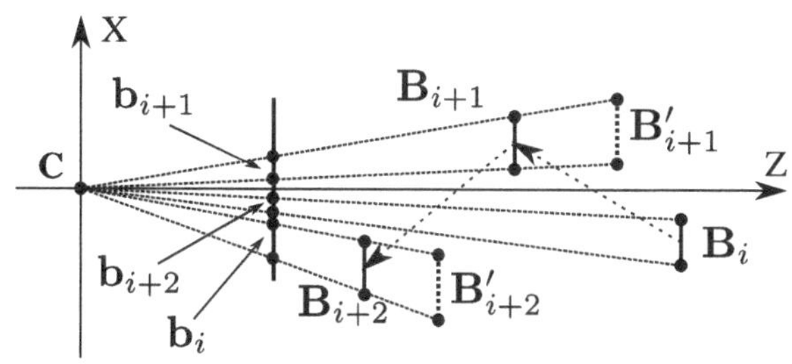

Fig. 11 Scale correction for the second segment to build a continuous trajectory

must be computed, so that the coordinates for that segment will be $\hat{\mathbf{B}}''_{i+2} = k_i^{\text{s}} \cdot k_{i+1}^{\text{s}} \cdot \mathbf{B}''_{i+2} = k_i^{\text{s}} \cdot \mathbf{B}'_{i+2}$, and so on.

3.4 Three-dimensional case

Extrapolating the problem to three dimensions is not straightforward. A first approach could be to perform the reconstruction independently for the horizontal and vertical coordinates, but this would yield different depth values (Z_i and ΔZ_i) for the horizontal and vertical reconstructions. Although we can adjust the scale factor k_s of one of the coordinates to make the depth values Z_i equal for both coordinates, the other value, ΔZ_i, will, in general, be different and so will be the object depth for $t = t_{i+1}$, but, obviously, the depth position should be the same for both coordinates along the whole path, since it is the same point.

The depth values for the horizontal and vertical coordinates will only be equal when the bounding box aspect ratio is kept constant between annotations. Thus, to improve the performance, we have to correct the aspect ratio variation between key frames. To see this, suppose given an object annotation for $t = t_i$ and $t = t_{i+1}$ as in Fig. 12. As the object moves away from the camera, its size will decrease and inversely. If the aspect ratio of the annotated rectangle was kept constant between the key frames, this would be the only variation on both height and width.

This can easily be checked if we look at the equations in (21) for the computation of Z and ΔZ. Applying the

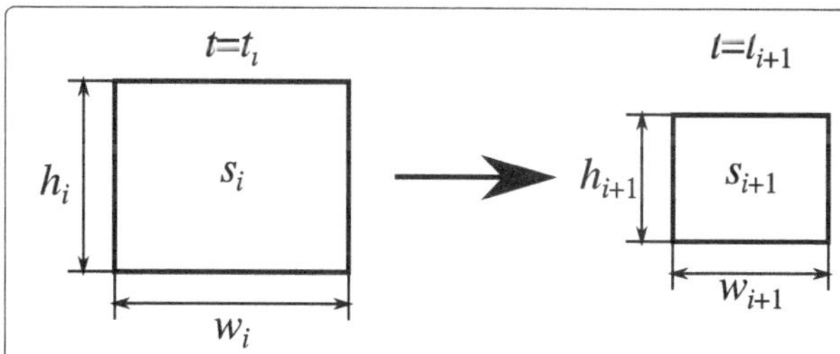

Fig. 12 Horizontal and vertical variables for an object in two consecutive annotations

equation for the computation of the depth information, substituting $s \to w$ for the horizontal coordinate, and $s \to h$ for the vertical coordinate, we have $Z_i^x = w_{i+1}$ and $\Delta Z_i^x = w_i - w_{i+1}$ for the x coordinate and $Z_i^y = h_{i+1}$ and $\Delta Z_i^y = h_i - h_{i+1}$ for the y coordinate. We can use the scale correction factor k_s introduced before to have $\hat{Z'}_i^y = k_i^s \cdot Z_i^y = Z_i^x$, that is, the depth position at the beginning of the segment will be the same for the horizontal and vertical reconstructions. In this situation, the depth shift for the vertical reconstruction will be

$$\Delta \hat{Z'}_i^y = k_i^s \cdot \Delta Z_i^y = \frac{Z_i^x}{Z_i^y} \cdot \Delta Z_i^y = \frac{w_{i+1}}{h_{i+1}}(h_i - h_{i+1}). \quad (24)$$

If both rectangles have the same aspect ratio,

$$\frac{h_{i+1}}{h_i} = \frac{w_{i+1}}{w_i} \quad (25)$$

and after a little calculation, we obtain

$$\Delta \hat{Z'}_i^y = \frac{w_i}{h_i}(h_i - h_{i+1}) = w_i - w_{i+1} = \Delta Z_i^x, \quad (26)$$

that is, the computed depths for the x and y coordinates are the same, as desired. However, if the aspect ratio is not maintained, the last equation does not hold, and so, the depth shifts for the horizontal and vertical reconstructions do not coincide.

The aspect ratio of an imaged object is only maintained when considering planar rigid objects that move without rotating. But even in this hypothetical case, errors in object annotations will generally result in the bounding box aspect ratio not being maintained. When annotating non-rigid objects, the problem gets worse. For instance, consider a person standing up in one key frame, and the same person crouching in the next key frame. In that situation, the bounding box height will halve, whereas the width will be the same, or even increase.

In order to improve performance, we need to take such aspect ratio variations into account. Again, consider the object annotations given in Fig. 12. As we saw before, one of the factors that make the height and width of the object change is depth variation. If we denote by k_i^{d} the size variation from t_i to t_{i+1} due to the change in depth and compute it from the geometrical mean of its width and height,

$$k_i^{\mathrm{d}} = \sqrt{\frac{h_{i+1} \cdot w_{i+1}}{h_i \cdot w_i}}, \quad (27)$$

the variation in height will be

$$h_{i+1} = k_i^{\mathrm{d}} \cdot k_i^{\mathrm{h}} \cdot h_i, \quad (28)$$

and similarly for its width. In the last equation, k_i^{h} (and k_i^{w} for the width) is the variation in height not included in the size variation due to the change in depth k_i^{d}. This factor can be computed by

$$k_i^{\mathrm{h}} = \frac{h_{i+1}}{k_i^{\mathrm{d}} \cdot h_i} = \sqrt{\frac{w_i \cdot h_{i+1}}{w_{i+1} \cdot h_i}} \quad (29)$$

for the height, and

$$k_i^{\mathrm{w}} = \frac{w_{i+1}}{k_i^{\mathrm{d}} w_i} = \sqrt{\frac{w_{i+1} \cdot h_i}{w_i \cdot h_{i+1}}} = \frac{1}{k_i^{\mathrm{h}}}. \quad (30)$$

That is, the factors are inversely proportional. Thus, to complete the system, we only have to take into account the last variable k_i^{h} along the trajectory, modifying the width and height of the bounding box of the annotated rectangle correspondingly. Since its value can change along the different segments of the trajectory, interpolation of this variable between key frames needs to be considered to eliminate abrupt size changes between segments.

3.5 Bounding box interpolation

The process described so far allows us to perform, up to a scale transformation, a three-dimensional reconstruction of the annotated trajectory. The last step is the design of the interpolation schema to compute the bounding box coordinates for any frame. First of all, thanks to the reconstruction process described, the interpolation is not performed over the image plane coordinates but over the reconstructed spatial coordinates. Once we have the interpolated 3D coordinates, we project them over the image plane to get the new bounding box coordinates. In Section 3.1, we chose the camera to have its center of projection at the origin of the system of coordinates, with its principal axis the Z axis and with focal length equal to 1. Since this is the camera employed in the reconstruction, it will be the one to project the interpolated coordinates to the image plane.

In order to model the actual object motion, instead of using linear interpolation between key frames, we use cubic spline interpolation, as can be seen in Fig. 13. Since we are using spatial coordinates, the three coordinates X, Y, and Z need to be interpolated. Furthermore, to eliminate abrupt changes in object size, the aspect ratio correction k_i^h also needs to be interpolated. So, a four cubic spline interpolators need to be implemented. Algorithm 1 summarizes the whole process needed to interpolate bounding box coordinates from a sparse set of annotations.

Algorithm 1: Algorithm for the interpolation of bounding box coordinates combining 3D reconstruction and cubic splines.

function Interpolation (T,j);
Input : $T = [\mathbf{b}_0, \mathbf{b}_1, \ldots \mathbf{b}_N]$ where $\mathbf{b}_i = [x_i, y_i, w_i, h_i]$: list of $N+1$ bounding box annotations for a given trajectory
j: Required interpolation time
Output: $\mathbf{b}_j$: bounding box coordinates for time j
if $\mathbf{b}_j \in T$ **then**
  return $\mathbf{b}_j$
else
  Perform 3D reconstruction to obtain $\mathrm{T}' = [\mathbf{B}_0, \mathbf{B}_1, \ldots \mathbf{B}_N]$, where $\mathbf{B}_i = [X_i, Y_i, Z_i.k_i^{\mathrm{h}}]$
  Compute $\mathbf{B}_j = [X_i, Y_i, Z_i]$ from $\mathbf{T}'$ using a 4 cubic spline interpolator
  Project $\mathbf{B}_j$ to the image plane to obtain $\mathbf{b}_j$, using projection matrix $\mathrm{P} = [\mathrm{I}|0]$
  return $\mathbf{b}_j$
end

Although cubic splines can accurately model general object motion, they are not able to model abrupt motion changes, like a ball bouncing on the floor. This can easily be solved by allowing the user to insert *trajectory breaks* at these points, so that the interpolation is computed independently at either side of the break. Figure 14 shows the improvement when breaks are included in the trajectory annotation.

4 Results and discussion

In order to test the interpolation schema described in this paper, the proposed algorithm has been implemented in Python, using OpenCV. For the cubic spline interpolation, we used the functions implemented in the SciPy library [19]. For comparison purposes, the following algorithms have been implemented and tested:

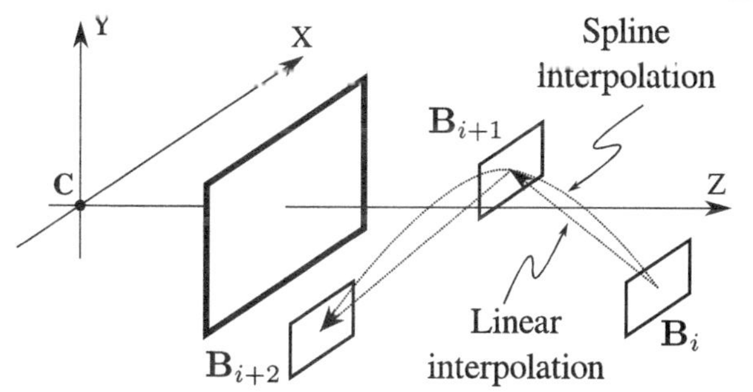

Fig. 13 Comparison between linear and spline interpolation over the reconstructed spatial coordinates

1. *Geometric cubic spline interpolation* (GC): This is the algorithm described in this paper.
2. *Geometric interpolation* (GI): A simplified version of the algorithm described, without using cubic spline interpolation.
3. *Linear interpolation* (LI): A linear interpolator, as defined in Eq. (2). This is the interpolation used in ViPER [12].
4. *LabelMe* (LM): Although the original work was designed for general polygonal annotations [14], the algorithm can be easily adapted to work with rectangles. For the computation of the vanishing point p_{v}, we used the cross point between the lines defined by the top-left and the bottom-right corners of consecutive rectangles. Velocity and point coordinates are computed according to (4) and (3).
5. *Cubic spline interpolation* (CS): Following the original work by Lee et al. [15], the center coordinates of the bounding box and the rectangle size in the image plane as well are interpolated between key frames using the cubic spline functions implemented in the SciPy library.

To test the performance of the interpolation algorithms, 19 short (a few seconds long) video clips were selected and manually annotated. The clips were extracted from surveillance and broadcasted video and personal cameras. All the videos, annotations, source code, and a Python application can be downloaded from the project web page [20]. The objects were carefully annotated in every frame (which can involve more than one hundred frames each). Table 1 summarizes the main properties of the videos selected for the test. The selection includes different object and/or camera motion types (see column *motion type* in this table), to check the algorithms against a number of situations. These include the following:

(a) Rigid object moving in straight path, static camera
(b) Rigid object moving in curved path, static camera
(c) Static object, with camera panning
(d) Static object, with camera translation
(e) Static object, with camera zoom
(f) Non-rigid objects

Also, Fig. 15 shows some sample images of the video data set.

The accuracy of the bounding box interpolation has been evaluated by a decimation followed by a further interpolation of the decimated samples, according to the following process. Once a complete trajectory is read, one half of the samples (the odd or the even samples) are removed, and their coordinates are interpolated using the implemented algorithms. For each computed bounding box $\mathbf{b}_i$, we calculate the *overlap error* as

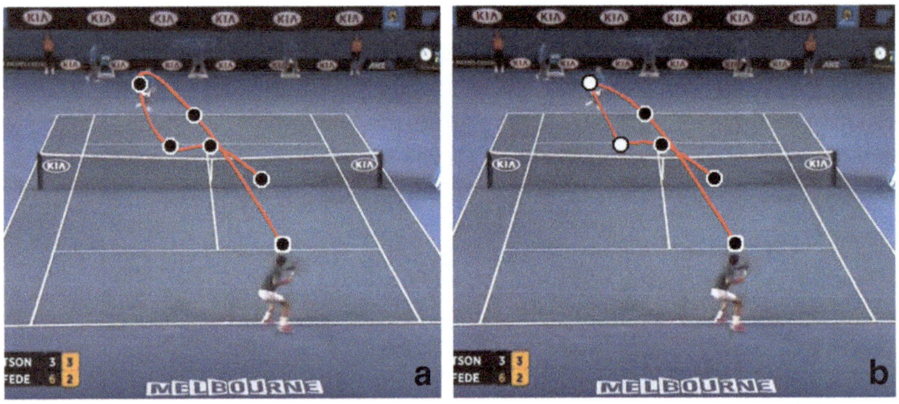

Fig. 14 Comparison of trajectories using cubic spline interpolation **a** without and **b** with *trajectory breaks*. Normal key frames are marked with *black circles*, while breaks are marked with *white circles*

$$e_i = \left(\mathbf{b}_i^p \cup \mathbf{b}_i^g\right) - \left(\mathbf{b}_i^p \cap \mathbf{b}_i^g\right) \tag{31}$$

where the first term denotes the union of the interpolated bounding box with the annotated one (or ground truth) and the second their intersection. For a given trajectory, the interpolation error is computed as the average of all partial overlap errors:

$$e_n = \frac{1}{M} \sum_{M} e_i \tag{32}$$

Table 1 Video data set summary

Motion type	Resolution ($W \times H$) (pixels)	Trajectory length (frames)	Bounding box size ($h \cdot w$) Min (pixels)	Max (pixels)
a	352 × 240	41	560	12,212
a	352 × 240	28	616	4888
a	352 × 240	53	660	1856
b	352 × 240	55	1440	5508
b	352 × 240	76	1392	4800
b	352 × 240	52	7488	16,020
b	352 × 240	188	2912	17,472
c	640 × 360	34	6030	7056
c	960 × 540	29	12,920	15,675
c	640 × 360	17	3942	5270
d	720 × 576	89	240	4640
d	640 × 480	164	1176	6360
d	640 × 360	47	432	2480
d	640 × 480	97	440	2808
e	640 × 360	67	680	3936
e	640 × 480	33	13,720	57,200
f	640 × 360	69	2072	3192
f	640 × 360	69	2184	4292
f	640 × 360	198	9396	33,200

See text for the description of the column motion type

where the subscript n indicates the interval between a pair of key frames. In this case, since we removed one half of the samples of the trajectory, $n = 1$ sample. Note that since we have to remove one half of the samples, we can create one test trajectory by removing the odd samples and a second one removing the even samples. So, the computation of e_1 is finally obtained as the mean value of e_1^0 and e_1^1, where the superscript 0 means the error when the test trajectory is obtained by keeping the even samples (that is, we start keeping the sample in 0, remove the sample in 1, keep the sample in 2, and so on), and similarly for the superscript 1. Likewise, e_2 is computed by constructing a new test trajectory removing $n = 2$ out of three samples. In this case, we can have e_2^0, e_2^1, and e_2^2, and so the error e_2 is computed as the mean of these three values. We performed this computation until the interval $n = 20$.

In Fig. 16, we can see the overall performance of the interpolators evaluated as a function of the interval n. In this case, the error has been computed as the average of the error e for all the videos used in this experiment. Note that, for $n = 1$, an error is present for all the interpolators. This error is mainly due to the process employed to annotate the videos manually, which adds a random shift to the object trajectory. Obviously, as the interval n increases, the average error also increases. However, the best performance is achieved for the interpolators based on depth recovery to compute the interpolated bounding boxes, since many of the objects annotated exhibit great variations of depth. In this sense, the algorithm proposed in this paper (GC) yields the best performance. The algorithm proposed by Yuen et al. (LM) and the simplified version of the algorithm proposed in this paper without cubic spline interpolation (GI) obtain worse results. In fact, for most of the cases, both algorithms are quite similar, since they are based on depth recovery and straight interpolation between key frames. Lastly, those algorithms not

Fig. 15 a–f Sample images of the video data set used in the experiments. See text for description

based on depth recovery but on interpolation from image coordinates, (LI) and (CS), yield the worst results. Table 2 shows the numerical results for the overall performance of the interpolators for different values of the interval *n*.

In order to make a more detailed analysis of the performance of the interpolators, we next present partial results considering the different types of object and camera motions. Figure 17 shows the mean error using only the trajectories of rigid objects moving in a straight path with great depth variations. In this case, the algorithms using depth recovery performed similarly and better than those using interpolation from the image coordinates.

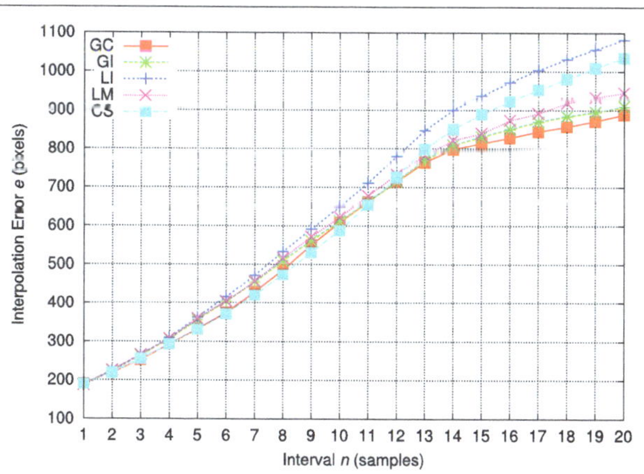

Fig. 16 Overall interpolation error

However, when the object follows a curved path, algorithms using cubic splines can model the object motion better. As can be seen in Fig. 18, the inclusion of a cubic spline interpolation improves the performance. In this case, the algorithm proposed in this paper (GC) yields the best results, since the objects also exhibits depth variations. The algorithms which compute straight paths between key frames decrease their performance.

When considering static objects but non-static cameras, the results are quite similar. If we analyze the trajectory on the image plane of a static object recorded with an on-board camera, we get the results shown in Fig. 19. In this case, the situation is equivalent to an object in front of a static camera moving in the opposite direction, so that the same comments made before are valid here, that is, the algorithms using depth recovery perform better than those interpolating from image coordinates. However, since the videos are recorded with on-board cameras, camera vibration adds a random shift to the object trajectories, which degrades the performance of all the interpolators.

Table 2 Overall interpolation error e_n (pixels)

Interval *n* (samples)	GC	GI	LI	LM	CS
5	332.9	356.0	362.5	360.3	331.3
10	610.2	612.5	649.7	623.8	588.2
15	815.5	832.0	937.9	840.8	889.4
20	890.4	910.9	1083.2	946.4	1035.6

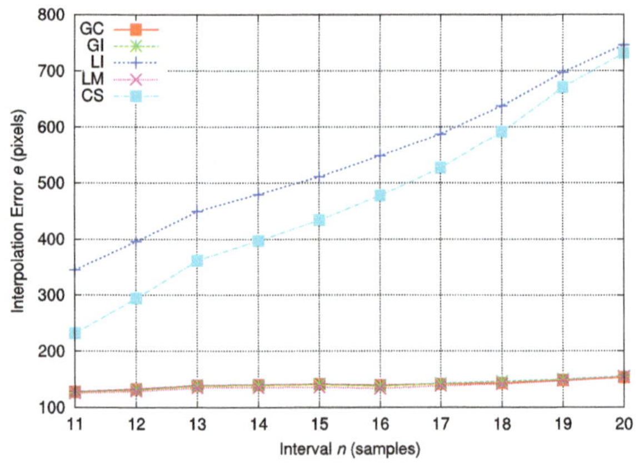

Fig. 17 Results for straight paths with depth variations, static camera

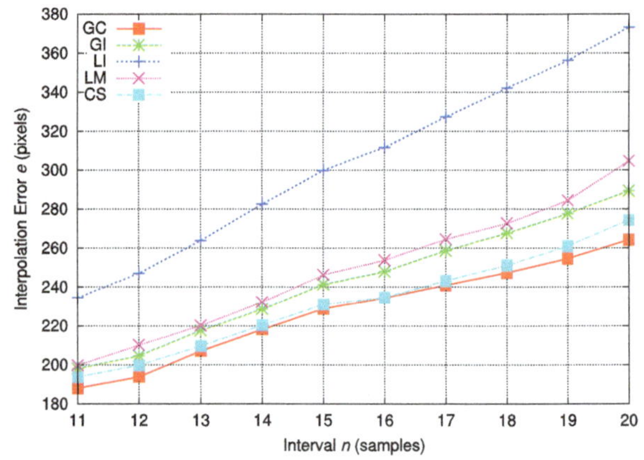

Fig. 19 Results for static object, with camera translation

Next, we obtained results for a static object while the camera is panning. In this case, since neither the object depth nor the size changes significantly, the algorithms based on image plane interpolation perform better than those based on depth recovery, as can be seen in Fig. 20. The main reason that makes the algorithms based on depth recovery decrease their performance is the nature of the annotations. In this case, although the object size in the image plane does not change, the annotations do, because they were done manually. This random change degrades the performance of those algorithms.

Next, we analyze the performance when the camera is zooming. In this case, the variation in the image plane of the object size due to the change in the camera's focal length is not equivalent to the change in depth. In fact, an alternative study to the one developed in this paper can be done to model this type of trajectory. Nevertheless, Fig. 21 shows that it can be modeled accurately with the algorithms which use depth recovery (GC, GI, and LM).

The algorithm proposed in this paper is mainly intended to work with rigid objects. On the other hand, when working with non-rigid objects, like humans, geometric assumptions do not hold, and the results get worse. This is due to the fact that a change in object size is erroneously interpreted as a change in object depth, and non-rigid objects can change their size in the image without changing their depth in space. For instance, the image of a man while crouching can reduce its bounding box height while maintaining the same distance from the camera. In order to analyze the performance for non-rigid object annotation, Fig. 22 shows the results for this kind of object. As can be seen, the results are a little worse than before, although all interpolators analyzed perform similarly, with a small advantage for the algorithms based on cubic splines.

5 Conclusions

In this paper, we have presented an alternative method that interpolates the bounding box annotations between

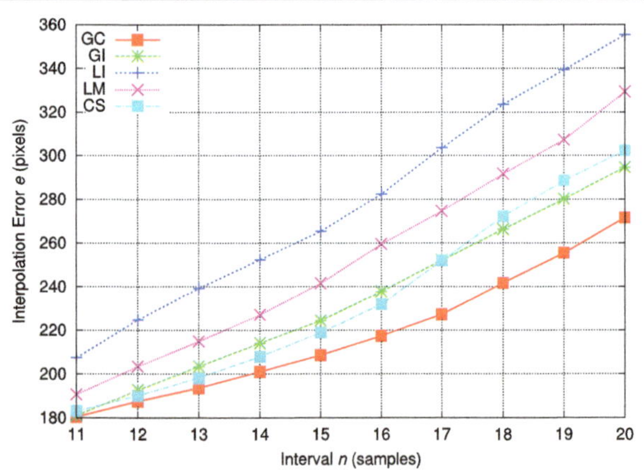

Fig. 18 Results for curved paths, static camera

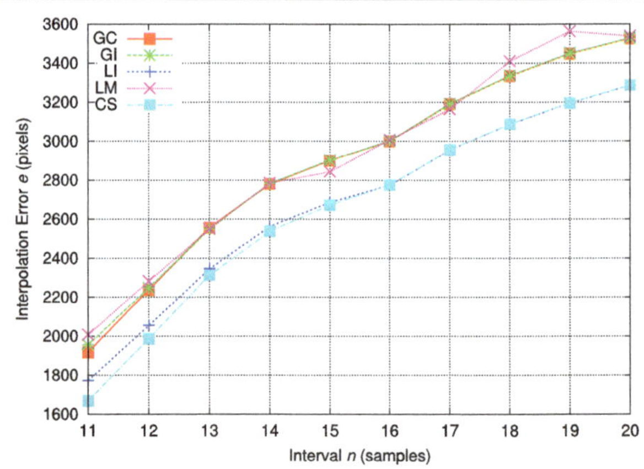

Fig. 20 Results for static object, with camera panning

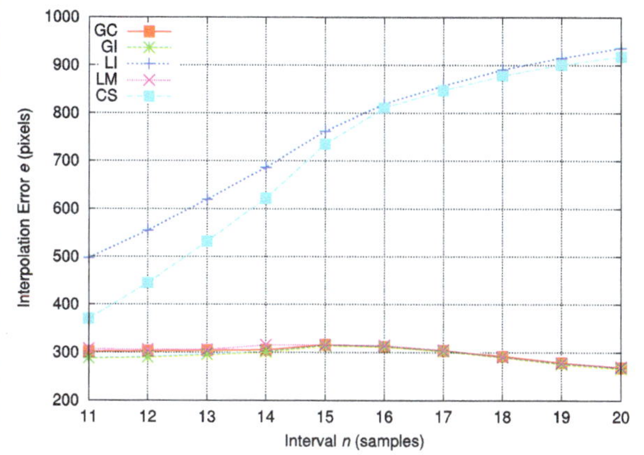

Fig. 21 Results for static object, with camera zoom

key frames for video labeling. The method is based on a 3D reconstruction of the bounding box using the provided annotations, based on the geometric properties of the elements involved. Once the 3D coordinates are computed, the interpolation of the bounding box for the remaining frames is performed in spatial coordinates, using cubic spline interpolation, and finally projected onto the image plane. The algorithm has been evaluated using a selected set of video clips that includes different types of object and camera motions and compared with other interpolation algorithms proposed in the literature. The results show a good performance, especially when considering rigid objects moving in trajectories with great variation in depth, where the accuracy of the interpolated bounding boxes is higher than the other evaluated algorithms. Since bounding box interpolation is a specific part of any annotation tool, we believe that the proposed algorithm can be a good alternative for all these tools when accurate object annotations are required.

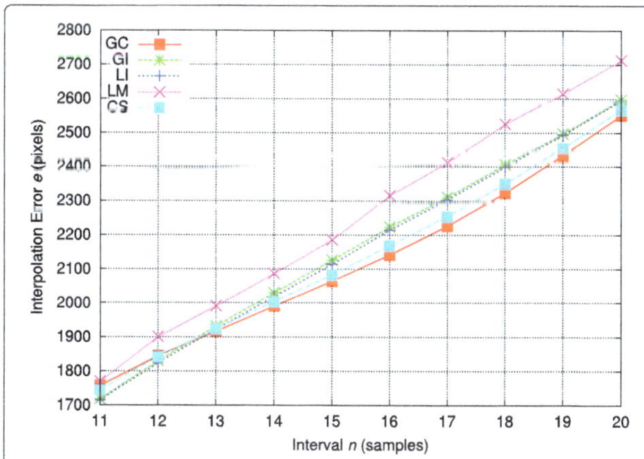

Fig. 22 Results for non-rigid objects

Competing interests

The authors declare that they have no competing interests.

Acknowledgements

This research was supported by projects CCG2014/EXP-055 and TEC2013-45183-R.

References

1. M Everingham, LV Gool, C Williams, J Winn, A Zisserman, The PASCAL visual object classes (VOC) challenge. Int. J. Comput. Vis. **88**, 303–338 (2010)
2. J Deng, W Dong, R Socher, LJ Li, K Li, L Fei-Fei, in *IEEE Conf. Computer Vision and Pattern Recognition*. ImageNet: a large-scale hierarchical image database (IEEE, New York, 2009)
3. B Russell, A Torralba, K Murphy, W Freeman, LabelMe: a datadata and web-based tool for image annotation. Int. J. Comput. Vis. **77**(1), 157–173 (2008)
4. L von Ahn, L Dabbish, in *Proceedings of the Conference on Human Factors in Computing Systems*. Labeling Ilmage with a computer game (ACM, New York, 2004)
5. A Sorokin, D Forsyth, in *IEEE Conference on Computer Vision and Pattern Recognition Workshops*. Utility data annotation with Amazon Mechanical Turk (EEE, 2008), pp. 1–8
6. M Kipp, in *Proceedings of the 7th European Conference on Speech Communication and Technology*. Anvil - A generic annotation tool for multimodal dialogue (INTERSPEECH, Aalborg, 2001), pp. 1367–1370
7. C Vondrick, DCR Patterson, Efficient scaling up crowdsourced video annotation. Int. J. Comput. Vis. **101**(1), 184–204 (2012)
8. V Badrinarayanan, F Galasso, R Cipolla, in *IEEE Conf. on Computer Vision and Pattern Recognition*. Label propagation in video sequences (IEEE, New York, 2010)
9. C Liu, WT Freeman, EH Adelson, Y Weiss, in *IEEE Conference on Computer Vision and Pattern Recognition*. Human-assisted motion annotation (IEEE, New York, 2008)
10. A Agarwala, A Hertzmann, DH Salesin, SM Seitz, in *ACM Transactions on Graphics*. Keyframe-based tracking of rotoscoping and animation (ACM, New York, 2004)
11. Z Kalal, K Mikolajczyk, J Matas, Tracking-learning-detection. Pattern Anal. Mach. Intell. **34**(7), 1409–1422 (2012)
12. D Doermann, D Mihalcik, in *Int. Conf. on Pattern Recognition. vol. 4*. Tools and techniques for video performance evaluation (IEEE, New York, 2000)
13. MA Serrano, J García, MA Patricio, JM Molina, in *Distributed Computing and Artificial Intelligence. vol. 79*. Interactive video annotation tool (Springer-Verlag, Berlin - Heidelberg, 2010), pp. 325–332
14. J Yuen, B Russell, C Liu, A Torralba, in *IEEE Int. Conf. Computer Vision*. LabelMe video: building a video database with human annotations (IEEE, New York, 2009)
15. JH Lee, KS Lee, GS Jo, in *Proceedings of the International Conference on Information Science and Applications (ICISA)*. Representation method of the moving object trajectories by interpolation with dynamic sampling (IEEE, New York, 2013), pp. 1–4
16. KS Lee, AN Rosli, IA Supandi, GS Jo, Dynamic sampling-based interpolation algorithm for representation of clickable moving object in collaborative video annotation. Neurocomputing. **146**, 291–300 (2014)
17. M Unser, Splines: a perfect fit for signal and image processing. Signal Process. Mag. **16**(6), 22–38 (1999)
18. R Hartley, A Zisermann, *Multiple View Geometry in Computer Vision*, 2nd ed. (Cambridge University Press, Cambridge, 2003)
19. E Jones, T Oliphant, P Peterson, et al., SciPy: Open source scientific tools for Python (2001). http://www.scipy.org/. Accessed 15 Feb 2016
20. P Gil-Jiménez, TrATVid annotation tool. http://agamenon.tsc.uah.es/Investigacion/gram/papers/Annotation. Accessed 16 Nov 2015

4

Audio bandwidth extension using ensemble of recurrent neural networks

Xin Liu and Chang-Chun Bao*

Abstract

In audio communication systems, the perceptual audio quality of the reproduced audio signals such as the naturalness of the sound is limited by the available audio bandwidth. In this paper, a wideband to super-wideband audio bandwidth extension method is proposed using an ensemble of recurrent neural networks. The feature space of wideband audio is firstly divided into different regions through clustering. For each region in the feature space, a specific recurrent neural network with a sparsely connected hidden layer, referred as the echo state network, is employed to dynamically model the mapping relationship between wideband audio features and high-frequency spectral envelope. In the following step, the outputs of multiple echo state networks are weighted and fused by means of network ensemble, in order to further estimate the high-frequency spectral envelope. Finally, combining the high-frequency fine spectrum extended by spectral translation, the proposed method can effectively extend the bandwidth of wideband audio to super wideband. Objective evaluation results show that the proposed method outperforms the hidden Markov model-based bandwidth extension method on the average in terms of both static and dynamic distortions. In subjective listening tests, the results indicate that the proposed method is able to improve the auditory quality of the wideband audio signals and outperforms the reference method.

Keywords: Audio bandwidth extension, Ensemble of recurrent neural networks, Echo state network, Spectral translation

1 Introduction

Due to the restrictions of network transmission rate and data processing capability, the effective bandwidth of audio signals available for the existing audio communication systems is commonly limited. To improve coding efficiency [1], a practical perceptual audio coding framework would encode and transmit only the low-frequency (LF) components of the audio. However, state-of-the-art audio communication systems are not satisfied with the auditory quality achieved by the wideband (WB) audio reproduction, which are sampled at 16 kHz corresponding to the bandwidth of 50~7000 Hz. Contemporary audio communication systems are looking forward to explore the technique to ensure high-quality audio service with more brightness and expressiveness. Therefore, it is desirable to allow existing WB audio systems to achieve the auditory quality of super-wideband (SWB) audio systems, which are sampled at 32 kHz with the bandwidth of 50~14,000 Hz.

Among the methods to enhance the perceptual quality of the WB audio signals, blind bandwidth extension (BWE) is designed to analyze the statistical relationship between the low-frequency and high-frequency components of WB audio signals and artificially restore the missing high-frequency (HF) components in the frequency range of 7000~14,000 Hz from the decoded WB signals at the decoder end. The advantage of this approach is to avoid any modifications inside the source coding and the network transmission process [2]. In recent decades, many blind BWE solutions have been developed for speech and audio signals, and these typical BWE methods can be summarized to perform two main tasks, namely, the estimation of the spectral envelope and the extension of the fine spectrum [1, 2]. Informal listening test results indicate that the estimation accuracy of the HF spectral envelope is crucial to the improvement of auditory quality for the reproduced signals [3]. Therefore, most BWE approaches have concentrated on modelling the mapping relationship between LF and

* Correspondence: baochch@bjut.edu.cn
Speech and Audio Signal Processing Lab, School of Electronic Information and Control Engineering, Beijing University of Technology, 100124 Beijing, China

HF spectral coefficients based on statistical learning methods and further estimated the HF spectral envelope under some error criterions. In 1994, a statistical recovery method [4] was proposed to predict the HF spectrum and achieved an improved perceptual quality of the reproduced signals. In the same year, Carl and Heute proposed the spectral envelope estimation method based on codebook mapping (CBM) [5]. This method explored the joint codebook between LF features and the HF spectral envelope and modelled their bijective mapping relationship by training. Further, the improved methods with interpolation, soft decision, and split codebook were proposed to decrease the spectral distortion caused by the single codebook [6–8]. In 2000, the Gaussian mixture model-based spectral envelope estimation method was proposed [9]. The Gaussian mixture model (GMM) is adopted to mimic the joint probability density between LF and HF spectral coefficients, and the HF spectral envelope is estimated under the minimum mean square error criterion. The GMM method builds a soft clustering-based statistical model to restrain the spectral discontinuity of the audio signals reproduced by CBM and achieves better performance in both subjective and objective tests. In addition, feed-forward neural networks (FNNs) are also adopted to extend the spectral envelope [10, 11]. Iser and Schmidt made a comparison between FNN and CBM in the context of the BWE application [12]. The results show that there was no significant difference in auditory quality of the extended signals among these two methods but the FNN method required less computational complexity in comparison with the CBM method.

The abovementioned methods lay stress on the relationship between LF and HF spectral coefficients in the current frame and exhibit static characteristic of the audio spectrum. Jax and Vary brought the inter-frame correlation of audio features into extension of the spectral envelope and employed a hidden Markov model (HMM) to represent the temporal evolution of the spectral envelope for audio signals [13, 14]. Some other research results also show that the HMM-based BWE method obtained better performance in terms of both subjective and objective quality [15, 16]. In our previous research work [17], we adopted the HMM-based estimator to extend the bandwidth of WB audio to SWB and obtained good subjective and objective quality of the reproduced audio signals. While improved results are achieved in [17], some dynamic distortion is still perceived. This is caused by the fact that only discrete states are adopted to piecewise approximate the actual evolution of the audio spectrum over time. Thus, we further introduce a continuous state-space model into the spectral envelope extension and propose the blind BWE method based on the ensemble of recurrent neural networks in this paper. The feature space of wideband audio is firstly divided into different regions through clustering. For each region in the feature space, a specific recurrent neural network with a sparsely connected hidden layer, referred as the echo state network, is employed to dynamically model the mapping relationship between wideband audio features and high-frequency spectral envelope. Different from the traditional fully connected recurrent neural networks, an echo state network (ESN) adopts the sparsely connected hidden layer in which the connectivity and weights of hidden neuron nodes are fixed and randomly assigned. The recurrent nature of the connections turns the time varying WB audio features into specific temporal patterns in high dimension. Then, a simple linear transformation is used to map the state of hidden neuron nodes to the desired HF spectral envelope. Given a rich collection of recurrently connected nodes, the ESN is able to effectively fit the nonlinear mapping function between WB audio features and the HF spectral envelope. Next, the outputs of the parallel ESNs for different regions in the feature space are weighted and fused by means of network ensemble techniques, so as to form ensemble echo state networks (EESNs) for further estimating the HF spectral envelope. Finally, combining with the HF fine spectrum extended by spectral translation, the proposed method can effectively extend the bandwidth of the WB audio to SWB and upgrade the subjective and objective quality of the WB audio.

2 Audio bandwidth extension using ensemble echo state network

2.1 Overview

The block diagram of the proposed blind BWE method is shown in Fig. 1. The input signals are the WB audio signals sampled at 16 kHz with a bandwidth of 7 kHz. Through up-sampling and low-pass filtering, the audio signals sampled at 32 kHz with the bandwidth of 7 kHz are separated into 32-ms frames with 16-ms overlap. Then, the audio signals are windowed with the Hamming window and are transformed into the frequency domain by using fast Fourier transform (FFT). The audio amplitude spectrum $A(k)$ in the frequency range of 64~7000 Hz is mapped onto the perceptually correlated equivalent rectangular bandwidth (ERB) scale by 20-channel triangular overlapping windows, in order to extract the 20-dimensional Mel frequency cepstral coefficient (MFCC) vector F_{MFCC}. Next, the extracted MFCCs are normalized to the range between 0 and 1 via min-max normalization. The normalized MFCCs are fed to the pre-trained EESN for estimating the HF spectral envelope. Here, the HF spectral envelope, $F_{\text{RMS}}(i)$, $i = 0,\ldots, 3$, is represented by the root mean square (RMS) values of the audio

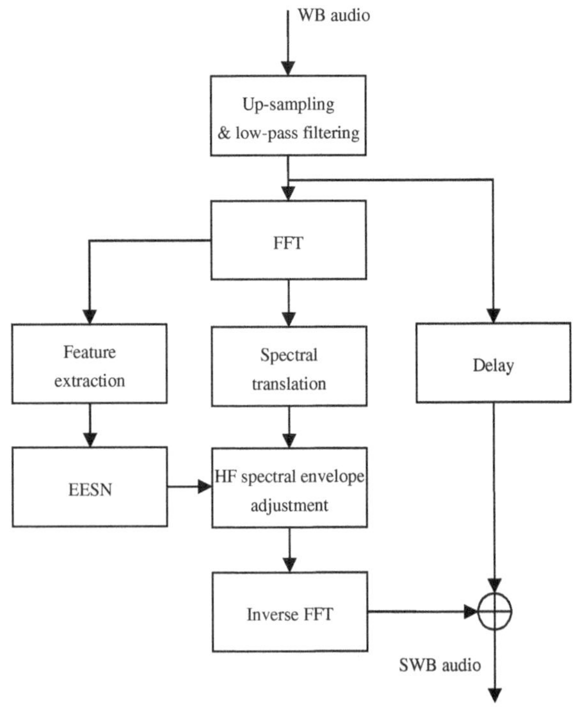

Fig. 1 Block diagram of the proposed blind bandwidth extension method for audio signals

amplitude spectrum in four nonuniform sub-bands in the frequency band of 7~14 kHz as follows:

$$F_{\mathrm{RMS}}(i) = \sqrt{\frac{1}{h(i)-l(i)+1}\sum_{k=l(i)}^{h(i)} A^2(k)}, \quad i = 0, 1, ..., 3 \tag{1}$$

where $A(k)$ is the spectral magnitude of audio signals and $h(i)$ and $l(i)$ correspond to the lower and upper boundaries of frequency bins in the ith sub-band, respectively. Note that there is no overlap between the adjacent sub-bands. The central frequencies of HF sub-bands are arranged at 8470, 9338, 11,653, and 13,657 Hz, according to the perceptually correlated ERB scale.

The extension of the fine spectrum is less critical for the subjective auditory quality of the extended audio signals, in comparison with the extension of the spectral envelope [2]. Thus, in order to reduce algorithm complexity, the HF fine spectrum is extended by using the simple spectral translation method [2]. The LF components are replicated into the HF band in the range of 7~14 kHz. The spectral envelope of the extended HF components is further adjusted by employing the estimated RMS values $\hat{F}_{\mathrm{RMS}}(i)$ of HF sub-bands. Finally, the reconstructed HF components are converted to the time domain by the inverse FFT, windowing, overlapping, and addition operations. By combining the WB audio signals with appropriate delay, the proposed method can reconstruct the SWB audio signals sampled at 32 kHz with the bandwidth of 14 kHz. Details of the proposed BWE method are discussed below.

2.2 Estimation of spectral envelope based on EESN

Let $F_X(m)$ denote the MFCCs of WB audio signals in the mth frame, whose dimension is set to $d_X = 20$. $F_Y(m)$ is the RMS value of HF sub-bands in the mth frame, and its dimension is set as $d_Y = 4$. Through an unknown mapping function $H(\bullet)$, the WB features $F_X(m)$ can be converted into the HF spectral envelope $F_Y(m)$ as follows:

$$F_Y = H(F_X) \tag{2}$$

We assume that the unknown mapping function can be modelled by ensembling several different state-space models. If a kth state-space model is applied, the process of regenerating the HF spectral envelope is approximated by

$$S_{\mathrm{hidden}}(m,k) = H_{\mathrm{state}}{}^{(k)}(S_{\mathrm{hidden}}(m-1,k), F_X(m-1), U(m,k)) \tag{3}$$

$$\hat{F}_Y\left(m|k\right) = H_{\mathrm{observation}}{}^{(k)}(S_{\mathrm{hidden}}(m,k), F_X(m), V(m,k)) \tag{4}$$

where $S_{\mathrm{hidden}}(m, k)$ with the dimension d_S is the hidden state vector of the kth state-space model in the mth frame, and its temporal evolution determines the dynamic characteristics of state space; $\hat{F}_Y(m|k)$ is the HF spectral envelope estimated by the kth state-space model in the mth frame; $H^{(k)}_{\mathrm{state}}(\bullet)$ is referred to as the state updating equation of the kth state-space model and describes the evolution process of hidden states given the input vector $F_X(m-1)$; $H^{(k)}_{\mathrm{observation}}(\bullet)$ is referred to as the observation equation of the kth state-space model and denotes the mapping process from $F_X(m)$ to $\hat{F}_Y(m|k)$ controlled by the hidden state $S_{\mathrm{hidden}}(m, k)$; and $U(m, k)$ and $V(m, k)$ represent the error caused by the state updating and observation processes, respectively.

By integrating different state-space models, we can estimate the HF spectral envelope $F_Y(m)$ from the WB audio feature $F_X(m)$. Therefore, there remain two important issues, namely, how to build an effective state-space model for approximating the true mapping process and how to choose the suitable state-space models among candidate models at each time.

2.2.1 Model structure of ESN

$H^{(k)}_{\mathrm{state}}(\bullet)$ and $H^{(k)}_{\mathrm{observation}}(\bullet)$ can be separately represented by two FNNs to build up a state-space-based recurrent neural network [18] for fitting the true mapping process.

Motivated by this, ESN [19, 20] is employed as the state-space model in this paper. Different from traditional fully connected recurrent neural networks, ESN adopts the sparsely connected hidden layer in which the connectivity and weights of hidden neuron nodes are randomly assigned and unchanged during training. The recurrent nature of the connections turns the time varying WB audio features into specific temporal patterns in high dimension. From the perspective of dynamic system theory, these states of hidden neuron nodes serve as a memory of the input audio feature and are termed as echoes of the input history. This is where ESN draws its name from. Given a rich collection of recurrently connected nodes with nonlinear activation functions, such a hidden layer, being an input-driven dynamical model, could provide a rich and relevant enough feature space, such that the desired HF spectral envelope could be obtained by linear combination from the state of hidden neuron nodes.

In the ESN-based state-space model, the updating process of hidden states for the kth ESN is represented by a leaky-integrated nonlinear activation unit [21] as follows:

$$\tilde{S}_{\text{hidden}}(m,k) = \tanh\left(W^{\text{in}}(k)\begin{bmatrix} 1 \\ F_X(m-1) \end{bmatrix} + W^{res}(k) S_{\text{hidden}}(m-1,k) \right) \quad (5)$$

$$S_{\text{hidden}}(m,k) = (1-\alpha) S_{\text{hidden}}(m-1,k) + \alpha \tilde{S}_{\text{hidden}}(m,k) \quad (6)$$

where $\tilde{S}_{\text{hidden}}(m, k)$ is the updating state of $S_{\text{hidden}}(m, k)$ in the mth frame. Leaking rate α of the leaky-integrated nonlinear unit represents the dynamic updating speed of hidden states. When the value of α is small, the dynamic change of hidden states may be slow. If α is close to 1, the nonlinear unit approximates a tanh function. $W^{in}(k)$ is the input weight matrix with the dimension of $d_S \times (d_X + 1)$, and the elements of $W^{in}(k)$ are uniformly distributed in $[-a_{in}, a_{in}]$. The scaling factor a_{in} of W^{in} decides the nonlinearity of the activation units. When a_{in} is close to 0, the leaky-integrated nonlinear activation units approach linear functions. For a larger a_{in}, the state updating process driven by $F_X(m-1)$ exhibits more nonlinearity. $W^{res}(k)$ is the recurrent weight matrix with the dimension of $d_S \times d_S$, and its spectral radius a_{res} can be artificially adjusted for controlling network stability in the actual applications. It is noteworthy that a_{res} and a_{in} jointly decide the relative importance between $F_X(m-1)$ and $S_{\text{hidden}}(m-1, k)$ in the state updating process. If a_{res} is larger, the previous hidden states own more effects on the current hidden states and the recurrent network would show more long-term correlations. Otherwise, the input vectors become the key factor to derive the hidden states. In addition, some research works indicated that the sparsely connected hidden states could slightly improve the estimation accuracy of ESN and accelerate the speed of model training [19, 20]. So, we can change the percentage of nonzero elements relative to the total amount of elements in W^{res} to adjust the sparsity f_{sparsity} of the recurrent weight matrix.

ESN takes advantage of a simple linear mapping function to represent the observation equation $H^{(k)}_{\text{observation}}(\bullet)$ [19] in the following way:

$$\hat{F}_Y(m|k) = W^{\text{out}}(k)\begin{bmatrix} 1 \\ F_X(m) \\ S_{\text{hidden}}(m,\ k) \end{bmatrix} \quad (7)$$

where $W^{\text{out}}(k)$ is the output weight matrix with the dimension of $d_S \times (1 + d_X + d_S)$ for the kth ESN.

From the perspective of network structure, ESN includes an input layer (F_X), a hidden layer (S_{hidden}), and an output layer (F_Y), as shown in Fig. 2. Here, the hidden layer, also referred to as the reservoir in ESN, consists of a large number of sparse-connected neuron nodes. With the help of the nonlinearity, sparsity, and feedback structure of neurons, the reservoir expands the audio features into a nonlinear high-dimensional feature space and further provides a more temporal context, so as to make the recurrent neural network exhibit rich dynamic characteristics [22]. The conclusions in relevant studies on recurrent learning algorithms [23] show that the weight dynamics favor a functional partition of the recurrent network into a fast output layer and a slower dynamical reservoir, whose rates of weight change are closely coupled. Therefore, ESN randomly generates the

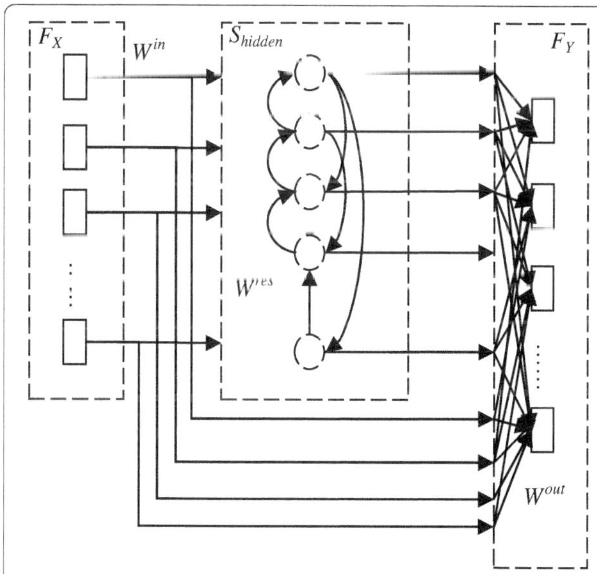

Fig. 2 Network structure of a single echo state network

input and recurrent weight matrices in network initialization and fixes them during network learning. Then, the network is optimized by adjusting the output weight matrix to fit the unknown nonlinear mapping function $H(\bullet)$.

2.3 Model training for ESN

The WB audio feature $F_X(m)$ and the HF spectral envelope $F_Y(m)$ are separately extracted frame by frame from the aligned WB and SWB audio signals, which are available for the training procedure, so as to form the training sample data $[F_X(m)^T, F_Y(m)^T]^T$, where $m = 0,1,\ldots, N_{\text{train}} - 1$ denotes the index of audio frame. In the model training of ESN, the hidden state $S_{\text{hidden}}(-1, k)$ of the kth ESN model is first initialized as 0. Next, $F_X(m)$ is fed into multiple pre-defined ESNs for driving the reservoirs. Then, by constantly updating the hidden states, the reservoirs exhibit different dynamic characteristics. Finally, the output weight matrix of the kth ESN is computed according to the information collected from the sequences of $F_X(m)$, $F_Y(m)$, and $S_{\text{hidden}}(m, k)$. A more detailed description about model training is provided below.

1. Initializing the reservoir parameters
   The initialization of reservoir parameters uses a heuristic method [19] as follows.
   (a) A matrix with the dimension of $d_S \times d_S$ is randomly created, and its element values are uniformly distributed within [−1, 1]. A fraction of the small element values within the random matrix are set to 0, in order to form a sparse matrix W. The sparsity f_{sparsity} is used to adjust the percentage of nonzero elements relative to the total amount of elements in W. Then, W is normalized according to its spectral radius $\lambda_{\max}$, i.e., $W/\lambda_{\max}$.
   (b) The resulting sparse matrix is scaled to obtain the recurrent weight matrix, i.e., $W^{\text{res}} = a_{\text{res}} W/|\lambda_{\max}|$. Here, $a_{\text{res}} < 1$ is the spectral radius of W^{res} and can be experimentally adjusted.
   (c) The input weight matrix W^{in} is randomly created according to a uniform distribution.
2. Driving the reservoir
   The hidden states $S_{\text{hidden}}(m, k)$ is updated in the light of two model parameters $\{W^{\text{res}}(k), W^{\text{in}}(k)\}$. But the update process is impacted by the initial hidden state $S_{\text{hidden}}(-1, k)$ [19]. Therefore, we set a network stability threshold $N_T = 13$ and assume that the whole network is approaching stability after the first N_T state updates.
   Starting with the $N_{T\text{th}}$ updates, the sample data $F_X(m)$ of the WB audio feature, the hidden states $S_{\text{hidden}}(m)$, and the sample data $F_Y(m)$ of the HF spectral envelope are collected for establishing the state collection matrix B and the expectation output matrix Q. The column vectors of the state collection matrix $B(k)$ for the kth ESN are represented as $[1, F_X(m)^T, S_{\text{hidden}}(m, k)^T]^T$, so the dimension of $B(k)$ is $(1 + d_X + d_S) \times (N_{\text{train}} - N_T)$. The column vectors of the expectation output matrix $Q(k)$ for the kth ESN are represented as $[F_Y(m)]$, so the dimension of $Q(k)$ is $(d_Y) \times (N_{\text{train}} - N_T)$. Through the abovementioned method of data collection, the negative effect of $S_{\text{hidden}}(-1, k)$ on ESN is effectively eliminated, and the mapping relationship between the WB audio feature $F_X(m)$ and the HF spectral envelope $F_Y(m)$ can be properly reflected according to the model trained from $B(k)$ and $Q(k)$.
3. Computing the output weight matrix

The output weight matrix $W^{\text{out}}(k)$ for the kth ESN is computed according to $B(k)$ and $Q(k)$ to minimize the error between the network output $\hat{F}_Y(m|k)$ and the true HF spectral envelope $F_Y(m)$. Informal evaluation indicates that, when the element value of $W^{\text{out}}(k)$ is large, $W^{\text{out}}(k)$ may amplify the slight difference of $F_X(m)$ and over-fitting may occur during network training. Therefore, the ridge regression method is employed in order to avoid over-fitting as follows:

$$W^{\text{out}}(k) = \underset{W^{\text{out}}(k)}{\arg\min}\left(\left[\frac{1}{N_{\text{train}}-N_T}\sum_{m=0}^{N_{\text{train}}-N_T-1}\|F_Y(m) - \hat{F}_Y(m|k)\|^2\right] + \beta\|W^{\text{out}}(k)\|^2\right) \tag{8}$$

where $\beta\|W^{\text{out}}(k)\|^2$ is the regular item for penalizing the large elements in $W^{\text{out}}(k)$ and the relative importance between two items in the above error function can be controlled by adjusting β.

Finally, the resulting $W^{\text{out}}(k)$ can be given as

$$W^{\text{out}}(k) = Q(k)B(k)^T\left[B(k)B(k)^T + \beta I\right]^{-1} \tag{9}$$

where I is a unit matrix.

The connection weight matrices $W^{\text{res}}(k)$, $W^{\text{in}}(k)$, and $W^{\text{out}}(k)$ for the kth ESN are obtained by using the training method above. In the actual extension, the frame-wise MFCCs are fed into the ESNs. The hidden states $S_{\text{hidden}}(k)$ are updated with the change of F_{MFCC} in a frame-by-frame basis. The dynamic characteristics of WB audio features are brought into the network model. Further, the HF spectral envelope $F_Y(m)$ is effectively estimated by fusing the outputs $\hat{F}_Y(m|k)$ of multiple ESNs.

2.3.1 Network ensemble fusion

According to the network ensemble theory, frequently, an ensemble of models performs better than any individual model. This is because the various errors of the models average out [24]. In this paper, a network ensemble framework based on the statistical distribution of input feature space is adopted to integrate multiple ESNs for further forming ensemble echo state network.

Figure 3 shows the block diagram of the EESN-based spectral envelope estimator. In the model training procedure of EESN, we first selected the GMM-based clustering method to separate the feature space of the input WB audio feature F_X into N_{ESN} regions. The mean vector of each Gaussian component corresponds to a clustering center of the feature space. Accordingly, the training sample data $[F_X(m)^T, F_Y(m)^T]^T$, $m = 0,1,\ldots, N_{\text{train}} - 1$ could be segmented into N_{ESN} groups on the basis of the separation of feature space. By assuming that the statistical properties of each group of input features were similar, N_{ESN} ESNs with the same network structure were built up to model the local mapping relationship between F_X and F_Y within the corresponding N_{ESN} regions of the feature space. It is worth mentioning that the WB audio features extracted from the training data can be simultaneously fed into all the ESNs in order to continuously drive their state updating equations since their input and recurrent weight matrices are unchanged during model training, while the output weight matrices of different ESNs need to be respectively trained according to their corresponding regions in the feature space. Therefore, the feature space separation-based network ensemble framework does not actually affect temporal modelling of ESNs.

In the actual bandwidth extension procedure, the mth-frame audio feature $F_X(m)$ is fed into all of these ESNs, so as to obtain the corresponding estimated values $\hat{F}_Y(m|k)$ of the HF spectral envelope for the kth ESN. Meanwhile, according to the probability distribution of the WB audio feature modelled by GMM, a posteriori probability $p(k|F_X(m))$ of the kth region in the feature space is computed and adopted as the weights for guiding us on integrating the HF spectral envelope estimated by different ESNs.

Within the GMM framework, $p(k|F_X(m))$ is computed as follows:

$$p(k|F_X(m)) = \frac{p(F_X(m)|k)p(k)}{\sum_{i=0}^{N_{\text{ESN}}-1} p(F_X(m)|i)p(i)} \tag{10}$$

where $p(k)$ is the a priori probability of the kth region in the feature space, and it needs to match the following condition:

$$\sum_{k=0}^{N_{\text{ESN}}-1} p(k) = 1, \quad 0 \le p(k) \le 1 \tag{11}$$

$p(F_X(m)|k)$ is the probability of $F_X(m)$ given the kth region, and it is represented as

$$p(F_X(m)|k) = N(F_X(m); m_k, C_k) = \frac{1}{(2\pi)^{\frac{D_X}{2}}|C_k|^{\frac{1}{2}}} \exp\left[-\frac{1}{2}(F_X(m)-m_k)^T C_k^{-1}(F_X(m)-m_k{}^k)\right] \tag{12}$$

where $N(\bullet)$ represents the Gaussian probability density function. m_k and C_k are the mean and covariance of the kth Gaussian component, respectively, and can be trained according to the standard EM algorithm. Finally,

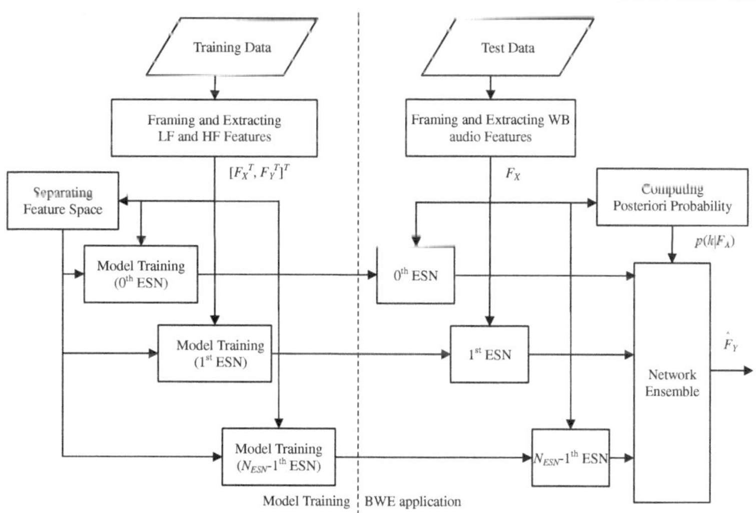

Fig. 3 Block diagram of the EESN-based spectral envelope estimator

in the GMM-based soft-decision ensemble framework, multiple ESNs can be linearly combined by taking advantage of the posterior probabilities $p(k|F_X(m))$, and the estimated HF spectral envelope $\hat{F}_Y(m)$ is given by

$$\begin{aligned}\hat{F}_Y(m) &= \sum_{k=0}^{N_{\mathrm{ESN}}-1} p(k|F_X(m))\hat{F}_Y(m|k) \\ &= \sum_{k=0}^{N_{\mathrm{ESN}}-1} p(k|F_X(m))W^{\mathrm{out}}(k)\begin{bmatrix}1\\ F_X(m)\\ S_{\mathrm{hidden}}(m,k)\end{bmatrix}\end{aligned} \tag{13}$$

2.4 High-frequency component reconstruction

In general, the LF and HF components have the similar spectral structure. The slightly different fine structure for the audio spectrum would not significantly affect the auditory quality of the bandwidth-extended audio signals, if the HF spectral envelope is well restored. So, in the proposed blind BWE method, spectral translation is adopted to directly copy the fine spectrum in the frequency range of 0~7 kHz to the HF band of 7~14 kHz. Here, the LF fine spectrum is represented by the normalized spectral magnitude $A_{\mathrm{norm}}(k)$, $k = 0,\ldots, 223$, as follows:

$$A_{\mathrm{norm}}(k) = \frac{A(k)}{F_{\mathrm{RMS_{WB}}}(i)}, \quad i = \frac{k}{N_{\mathrm{subband}}} \tag{14}$$

where $A(k)$ is the spectral magnitude of audio signals. $F_{\mathrm{RMS\_WB}}(i)$ is the RMS value of the LF sub-bands for describing the LF spectral envelope of audio signals. Here, in order to maintain the spectral flatness of $A_{\mathrm{norm}}(k)$, the LF spectrum is uniformly divided into 14 sub-bands in the frequency scale and each LF sub-band contains $N_{\mathrm{subband}} = 16$ frequency bins. Then, the HF fine spectrum extended by spectral translation is given by

$$A_{\mathrm{norm}}(k) = A_{\mathrm{norm}}(k-224), \quad k = 224, \ldots .447 \tag{15}$$

Since the HF spectral envelope is analyzed in the ERB scale, the HF fine spectrum is also divided into four non-overlapping sub-bands. By combining the estimated HF sub-band RMS values $\hat{F}_{\mathrm{RMS}}(i)$, $i = 0,\ldots, 3$, the extended HF spectrum $A_{\mathrm{swb}}(k)$, $k = 224,\ldots, 447$, can be presented as

$$A_{\mathrm{swb}}(k) = A_{\mathrm{norm}}(k)\hat{F}_{\mathrm{RMS}}(i), \quad i = \mathrm{Subband}(k) \tag{16}$$

in which Subband(k) represents the index of the HF sub-band corresponding to the kth frequency bin.

The phase of HF spectrum $\theta(k)$, $k = 224,\ldots, 447$, is also obtained by spectral translation as follows:

$$\theta(k) = \theta(k-224), \quad k = 224, \ldots .447 \tag{17}$$

Finally, the HF spectrum is converted into the time domain by inverse FFT. The resulting HF signals are combined with the properly delayed WB signals to reconstruct the SWB audio signals.

3 Parameter selection for EESN

In this section, we made some preliminary experiments in order to select the appropriate model parameters according to the objective performance of spectral envelope extension. The SWB audio data used for parameter selection came from live concert recordings with the length of about 6 h and contain dialog, pop music, singing, and live background sounds. These signals were sampled at 32 kHz with the bandwidth of 14 kHz and were stored by 16-bit PCM. After low-pass filtering and time alignment, the parallel WB training dataset was built up. Twenty-dimensional MFCCs and four-dimensional HF sub-band RMS values were extracted from the WB and SWB datasets as the input vectors F_X and the expectation output vectors F_Y for EESN, respectively. Seventy-five percent of these sample data were randomly selected for training the EESN model, and others were used as the validation set for parameter selection.

In this section, the mean square error between the original and estimated spectral envelope in the frequency range of 7~14 kHz was selected as the objective measurement for evaluating the performance of the EESN method. The estimation error of HF spectral envelope can be defined as

$$e_{\mathrm{MS}}(i) = \frac{1}{4}\sum_{n=0}^{3}\left(F_{\mathrm{RMS}}(i,n) - \hat{F}_{\mathrm{RMS}}(i,n)\right)^2 \tag{18}$$

where $e_{\mathrm{MS}}(i)$ is the mean square error of the HF spectral envelope for the ith frame, $F_{\mathrm{RMS}}(i, n)$ and $\hat{F}_{\mathrm{RMS}}(i,n)$ are the true and estimated RMS values of the nth HF sub-band in the ith audio frame, respectively, and $n = 0, 1, 2, 3$ is the index of HF sub-band. Before computing the mean square error, all the data needed to be resampled to 32 kHz and were temporally aligned with the original SWB audio. Then, the resulting error values were averaged over all the frames for each audio signal in the validation set, in order to be used as the distortion measure of the spectral envelope estimation.

The extension performance of the EESN method is mainly affected by seven model parameters: the scaling factor of W^{in} (a_{in}), spectral radius of W^{res} (a_{res}), sparsity of W^{res} (f_{sparsity}), leaking rate (α), regular factor for ridge regression (β), dimension of hidden state (d_S), and number of ESNs (N_{ESN}). In EESN, different ESNs are generally independent of each other, so we can first reduce the EESN model to a single ESN model by setting $N_{ESN} = 1$ and optimize the ESN model parameters (a_{in}, a_{res}, $f_{sparsity}$, α, β, and d_S), by using grid search. Next, several ESNs are constructed with the "optimal" ESN parameters and are integrated for further forming EESN, and then, the optimal value of N_{ESN} can be manually determined, guided by evaluation on the validation set.

3.1 ESN model parameters

In order to improve search efficiency, a two-stage grid search was conducted to optimize the ESN model parameters, while the number of ESN was preset as 1.

In the first-stage search, we manually set the bounds for each model parameter, according to their actual characteristics in the ESN model. Then, a finite set of reasonable values for each parameter were pre-selected, as shown in Table 1, so as to perform the coarse grid search over wider parameter intervals. Next, an ESN was trained with each candidate in the Cartesian product of these six parameter sets. Finally, their performance was evaluated on the validation set in terms of the estimation error of the HF spectral envelope. According to the experiment results, the parameter setting that achieves the lowest estimation error (12.7824) in the validation procedure is obtained as $\{a_{in} = 1/8,\ a_{res} = 1/2,\ f_{sparsity} = 1/4,\ \alpha = 1,\ \beta = 3,\ d_S = 400\}$. The resulting parameter setting can be used to identify the promising regions for each model parameter: $a_{in} \in (1/16, 1/4)$, $a_{res} \in (1/4, 3/4)$, $f_{sparsity} \in (1/8, 1/2)$, $\alpha \in (4/5, 1)$, $\beta \in (2, 4)$, and $d_S \in (200, 600)$.

The second-stage gird search was further adopted with smaller steps, so as to determine the "optimal" ESN model parameters. For each model parameter, five grid points were equally spaced within the corresponding promising region. Then, all the 5^6 possible combinations of the six ESN model parameters were evaluated, and the lowest estimation error (12.4961) of the HF spectral envelope was obtained at $\{a_{in} = 0.1094,\ a_{res} = 0.625,\ f_{sparsity} = 0.2188,\ \alpha = 1,\ \beta = 3,\ d_S = 500\}$.

In addition, we should note that there are no evident differences for different leaking rates of the leaky-integrated nonlinear activation unit in terms of the estimation error values of the HF spectral envelope, and $\alpha = 1$ provides a smaller estimation error than other α values, according to the evaluation results in the two-stage grid search process. This suggests that the long-term memory of hidden states does not clearly improve the performance of the ESN model.

Table 1 Candidates of the ESN model parameters for coarse grid search

a_{in}	{1/16, 1/8, 1/4, 1/2, 1}
a_{res}	{1/4, 1/2, 3/4, 1, 5/4}
$f_{sparsity}$	{1/16, 1/8, 1/4, 1/2, 1}
α	{1/5, 2/5, 3/5, 4/5, 1}
β	{0, 1, 2, 3, 4}
d_S	{200, 400, 600, 800, 1000}

3.2 Number of ESNs

EESN adopts the a posteriori probabilities of each region in the feature space as weights to integrate the outputs of different ESNs, which model the local mapping functions between LF and HF components within the corresponding regions of the feature space. So, the number of ESNs is an important factor for the proposed EESN-based BWE method. According to the results of the two-stage grid search, the model parameters for each ESN are set as $a_{in} = 0.1094$, $a_{res} = 0.625$, $f_{sparsity} = 0.2188$, $\alpha = 1$, $\beta = 3$, and $d_S = 50$. Then, we experimented with several values of N_{ESN} to form different EESNs and evaluated them under the estimation error of the HF spectral envelope. The mean and standard deviation of the spectral envelope estimation error are shown in Table 2. It is noticeable that the performance of the EESN-based spectral envelope estimator improves quickly with the increase of N_{ESN} and the mean estimation error of HF spectral envelope is reduced down to about 11.3. With the further consideration of computation complexity and memory demands, the number of ESN can be set as 8.

As a consequence of the above, the EESN model parameters are finally selected as $a_{in} = 0.1094$, $a_{res} = 0.625$, $f_{sparsity} = 0.2188$, $\alpha = 1$, $\beta = 3$, $d_S = 50$, and $N_{ESN} = 8$.

4 Performance evaluation

In this section, log spectral distortion, cosh measure, and differential log spectral distortion (DLSD) were adopted as objective measurements for evaluating the performance of the proposed BWE method in comparison with the HMM-based reference method. In addition, the auditory quality improvement of the extended audio signals was further discussed in terms of subjective preference tests.

Table 2 LSD of EESN model with different N_{ESN}

N_{ESN}	Spectral envelope estimation error	
	Mean	Standard deviation
1	12.4961	3.5859
2	12.0504	3.7867
4	11.6919	3.0442
6	11.4871	4.9626
8	11.3662	3.6441
16	11.3384	4.7234
32	11.3074	3.5859

4.1 Reference method and audio data

Except for the spectral envelope estimation module, the HMM-based reference method follows the similar principle shown in Fig. 1. The MFCCs were extracted from the WB audio and fed into the HMM-based Bayesian estimator of the HF spectral envelope under the minimum mean square error criterion, while the HF fine spectrum was extended by spectral translation. In the HMM-based reference method, the HMM states were defined by vector quantization of the joint space between LF and HF coefficients. For each state of the HMM, the state-dependent conditional expectation of the HF spectral envelope was calculated based on a GMM of the joint probability density of the WB audio feature and HF spectral envelope. The respective a posteriori probabilities of the corresponding states of the HMM were used to weight the conditional expectation, in order to obtain the estimation of the HF spectral envelope. Here, the HMM parameters were learned by using the hybrid approach suggested by Jax and Vary [13], and the state number of HMM and mixture number for each GMM were also selected in terms of the estimation error of the HF spectral envelope. The experiment results indicated that the estimation error was slightly decreased with an increase of the number of hidden states and Gaussian mixtures. Finally, taking into consideration algorithm complexity and auditory quality, we set the number of hidden states as 16 and required each GMM to have 32 mixtures and full covariance matrices in the HMM-based reference method.

The audio data of the training and validation set for the proposed and reference methods were from the lossless audio recorded in a live concert with the length of about 6 h. They contained dialog, music, singing, and background sound. Through resampling and low-pass filtering, the parallel WB and SWB audio datasets were obtained, and the level of all the signals needed to be normalized to −26 dBov before further processing. In addition, 15 SWB audio signals were selected from the MPEG audio quality listening test dataset as the test data and contained pop music, solo instrumental music, symphony, and speech. These signals were sampled as 32 kHz with the bandwidth of 14 kHz and were limited in the length of 10~20 s. Then, they were converted into the WB audio through low-pass filtering and downsampling, and the level of the resulting signals were normalized to −26 dBov as the input of the BWE methods. Since disjoint speakers and disjoint music pieces were used in training, validation, and testing, the accuracy and generalization of the proposed EESN method could be evaluated more objectively and fairly. A detailed analysis of the subjective and objective quality for the extended signals is given below.

4.2 Objective measurement

The objective quality of the SWB signals reproduced by different methods was evaluated in terms of log spectral distortion (LSD), cosh, and DLSD measurements.

4.2.1 LSD

The LSD [25, 26] between the original SWB audio and the extended audio signals in the frequency range of 7~14 kHz was selected as the objective measurement for comparing the extension performance between the EESN method and HMM method. LSD measurement is computed directly from the FFT power spectra as

$$d_{\mathrm{LSD}}(i) = \sqrt{\frac{1}{N_{\mathrm{high}} - N_{\mathrm{low}} + 1} \sum_{n=N_{\mathrm{low}}}^{N_{\mathrm{high}}} \left[10 \log_{10} \frac{P_i(n)}{\hat{P}_i(n)} \right]^2} \tag{19}$$

where $d_{\mathrm{LSD}}(i)$ is the LSD value of the ith frame and P_i and $\hat{P}_i$ are the FFT power spectra of the original SWB audio signals and the audio signals processed with BWE methods, respectively. N_{high} and N_{low} are the indices corresponding to the upper and lower bounds of the HF band with the frequency range of 7~14 kHz. Before computing the LSD values, all the data needed to be resampled to 32 kHz and were temporally aligned with the original SWB audio. Then, the LSD values were computed only for the HF band in the range of 7~14 kHz. The resulting LSD values were averaged over all the frames for each test signal, and the mean LSD was used as the distortion measure.

Table 3 shows the LSD values of the signals extended by the proposed and reference BWE methods. On average, the EESN method outperforms the HMM method in terms of LSD measure. For rock music, violin solo, and speech, the LSD difference between two methods is with the range of ±0.3 dB. But for jazz music, the LSD difference is relatively large, because this type of music has rich HF components and obvious transients over time. In comparison with the HF spectra extended by the HMM method, the spectra extended by the

Table 3 LSD of the signals reproduced by different BWE methods

Data type	HMM	EESN
Country music	6.6788	7.5653
Jazz music	12.6068	8.4862
Rock music	6.3262	6.4720
Violin solo	3.8470	3.5376
Symphony	3.9713	5.2903
Speech	6.1027	5.8650
Average	6.5888	6.2027

proposed EESN method are more similar to the original ones and obtain a lower LSD value. For symphony and country music, the energy of the HF components extended by the EESN method is slightly high, so the LSD value of the EESN method is higher than that of the HMM method, though the audio extended by EESN is brighter.

4.2.2 Cosh measure

In comparison to LSD measure, Itakura-Saito distortion gives more heavy weights on the peaks of the audio spectrum and is more relative to the true subjective auditory quality [26]. Itakura-Saito distortion is defined as

$$d_{\mathrm{IS}}\left(P_i,\hat{P}_i\right) = \frac{1}{N_{\mathrm{high}}-N_{\mathrm{low}}+1}\sum_{n=N_{\mathrm{low}}}^{N_{\mathrm{high}}}\left[\frac{P_i(n)}{\hat{P}_i(n)} - \log_{10}\frac{P_i(n)}{\hat{P}_i(n)} - 1\right] \tag{20}$$

But, Itakura-Saito distortion is asymmetrical as distance metric, so cosh measure is selected as the modified measurement to describe the perceptual distortion of the reproduced audio. The cosh measure [26] is defined as

$$d_{\mathrm{COSH}}(i) = \frac{1}{2}\left[d_{\mathrm{IS}}\left(P_i,\hat{P}_i\right) + d_{\mathrm{IS}}\left(\hat{P}_i,P_i\right)\right] \tag{21}$$

In this paper, the cosh measure was only computed within the frequency range of 7~14 kHz. The resulting cosh values were averaged over all the frames for each test signal, and the mean cosh value was used as the distortion measure. Table 4 gives the comparison results of the COSH measure of the signals extended by two methods. For jazz music and speech, the EESN method gains a better quality than the HMM method, but for the country music, the EESN method has more distortion since the reconstructed HF spectrum owns higher energy. On average, the SWB audio reproduced by EESN can achieve a better objective quality in terms of the cosh measure.

4.2.3 DLSD

The continuity of the audio spectrum over time is as perceptually important as the accuracy of the spectrum reconstruction. Here, DLSD is utilized as a dynamic distortion measure for evaluating the smoothness of temporal evolution for the spectral envelope of the extended audio signals [27]. If the DLSD value of the reproduced audio is small, it indicates that the spectrum evolves smoothly over time and is helpful for the overall subjective auditory quality of the reconstructed audio. DLSD is defined as

$$d_{\mathrm{DLSD}}(i) = \sqrt{\frac{1}{2(N_{\mathrm{high}}-N_{\mathrm{low}}+1)}\sum_{n=N_{\mathrm{low}}}^{N_{\mathrm{high}}}\left[10\log_{10}\frac{P_i(n)}{P_{i-1}(n)} - 10\log_{10}\frac{\hat{P}_i(n)}{\hat{P}_{i-1}(n)}\right]^2} \tag{22}$$

where P_{i-1} and $\hat{P}_{i-1}$ are the FFT spectra of the original and extended SWB audio signals in the previous frame.

Table 5 shows the DLSD values of the signals extended by two BWE methods. The dynamic distortion of violin solo and symphony is similar, because their energy of the HF components is low and evolves smoothly over time. For country, jazz, and rock music, the EESN method can effectively restore transients and achieve an improvement in DLSD over the HMM method on the average. For speech, the DLSD value of EESN is about 3.89 dB. In some unvoiced frames, the HF energy of the extended signals has some difference from that of the original signals, and it causes some dynamic distortion.

4.3 Subjective preference test

The subjective quality of two BWE methods was assessed by using the pair-wise subjective preference tests. The subjective tests were arranged in a quiet room, and 15 listeners were invited to take part in the tests. Five test signals were selected from the MPEG audio dataset. The listeners were asked to choose which audio they preferred from two presented test items or to indicate no preference in each test case, and they were also allowed to repeat the testing data with no time limitation before giving answers.

Table 4 cosh measure of the signals reproduced by different BWE methods

Data type	HMM	EESN
Country music	29.3467	31.6246
Jazz music	24.4898	18.1792
Rock music	19.4905	19.4480
Violin solo	2.3990	2.3325
Symphony	3.1477	3.0678
Speech	40.2552	34.6777
Average	19.8548	18.2216

Table 5 DLSD of the signals reproduced by different BWE methods

Data type	HMM	EESN
Country music	4.4466	3.9017
Jazz music	6.7463	4.8963
Rock music	3.9707	3.2137
Violin solo	2.1191	2.0595
Symphony	3.3856	3.5123
Speech	3.9606	3.8925
Average	4.1048	3.5793

Table 6 Results of the subjective preference tests

	Prefer the former (%)	Prefer the latter (%)	No preference (%)
WB vs original SWB	6.7	76.0	17.3
EESN vs WB	46.5	25.3	28.2
HMM vs WB	44.6	29.3	26.1
EESN vs HMM	28.7	24.1	47.2
EESN vs original SWB	26.1	31.1	42.8
Original SWB vs HMM	35.1	20.9	44.0

The examples of demmostation are given in the Additional file 1.

Six groups of tests were presented to each listener. They were the comparison between the original WB and SWB signals, the comparison between the extended signals by the EESN method and the WB signals, the comparison between the extended signals by the HMM method and the WB signals, the comparison between the extended signals by the EESN method and the HMM method, the comparison between the extended signals by the EESN method and the original SWB signals, and the comparison between the original SWB signals and the extended signals by the HMM method. The results of the subjective tests are shown in Table 6. It is found that the extended signals by two BWE methods are preferable over the WB signals in terms of subjective auditory quality and the original SWB signals gain the best performance in comparison with the extended signals and the WB signals. In addition, the subjective audio quality of the proposed EESN method is slightly better with a little bit less of artifacts, compared to the HMM reference method, and is subjectively comparable to the original SWB signals. According to the comments of listeners, more dynamic contents in country, jazz, and rock music are restored by using the EESN methods, and it leads to a preference for EESN in terms of subjective listening quality, in comparison with the reference method. For symphony and violin solo, the audio spectrum changes smoothly over time, so no significant difference can be perceived between the audio signals extended by different BWE methods. In addition, some bandwidth-extended audio signals of country and jazz music own higher energy than the original SWB audio signals, so some listeners subjectively prefer the extended audio signals to the original ones.

5 Conclusions

A bandwidth extension method for audio signals based on ensemble echo state network is proposed in this paper. For each region in the feature space, a specific echo state network with recursive structure is utilized to dynamically model the mapping relationship between the LF and HF coefficients in the light of the continuous state updating equation. And, the outputs of multiple ESNs are fused by means of the GMM-based network ensemble techniques, in order to further estimate the HF spectral envelope. Subjective and objective quality test results show that the EESN method achieves improvement in terms of both static and dynamic distortions in comparison with the HMM-based reference method on the average and the auditory quality of the reproduced signals is close to that of the original SWB audio.

Additional file

Additional file 1: Demonstration for subjective listening tests. (PPT 10166 kb)

Competing interests

The authors declare that they have no competing interests.

Acknowledgements

This work was supported by the National Natural Science Foundation of China under Grants 61072089 and 61471014.

References

1. P Vary, R Martin, *Digital speech transmission—enhancement, coding and error concealment* (Wiley, UK, 2006)
2. E Larsen, MR Aarts, *Audio bandwidth extension—application of psychoacoustics, signal processing and loudspeaker design* (Wiley, UK, 2004)
3. C Avendaño, H Hermansky, EA Wan, *Beyond Nyquist towards the recovery of broad bandwidth speech from narrow-bandwidth Speech* (Proc European Conference on Speech and Language Processing, Madrid, 1995), pp. 165–168
4. YM Cheng, D O'Shaughnessy, P Mermelstein, Statistical recovery of wideband speech from narrowband speech. IEEE Trans. Speech Audio Process **2**(4), 544–548 (1994)
5. H Carl, U Heute, *Bandwidth enhancement of narrow-band speech signals* (Proc 7th European Signal Processing Conference, Edinburgh, 1994), pp. 1178–1181
6. J Epps, WH Holmes, *A new technique for wideband enhancement of coded narrowband speech* (Proc IEEE Workshop on Speech Coding, Porvoo, 1999), pp. 174–176
7. IY Soon, KY Chai, *Bandwidth extension of narrowband speech using soft-decision vector quantization* (Proc Fifth International Conference on Information, Communications and Signal Processing, Bangkok, 2005), pp. 734–738
8. U Kornagel, Techniques for artificial bandwidth extension of telephone speech. Signal Process. **86**(6), 1296–1306 (2006)
9. KY Park, HS Kim, *Narrowband to wideband conversion of speech using GMM based transformation* (Proc IEEE International Conference on Acoustics Speech and Signal Processing, Istanbul, 2000), pp. 1843–1846
10. CV Botinhao, BS Carlos, LP Caloba, MR Petraglia, *Frequency extension of telephone narrowband speech signal using neural networks* (Proc IMACS Multiconference on Computational Engineering in Systems Applications (CESA), Beijing, 2006), pp. 1576–1579
11. VP Tuan, F Schaefer, G Kubin, *A novel implementation of the spectral shaping approach for artificial bandwidth extension* (Proc 3rd International Conference on Communications and Electronic, Nha Trang, 2010), pp. 262–267
12. B Iser, G Schmidt, *Neural networks versus codebooks in an application for bandwidth extension of speech signals* (Proc European Conference on Speech and Language Processing, Geneva, 2003), pp. 565–568
13. P Jax, P Vary, *Wideband extension of telephone speech using a hidden Markov model* (Proc 7th IEEE Workshop on Speech Coding, Delavan, 2000), pp. 133–135
14. P Jax, P Vary, On artificial bandwidth extension of telephone speech. Signal Process. **83**(8), 1707–1719 (2003)
15. GB Song, P Martynovich, A study of HMM-based bandwidth extension of speech signals. Signal Process. **89**(10), 2036–2044 (2009)
16. C Yagli, MAT Turan, E Erzin, Artificial bandwidth extension of spectral envelope along a Viterbi path. Speech Comm. **55**(1), 111–118 (2013)

17. X Liu, CC Bao, Blind bandwidth extension of audio signals based on non-linear prediction and hidden Markov model. APSIPA Transactions on Signal and Information Processing **3**(e8), 1–16 (2014)
18. G Lear, A state-space-based recurrent neural network for dynamic system identification. J. Syst. Eng. **6**(3), 186–193 (1996)
19. M Lukosevicius, A practical guide to applying echo state networks, in *Neural networks: tricks of the trade*, ed. by G Montavon, GB Orr, KR Müller, 2nd edn. (Springer, Heidelberg, 2012), pp. 659–686
20. M Lukosevicius, H Jaeger, Reservoir computing approaches to recurrent neural network training. Comput. Sci. Rev. **3**(3), 127–149 (2009)
21. H Jaeger, M Lukosevicius, D Popovici, U Siewert, Optimization and applications of echo state networks with leaky-integrator neuron. Neural Netw. **20**(3), 335–352 (2007)
22. H Jaeger, H Haas, Harnessing nonlinearity: predicting chaotic systems and saving energy in wireless communication. Science **304**(5667), 78–80 (2004)
23. UD Schiller, JJ Steil, Analyzing the weight dynamics of recurrent learning algorithm. Neucomputing **63**, 757–779 (2005)
24. LK Hansen, P Salamon, Neural network ensembles. IEEE Trans. Pattern Anal. Mach. Intell. **12**(10), 993–1001 (1990)
25. H Pulakka, L Laaksonen, M Vainio, J Pohjalainen, P Alku, Evaluation of an artificial speech bandwidth extension method in three languages. IEEE Trans. Audio Speech Lang. Process. **16**(6), 1124–1137 (2008)
26. AH Gray, JD Markel, Distance measures for speech processing. IEEE Trans. Audio Speech Lang. Process. **24**(5), 380–391 (1976)
27. F Norden, T Eriksson, Time evolution in LPC spectrum coding. IEEE Trans. Speech Audio Process **12**(3), 290–301 (2004)

5

BM3D filter in salt-and-pepper noise removal

Igor Djurović

Abstract
There is a significant recent advance in filtering of the salt-and-pepper noise for digital images. However, almost all recent schemes for filtering of this type of noise are not taking into an account the shape of objects (in particular edges) in images. We have applied the block-matching and 3D filtering (BM3D) scheme in order to refine the output of the decision-based/adaptive median techniques. Obtained results are excellent, surpassing current state-of-the-art for about 2 dB for both grayscale and color images.

Keywords: Filtering, Salt-and-pepper noise, Block-matching filters, Median filters

1 Introduction

One of the simplest noise model for digital images is the salt-and-pepper noise. For such environments, pixels exhibit maximal or minimal values with some predefined probability. Appearance of such impulses is rather common and it can be caused by various effects including existence of dead pixels in images or failures in acquisition, hardware, and transmission. In addition, this model is almost the same as the model of images with small percentage of available samples (sometimes considered as a part of wider compressive sensing framework). As for other types of impulsive noise environments, nonlinear filters from the median family are common tool for treating such disturbances. Several effective techniques have been proposed for elimination/reduction of the salt-and-pepper noise influence [1–15]. Quite simple solution is a technique proposed in [1] where pixels recognized as corrupted are replaced by the median of the uncorrupted ones from neighborhood of 3×3 size. The median filter has rather good ability to preserve image edges. However, its lowpass filtering characteristic causes errors for pixels close to edges that are more emphatic for high percentage of noise. In [1], relatively small neighborhood is used meaning that for high percentage of pulses it could happen that there is no any reliable pixel in the neighborhood. Nevertheless, this simple technique motivated numerous recent developments. In these techniques, the neighborhood is gradually enlarged until some uncorrupted pixels are found and then the median filtering is performed only to those pixels. These techniques are surveyed in [14] and it seems that a technique called the decision-based median filter (DBMF) from [15] produces the best results. The second candidate for the best technique is a statistics-based scheme proposed in [3]. Note that just minor differences in implementation of these techniques significantly affect accuracy. Some of the main ingredients in these techniques are discussed in Section 2. There are several alternatives related to weighted median techniques, sharpening filters, and variants of total variation criterion embedded with nonlinear filtering. Some of these strategies are presented in [9–13]. The biggest issue is handling environments with extremely high percentage of impulses where filters with wide neighborhood can select output from region from a wrong edge side. It causes large errors, broken edges, and other visually impaired effects.

There are several strategies in nonlinear filtering for improving results related to edges and details preserving in the case of environments with high percentage of impulse noise. A particular interesting strategy is described in [7] with sharpening filters. However, these filters are applied only to the part of available pixels in attempt to preserve edges and similar details. For high-density salt-and-pepper noise, it is difficult to have large number of uncorrupted pixels within the local neighborhood to perform such operations. An alternative solution for this problem is described in [6, 11], with variants of the

Correspondence: igordj@ac.me
Electrical Engineering Department, University of Montenegro, Podgorica, Montenegro

total variation filters used in order to stabilize output of the filtering scheme preserving edges and details. Again, these strategies are not robust enough to a high-density salt-and-pepper noise. In addition, they require elaborate algorithm for implementation meaning they are relatively inefficient.

In this paper, we perform postprocessing (post-filtering) of the salt-and-pepper denoising filters class considered in [1, 3, 4, 14, 15]. For postprocessing, we have used the block matching and 3D filtering (BM3D) [16]. It is the filter from the non-local means class that is looking for local neighborhoods of similar shapes [17]. These similar shapes are put into a 3D matrix. Filtering is performed in transform domain employing appropriate threshold. In this way, we are able to achieve two goals for the salt-and-pepper noise environment in the considered scheme. Firstly, for low-density noise, this scheme is able to reduce residual noise caused by changes in the local neighborhood (as an output of the filter, we are selecting one of the pixels in the local neighborhood). For a high-density impulse noise, an output of the decision-based/adaptive median filters is corrupted by artifacts and edges are disturbed. The BM3D looks for similar patches in such images and filters them together, reducing both effects of the artifacts and errors related to the edges [18]. The proposed technique can also be considered as an extension of the BM3D algorithm that is able to work with images corrupted by the salt-and-pepper noise. This is a powerful demonstration of how legacy of existing well-developed filtering technique can be applied for improving results in digital image filtering even to problems previously not considered. This is an attempt in direction where huge legacy of developed digital image processing methods can be used for a different purpose with respect to the original one.

The paper is organized as follows. Section 2 gives background information on noise model, techniques for a high-density salt-and-pepper noise removal, and the BM3D algorithm. The proposed denoising scheme is described in Section 3. Numerical examples demonstrating ability of the proposed technique are given in Section 4. Concluding remarks are given in Section 5.

2 Background information

For notational simplicity, we are going to describe a model for single channel—grayscale images while all results can straightforwardly be extended to color images.

2.1 Noise model

Denote an image as $f(n,m)$, where $n \in [1,N]$ and $m \in [1,M]$, and $N \times M$ is the image size. The domain of image luminance is defined as unsigned integers within interval $f(n,m) \in [f_{\min}, f_{\max}]$ (commonly $f_{\min} = 0$ while $f_{\max} = 2^p - 1$ where p is number of bits used for memory representation of images). An image corrupted by the salt-and-pepper noise is given as

$$x(n,m) = \begin{cases} f_{\max} & \text{with probability } p_1, \\ f(n,m) & \text{with probability } 1 - p_1 - p_2, \\ f_{\min} & \text{with probability } p_2. \end{cases} \quad (1)$$

Usually, it is assumed that $p_1 = p_2$; but for our technique, it is not of critical importance.

For color images, we assume that corrupted image has output equal to the original image $x(n,m;r) = f(n,m;r)$ with percentage $1 - p_1 - p_2$ where r is the channel of a color (or multispectral) image. With probability $p_1 + p_2$ at least in one of the channels, pixels take values that are equal to $f_{\max}$ or $f_{\min}$ for the corresponding channel.

2.2 Review of decision-based/adaptive filters

Instead of various elaborated techniques for rejection of the salt-and-pepper noise, recently, a rather simple scheme has been proposed in [1] called the modified decision-based unsymmetric trimmed median filter. This scheme consists of the following steps.

Algorithm. *Modified decision-based unsymmetric trimmed median filter.*
Consider pixel $x(n,m)$.

1. If $x(n,m) \neq f_{\min}$ and $x(n,m) \neq f_{\max}$, take that the filter output is $\hat{f}(n,m) = x(n,m)$.
2. Else, create 3×3 local neighborhood and eliminate all pixels that are equal to $f_{\max}$ or $f_{\min}$, $D'(n,m) = \{x(q,l) | x(q,l) \neq f_{\min} \wedge x(q,l) \neq f_{\max}, q \in [n-1, n+1], l \in [m-1, m+1]\}$.
3. If there is at least one pixel in the set $D'(n,m)$, a filter output is $\hat{f}(n,m) = \text{median}\{D'(n,m)\}$.
4. Otherwise, filter output is selected as the mean of all pixels in the local neighborhood $D(n,m) = \{x(q,l) | q \in [n-1, n+1], l \in [m-1, m+1]\}$, $\hat{f}(n,m) = \text{mean}\{D(n,m)\}$.

This is a very simple strategy that is using only uncorrupted pixels in the narrow local neighborhood of size 3×3 and performs median filtering of these samples. For large percentage of corrupted pixels, it can happen that there are no uncorrupted pixels in the local neighborhood and this filter does not have any strategy for handling such pixels. However, excellent results achieved with this technique with respect to other strategies motivated numerous other recent developments.

Here, we are going to describe the main ingredients in all adaptive/decision-based filters for the removal of the salt-and-pepper noise. Description is given for grayscale images while simple extension can be given for color (multichannel) images. This description covers more or less all filters considered in [1, 3, 4, 14, 15]. It is applied as initial stage of the proposed filtering approach with design parameters explained in the next section.

Algorithm. *Decision-based/adaptive median filter.* Consider pixel $x(n, m)$.

1. If $x(n, m) \neq f_{\min}$ and $x(n, m) \neq f_{\max}$, take that the filter output is $\hat{f}(n, m) = x(n, m)$.
2. Else, set $\Delta = 1$ and proceed with the next steps.
3. From the local neighborhood, eliminate all pixels that are equal to $f_{\max}$ or $f_{\min}$, $D'(n, m) = \{x(q, l) | x(q, l) \neq f_{\min} \wedge x(q, l) \neq f_{\max}, q \in [n - \Delta, n + \Delta], l \in [m - \Delta, m + \Delta]\}$.
4. If there is at least k uncorrupted pixels in the local neighborhood, we can take as an output of the filter $\hat{f}(n, m) = \text{median}\{D'(n, m)\}$.
5. If all pixels in local neighborhood are corrupted $D(n, m) = \{x(q, l) | q \in [n - \Delta, n + \Delta], l \in [m - \Delta, m + \Delta]\}$ and equal to $f_{\max}$ or $f_{\min}$ depending on the neighborhood size, we can select that the filter output is equal to $\hat{f}(n, m) = f_{\max}$ or $\hat{f}(n, m) = f_{\min}$ (if all pixels in the neighborhood are equal to $f_{\min}$ then set to this value and the same for all pixels equal to $f_{\max}$).
6. If $\Delta = \Delta_{\max}$ (maximal size of the neighborhood), go to 7; otherwise, $\Delta = \Delta + 1$ and go to 3.
7. The remaining question is what to do if maximal size of the local neighborhood is reached and all pixels in the neighborhood are indicated as corrupted. There are several alternatives considered in the literature. An alternative is to take the median or mean of the largest local neighborhood while the other is to take the last processed pixel as the filter output. The last processed pixel could be in various directions (related to direction of processing of images) or some combination of such pixels. According to our experiments, step 5 should be skipped if neighborhood is smaller than 5×5 ($\Delta < 2$). Note that some alternative statistics could be used instead of median in step 4.

Minor differences in the algorithm setup (size of neighborhood, filter in step 4, number k of uncorrupted pixels in the neighborhood to perform filtering, strategy for handling unprocessed pixels, etc.) contribute to significant changes in output that can be measured with several decibels. This is of importance in corresponding applications so it is still rather a hot research topic with numerous recent developments. Also, it has potential to contribute to the compressive sensing research for digital images to great extend [19].

2.3 BM3D

Recently, the rather attractive fields of the image processing research are non-local algorithms for image processing [17]. Instead of filtering in the local neighborhood, similar image patches are recognized. Commonly, these similar patches are selected to be rectangular but also other shapes are possible. An example of these similar patches is shown in Fig. 1 with test image "baboon" (size 512×512). In the BM3D algorithm, similar patches form the 3D matrix and this matrix is filtered in the transform domain with appropriately selected threshold [16]. Different patches have smaller correlation of noise than local neighborhoods giving better results than the local neighborhood-based filtering schemes. This technique is thoroughly studied recently; and in [20], it has been shown that it produces performance close to the achievable. We are using here the BM3D to filter small mistakes caused by taking as a filter output values from the local neighborhood different from the true one. Also, the BM3D is able to filter out (smooth) artifacts in the output from decision-based/adaptive median filters that appear in images with high percentage of impulses. These artifacts are caused by spreading values of rare uncorrupted pixels over wide neighborhood in such images. In addition, for an environment with high percentage of impulses, the BM3D is able to improve filtering around edges, objects, and other details by looking for similar patches and repetitive details. The similarity check cannot be applied directly to corrupted images due to high density of impulses while it is possible to apply it on the images filtered by the decision-based/adaptive median filter described in the

Fig. 1 Test image "baboon" with single set of similar blocks recognized in the BM3D algorithm

previous subsection. Details related to ability of the BM3D to the artifact removal can be found in [18] and references therein.

The BM3D algorithm consists of the coarse and fine algorithm runs [16, 21, 22] (http://www.cs.tut.fi/~foi/GCF-BM3D/). Here, the BM3D algorithm is briefly summarized.

Algorithm. *BM3D filter.*
Consider an image region D centered around the pixel $x(n, m)$.

1. Search for regions D' that are similar to the considered region D. Note that this similarity check is performed in the transform domain (commonly Hadamard/Walsh transform). These transform coefficients are compared with threshold and those below the threshold are set to 0. Similar patches are selected to be relatively close to the considered patch in order to simplify search procedure.
2. Similar patches are put into the 3D matrix. 3D discrete linear transform of the 3D matrix is evaluated. Obtained transform coefficients are thresholded and all coefficients below the threshold are removed. Temporary filtered blocks are obtained using inverse 3D transform. Note that in practice, separated 2D/1D transforms are used instead of the 3D transform due to efficiency reasons.
3. These filtered blocks are returned back to the image. The procedure is performed for each pixel. Pixels belong to various number of different blocks. Therefore, an aggregation of the filtered blocks should be performed. Weighted coefficients in aggregation are calculated based on number of transform coefficients that are above the threshold in the 3D transforms. More coefficients above the threshold mean that there is more residual noise in the blocks and vice versa.
4. Similar blocks are found for the filtered image from the previous step.
5. Similar blocks form the 3D matrix.
6. The 3D transform is calculated for the 3D matrix.
7. 3D transform coefficients are filtered using the Wiener filter.
8. Inverse 3D filtering is performed to obtain the final version of filtered patches.
9. These filtered patches are again aggregated to obtain the final estimate.

Steps 1–3 represent the first (coarse) run of the algorithm, while steps 4–9 represent the fine algorithm run applied to the output from the first three steps. For all details of this algorithm realization, interested readers can refer to [16, 21, 22], while software realization is available at (http://www.cs.tut.fi/~foi/GCF-BM3D/).

Detailed calculation complexity analysis of the BM3D algorithm is considered in [16] with several algorithm "profiles." In this research, we are using the so-called "normal profile" requiring less than 1 s for the largest image considered in our study (color image "boats" of size 576×787) on a standard laptop computer. In all experiments, we have used a default BM3D setup in realization from (http://www.cs.tut.fi/~foi/GCF-BM3D/).

3 Proposed strategy

The output of the decision-based/adaptive median filter does not take care about particular image features such as edges and other details. This is critical in high-density noise where several pixels from the other side of the edge can exist in wide local neighborhoods. It can cause significant degradation of the filtering output quality. In addition, for such environments, some uncorrupted pixels could be spread over wide neighborhood. In this way, unpleasant artifacts are obtained, also disturbing quality of edges, other details, and overall visual quality of filtered images. Therefore, it is required to smooth these artifacts and in the same time to preserve edges and other repeating elements appearing in natural images. Both of these requirements can be met with the BM3D filter. The BM3D smooths artifacts caused by spreading the same values of uncorrupted pixels to the wide neighborhood in highly corrupted images [18]. Even more importantly, filtering together similar patches attenuates these artifacts and helps in preserving of image edges and other visually important elements. It is possible since recognized similar patches have different forms of artifacts and they are smoothed/attenuated for recognized repetitive patterns.

The proposed strategy can be briefly summarized as follows:

1. Application of the decision-based/adaptive median filter (Section 2.2). Denote output of this filter as $h^{(0)}(n, m) = \hat{f}(n, m)$, $i = 0$.
2. Apply the BM3D algorithm on the filtered image (Section 2.3):

$$g^{(i)}(n, m) = BM3D(h^{(i)}(n, m)). \tag{2}$$

Output image is calculated as

$$h^{(i+1)}(n, m) = \begin{cases} f(n, m) & (n, m) \text{ uncorrupted pixel of original image,} \\ g^{(i)}(n, m) & \text{other pixels.} \end{cases} \tag{3}$$

Step 2 can be repeated several times. In each stage, filtering is performed using previous output of this procedure. The proposed algorithm is summarized in Fig. 2.

In our research, we have used the following setup of the decision-based/adaptive median filter described in

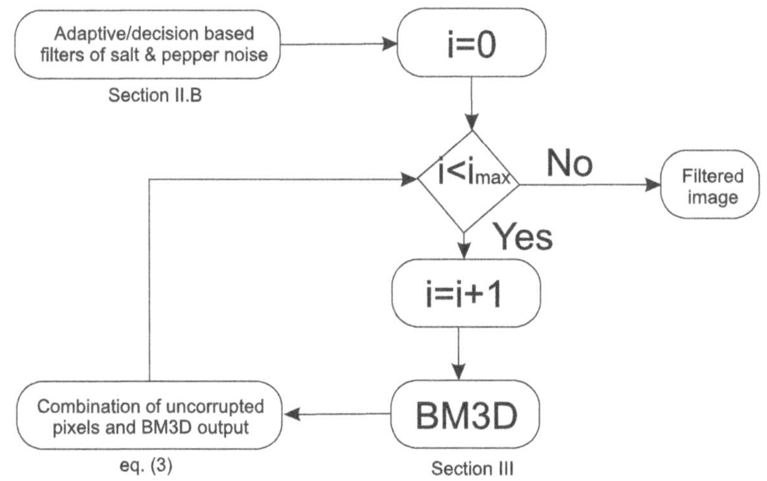

Fig. 2 Flow chart of the image filtering algorithm

Section 2.2. Minimal number of pixels in the local neighborhood required for performing filtering is set to $k = 1$. The median is applied on the obtained pixels for grayscale or vector median for color images. Maximal size of the neighborhood is selected to be rather large $\Delta_{max} = 9$. This large neighborhood is used to minimize probability of unprocessed pixels after this stage. Therefore, the need for handling unprocessed pixels (last good processed pixel and other strategies) is significantly reduced. Only for 5 % or less uncorrupted pixels it can happen with small probability that there are still unprocessed pixels (for 95 % of impulses it is less than 1×10^{-8} , while for 97.5 % of pixels it is about 1×10^{-4}). In that case, the median of the local neighborhood is calculated. The reason is in the fact that in some fields of image processing, large portion of image could be black (for example, in images coming from astronomy) or white and it is better to take dominant pixels from the neighborhood since sometimes we are wrongly deciding that these pixels are corrupted. In the case when we are sure that there are no many pixels in recording that can be misidentified as salt-and-pepper noise, then previous or the closest processed pixel can be used (or mean of pixels in the considered neighborhood as in [1]), etc. However, we wanted that our technique has no problems with rather dark recordings from astronomical images so unprocessed pixels are filtered with median of the widest considered neighborhood. In general, such situation is extremely rare and also requirement for employing wide neighborhoods is also rare (for 95 % of pixels wider than 9×9 is required only with probability of less than 1.6 %).

The BM3D has been originally introduced for filtering Gaussian noise. However, it can also be applied in this case when the decision-based/adaptive median filter in the initial stage removes impulsive noise. Then, residual noise is not of the impulsive nature and we can apply the BM3D accurately as in the case of the Gaussian noise environments. In this manner, we have extended the application range of the BM3D since relatively simple pre-processing strategy allows application of the BM3D to the salt-and-pepper noise.

4 Numerical study

We have considered numerous grayscale and color images. For brevity reasons, results for 9 grayscale and 9 color $p = 8$-bit depth images given in Fig. 3 are only presented. Images' sizes are given in Table 1. We have compared the results of the procedures in terms of the pseudo signal-to-noise ratio (PSNR) defined as

$$\text{PSNR} = 10 \log_{10} \frac{f_{max}^2}{\frac{1}{NM} \sum_{n,m} [h^{(i)}(n,m) - f(n,m)]^2} \tag{4}$$

where it is selected that $f_{max} = 2^p - 1 = 255$ for all images, $f(n,m)$ is the original image, and $h^{(i)}(n,m)$ is the image filtered with the proposed procedure. For $i = 0$, the image filtered only with the decision-based/adaptive median filter follows; while for $i > 0$, the BM3D procedure (2), (3), is applied i times. Figure 4 depicts results achieved in filtering of grayscale images "clock" and "Lena" for single trial in cases of 80, 90, and 95 % of impulses. For each case, the upper left image is corrupted by corresponding percentage of the salt-and-pepper noise, the upper right row is image filtered by the decision-based/adaptive median filter, the lower left corner depicts image filtered with the proposed scheme with single iteration of the BM3D filter, and lower right corner is image filtered with five iterations of the BM3D filtering. For color images "baboon" and "monarch," results of simulations in a single trial are

Fig. 3 Test images. **a** Grayscale images. **b** Color images

presented in Fig. 5. The PSNR is given next to images in both Figs. 4 and 5. For improved visualization of results achieved with the proposed technique, a part of the test image "monarch" is enlarged for 80, 90, and 95 % of corrupted pixels in Fig. 6. The first column represents the noiseless image, the second column is the result achieved with the decision-based/adaptive median filters, the third column represents the results after single iteration of the BM3D algorithm, while the fourth column represents the images after five iterations of the BM3D. Comparing the second and the fourth column, it is obvious improvement achieved with the proposed procedure in elimination of artifacts from the decision-based/adaptive median filters. The BM3D takes similar patches in various parts and filters and significantly attenuates artifacts. The results of the Monte Carlo simulations for grayscale and color test images for several percentages of impulses are given in Tables 2 and 3. In each cell, the upper row represents the PSNR of the decision-based/adaptive median filter, the middle row represents the PSNR for the filter after single iteration of the BM3D algorithm, while the bottom row represents the PSNR after five iterations of the BM3D algorithm. Obviously, the largest gain is achieved with the first iteration of the BM3D algorithm while only moderate gain is achieved with additional four iterations.

The BM3D can be applied with similar improvement to all filters from this decision-based/adaptive median filter class [1, 3, 4, 14, 15]. Just to illustrate the improvement that is achieved in the last several years, we compare the obtained results with those from the initial paper [1] for test image "Lena" and several percentages of the salt-and-pepper noise. For example, the method from [1] for test image "Lena" corrupted by 20 % of impulses achieves PSNR = 34.78 dB, while the proposed technique as seen in Table 2 achieves PSNR = 39.49 dB. For 40 % of the impulses, these two filters achieve PSNRs 30.32 dB/35.70 dB (more than 5-dB gain); for 60 % of the impulses, it is 26.43 dB/32.68 dB (more than 6-dB gain); for 80 % of the impulses, the obtained PSNRs are 20.70 dB/29.78 dB (more than 9-dB gain); and for 90 % of the impulses, it is 18.10 dB/27.74 dB (more than 9-dB gain). It can be observed the significant improvement achieved in a series of small steps for just a couple of years to which we are contributing with additional gain [1, 3, 4, 14, 15].

Table 1 Test images

Grayscale images	Size	Color images	Size
Cameraman	256 × 256	Peppers	384 × 512
Mandi	504 × 760	Autumn	206 × 345
Moon	537 × 358	F16	512 × 512
Pout	291 × 240	Baboon	512 × 512
Saturn	500 × 400	Lena	512 × 512
West Concord	366 × 364	House	256 × 256
Lena	512 × 512	Boats	576 × 787
Baboon	512 × 512	Monarch	512 × 768
Clock	256 × 256	Jelly	256 × 256

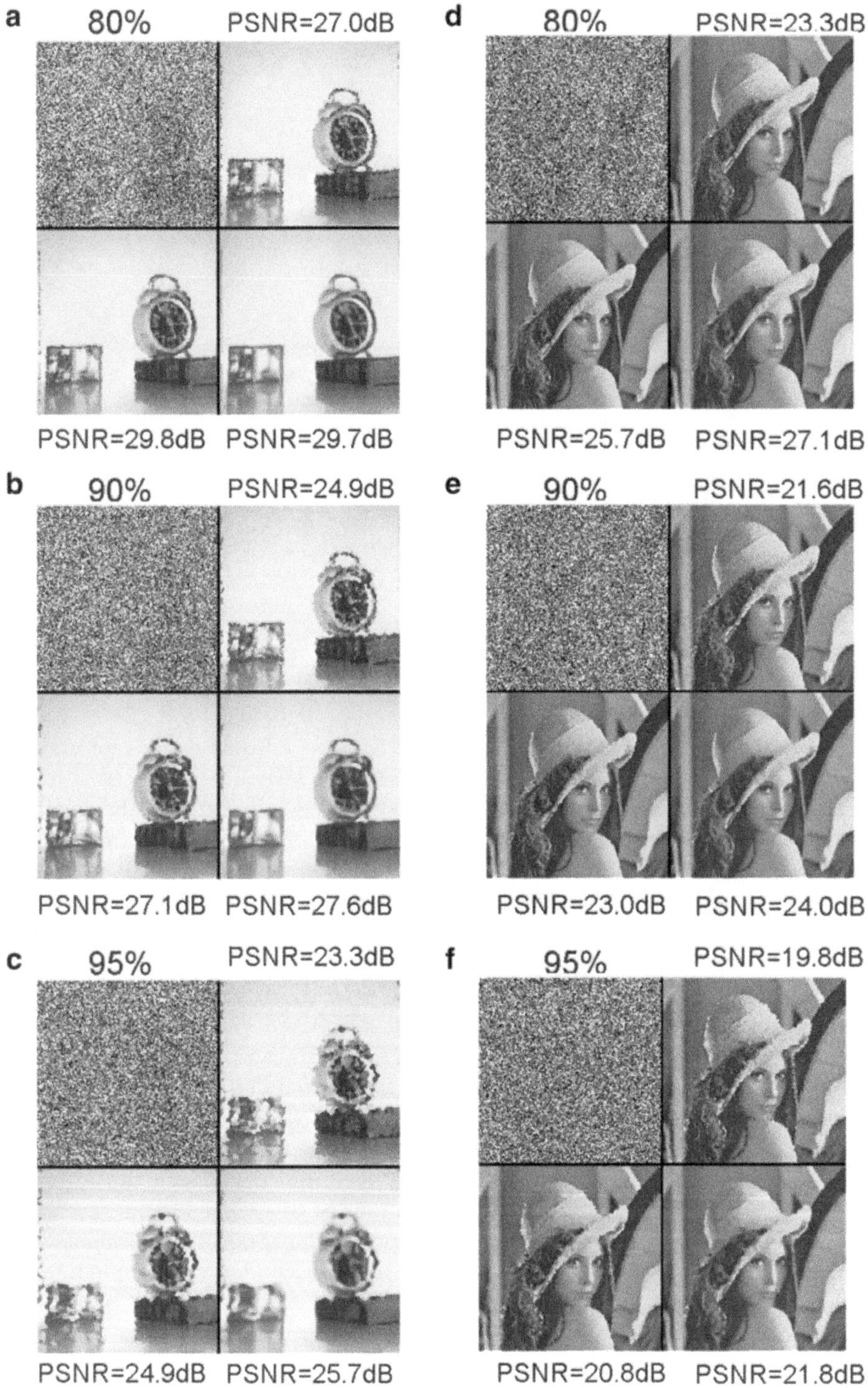

Fig. 4 Test trials for filtering grayscale images. *Left column* - image "clock;" *right column* - image "Lena." *First row* - 80 % of impulses (**a**, **d**); *second row* - 90 % of impulses (**b**, **e**); *third row* - 95 % of impulses (**c**, **f**). For each image and noise percentage: *upper left corner* - corrupted image; *upper right corner* - filtered with decision-based/adaptive median filter; *lower left corner* - after single iteration with BM3D; *lower right corner* - after five iterations with BM3D

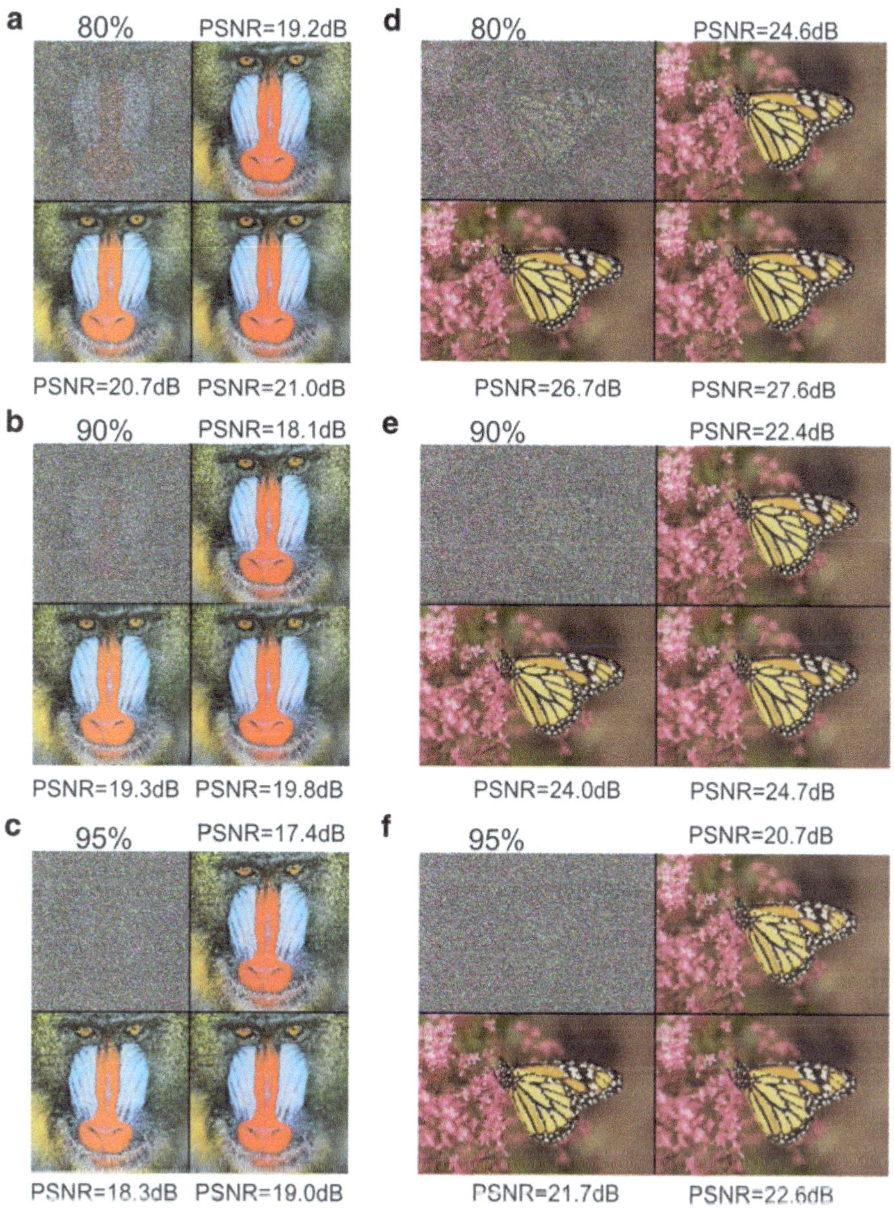

Fig. 5 Test trials for filtering color images. *Left column* - image "baboon;" *right column* - image "monarch." *First row* - 80 % of impulses (**a**, **d**); *second row* - 90 % of impulses (**b**, **e**); *third row* - 95 % of impulses (**c**, **f**). For each image and noise percentage: *upper left corner* - corrupted image; *upper right corner* - filtered with decision-based/adaptive median filter; *lower left corner* - after single iteration with BM3D; *lower right corner* - after five iterations with BM3D

In order to quantify this gain, Fig. 7 presents the average improvement in the signal-to-noise ratio (ISNR) for all considered test images

$$\mathrm{ISNR}_{i,j} = 10\log_{10}\frac{\frac{1}{NM}\sum_{n,m}[h^{(j)}(n,m)-f(n,m)]^2}{\frac{1}{NM}\sum_{n,m}[h^{(i)}(n,m)-f(n,m)]^2}. \quad (5)$$

Three values are calculated: $\mathrm{ISNR}_{0,1}$ (improvement in the SNR achieved for the first iteration), $\mathrm{ISNR}_{1,5}$ (improvement in the SNR in the next four iterations), and $\mathrm{ISNR}_{0,5}$ (overall improvement in five iterations). Gain achieved in the first iteration for 10 % of impulses is about 1 dB then gradually increases up to 2.5 dB for 50 % of impulses and after that decreases to about 0.8 dB for 97.5 % of impulses. Additional gain achieved in the next four iterations is moderate from 0.15 dB for 10 % of

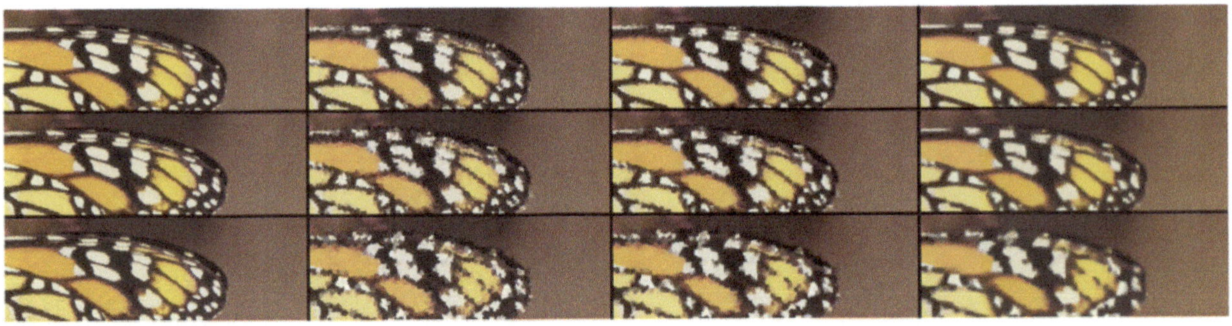

Fig. 6 Enlarged part of test trials for test image "monarch." *First row* - 80 % of impulses; *second row* - 90 % of impulses; *third row* - 95 % of impulses. *First column* - original image; *second column* - filtered by the decision-based/adaptive median filter; *third column* - after single iteration with BM3D; *fourth column* - after five iterations with BM3D

Table 2 PSNR in filtering of grayscale images for various percentage of impulses

	20 %	40 %	60 %	80 %	90 %	95 %	97.5 %
	30.04	25.92	23.27	21.11	19.68	18.53	17.02
Cameraman	33.12	28.71	25.47	22.78	20.88	19.41	17.59
	35.06	30.88	27.50	23.78	22.08	20.47	18.41
	38.95	34.78	32.15	29.90	28.00	26.28	24.70
Mandi	40.61	37.38	35.22	32.98	30.55	28.13	25.96
	40.51	37.16	34.76	45.14	30.49	28.61	26.80
	40.95	36.03	32.71	30.34	28.30	25.80	23.47
Moon	42.97	39.17	36.03	32.98	30.25	26.96	24.14
	42.79	39.17	36.63	34.19	32.30	28.70	25.25
	43.94	39.01	36.05	33.58	31.30	29.36	27.46
Pout	44.26	41.03	38.25	36.33	33.56	31.11	28.78
	43.85	40.33	36.97	34.10	32.16	30.90	29.22
	41.16	37.08	34.52	32.00	25.86	16.71	13.31
Saturn	44.47	41.90	39.28	36.15	27.23	17.04	13.51
	44.96	42.68	40.08	37.05	32.02	17.59	14.81
	28.82	24.49	21.80	19.68	18.02	16.67	15.47
West Concord	30.31	26.53	24.00	21.60	19.50	17.74	16.18
	30.41	46.57	24.01	21.37	19.84	18.44	16.88
	36.94	32.26	29.44	26.87	25.00	23.16	21.54
Lena	39.46	35.51	32.81	29.76	27.31	24.79	22.58
	39.49	35.70	32.68	29.78	27.74	25.78	23.58
	27.73	23.76	21.33	19.44	18.40	17.66	16.92
Baboon	29.63	25.75	23.27	21.07	19.70	18.67	17.64
	30.06	26.33	23.68	21.47	20.20	19.35	18.52
	32.61	28.16	23.27	23.27	21.64	20.28	18.94
Clock	37.41	32.40	28.97	25.53	23.28	21.42	19.58
	37.91	34.33	30.88	26.75	24.30	22.23	20.04

Upper row- decision-based/adaptive median filers; *middle row* - proposed scheme after single iteration by BM3D filter; *bottom row* - proposed scheme after five iterations by BM3D filter

Table 3 PSNR in filtering of color images for various percentage of impulses

	20 %	40 %	60 %	80 %	90 %	95 %	97.5 %
	29.74	28.48	27.13	25.21	22.31	20.58	19.01
Peppers	29.95	29.09	28.06	26.30	23.12	21.27	19.52
	29.72	29.14	28.16	27.32	25.07	22.52	19.71
	31.77	27.35	24.82	22.72	21.21	19.89	19.06
Autumn	35.62	31.00	27.89	25.01	22.81	20.98	19.81
	36.33	32.36	29.53	26.18	23.63	22.02	20.34
	31.54	28.98	26.91	24.72	23.08	21.61	20.28
F16	32.70	30.78	28.91	26.52	24.56	22.68	21.00
	32.82	31.32	29.54	27.13	25.12	23.40	21.79
	27.24	23.58	21.14	19.19	18.08	17.22	16.67
Baboon	28.97	25.32	22.89	20.73	19.33	18.20	17.36
	29.42	25.82	23.25	21.02	19.84	18.93	18.02
	36.60	32.21	29.30	26.75	24.94	23.36	21.83
Lena	38.52	34.80	32.13	29.36	27.06	24.88	22.84
	38.63	34.93	32.10	29.44	27.57	25.69	24.04
	29.78	27.98	26.21	24.46	22.87	21.53	20.36
House	30.45	29.57	28.26	26.61	24.58	22.70	21.20
	30.44	29.80	28.97	27.73	25.24	23.67	21.60
	29.04	27.63	25.81	23.59	21.90	19.43	17.65
Boats	29.41	28.62	27.24	25.05	23.14	20.21	18.16
	29.45	28.62	27.35	25.25	23.26	20.50	18.26
	35.24	30.43	27.26	24.44	22.43	20.56	18.98
Monarch	37.96	33.45	30.12	26.62	23.99	21.59	19.65
	38.32	34.42	31.06	27.43	24.81	22.53	20.29
	40.25	35.09	31.74	28.53	26.18	24.50	22.57
Jelly	42.23	37.71	34.79	31.50	28.20	25.91	23.43
	41.51	37.09	33.85	30.89	28.85	26.64	23.76

Upper row - decision-based/adaptive median filers; *middle row* - proposed scheme after single iteration by BM3D filter; *bottom row* - proposed scheme after five iterations by BM3D filter

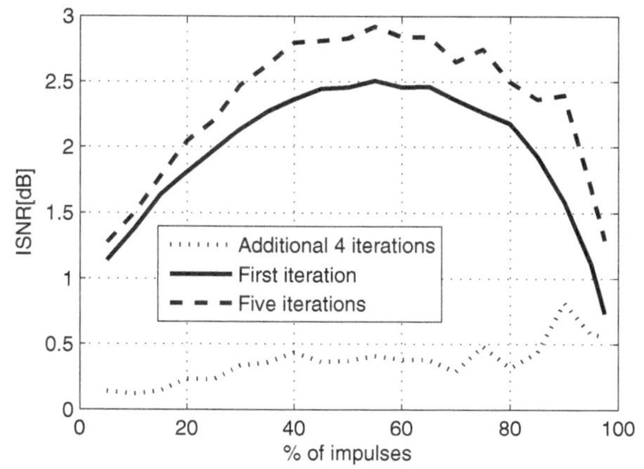

Fig. 7 Average improvement in the signal to noise ratio. *Solid line* - gain in the first iteration; *dashed line* - gain after five iterations; *dotted line* - gain in four iterations after the first one

impulses and gradually increases up to 0.8 dB for 90 % of impulses and after that decreases to 0.6 dB for 97.5 % of impulses. Total improvement is larger than 2 dB in the range of 20–93 % of impulses and in the entire considered range average ISNR is more than 1.2 dB with respect to the best decision-based/adaptive median filter.

It can be seen that four additional iterations of the BM3D filter brings relatively small gain; and if calculation complexity is critical, it can be used as a single iteration of the BM3D algorithm. Alternatively, some adaptation criterion can be employed to decide the number of the BM3D iterations. A simple criterion can be

$$10\log_{10}\frac{\frac{1}{NM}\sum_{n,m}[h^{(i+1)}(n,m)-x(n,m)]^2}{\frac{1}{NM}\sum_{n,m}[h^{(i)}(n,m)-x(n,m)]^2}\leq\varepsilon \qquad (6)$$

where it can be selected $\varepsilon \approx 0.1$ dB-0.2 dB.

In addition, we have compared considered filters using other quality measures of digital image filtering and reconstruction methods. In particular, we have considered the image enhancement factor (IEF) [23], structural similarity index (SSIM) [24], and edge-preserving index (EPI) [25]. The IEF is defined as

$$\mathrm{IEF}=\frac{\frac{1}{NM}[x(n,m)-f(n,m)]^2}{\frac{1}{NM}[h(n,m)-f(n,m)]^2}. \qquad (7)$$

The SSIM is proposed in [24] as universal image quality index between original $f(n,m)$ and corrupted and/or processed image $h(n,m)$:

$$\mathrm{SSIM}=\frac{(2\mu_h\mu_f+c_1)(2\sigma_{hf}+c_2)}{\left(\mu_h^2+\mu_f^2+c_1\right)\left(\sigma_h^2+\sigma_f^2+c_2\right)} \qquad (8)$$

where μ_h, μ_f are the mean values of corresponding images, σ_h^2, σ_f^2 are the image variances, while σ_{hf} is the covariance:

$$\mu_h=\frac{1}{NM}\sum_{n,m}h(n,m) \qquad (9)$$

$$\sigma_h^2=\frac{1}{NM}\sum_{n,m}[h(n,m)-\mu_h]^2 \qquad (10)$$

$$\sigma_{hf}=\frac{1}{NM}\sum_{n,m}[h(n,m)-\mu_h][f(n,m)-\mu_f]. \qquad (11)$$

Values c_1 and c_2 are introduced to stabilize division for small denominator and commonly they are selected as $c_1 = 10^{-4}f_{\max}^2$ and $c_2 = 9\cdot 10^{-4}f_{\max}^2$. Values closer to 1 correspond to higher structural similarity (better filtering). The EPI index is proposed in [25]. This index is very interesting here since our approach is developed to improve the shape of objects and in particular edges for images corrupted by high percentage of the salt-and-pepper noise. The EPI is defined as

$$\mathrm{EPI}=\frac{\sigma_{h_H f_H}}{\sqrt{\sigma_{h_H}^2\sigma_{f_H}^2}} \qquad (12)$$

where $\sigma_{h_H}^2$ and $\sigma_{f_H}^2$ are variances, while $\sigma_{h_H f_H}$ is the covariance of highpass version of filtered $h_H(n,m)$ and original $f_H(n,m)$ images. The Laplacian operator is used as a highpass filter for emphasizing high-frequency content, i.e., edges and details:

$$\begin{aligned}h_H(n,m)=&h(n-1,m)+h(n+1,m)+h(n,m+1)\\&+h(n,m-1)-4h(n,m).\end{aligned} \qquad (13)$$

Note that the EPI closer to 1 indicates better results as in the case of the SSIM. For brevity reasons, we have given only the results for test image "clock" in Table 4 while similar results as in Tables 2 and 3 are achieved for other test images.

Table 4 Alternative filtering quality measures for grayscale color image "clock" (IEF, SSIM, and EPI)

	20 %	40 %	60 %	80 %	90 %	95 %	97.5 %
	128.57	91.87	73.80	61.95	46.19	33.81	27.59
IEF	396.87	244.12	156.46	107.60	66.77	42.51	32.21
	497.86	358.55	254.93	158.33	88.00	52.42	37.76
	0.98	0.94	0.89	0.83	0.78	0.72	0.67
SSIM	0.99	0.97	0.93	0.89	0.84	0.77	0.72
	0.99	0.97	0.94	0.90	0.84	0.78	0.74
	0.90	0.74	0.60	0.47	0.42	0.40	0.54
EPI	0.97	0.90	0.81	0.67	0.61	0.56	0.54
	0.97	0.94	0.88	0.77	0.69	0.64	0.61

Upper row - decision-based/adaptive median filters; *middle row* - proposed scheme after single iteration by BM3D filter; *bottom row* - proposed scheme after five iterations by BM3D filter

5 Conclusions

In this paper, we have proposed an application of the BM3D algorithm to the decision-based/adaptive median filter output for filtering salt-and-pepper noise. The BM3D filter with the proposed switching rule brings quite visible and measurable gain with respect to the decision-based/adaptive median filters. This research is a step in application of non-local and shaping filters to salt and pepper and other types of impulsive noises from digital images. Also, the proposed technique can be considered as an extension of the BM3D to challenging salt-and-pepper noise environment. Similar extensions and combinations of the BM3D with other filters are probably possible but they remain for further research. In any case, we strongly believe that one of the most important directions in an actual digital image processing field is designing procedures for combining different, well-developed, filters and their application to other problems outside of the originally considered area.

Competing interests

The author declares that he has no competing interests.

Acknowledgements

This research is supported in part by the Ministry of Science of Montenegro through the national project "Intelligent search techniques for parametric estimation in communication and power engineering" and by the EU with FP7 Project Foremont.

References

1. S Esakkirajan, T Veerkumar, AN Subramanyam, CH PremChand, Removal of high density salt and pepper noise through modified decision based unsymmetric trimmed median filter. IEEE Signal Process. Lett. **18**(5), 287–290 (2011)
2. RH Laskar, K Banerjee, D Basak, in *Proc. of International Conference on Circuits, Power and Computing Technologies (ICCPCT)*. Removal of high density salt and pepper noise from color images through variable window size, (2013), pp. 1132–1136. doi:10.1109/ICCPCT.2013.6528868
3. MM Rahman, F Ahmed, MI Jubair, SA Priom, IM Ziko, An enhanced non-linear adaptive filtering technique for removing high density salt-and-pepper noise. Int. J. Comput. Appl. **38**(11), 7–12 (2012)
4. AR Telepatil, SA Patil, VP Parama, A survey on median filters for removal of high density salt & pepper noise in noisy image. IOSR Journal of Electronics and communication Engineering (IOSR-JECE). **1**, 22–26 (2013). ISSN: 2278-2834, ISBN: 2278-8735
5. P Rodrigez, Total variation regularization algorithms for images corrupted with different noise models: A review. J. Electr. Comput. Eng. **2013**(Article ID 2170 21), 18
6. RH Chan, C-W Ho, M Nikolova, Salt-and-pepper noise removal by median-type noise detector and detail-preserving regularization. IEEE Trans. Image Process. **14**(10), 1479–1485 (2005)
7. R Lukac, B Smolka, KN Plataniotis, Sharpening vector median filters. Signal Process. **87**, 2085–2099 (2007)
8. J Astola, P Kuosmaneen, *Fundamental of Nonlinear Digital Filtering*. (CRC, Boca Raton, 1997)
9. S Zhang, MA Karim, A new impulse detector for switching median filters. IEEE Signal Process. Lett. **9**(11), 360–363 (2002)
10. KS Srinivasam, D Ebenezer, A new fast and efficient decision based algorithm for removal of high density impulse noise. IEEE Signal Process. Lett. **14**(3), 189–192 (2007)
11. M Nikolova, A variational approach to remove outliers and impulse noise. J. Math. Imaging Vis. **20**, 7 (2014)
12. R Lukac, B Smolka, KN Plataniotis, AN Venetsanopulos, Vector filtering for color imaging. IEEE Signal Process. Mag. **22**(1), 74–86 (2005)
13. S Morrilas, V Gregori, G Peris-Fajarnes, P Latorre. A fast impulsive noise color image filter using fuzzy metrics.Real-time Imaging. **11**(5-6), 421–428 (2005)
14. R Kaur, YS Randhawa, Comparative analysis of mean and median filter for high density salt and pepper noise removal. Int. J. Adv. Res. Comput. Sci. Softw. Eng. **4**(6), 1394–1401 (2014)
15. A Bhatia, in *Proc. of international conference on ITR*. Decision based median filtering technique to remove salt and pepper noise in images (Bhubaneswar, India, 2013), pp. 9–12. ISSN: 978-93-82702-26-9
16. K Dabov, A Foi, V Katkovnik, K Egiazarian, Image denoising by sparse 3D transform-domain collaborative filtering. IEEE Trans. Image Process. **16**(8), 2080–2095 (2007)
17. A Buades, B Coll, JM Morel, in *Proc. of IEEE Computer Society Conference on Computer Vision and Pattern Recognition*. A non-local algorithm for image denoising, vol. 2 (CVPR, 2005), pp. 60–65. doi:10.1109/CVPR.2005.38
18. J Zhang, R Xiong, C Zhao, Y Zhang, S Ma, W Gao, CONCOLOR: constrained non-convex low-rank model for image deblocking. IEEE Trans. Image Process. **25**(3), 1246–1259 (2015)
19. EJ Candes, KJ Romberg, T Tao, Stable signal recovery from incomplete and inaccurate measurements. Commun. Pure Appl. Math. **58**(Np. 6), 1207–1223 (2006)
20. P Chatterjee, P Milanfar, Is denoising dead? IEEE Trans. Image Process. **19**(4), 895–911 (2010)
21. S Sarjanoja, *BM3D image denoising using heterogeneous computing platforms*. (University of Oulu, M.Sc.Thesis, 2015)
22. M Lebrun, An analysis and implementation of the BM3D image denoising method. Image Process Line. **2**, 175–213 (2012). http://dx.doi.org/10.5201/ipol.2012.l-bm3d
23. S Malviya, H Ahmia, Image enhancement using improved mean filter at low and high noise density. Int. J. Emerging Eng. Res. Technol. **2**(3), 45–52 (2014)
24. Z Wang, AC Bovik, HR Sheikh, EP Simoncelli, Image quality assessment: from error visibility to structural similarity. IEEE Trans. Image Process. **13**(4), 600–612 (2004)
25. F Sattar, L Floreby, G Salomonsson, B Lovstrom, Image enhancement based on a nonlinear multiscale method. IEEE Trans. Image Process. **6**(6), 888–895 (1997)

6

Adaptive error protection coding for wireless transmission of motion JPEG 2000 video

Giuseppe Baruffa* and Fabrizio Frescura

Abstract

The delivery of video over wireless, error-prone transmission channels requires careful allocation of channel and source code rates, given the available bandwidth. In this paper, we present a theoretical framework to find an optimal joint channel and source code rate allocation, by considering an intra-coded video compression standard such as Motion JPEG 2000 and an error-prone wireless transmission channel. Lagrangian optimization is used to find the optimal code rate allocation, from a PSNR perspective, starting from commonly available source coding outputs, such as intermediate rate-distortion traces. The algorithm is simple and adaptive both on the available bandwidth and on the transmission channel conditions, and it has a low computational complexity. Simulation results, using Reed-Solomon (R-S) coding, show that the achieved performance, in terms of PSNR and MSSIM, is comparable with that of other methods reported in literature. In addition, a simplified and sub-optimal expression for determining the channel code assignment is also provided.

Keywords: Motion JPEG 2000, Wireless video transmission, Lagrangian optimization, Unequal error protection, Forward error correction

1 Introduction

Many multimedia devices are being turned into complete entertainment centers, also by taking profit of wireless transmission. There exist several industry-backed liaisons aimed at transmitting wireless audio/video contents between multimedia home appliances using either the 60-GHz band [1–3] or the 2.4–5.0-GHz unlicensed spectrum [4, 5], using for this purpose techniques such as UltraWideBand, orthogonal frequency division multiplexing, and multi-antenna links.

Moreover, JPEG 2000 [6] is rapidly spreading as a valuable intra-coding scheme for video contribution applications [7] due to the high compression efficiency, wide coverage of encoding profiles from lossless to lossy, and the low latency. Recently the International Organization for Standardization (ISO), jointly with the International Electrotechnical Commission (IEC) and the International Telecommunications Union (ITU), added new profiles, to the JPEG 2000 standard, for broadcast video contribution and distribution with an amendment to the JPEG 2000 core coding system [8]. This amendment defines three new profiles, aimed at studio contribution links, specifying encoding parameters and rate limits over seven operating levels for video encoded with JPEG 2000. Even JPEG 2000 over MPEG-2 Transport Stream is a recently standardized method suited to this scenario [9]. In this kind of application, wireless cameras may produce a video contribution that has to adapt, in real time, to time-varying transmission channel profiles. In such case, both the available bandwidth and the wireless link bit error rate (BER) may be considered slowly variable with respect to the video frame rate [10].

The streaming of video either directly over the physical layer or using IP packets, is subject to transmission errors. Retransmission of lost/corrupted data or packets is viable, but decreases interactivity and real-time requirements. Thus, forward error correction (FEC) is generally adopted, and the channel code rate may be matched to compressed data error sensitivity, performing unequal error protection (UEP) [11–13].

*Correspondence: giuseppe.baruffa@unipg.it
Department of Engineering, University of Perugia, via G. Duranti 93, I-06125 Perugia, Italy

Many researchers have investigated optimal methods for the protection of intra-coded video streams. For instance, JPEG 2000 for wireless (JPWL) [14] has been standardized for this purpose, and several works have shown its good performance either when used on IP networks [15–17] or directly over the physical layer [18–23].

In [24, 25], the authors addressed similar problems showing how rate-distortion optimized (RaDiO) audio/video streaming over packetized networks can be achieved, and they solved this problem using Lagrangian optimization. Cataldi et al. proposed a technique based on raptor codes and sliding windows, where different H.264 [26] code rates are associated to each quality layer [27]. In [28], the authors proposed Wyner-Ziv coding for the protection of a coarse version of the video, where side information is provided by a primary H.26x decoder. Ahmad et al. proposed a UEP system using fountain codes [29]. In general, many of the solutions reported in literature for searching optimal UEP strategies are based on heuristic methods or use optimization algorithms [30–33]. It should be noted, however, that such solutions are based on search strategies characterized by a variable amount of computational complexity, which could prevent their use in real-time and bandwidth-adaptive video transmission.

1.1 Review of recent works

In recent years, several researchers have investigated techniques capable to apply differentiated FEC levels to wavelet-based image/video compressors, when the multimedia stream is delivered over unreliable or wireless channels.

In [34], a motion-compensated temporal filtering discrete wavelet transform (DWT) video coder is coupled with double binary turbo codes: the joint source-channel coding strategy is based on distortion profiling and code statistics. Also, Ho and Tsai [35] used 3D-wavelets, data interleaving, and Reed-Solomon (R-S) codes for their UEP streaming system. In [36], the authors investigate the performance of MJPEG 2000 and rate-compatible punctured convolutional codes for streaming over a time-varying binary symmetric channel (BSC); in their work, rate-distortion tradeoff of the coding units adapts the error correction code to the bandwidth and error characteristics. Schwartz et al. [37] adopted the DWT-based compression and convolutional coding FEC of CCSDS 122.0-B-1 and 131.0-B-1 satellite standards. Their results show that wavelet coefficient UEP outperforms the equal error protection (EEP) method over the simulated AWGN channel.

In [15, 38], the authors focus on JPEG 2000, R-S coding, and interleaving over wireless channels, simulated by a time-varying BSC with Gilbert-Elliot (GE) model. UEP is performed by variable FEC rates defined by solving a convex optimization problem. Based on interleaving effects, they derive a lower bound for successful image decoding rate in wireless environments. In [39], several UEP schemes are compared and layered JPWL streaming with RTP packetization on wireless channels is studied; the FEC allocation method is comparably faster and less complex than others, although yielding comparable quality.

JPWL has been shown to be both flexible and reactive to variable channel status. In [21, 40], streaming performance is simulated over realistic wireless channels, such as multiple-input multiple-output or Rayleigh fading ones. In [17], the authors conjugate JPWL with a dynamic bandwidth estimation scheme in order to provide the best layering, scaling, and protection of video streams. Even if source distortion is coarsely estimated, it has been shown that it can be effectively used to find an optimal rate allocation that outperforms EEP [23].

In our previous papers [16, 41], we used JPWL and interleaving over lossy packet networks. The UEP solution, found by means of a recursive, dichotomic search algorithm, was shown to always outperform EEP, and a low complexity interleaving strategy was devised for JPWL implementation on a DSP device. Iqbal et al. [42, 43] devised a family of dynamic programming code allocation methods for FEC protected wireless video streaming. Their protection assignment can provide variable tradeoff between performance and implementation complexity.

A different view was adopted by Bahmani et al. [44]. The method devised by the authors operates mainly at the decoder side, leaving the particular UEP implementation open. By leveraging the error resilience features of JPEG 2000, their method guesses the erased received symbols and improves the error correction capability.

Ouaret et al. [45] compared R-S coded JPEG 2000 to the Slepian-Wolf/Wyner-Ziv distributed video coding (DVC) approach. Their results show that JPEG 2000 results in better quality at high error rates, even if only an EEP scheme was used, whereas DVC performs better at lower packet loss rates. In [46], JPEG 2000 and H.264 streams are protected with UEP and transmitted over lossy packet networks. A performance comparison with multiple description coding shows that UEP achieves better quality. Chen et al. [47], by using progressive digital fountain codes, allowed different users to receive broadcast video at different qualities, depending on the reliability of their UDP-based WiFi link. In [48], the authors describe the emerging MPEG multimedia transport standard for delivering high bit rate video over packet-lossy networks, using a low-density generator-matrix FEC. They show the effectiveness of their method on the streaming of JPEG 2000 digital cinema.

1.2 Proposed contributions

In this paper, we present a simple mathematical and algorithmic solution for finding an optimal UEP channel code rate allocation strategy. The method used for solving the problem is based on Lagrangian optimization, which is known in the literature and has been applied in several works of other authors [25, 34, 36]. Our proposed solution can be calculated with a closed-form expression, directly from the knowledge of the data units rate-distortion and of the transmission channel characteristics.

Differently from other existing strategies, such as those reviewed in Sec. 1.1, our method has a closed-form representation for the solution to the UEP problem, and it does not require iterative or dynamic programming-based strategies. Moreover, the presence of data interleaving enables optimal video quality when channel conditions are time-varying, and data- and channel-adaptivity can be fulfilled.

The devised method does not rely on a particular channel coding technique, since it can be applied universally to all block-based FEC schemes. The algorithm itself is also lightweight, as it gives a closed-form expression of the optimal channel code allocation strategy, which can be computed in real time and adaptively respond to changes in the available transmission bandwidth and experienced channel BER. This makes the technique suitable for channels with unknown and slowly changing error rate and, even, available bandwidth; the video stream receiver should communicate the experienced error rate to the sender side, which in turn would change the UEP solution accordingly. We also show how the algorithm can be practically applied, using JPEG 2000 source coding and R-S channel coding, and present some performance results expressed in terms of either PSNR (peak signal-to-noise ratio) or MSSIM (mean structural similarity index metric).

Some mathematical derivations and performance results shown in this paper have already been partially presented in [49]. However, in this paper, we present additional derivations and novel simulation results. First, we describe in detail a method for assigning codewords of a channel code and implementing the designed protection profile. Furthermore, we work out a formula that allows approximating the optimal protection profile without knowledge of the image content, but only by means of a statistical entropy approach. Finally, we also present some results on the simulated transmission of a complete video sequence.

The paper is structured as follows. First, the theoretical framework for optimized unequal error protection is presented in Sec. 2 and a Lagrangian optimization strategy is shown to be able to find a UEP solution, both for particular and general cases. Then, a practical UEP code rate assignment using R-S codes is presented in Sec. 3. In Sec. 4, the Monte Carlo simulation setup is described, and the results of several simulated scenarios are presented and discussed. Eventually, conclusions are drawn, followed by an Appendix that contains proofs to assumptions and lemmas.

2 Optimized unequal error protection

In this paper, we model the distortion as a function of the combined channel characteristics/channel code performance in terms of probability of data loss. This approach is commonly used in the reference literature [16, 30]. We use Motion JPEG 2000 as source coding algorithm and R-S as channel coding algorithm, but the same optimization technique can be applied to other source/channel coding combinations as well. Table 1 summarizes the mathematical notation that will be used throughout the following sections and in the appendices.

Since we are considering an intra-video coding method, the video stream is regarded as a sequence of frames that are compressed independently of each other [50]. Each frame is compressed using JPEG 2000, and the size of the compressed bitstream is of B bits or, equivalently, the source code rate is r_s bits/pixel. The compressed bitstream is then divided in N_k pieces, each one of $(k - 16)$ bits; a two-byte cyclic redundancy check (CRC) codeword, used to test for the error-free condition, is appended at the end of each piece (see Fig. 1). The CRC codeword is considered in our method, even if it is not necessary when R-S codes are used, since they are able to detect the presence of residual errors. However, for different types of channel codes (e.g., convolutional codes), the CRC codeword is required. In this work, the detection of eventual erasures is not managed differently than the detection of errors. The message words of the k-bit long pieces are then R-S encoded, and the codewords are assembled in n_i-bit long pieces, $0 \leq i < N_k$. If the average channel code rate is $\bar{r}$, the combined source/channel code rate for the whole codestream will be $R = r_s/\bar{r}$.

The packetized codestream is then transmitted over a Q-ary symmetric transmission channel, where Q is the number of available different transmission symbols, that is, the pieces are assembled in symbols of $\log_2 Q$ bits. We define p_i as the probability that no errors occur up to piece i, and M_i as the mean square error (MSE) distortion of the decompressed image using all pieces up to piece i. The combined transmission channel performance and channel coding correction capability results in the probability $h(n_l)$ that a piece of n_l bits (after channel decoding) is received with errors (residual error rate); many other researchers, in the past, have already used this model (e.g., [30]). In this paper, we assume that this probability is approximated by a log-linear model [51], such as

$$h(n_l) \cong Ce^{-s(n_l-d)}, \tag{1}$$

Table 1 Summary of mathematical notation

Symbol	Meaning	Symbol	Meaning
B	Size of the compressed bitstream, in bits	m_i	$m_i = \Sigma_{j=i}^{N_k-1} M_j$, complementary cumulative distortion (CCD)
N_k	Number of pieces the bitstream is split into	$\hat{m}$	$\hat{m} = (\Pi_{i=0}^{N_k-1} m_i)^{1/N_k}$, geometric mean of the CCD
k	Size of each piece (with 2-byte CRC) before channel coding, in bits	$\bar{r}_{\min}$, $\bar{n}_{\min}$	Minimum average channel code rate and piece length after channel coding, in bits or bytes
n_i	Size of the i-th piece after channel coding, in bits	σ^2	Variance of the compressed frame coefficients
p_i	$p_i\,(n_0, n_1, \ldots, n_i)$, probability that no transmission errors occur up to piece i	$\Delta\rho$	MSE profile sampling step width, in bits/pixel
M_i	MSE distortion of reconstructed frame using the first i pieces	H	Differential entropy of the frame
Q	Number of available transmission alphabet symbols in the symmetric channel	$\tilde{N}$	Number of *mother* R-S codes
P_S	Symbol error rate of the Q-ary symmetric channel	$\tilde{r}_l$	$\tilde{k}_l/\tilde{n}_l$, channel code rate of the l-th R-S code
$\tilde{P}_S$	Q-ary symbol error rate estimated at the receiver	w_i	Index of the R-S code used in the i-th piece
P_b	Effective bit error rate of the Q-ary symmetric channel	α_i, β_i	Fractional number of codewords, in piece i, with R-S channel code rate $\tilde{r}_{w_i}$ and $\tilde{r}_{w_i+1}$, respectively
γ_b	Average signal-to-noise ratio of the AWGN or Rayleigh fading channels	α_i', β_i'	Integer number of codewords, in piece i, with R-S channel code rate $\tilde{r}_{w_i}$ and $\tilde{r}_{w_i+1}$, respectively
$h\,(n_l)$	Probability that a piece of n_l bits is channel-decoded without residual errors	$\delta_{\alpha_i'}$	Amount of rate $\tilde{r}_{w_i}$ codewords normalized to the total number of codewords
C, s, d	Amplitude, decay, and offset constants of the log-linear model for the combined transmission channel/channel code performance	N_{row}, N_{col}	Dimensions (rows and columns) of the block interleaver
f_D	Doppler spread of the Rayleigh fading channel	λ	Lagrange multiplier used for the optimization step
r_s	Source code rate, in bits/pixel	$\bar{r}$	Channel code rate
R	Combined source and channel rate, in bits/pixel	N_F	Number of frames in the YUV video sequence
R_b	Transmitted bit rate, after source and channel coding	$\epsilon_i, \bar{\epsilon}$	Per-frame and average MSE of the received video sequence
$\bar{n}$	Average piece length after channel coding, in bits or bytes	$\Gamma_i, \bar{\Gamma}$	Per-frame and average PSNR (in dB) of the received video sequence
$\iota_i, \bar{\iota}$	Per-frame and average MSSIM of the received video sequence	$F_E(\epsilon)$	*a posteriori* cumulative distribution function of the per-frame MSE ϵ_i

where C, s, and d represent the combined transmission channel/channel code performance characteristics. In particular, the parameter s is a decay connected to the correcting performance of the code, d is an offset to n_l used to improve the validity of this model, and C is an amplitude normalization constant. Further details on this approximation are given in Appendix A.1. This approximation is correct when the channel error probability is lower than 10^{-1}, as for higher values, the log-linear relationship does not prove to be valid; however, in such cases, any practical rate allocation/FEC method is hardly operational without increasing the channel coding redundancy to a limit beyond which the video quality is severely impaired by the high decoding latency and the low source code rate. In particular, for the model presented in (1), it is found that a unique parameter, s, can be used to represent the transmission scenario, and its value increases as the combined transmission channel/channel code performance improves (see Appendix A.1). At the receiving side, the probability of having no errors up to piece i depends on the chosen sequence of code word lengths (up to piece i) as $p_i = p_i\,(n_0, n_1, \ldots, n_i)$. The optimization objective is that of minimizing the average MSE distortion, given that a constant amount of combined source and channel coding bits is sent on the channel, i.e.,

$$\min_{\{n_i\}} \sum_{i=0}^{N_k-1} M_i p_i \quad \text{s.t.} \quad \sum_{i=0}^{N_k-1} n_i = \frac{B}{\bar{r}} \,. \tag{2}$$

Lemma 1. *The optimal UEP profile, i.e., the set of codeword lengths $\{n_i\}$ to be used for the pieces, which minimizes the overall distortion on the received image, is given by*

$$n_i = \bar{n} + \frac{1}{s} \ln \frac{m_i}{\hat{m}}, \; 0 \le i < N_k \,, \tag{3}$$

where $\bar{n} = k/\bar{r}$ is the average protected piece length, m_i is a complementary cumulative distortion (CCD),

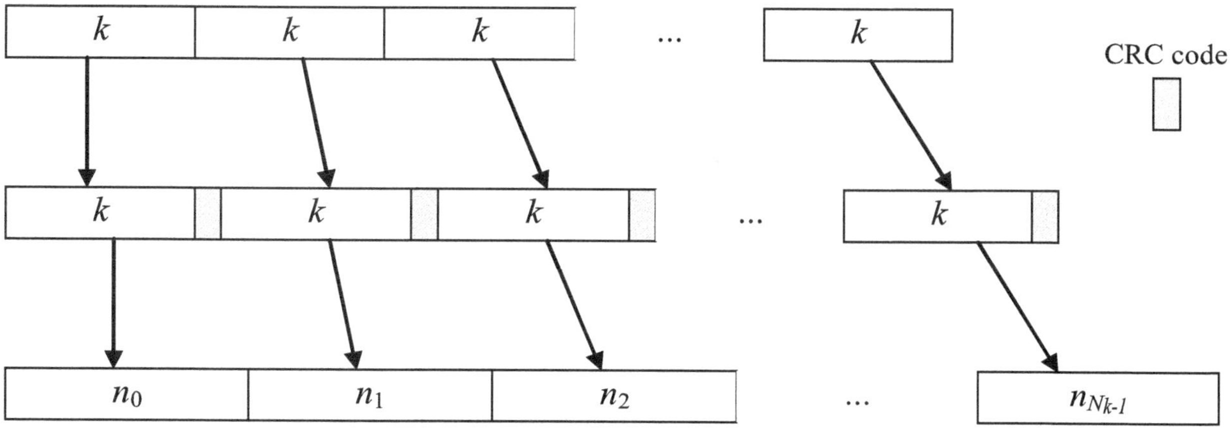

Fig. 1 Organization of the protected bitstream after division in N_k pieces

$m_i = \Sigma_{j=i}^{N_k-1} M_j$, *and* $\hat{m}$ *is the CCD geometric mean* $\hat{m} = \left(\Pi_{i=0}^{N_k-1} m_i\right)^{1/N_k}$.

Appendix A.2 shows how the closed-form expression (3), mathematical solution of (2), can be obtained. The relationship in (3) can be commented as follows:

- The protection rate for piece i depends on the average protection rate $\bar{n}$ plus a modification term.
- The modification term logarithmically weights the CCD at piece i, normalized by the geometric mean of the CCD.
- If the channel code has high-error correction performance and/or the channel conditions are good, the parameter s is large, which gives a small modification term.
- The protection profile depends on the equivalent transmission channel conditions only by means of the parameter s, not C and d.
- Since m_i is monotonically not increasing and $\ln(\cdot)$ is a monotonic function, the modification term is monotonically not increasing, i.e., pieces at the beginning of the bitstream are more protected than those at the end.
- The protection level at piece i depends on the cumulative amount of distortion of all following pieces.
- The shape of the protection profile is determined by the CCD. Ordinate extrema are defined only by the channel/code combined performance.

With respect to other similar solutions presented in the literature, and described in Sec. 1.1, the main advantage of our method is that a closed-form solution to the UEP problem is readily available, without needing iterative or dynamic programming-based solving strategies. The proposed solution is data and channel adaptive; regular, low bit rate feedback from the receiver lets the transmitter modify the UEP strategy, which, considering also the presence of data interleaving, enables optimal decompressed video quality when channel conditions change with time. Moreover, the side information produced at no cost during the compression process allows implementing a well-crafted protection profile, which minimizes the expected amount of distortion due to missing or corrupted data at the receiver. Eventually, we also want to outline that when stringent real-time requirements are needed, and the wireless channel is time-varying, deep interleaving matrices are necessary to intersperse the symbol losses occurring in the channel far away (for high Doppler spread f_D), and the decoding delay increases correspondingly.

Upon looking carefully at the solution (3) proposed in Lemma 1, it can be noticed that an additional condition to be satisfied is that $n_i \geq k$, $0 \leq i < N_k$, meaning that we cannot overprotect the pieces at the beginning, since there would be not enough bits to allocate for the last pieces, not even the source coding bits; especially at high symbol error probabilities, the protection profile, given the total bit budget, could be extremely unbalanced and might provide values lower than k.

Lemma 2. *The minimum average channel code rate* $\bar{r}_{\min}$, *expressed in terms of minimum average piece size after channel coding, is approximated by*

$$\bar{n}_{\min} = \max_{i<N_k}\left(k + \frac{\ln\left(\hat{m}/m_i\right)}{s}\right), \tag{4}$$

for a given equivalent channel error performance s.

Lemma 2 can be proved after some work on (3) and supposing that $\ln\left(m_i/\hat{m}\right) < 0$, for large i close to N_k.

In all cases where the exact rate-distortion curve of the compressed image is not known or cannot be calculated exactly, an approximation of (3) can be done, if the source coding process is expected to generate an ideal progressive codestream with a typical rate-distortion curve.

Lemma 3. *The approximated protection profile for progressively source encoded codestreams is given by*

$$n'_i \cong \bar{n} + \frac{\Delta\rho \ln 2}{s}(N_k - 1 - 2i),\ 0 \leq i \ll N_k\,, \qquad (5)$$

for N_k large, where $\Delta\rho$ is the bit rate sampling step of the MSE profile.

Proof of this lemma is given in Appendix A.3.

Our proposed UEP method has been devised in order to be as general as possible, with application scenarios that can extend also to other source and channel coding methods. For instance, considerations on the distortion reduction carried by data packets may be similarly done also for the network abstraction layer (NAL) units used in H.264 or H.265. In this case, NAL units naturally segment the video stream in pieces for which there is a correspondence with the pieces used in Fig. 1. The computation of the distortion profile can be achieved in several ways, if temporal and spatial frames intra/inter-dependency is maintained or not; layering methods for post-compression reordering of NAL units, similar to those adopted by the scalable video coding (SVC) extension of H.264 are then possible [52].

3 UEP profile generation using R-S coding

In the scenario adopted in this paper, when R-S coding is used, a strategy must be devised for the practical implementation of the optimal UEP profile found with (3). First, we consider a list of $\tilde{N}$ *mother* code rates $\{\tilde{r}_l\} = \{(\tilde{k}_l/\tilde{n}_l)\}$, $l = 0, 1, \ldots, \tilde{N} - 1$, ordered by decreasing code rate and not containing repeated code values, i.e., $\tilde{r}_l > \tilde{r}_{l+1}$, $j = 0, 1, \ldots, \tilde{N} - 2$.

Given the actual code rate for a generic piece i, $r_i = k_i/n_i$ (it was previously assumed that pieces are of fixed length, i.e., $k_i = k$, thus this is a generalization), an index w_i can be determined, such that $\tilde{r}_{w_i+1} \leq r_i \leq \tilde{r}_{w_i}$. The number α_i of R-S codewords with rate $\tilde{r}_{w_i}$, and β_i of codewords with rate $\tilde{r}_{w_i+1}$ is chosen in such way to achieve the target code rate r_i for the piece i. Due to the constraints

$$\alpha_i \tilde{k}_{w_i} + \beta_i \tilde{k}_{w_i+1} = k_i \quad \text{and} \quad \alpha_i \tilde{n}_{w_i} + \beta_i \tilde{n}_{w_i+1} = n_i\,,$$

the number of codewords for each code is

$$\alpha_i = \frac{k_i \tilde{n}_{w_i+1} - n_i \tilde{k}_{w_i+1}}{\tilde{k}_{w_i}\tilde{n}_{w_i+1} - \tilde{n}_{w_i}\tilde{k}_{w_i+1}}\,,$$

$$\beta_i = \frac{\tilde{k}_{w_i} n_i - \tilde{n}_{w_i} k_i}{\tilde{k}_{w_i}\tilde{n}_{w_i+1} - \tilde{n}_{w_i}\tilde{k}_{w_i+1}}\,. \qquad (6)$$

Positive-valued solutions always exists for (6), since the constraint $\tilde{r}_{w_i+1} \leq r_i \leq \tilde{r}_{w_i}$ gives

$$\tilde{k}_{w_i} n_i - k_i \tilde{n}_{w_i} > 0, \quad k_i \tilde{n}_{w_i+1} - n_i \tilde{k}_{w_i+1} > 0$$

and the constraint $\tilde{r}_{w_i} > \tilde{r}_{w_i+1}$ gives

$$\tilde{k}_{w_i}\tilde{n}_{w_i+1} - \tilde{k}_{w_i+1}\tilde{n}_{w_i} > 0\,.$$

The values of α_i and β_i obtained from (6) are fractional numbers. In order to best approximate them with integer numbers, we first compute α_i with rounding, and then use this result to compute β_i, as

$$\alpha'_i = \left\lfloor 1/2 + \frac{k_i \tilde{n}_{w_i+1} - n_i \tilde{k}_{w_i+1}}{\tilde{k}_{w_i}\tilde{n}_{w_i+1} - \tilde{n}_{w_i}\tilde{k}_{w_i+1}} \right\rfloor,$$

$$\beta'_i = \frac{k_i - \alpha_i \tilde{k}_{w_i}}{\tilde{k}_{w_i+1}}\,. \qquad (7)$$

4 System simulation and performance

4.1 Simulation setup

For purposes of assessing the performance of the technique presented in this paper, we have prepared a 512-frame video composed by the first 32 frames of each of the following 16 clips with CIF resolution (352 × 288, 30 frames/s) and YUV 4:2:0 format, combined in sequence: *akiyo, bus, coastguard, crew, flower, football, foreman, harbor, husky, ice, news, soccer, stefan, tempete, tennis,* and *waterfall* [53]. Only the luminance (Y) component of the video frames has been used to perform the optimization strategy and the transmission and reception over a simulated channel.

First, the partial distortions M_i for each frame in the video sequence have been calculated. To this purpose, JPEG 2000 compression has been performed using Kakadu 6.0 [54], with default parameters, no visual weighting, and the '-rate' option on every frame. The portion of each JPEG 2000 codestream located after the start-of-data (SOD) marker has been split into multiple pieces, each one with a size of $(k - 2)$ bytes (after CRC insertion, the piece will be of k bytes). Then, a new codestream has been constructed using the original header data, with an amended start-of-tile (SOT) marker to account for the new codestream length, a number i of codestream pieces, and the end-of-codestream (EOC) marker. The obtained codestream has been decompressed using Kakadu 6.0, and M_i has been calculated as the distortion of the reconstructed frame. Although this process of determining M_i is cumbersome, it should be said that the JPEG 2000 encoding process is able, *per se*, to provide such values; during the encoding procedure of JPEG 2000, an accurate rate-distortion estimation of the compressed frame is calculated, since the distortion values are gathered for the selection of the compressed wavelet coefficients with embedded block coding optimized truncation (EBCOT) of the bit-stream [55]. In this work, we have favored a direct calculation of the distortion values, in order to achieve more precise results.

When the piece boundaries are not coincident with the codestream interruption points decided by the JPEG 2000 compressor, we adopt a *continuum* hypothesis, i.e., we assume that the intermediate distortions at the piece boundaries can be calculated using linear interpolation from the nearest known, available distortions. This assumption is generally valid, since JPEG 2000 is a position-progressive encoder, and distortions are related to the way wavelet coefficients are truncated by EBCOT, in order to best satisfy the quality-rate constraints imposed on the compression process.

Moreover, in order to make a fair comparison among different channel code rates, we have kept fixed the total amount of data sent on the channel, i.e., the combined source and channel code rate R.

The transmission of the compressed stream has been simulated, using MATLAB, on three different types of channels. The first type is a Q-ary symmetric channel ($Q = 256$), characterized by symbol error rates P_S ranging from 10^{-3} to 10^{-1}. In this type of channel, errors occur at a symbol-level; since the bit errors are equiprobable and uniformly distributed over the symbol bits ($\log_2 Q = 8$ bits/symbol), there is a simple relationship between bit and symbol error rates when the number of bits per symbol is large, i.e., $P_b \cong P_S/2$ ([56] Section 4.4-1).

The second type of channel uses binary phase shift keying (BPSK) and additive white Gaussian noise (AWGN), in order to represent a transmission condition akin to physical level signaling on an actual, but ideal, channel. In this case, the performance depends on the signal-to-noise ratio (SNR) expressed by γ_b. Finally, the last type of simulated channel uses BPSK, AWGN, and Rayleigh-distributed flat fading, which represents a condition similar to that experienced on actual, wireless, non-line-of sight (NLOS) channels. In this case, the performance depends on the SNR γ_b and on the correlation degree among fades, expressed by the Doppler spread f_D. For channels using BPSK, the expressions used to determine the average BER (and the corresponding channel parameter s), given a certain value of γ_b, are

$$P_{b,\text{AWGN}} = Q\left(\sqrt{2\gamma_b}\right), \tag{8}$$

$$P_{b,\text{Rayleigh}} = \frac{1}{2}\left(1 - \sqrt{\frac{\gamma_b}{1+\gamma_b}}\right), \tag{9}$$

where $Q(\cdot)$ is the Gaussian tail function, and the Rayleigh channel BER is calculated for the maximum uncorrelated Doppler spread [56].

The UEP profile has been generated using (3), given the distortion profile M_i and the channel parameter s. Then, the codestream has been split into pieces that have been protected according to the determined UEP profile, using R-S coding with $\tilde{N} = 24$ mother code rates $\{\tilde{r}_l\} = \{32/36, 32/38, 32/40, \ldots, 32/80\}$, and adopting the codeword allocation strategy given by (7). The effect of the channel is simulated by randomly changing the transmitted bytes according to the simulation symbol error rate P_S. The error-affected codestream has been recomposed by terminating it at the last error-free received piece (thanks to the CRC codeword). In this way, any image reconstruction artifact due to wrong/erased codestream bytes has been eliminated, and the reconstructed image MSE is that used by the UEP allocation strategy. The JPEG 2000 header (about 300 bytes) has been considered as transmitted on a reliable channel, since it represents the most critical section of the codestream. At the receiving side, the JPEG 2000 header has been pre-pended to the JPEG 2000 bitstream bytes, and only the portion of the header carrying information on the bitstream size (*Psot* field of the SOT marker) has been changed accordingly. Performance has been evaluated as objective visual quality, and Y-PSNR has been used as objective quality indicator. In addition, we used MSSIM to faithfully represent the subjective evaluation by a human observer. The overall performance has been calculated by averaging the PSNR and MSSIM values calculated at each frame of the video sequence. The performance of the UEP method has been directly compared with that of an EEP method.

Additionally, comparisons with existing techniques in literature have been done using a static reference image, the 512 × 512 pixel grayscale version of *lena*, compressed at a total bit rate (joint source and channel code rate) of $R = 0.5$ bits/pixel. For each simulated channel error rate, at least 100 independent transmissions of the image have been repeated, and the results averaged. Since both the video sequence and the static image cases cover a standard definition application scenario, we have also used a static image frame from the high-definition 1920 × 1080 pixels *crowdrun* RGB sequence [53], in order to show some properties of the calculated UEP profiles.

4.2 Simulated performance results

4.2.1 Performance for static images on BSC

We first report the performance obtained with static images. Both images (*lena* and *crowdrun*) were compressed to a total rate of $R = 2.5$ bits/pixel, comprising both the source and the channel code bits.

Figure 2 (top) presents the UEP profile for *lena*, calculated with (3) at an average channel code rate of $\bar{r} = 32/44$, and a channel symbol error probability $P_S = 5 \times 10^{-2}$. The UEP profile, represented by the solid line, is shown in terms of the protected piece length n_i versus the piece index i; the average protected piece length $\bar{n}$, coincident with the EEP profile, is represented by the dashed line.

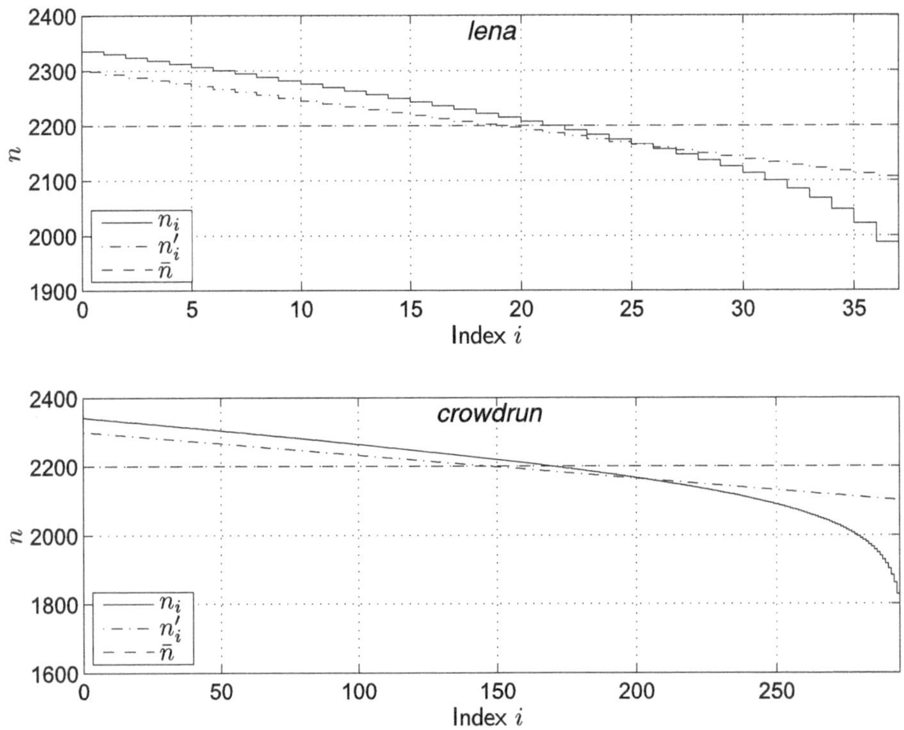

Fig. 2 UEP profiles for *lena* (*top*) and *crowdrun* (*bottom*), with code rate of $\bar{r} = 32/44$, for $k = 1\,600$ bytes and equivalent channel characterized by $s = 0.0127$ ($P_S = 5 \times 10^{-2}$). The values of k and n are expressed in bytes

Figure 2 (bottom) shows the equivalent case for *crowdrun*. In both cases, the piece length is $k = 1\,600$ bytes, and the channel parameter is $s = 0.0127$ (see Appendix A.1 and Table 3). The dot-dashed line in Fig. 2 represents the protection profile obtained using (5), at rate steps corresponding to the situation illustrated so far. The UEP profiles begin with a high protection level (low code rate), which then gradually decreases as the index of the piece increases, as expected.

Figure 3 shows how the UEP profile is practically generated, for *lena*, using a proper combination of the $\{\tilde{r}_l\}$ mother codes and with a number of codewords α'_i and β'_i, for each pair of codes, as given by (7). The top subplot shows the mother code rates $\tilde{r}_{w_i}$ and $\tilde{r}_{w_i+1}$ used in each i-th piece, expressed in terms of codeword size $\tilde{n}_{w_i}$. The bottom subplot describes the normalized amount $\delta_{\alpha'_i}$ of codewords with rate $\tilde{r}_{w_i}$ with respect to the total number of codewords used in the ith piece,

$$\delta_{\alpha'_i} = \frac{\alpha'_i}{\alpha'_i + \beta'_i} \,. \tag{10}$$

Clearly, $(1 - \delta_{\alpha,i})$ is the normalized amount of codewords with rate $\tilde{r}_{w_i+1}$.

Figure 4 shows the UEP profile n_i obtained for several values of the channel symbol error rate P_S (from 10^{-3} to 10^{-1}) and an average code rate $\bar{r} = 32/46$, for *lena*. The used message word size is $k = 1\,600$ bytes, which corresponds to the floor of the plot. For low values of error rates, the protection profile is almost linear, meaning that an EEP solution is optimal. On the other side, at higher error rates, the profile climbs over, at the beginning, and submerge below, toward the ending, the average protection level. At even higher error rates, the average protection is not sufficient for keeping the profile above the message size floor, and an increased protection rate is requested for correct operation of the algorithm.

Figure 5 shows the minimum required protected piece average size $\bar{n}_{\min}$ given by (4), plotted versus the channel symbol error rate P_S and the total bit rate R, using a message word size of $k = 1\,024$ bytes and the *crowdrun* image. The plot has been generated considering also the channel code rate in the total bit rate. We outline that, using this relationship, the system can adaptively respond to variations of the conditions of the transmission channel and of the available bandwidth, by keeping in all cases the received video quality at the optimal level. Given a channel error rate, the minimum average protection slightly increases as the available bandwidth increases, thus meaning that higher protection is required to achieve the optimal UEP profile for an error rate P_S. It is also evident how, when the available bandwidth increases, the

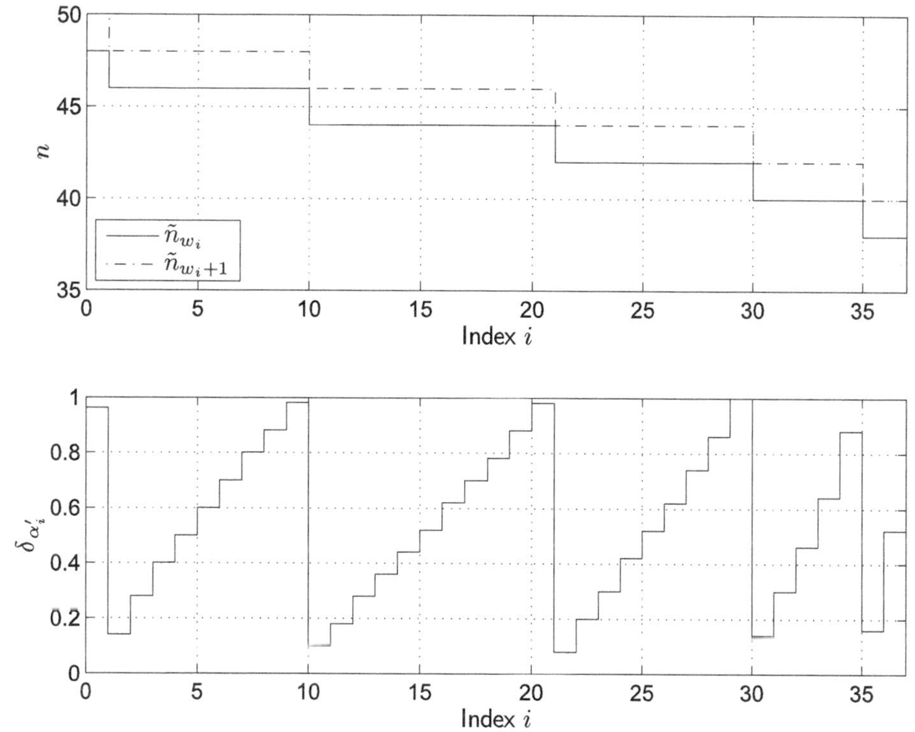

Fig. 3 Generation of the UEP profile for *lena* with an average channel code rate of $\bar{r} = 32/44$, $k = 1\,600$ and equivalent channel $s = 0.0127$ ($P_S = 5 \times 10^{-2}$). The used mother code rates are $\{\tilde{r}_l\} = \{32/36, 32/38, 32/40, \ldots, 32/80\}$. The values of k and n are expressed in bytes

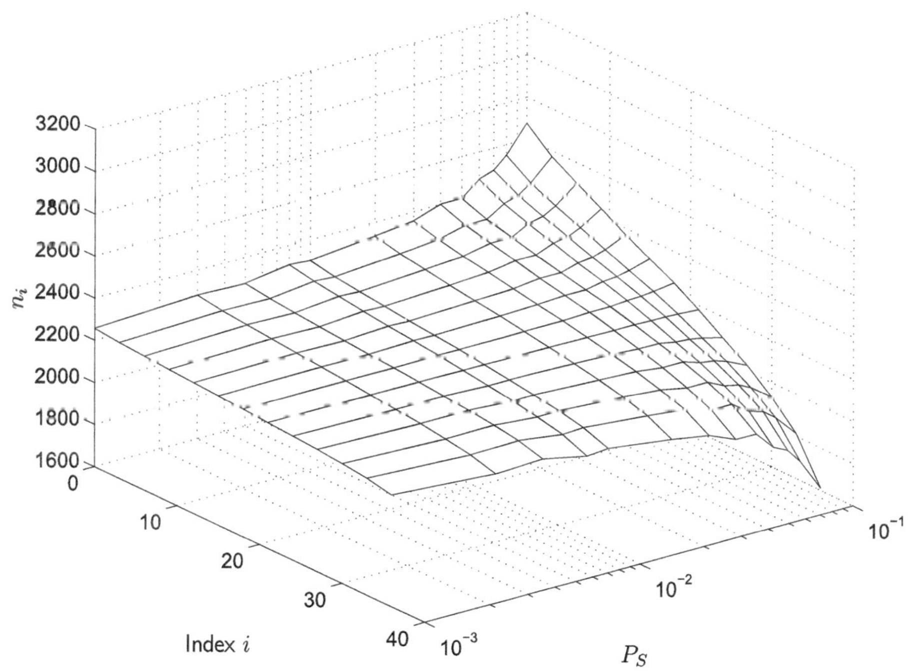

Fig. 4 UEP profiles for different channel symbol error rates for *lena* ($k = 1\,600$). The average protection code rate is $\bar{r} = 32/46$. The *floor* of the plot represents k. The k and n values are given in bytes

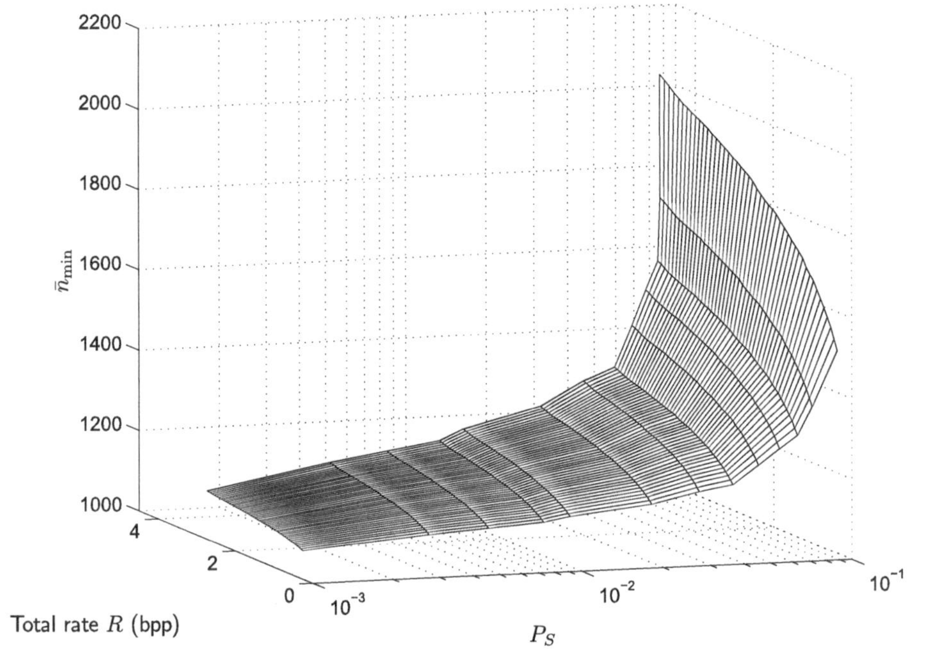

Fig. 5 Minimum average protection required for *crowdrun*, plotted versus the channel symbol error rate and the total bit rate, in case of $k = 1\,024$. The values of k and n are expressed in bytes

algorithm selects progressively higher levels of protection to maximize image quality. Clearly, in order to achieve the optimal profile for a given error rate, there must be some knowledge of the channel status even at the transmitter. Thus, the receiver must be able to calculate an estimate $\tilde{P}_S$ of the current channel symbol error rate and feed this information back to the transmission side. If channel conditions are slowly varying with respect to the information exchange rate, then this feedback may happen with minimum signaling requirements. Since our method employs the secure delivery of the JPEG 2000 header part (using a reliable transmission technique), these data can be repeated periodically (e.g., 1–2 times per second) on the same channel and used as pilot information for estimating $\tilde{P}_S$.

The performance of our method has been measured using the PSNR and MSSIM quality metrics. In order to compare the results with similar methods referenced in [16, 57, 58], Monte Carlo simulations have been done at a total bit rate (joint source and channel code rate) of $R = 0.5$ bits/pixel, and the metrics have been averaged. Figure 6 shows the achieved performance plotted versus the actual bit error probability P_b before channel decoding at the receiver side. The simulations have been done with a total of nine different configurations: three possible piece length values k (512, 1 024, and 1 600 bytes) and three different average channel code rates $\bar{r}$ (32/40, 32/44, and 32/48). In all the presented cases, it can be seen that shorter values of k allow achieving slightly higher PSNR/MSSIM values, for the same error rate on the channel.

The results summarized in Table 2, for the case $\bar{r} = 32/40$ and $k = 512$, show that the achieved PSNR is comparable or better than that obtained in [16, 57, 58]. There is an exception in the comparison with [58], which presents a higher PSNR. In that work, the authors used turbo codes with codewords longer than those used in our work, resulting in improved error correction capability. However, we want to outline that our algorithm is designed to find optimal UEP profiles, and using different error-correcting codes would result in different final PSNR values.

4.2.2 Performance for video sequence on BSC

As for the performance obtained on the video sequence, Fig. 7 shows the quality metric indicators history along the $N_F = 512$ frames of the test video. The piece length is $k = 1\,024$ bytes, the average code rate is $\bar{r} = 32/48$, and the total rate is $R = 2.5$ bits/pixel. Given the video frame rate (30 frames/s) and resolution (352 × 288), this corresponds to a transmitted bit rate of nearly $R_b = 7.6$ Mbit/s. The solid dark-green line (MAX) represents the maximum theoretically achievable PSNR at the receiver; it is due only to the compression artifacts introduced by the JPEG 2000 lossy source encoding. The performance indicators have been measured in the following way. First, the MSE ϵ_i of the luminance component of every decoded frame has been converted into logarithmic PSNR as $\Gamma_i =$

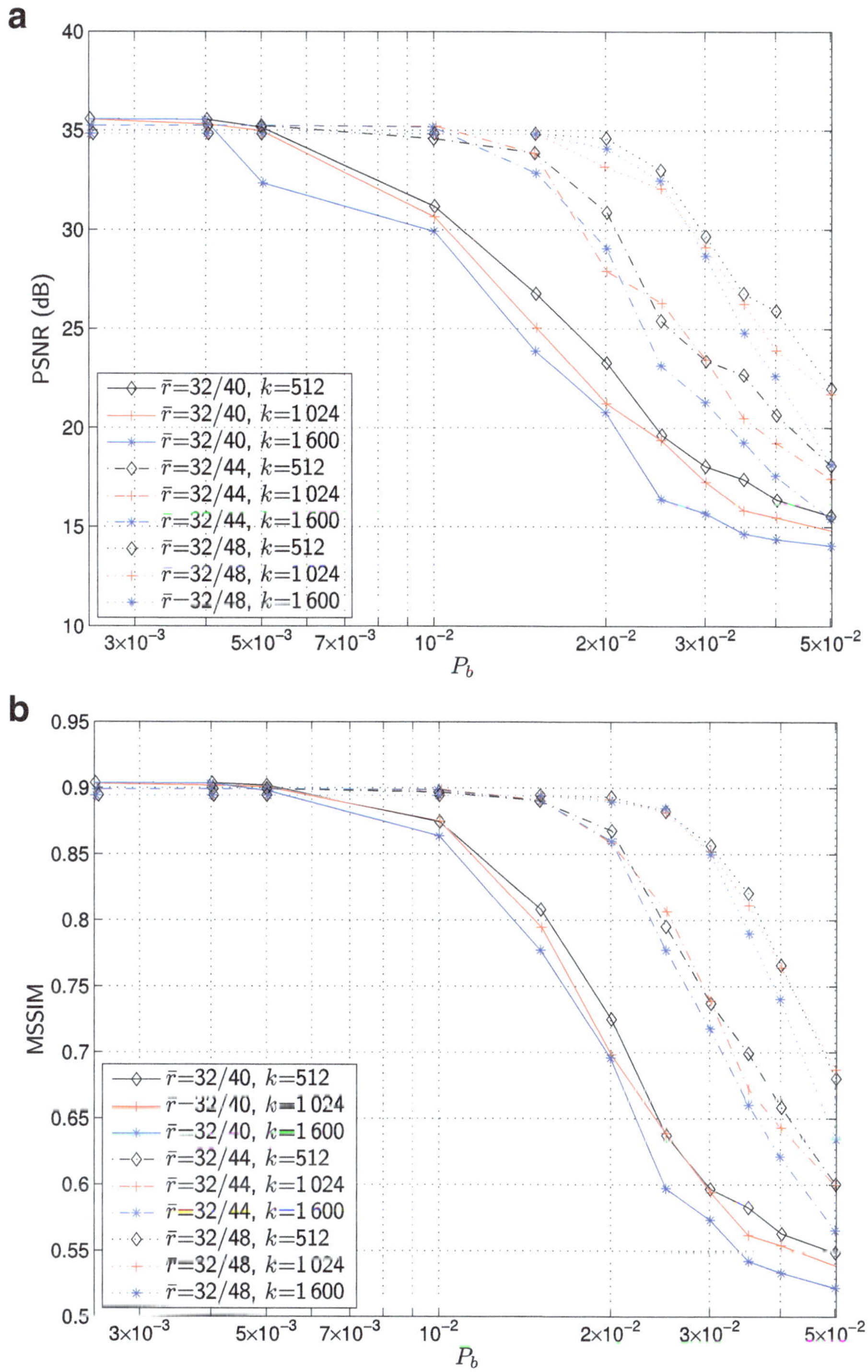

Fig. 6 Performance plotted in terms of PSNR (**a**) and MSSIM (**b**) versus channel bit error rate, for $\bar{r} = 32/40$ (*solid lines*), $\bar{r} = 32/44$ (*dashed lines*), $\bar{r} = 32/48$ (*dotted lines*), respectively, and $k = 512$ (*diamond*), $k = 1\,024$ (*plus*), $k = 1\,600$ (*asterisk*). The image is *lena*, compressed at 0.5 bpp

$10\log_{10}(1/\epsilon_i)$ and plotted versus the frame number i. Then, the average MSE has been calculated using the arithmetic mean of all MSEs, $\bar{\epsilon} = (1/N_F)\sum_{i=0}^{N_F-1}\epsilon_i$. Finally, the average PSNR is calculated from the average MSE as $\bar{\Gamma} = 10\log_{10}(1/\bar{\epsilon})$. For the MSSIM, the values ι_i obtained at each frame have been plotted versus the frame number i and arithmetically averaged as $\bar{\iota} = (1/N_F)\sum_{i=0}^{N_F-1}\iota_i$. Also in this case, the actual BER P_b, before channel decoding at the receiver side, is calculated and used to compare the performance.

Table 2 PSNR (dB) for *lena* at 0.5 bpp, $\bar{r} = 32/40$, and $k = 512$ bytes

P_b	1E-1	8E-2	3E-2	1E-2	1E-3	1E-4
[16]	16.5	16.5	20.0	23.0	33.0	37.0
[57]	–	–	–	16.04	22.11	29.02
[58]	32.76	33.20	35.15	35.67	–	–
Fig. 6	–	–	18.1	31.2	35.6	–

The PSNR and MSSIM histories Γ_i and ι_i plotted in Fig. 7 are obtained on a simulated channel with an error rate of $P_S = 7 \times 10^{-2}$, equivalent to a bit error rate of $P_b = 3.5 \times 10^{-2}$. The red line depicts the quality of the received UEP frames after decompression, with additional artifacts due to the errors introduced by the loss of pieces during transmission on the channel. For comparison purposes, we have also reported the quality of an EEP profile (blue line), having the same total rate R. The average performance of the UEP method results in a PSNR of $\bar{\Gamma} = 25.3$ dB, while the PSNR of the EEP method is of $\bar{\Gamma} = 19.6$ dB. Similarly, we have an MSSIM of $\bar{\iota} = 0.84$ and $\bar{\iota} = 0.66$, for the UEP and EEP methods, respectively.

Figure 8 shows the obtained average PSNR $\bar{\Gamma}$ and MSSIM $\bar{\iota}$ for different values of channel SER P_S, varying from 2×10^{-2} to 10^{-1} (P_b varying from 10^{-2} to 5×10^{-2}). The results are plotted both for the UEP and EEP cases. The curves show that the proposed UEP method outperforms the EEP method, in terms of PSNR, up to nearly 7 dB for $P_S = 10^{-1}$ ($P_b = 5 \times 10^{-2}$). It is also evident that the UEP method begins to provide better results than EEP starting from a SER of $P_S = 3 \times 10^{-2}$ ($P_b = 1.5 \times 10^{-2}$), while for values of SER lower than $P_S = 2 \times 10^{-2}$ ($P_b = 10^{-2}$) the two methods are equivalent. Similar considerations can be declared for the MSSIM. In this case, the quality index of UEP begins to improve over the EEP one for SER higher than $P_S = 4 \times 10^{-2}$ ($P_b = 2 \times 10^{-2}$); the protection advantage given by UEP over EEP is thus only slightly reduced when considering this more subjective-quality related metric.

Figure 9 shows the *a posteriori* cumulative distribution function (CDF) of the per-frame MSE, $F_E(\epsilon)$, defined as the computed probability that the MSE of a decompressed frame is lower than ϵ. The figure plots the CDFs for UEP and EEP cases, and for two different values of channel SER, $P_S = 5 \times 10^{-2}$ ($P_b = 2.5 \times 10^{-2}$) and $P_S = 8 \times 10^{-2}$ ($P_b = 4 \times 10^{-2}$). By setting, for instance, a threshold probability of 0.9, we find that, for the higher SER, the UEP MSEs are lower than 0.009 whereas for EEP they are lower than 0.05. Moreover, for every threshold probability, the UEP curves always give lower MSEs than the UEP curve. Similarly, at the lower SER, the 90 % threshold values are 0.0017 and 0.0025 for UEP and EEP, respectively. However, in this case, for a probability of 79 % and MSE of 0.0008, the two curves cross each other. This peculiar fact can be explained in the following way: in the EEP case, there are few values of ϵ_i that are much worse than the worst values obtained in the UEP case. On the contrary, UEP gives a lot of ϵ_i values that are somewhat lower than those of EEP ones, but never get much worse than that. This fact is mainly responsible for the improved average PSNR $\bar{\Gamma}$ exhibited by the UEP method over the EEP method in Fig. 8.

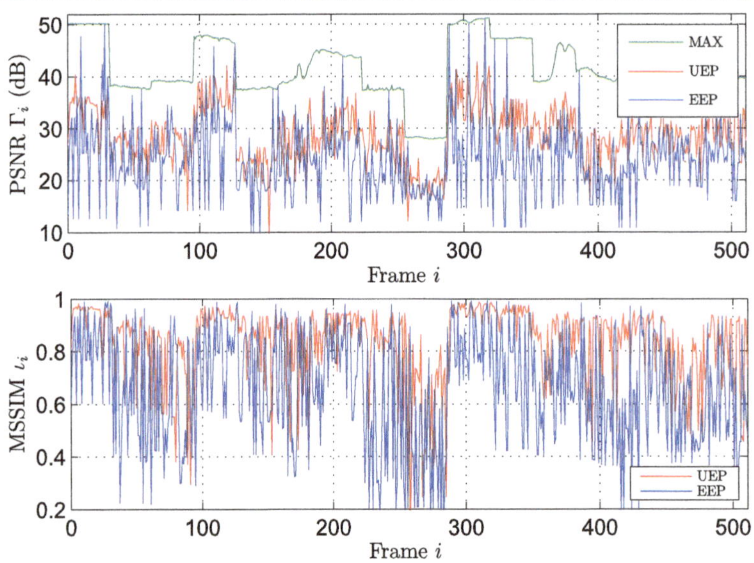

Fig. 7 PSNR and MSSIM history for the test video sequence on *Q*-ary channel. The piece length is $k = 1\,024$ bytes, the average code rate is $\bar{r} = 32/48$, and the total rate is $R = 2.5$ bits/pixel. The simulated channel has an error rate of $P_S = 7 \times 10^{-2}$

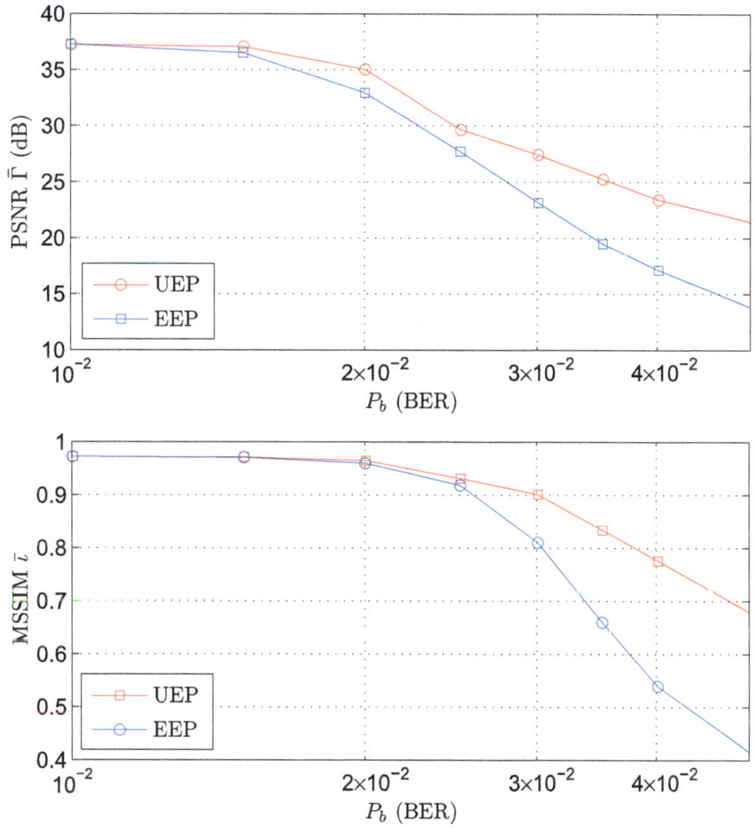

Fig. 8 Average PSNR and MSSIM for the test video sequence at different values of Q-ary channel BER. The piece length is $k = 1\,024$ bytes, the average code rate is $\bar{r} = 32/48$, and the total rate is $R = 2.5$ bits/pixel

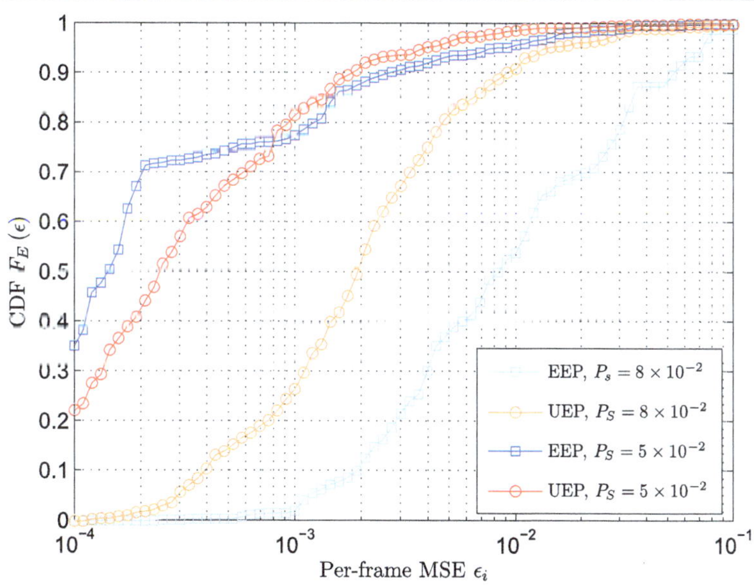

Fig. 9 CDF of MSE for the test video sequence at different values of Q-ary channel BER. The piece length is $k = 1\,024$ bytes, the average code rate is $\bar{r} = 32/48$, and the total rate is $R = 2.5$ bits/pixel

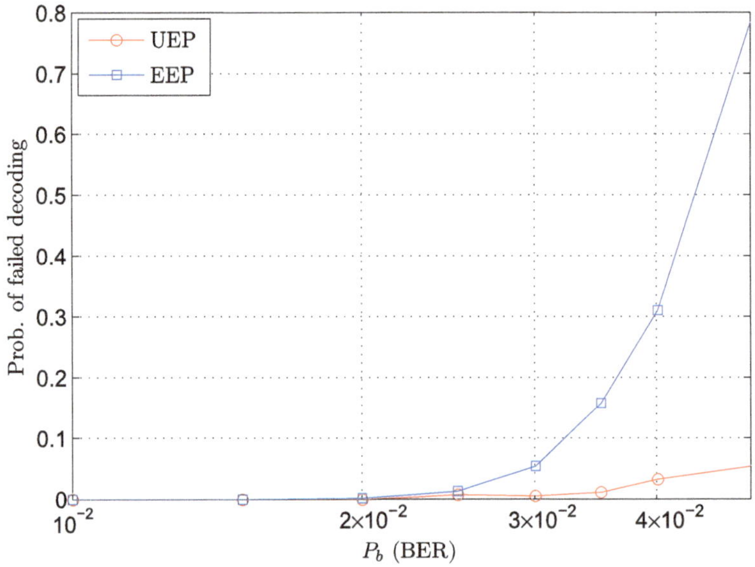

Fig. 10 Probability of failed decoding for the test video sequence at different values of Q-ary channel BER. The piece length is $k = 1\,024$ bytes, the average code rate is $\bar{r} = 32/48$, and the total rate is $R = 2.5$ bits/pixel

A proof that this phenomenon affects the decoded video quality is also given by Fig. 10. The figure shows the measured probability that no decoding happens at all in the received compressed video frame, due to the presence of uncorrectable errors in the first piece ($i = 0$). Both methods can successfully decode at least the first packet up to a SER of $P_S = 4 \times 10^{-2}$ ($P_b = 2 \times 10^{-2}$). For larger BERs, the UEP method attains a maximum of about 5 % probability of no decoding, whereas the EEP method failure probability is an order of magnitude higher and grows up to 80 %. Figure 11 shows two decompressed frames (frames no. 6 and 360) obtained during the transmission

Fig. 11 Comparison between decompressed frames using the EEP (**a–c**) and UEP (**b–d**) methods, for a Q-ary channel simulated error rate of $P_S = 7 \times 10^{-2}$. *Top row* contains frame no. 6, *bottom row* contains frame no. 390

on a channel with an error rate of $P_S = 7 \times 10^{-2}$ ($P_b = 3.5 \times 10^{-2}$). In this case, the error sequence has been exactly the same for the UEP and the EEP methods. Figure 11a and c display the frames obtained with EEP, while Fig. 11b and d present the frames obtained with UEP. Simulation results and samples of the original and decompressed video clips are available for download [59].

4.2.3 Performance on AWGN and Rayleigh channels

The performance of the proposed system has been also verified using the AWGN and uncorrelated ($f_D \approx R_b$) Rayleigh flat fading channels, for the video sequence only (without loss of generality, the results apply also to the static images case). Figure 12 shows the average PSNR and MSSIM versus the average channel SNR γ_b. For both types of channels, UEP outperforms EEP. This is not surprising, as the proper combination of interleaving depth and channel coding results in a equivalent BSC channel, which we have already simulated. In case of correlated Rayleigh fading ($f_D < R_b$), the bit interleaver size has to be chosen to span over an amount of bits such that, after deinterleaving, the fades are practically uncorrelated. If transmission on a channel with Doppler spread f_D adopts an $N_{\text{row}} \times N_{\text{col}}$ block interleaver, then, after deinterleaving, the equivalent Doppler spread becomes N_{row} times higher, $N_{\text{row}} f_D$ [60]. Thus, by properly choosing the interleaver dimension N_{row}, one can revert to the condition of uncorrelated Rayleigh fading, for which the performance is plotted in the right-side curves of Fig. 12. If, on the other side, the Doppler spread is so low (such as it happens on nearly static NLOS channels, $f_D \ll R_b$) that the interleaver size should exceed the available memory or the decoding delay bounds, then the periodic feedback from the receiver would allow the transmitter to adapt the protection profile and coding rate at the measured channel conditions. In this case, in the short term, the performance will be practically that of the AWGN case, for which the curves on the left side of Fig. 12 apply.

4.3 Computational complexity

The optimization problem requires the knowledge of the distortion profile of the image. Using JPEG 2000 compression, the partial distortions M_i (and the CCD m_i) can be easily obtained during the rate allocation step of the JPEG 2000 bitstream preparation [55]; thus, these values can be obtained virtually at no cost.

For the calculation of the C, s, and d coefficients, a look-up table (LUT) can be used to store the parameters,

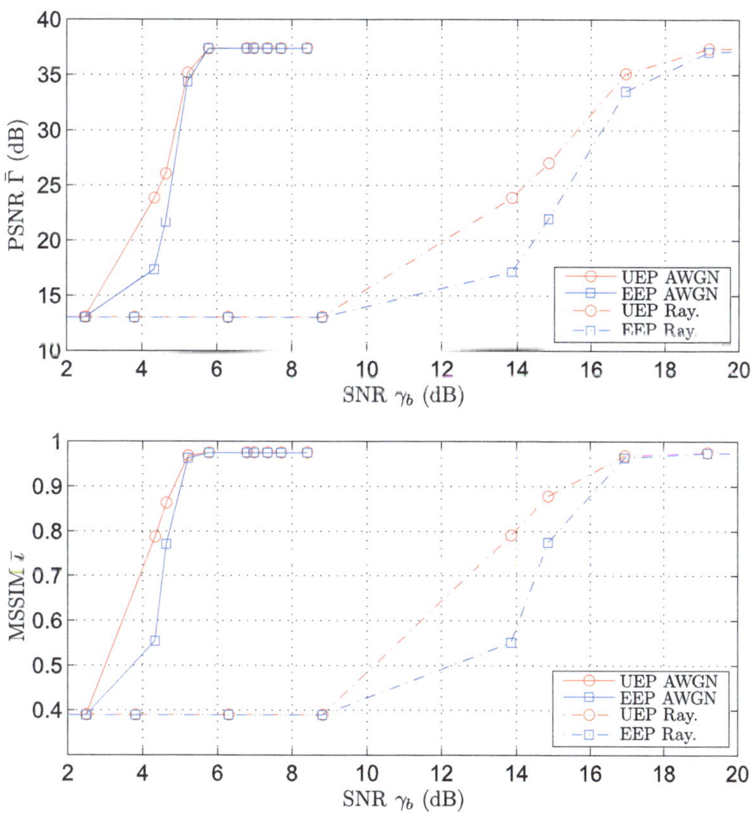

Fig. 12 Average PSNR and MSSIM for the test video sequence at different values of γ_b on the AWGN and Rayleigh fading channels. The piece length is $k = 1\,024$ bytes, the average code rate is $\bar{r} = 32/48$, and the total rate is $R = 2.5$ bits/pixel

for different values of the packet size N_k, of the channel bit/symbol error rate P_b/P_S, and possibly even for different channel coding algorithms (e.g., convolutional, binary R-S, low-density parity check codes). The LUT can then be accessed to provide the parameters that will be used in the protection profile generation, with a large saving with respect to storing the entire UEP profile, for each combination of the three variables. As for generating the coefficients stored in the LUT, they can be calculated off-line and smoothly interpolated to provide all the intermediate values that could be requested by the system.

The calculation of the optimal protection profile in (3) depends on the geometric mean of the CCD function. In order to avoid overflow or underflow problems due to floating point operations rounding during the computation, the geometric mean should be calculated logarithmically as

$$\hat{m} = e^{\frac{1}{N_k}\sum_{i=0}^{N_k-1}\ln m_i}, \qquad (11)$$

in which case it takes N_k logarithms, $(N_k - 1)$ additions, 1 division, and 1 exponentiation to be computed. Then, we need $(N_k - 1)$ additions for the computation of the CCD function, N_k multiplications for the logarithm operand (one division to obtain the inverse of the geometric mean), N_k logarithm operations, N_k multiplications for logarithm result scaling (one division to obtain the inverse of s, if not already saved in this form in the LUT), and N_k additions. In summary, to implement (3), a total of $(3N_k - 2)$ additions, $2N_k$ multiplications, $2N_k$ logarithms, 3 divisions, 1 exponentiation is needed. Assuming that natural logarithms and powers of e can be implemented by means of another LUT, with a sufficient precision once the dynamic ranges of the operands have been characterized, the asymptotic complexity becomes $\mathcal{O}(N_k)$. Differently, the solutions presented in [16] or in [58] require multiple evaluations of expressions similar to (2), which are more cumbersome to calculate from the computational viewpoint.

5 Conclusions

The transmission of video over error-prone wireless channels is a problem that can be solved by using an adequate error protection layer added to the streams, once the characteristics of the channel are known. In this work, we have presented a UEP strategy devised to allocate channel code bits, using an optimization algorithm that is computationally light during the search of the UEP profile. The general formulation of the problem has been solved using a Lagrangian minimization method. The discovered closed-form UEP expression requires already available data, such as the image rate-distortion curve, the average error protection code rate, the typical allowed packet size for transmission, and the channel error model (represented by one parameter). In addition, we have also presented a practical method for implementing the discovered UEP profile using R-S codes. The simulated performance of the proposed UEP strategy shows that the results outperform those obtained using an EEP strategy, that they are comparable with the UEP performance results of other methods presented in similar works, yet having a lower computational complexity, and that this UEP method can be used to effectively counteract the impairments introduced by an error-prone transmission channel.

Appendix A

A.1 Loglinear approximation of residual BER curves

The expression (1), derived from a more general expression found in [51], approximates the functions $h(n_l)$ with exponentials, at least in definite regions where $h(\cdot)$ is lower than 10^{-1}, which is a common requirement. In order to show the validity of this approximation, we have simulated the performance of R-S error coding applied to the devised packetization scheme. Each piece of k bytes has been split in message words of $\tilde{k} = 32$ bytes and a R-S code with rate $\tilde{n}/\tilde{k}$ has been applied to each word. Then, the resulting codewords have been concatenated, producing a piece of n bytes. Multiple pieces have been transmitted over an Q-ary channel ($Q = 256$) with a defined symbol error rate P_S, and the residual error probability after decoding, $h(n)$, has been measured. For a fixed P_S, the value of n has been increased and the measurement on the residual error rate performed again. This procedure has been repeated for several different values of P_S. Figure 13 shows the set of residual error probability curves obtained adopting a piece length $k = 1\,600$ bytes, for channel symbol error rates from $P_S = 10^{-3}$ ($P_b = 5 \times 10^{-4}$) to $P_S = 10^{-1}$ ($P_b = 5 \times 10^{-2}$). Similar sets of error rate curves have been obtained for different lengths of the pieces. The resulting curves have been fit using a nonlinear least-squares method in the logarithmic domain, thus providing the relevant model parameters; Table 3 lists the parameters C, s, and d for several piece lengths and channel symbol error probabilities P_S.

In Fig. 13, solid lines represent the results of simulations, whereas the dashed lines are obtained by evaluating (1) with the best-fit model parameters of Table 3, for every channel symbol error rate.

The following can be said on C, s, and d parameters:

- The amplitude C is generally much lower than 1.
- The decay s increases as the channel/channel code performance improves.
- The offset d is the point where the error rate curve becomes linear, and is higher than 10^{-2}.
- d is greater than the values of k that we have used.

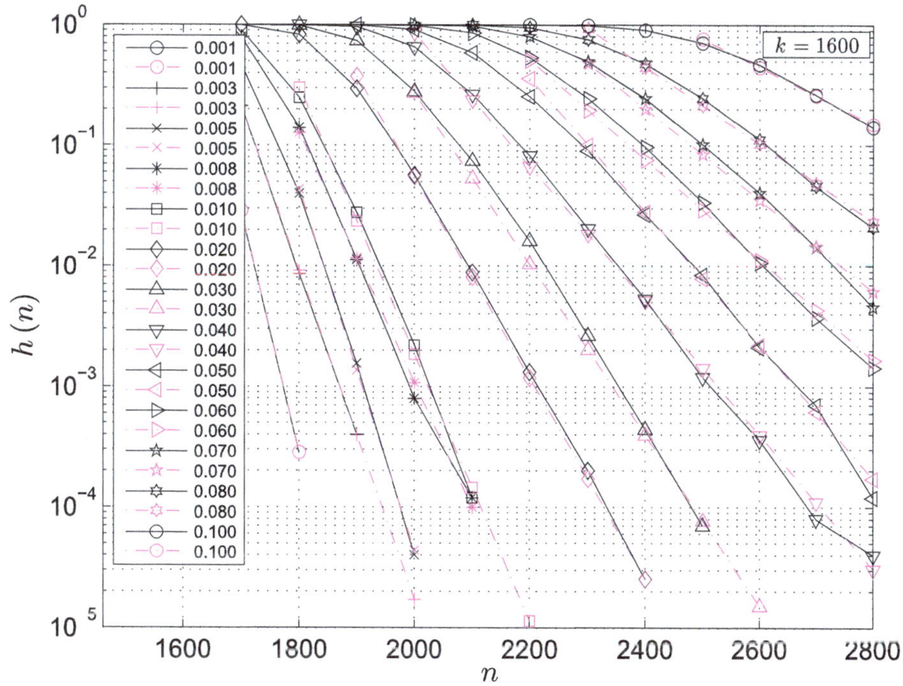

Fig. 13 Simulated (*solid line*) and modeled (*dashed line*) piece error rate performance for R-S coding, using a mother code from $\{\bar{r}_l\}$ and a piece size of $k = 1\,600$ (n and k are expressed in bytes). The channel symbol error rate P_S varies from 10^{-3} to 10^{-1} (*curve styles* are in the *left-side legend*)

For other types of transmission methods, such as, for instance, BPSK on AWGN or Rayleigh fading channels, if a bit/byte interleaver interspersing consecutive errors is present, then the results and comments discussed above are still valid.

A.2 Proof of Lemma 1

Proof. The constrained minimization problem expressed by (2) can be solved using the Lagrange multipliers method, as

$$\frac{\partial}{\partial n_i} \sum_{j=0}^{N_k-1} \left(M_j p_j \left(n_0, n_1, \ldots, n_j \right) - \lambda n_j \right) = 0 , \quad 0 \le i < N_k . \tag{12}$$

First, we simplify the probability of having no received errors up to piece i, $p_i = p_i(n_0, n_1, \ldots, n_i)$. We suppose that this probability is expressed by the product of correct reception probabilities for each piece, since they are independently decoded of each other, as

$$p_j \left(n_0, n_1, \ldots, n_j \right) = \prod_{l=0}^{j} \left(1 - h\left(n_l\right) \right) , \tag{13}$$

where $h(n_l)$ is the probability that a piece of k bits (n_l bits after channel encoding) is received with errors. Under such conditions, products in (13) are approximated as

$$\prod_{l=0}^{j} \left(1 - h\left(n_l\right) \right) \cong 1 - \sum_{l=0}^{j} h\left(n_l\right) , \tag{14}$$

Table 3 Fitting parameters *C*, *s*, and *d* found for the curves in Fig. 13

	k								
	512			1024			1600		
P_S	C	s	d	C	s	d	C	s	d
1E-3	0.0085	0.1893	544	0.0180	0.0783	1088	0.0282	0.0461	1700
3E-3	0.0000	0.0984	623	0.0009	0.0512	1186	0.0074	0.0314	1806
5E-3	0.0000	0.0886	660	0.0008	0.0453	1228	0.2666	0.0345	1747
1E-2	0.0000	0.0801	684	0.0008	0.0364	1298	0.0761	0.0255	1854
3E-2	0.0002	0.0451	772	0.0026	0.0248	1451	0.0237	0.0163	2148
5E-2	0.0003	0.0404	849	0.0027	0.0178	1628	0.1258	0.0127	2281
1E-1	0.0011	0.0174	1119	0.0246	0.0077	1996	0.2450	0.0054	2707

since products of $h(n_l)$ terms can be neglected. Supposing that all $h(n_l)$ have the same value, it can be found that when $h(n_l) < 2 \times 10^{-2}$, the approximation (14) is valid with an error lower than 10 %.

After substituting (14) in (13) and differentiating, we obtain

$$\frac{\partial p_j}{\partial n_i} = -\frac{\partial h(n_i)}{\partial n_i}\left(1 - \sum_{\substack{l=0 \\ l \neq i}}^{j} h(n_l)\right). \quad (15)$$

With the approximations (1) and (15), and neglecting C^2 (since $C \ll 1$), (12) becomes

$$e^{-sn_i} = \frac{\lambda}{sCe^{sd}\sum_{j=i}^{N_k-1} M_j},\ 0 \leq i < N_k. \quad (16)$$

The summation at the denominator of (16) is the CCD $m_i = \Sigma_{j=i}^{N_k-1} M_j$, which is a nonincreasing function in the i variable (as i increases, there are less M_j's to sum). Then, n_i is found to be

$$n_i = d + \frac{1}{s}\ln\left(\frac{sCm_i}{\lambda}\right),\ 0 \leq i < N_k. \quad (17)$$

In order to find the Lagrange multiplier λ, we use the constraint in (2). After some work, the constraint becomes

$$\lambda = sCe^{-\frac{s(T-D)}{N_k}}\hat{m}, \quad (18)$$

where $\hat{m} = (\Pi_{i=0}^{N_k-1} m_i)^{1/N_k}$ is the geometric mean of the CCD, $T = B(\bar{n} - k)/k$, and $D = N_k(d - k)$. We can substitute (18) into (17), and find the closed form solution to the minimization problem

$$n_i = \frac{k}{\bar{r}} + \frac{1}{sN_k}\ln\left(\frac{\left(\sum_{j=i}^{N_k-1} M_j\right)^{N_k-1}}{\prod_{\substack{l=0 \\ l \neq i}}^{N_k-1}\left(\sum_{j=l}^{N_k-1} M_j\right)}\right) \quad (19)$$

$$= \bar{n} + \frac{1}{s}\ln\frac{m_i}{\hat{m}},\ 0 \leq i < N_k.$$

□

A.3 Proof of Lemma 3

Proof. If the exact R-D curve of the image is not known, it is still possible to calculate an approximate UEP profile considering the MSE profile, sampled at $\Delta\rho$ bit/symbol steps, using the bounds

$$\frac{1}{2\pi e}2^{-2(i\Delta\rho - H)} < M_i < \sigma^2 2^{-2i\Delta\rho}, \quad (20)$$

$0 \leq i < N_k$, where the lower bound expresses the differential entropy H of the actual source, and the upper bound is calculated considering the hypothesis of Gaussian source (with encoded image coefficients that are Gaussian distributed with variance σ^2), respectively.

We start by expressing the CCD, for both bounds in (20), as

$$m_i = K\sum_{j=i}^{N_k-1}\left(2^{-2\Delta\rho}\right)^j = K\frac{2^{-2\Delta\rho i} - 2^{-2\Delta\rho N_k}}{1 - 2^{-2\Delta\rho}}, \quad (21)$$

where either $K = \left(2^{2H-1}/\pi e\right)$ or $K = \sigma^2$. Considering $i \ll N_k$ and N_k large, (21) can be approximated by $m_i \cong K2^{-2\Delta\rho i}/(1 - 2^{-2\Delta\rho})$. The geometric mean is expressed and approximated by

$$\hat{m} \cong \frac{K}{1 - 2^{-2\Delta\rho}}2^{-\frac{2\Delta\rho}{N_k}\sum_{i=0}^{N_k-1} i} = K\frac{2^{-\Delta\rho(N_k-1)}}{1 - 2^{-2\Delta\rho}},$$

when N_k is large. Eventually, the approximated protection profile n'_i is given by

$$n'_i \cong \bar{n} + \frac{1}{s}\ln\left(\frac{K2^{-2\Delta\rho i}}{\hat{m}\left(1 - 2^{-2\Delta\rho}\right)}\right)$$
$$= \bar{n} + \frac{\Delta\rho \ln 2}{s}(N_k - 1 - 2i),$$
$$0 \leq i \ll N_k,\ N_k \text{ large}.$$

□

Competing interests
The authors declare that they have no competing interests.

Authors' contributions
GB and FF worked jointly on the development of the theoretical model. GB performed the simulations and drafted the manuscript, FF carried out the statistical analysis of video and helped to draft the manuscript. All authors read and approved the final manuscript.

Acknowledgements
The authors thank Paolo Micanti and Barbara Villarini for the help on theoretical aspects and simulations carried out for this paper.

References
1. H Singh, J Oh, C Kweon, X Qin, H-R Shao, CY Ngo, A 60 GHz wireless network for enabling uncompressed video communication. IEEE Commun. Mag. **46**(12), 71–78 (2008). doi:10.1109/MCOM.2008.4689210
2. CJ Hansen, WiGiG: Multi-gigabit wireless communications in the 60 GHz band. IEEE Wireless Commun. **18**(6), 6–7 (2011). doi:10.1109/MWC.2011.6108325
3. S Yong, C-C Chong, An overview of multigigabit wireless through millimeter wave technology: potentials and technical challenges. EURASIP J. Wirel. Commun. Netw. **2007**(078907) (2007). doi:10.1155/2007/78907
4. G Lawton, Wireless HD video heats up. IEEE Computer. **41**(12), 18–20 (2008). doi:10.1109/MC.2008.509
5. S Srinivasan, in *Proc. of 20th Int. Conf. on Computer Commun. and Networks (ICCCN)*. An assessment of technologies for in-home entertainment (IEEE, Maui, Hawaii, 2011), pp. 1–6. doi:10.1109/ICCCN.2011.6005803
6. ISO/IEC, *JPEG 2000 image coding system – part 1: core coding system. ISO/IEC ISO/IEC 15444-1*. (Int. Standards Org./Int. Electrotech. Comm., Geneva, Switzerland, 2004)
7. F-O Devaux, C De Vleeschouwer, Parity bit replenishment for JPEG 2000-based video streaming. EURASIP J. Image Video Process. **2009**(1), 683820 (2009). doi:10.1155/2009/683820

8. ISO/IEC, *JPEG 2000 profiles for broadcast applications*. (ISO/IEC 15444-1:2004/Amd 3:2010, Int. Standards Org./Int. Electrotech. Comm., Geneva, Switzerland, 2010)
9. ISO/IEC, *ISO/IEC 13818-1:2007/FPDAM 5, Int. Standards Org./Int. Electrotech. Comm.*, (Geneva, Switzerland, 2010)
10. S Pejoski, V Kafedziski, in *Proc. of 5th European Conf. on Circuits and Sys. for Commun. (ECCSC)*. Causal video transmission over fading channels with full channel state information (IEEE, Belgrade, Serbia, 2010), pp. 294–297
11. M Etoh, T Yoshimura, Advances in wireless video delivery. Proc. IEEE. **93**(1), 111–122 (2005). doi:10.1109/JPROC.2004.839605
12. A Albanese, J Blomer, J Edmonds, M Luby, M Sudan, Priority encoding transmission. IEEE Trans. Inf. Theory. **42**(6), 1737–1744 (1996). doi:10.1109/18.556670
13. R de Albuquerque, D Cunha, C Pimentel, On the complexity-performance trade-off in soft-decision decoding for unequal error protection block codes. EURASIP J. Adv. Signal Process. **2013**(1), 28 (2013). doi:10.1186/1687-6180-2013-28
14. ISO/IEC, *JPEG 2000 image coding system-part 11: wireless*. (ISO/IEC WD2.0 15444-11, Int. Standards Org./Int. Electrotech. Comm., Geneva, Switzerland, 2003)
15. M Agueh, J-F Diouris, M Diop, F-O Devaux, C De Vleeschouwer, B Macq, Optimal JPWL forward error correction rate allocation for robust JPEG 2000 images and video streaming over mobile ad hoc networks. EURASIP J. Adv. Signal Process. **2008**(1), 192984 (2008). doi:10.1155/2008/192984
16. G Baruffa, P Micanti, F Frescura, Error protection and interleaving for wireless transmission of JPEG 2000 images and video. IEEE Trans. Image Process. **18**(2), 346–356 (2009). doi:10.1109/TIP.2008.2008421
17. C Mairal, M Agueh, in *Proc. of 2nd Int. Conf. on Advan. in Multimedia (MMEDIA)*. Smooth and scalable wireless JPEG 2000 images and video streaming with dynamic bandwidth estimation (IARIA, Athens, Greece, 2010), pp. 174–179. doi:10.1109/MMEDIA.2010.40
18. M Murroni, A power-based unequal error protection system for digital cinema broadcasting over wireless channels. Signal Process. Image Commun. **22**(3), 331–339 (2007). doi:10.1016/j.image.2006.12.006
19. MI Iqbal, H-J Zepernick, U Engelke, in *Proc. of 2nd Int. Conf. on Signal Process. and Commun. Sys. (ICSPCS 2008)*. Error sensitivity analysis for wireless JPEG 2000 using perceptual quality metrics (IEEE, Gold Coast, Australia, 2008), pp. 1–9. doi:10.1109/ICSPCS.2008.4813665
20. MI Iqbal, H-J Zepernick, U Engelke, in *Proc. of 6th Int. Symp. on Wireless Commun. Sys. (ISWCS 2009)*. Perceptual-based quality assessment of error protection schemes for wireless JPEG 2000 (IEEE, Siena, Italy, 2009), pp. 348–352. doi:10.1109/ISWCS.2009.5285217
21. W Xiang, A Clemence, J Leis, Y Wang, in *Proc. of 7th Int. Conf. on Inf., Commun. and Signal Process. (ICICS 2009)*. Error resilience analysis of wireless image transmission using JPEG, JPEG 2000 and JPWL (IEEE, Macau, China, 2009), pp. 1–6. doi:10.1109/ICICS.2009.5397742
22. KM Alajel, W Xiang, J Leis, in *Proc. of 4th Int. Conf. on Signal Proc. and Commun. Sys. (ICSPCS)*. Error resilience performance evaluation of H.264 I-frame and JPWL for wireless image transmission (IEEE, Gold Coast, Australia, 2010), pp. 1–7. doi:10.1109/ICSPCS.2010.5709766
23. C Bergeron, B Gadat, C Poulliat, D Nicholson, in *Proc. of 17th IEEE Int. Conf. on Image Process. (ICIP)*. Extrinsic distortion based source-channel allocation for wireless JPEG2000 transcoding systems (IEEE, Hong Kong, China, 2010), pp. 4469–4472. doi:10.1109/ICIP.2010.5651228
24. J Chakareski, PA Chou, Application layer error-correction coding for rate-distortion optimized streaming to wireless clients. IEEE Trans. Commun. **52**(10), 1675–1687 (2004). doi:10.1109/TCOMM.2004.836436
25. PA Chou, Z Miao, Rate-distortion optimized streaming of packetized media. IEEE Trans. Multimedia. **8**(2), 390–404 (2006). doi:10.1109/TMM.2005.864313
26. ISO/IEC, *Coding of audio-visual objects-part 10: Advanced video coding*. (ISO/IEC 14496-10, Int. Standards Org./Int. Electrotech. Comm., Geneva, Switzerland, 2010)
27. P Cataldi, M Grangetto, T Tillo, E Magli, G Olmo, Sliding-window raptor codes for efficient scalable wireless video broadcasting with unequal loss protection. IEEE Trans. Image Process. **19**(6), 1491–1503 (2010). doi:10.1109/TIP.2010.2042985
28. L Liang, P Salama, E Delp, Unequal error protection techniques based on Wyner-Ziv coding. EURASIP J. Image Video Process. **2009**(1), 474689 (2009). doi:10.1155/2009/474689
29. S Ahmad, R Hamzaoui, MM Al-Akaidi, Unequal error protection using fountain codes with applications to video communication. IEEE Trans. Multimedia. **13**(1), 92–101 (2011). doi:10.1109/TMM.2010.2093511
30. J Lu, A Nosratinia, B Aazhang, in *Proc. of Int. Conf. on Image Process. (ICIP 98)*. Progressive source-channel coding of images over bursty error channels, vol. 2 (IEEE, Chicago, Illinois, 1998), pp. 127–131. doi:10.1109/ICIP.1998.723331
31. VM Stankovic, R Hamzaoui, Z Xiong, Real-time error protection of embedded codes for packet erasure and fading channels. IEEE Trans. Circuits Syst. Video Technol. **14**(8), 1064–1072 (2004). doi:10.1109/TCSVT.2004.831964
32. N Thomos, NV Boulgouris, MG Strintzis, Optimized transmission of JPEG2000 streams over wireless channels. IEEE Trans. Image Process. **15**(1), 54–67 (2006). doi:10.1109/TIP.2005.860338
33. Y Zhang, S Qin, B Li, Z He, Rate-distortion optimized unequal loss protection for video transmission over packet erasure channels. Signal Process. Image Commun. **28**(10), 1390–1404 (2013). doi:10.1016/j.image.2013.05.009
34. N Ramzan, S Wan, E Izquierdo, Joint source-channel coding for wavelet-based scalable video transmission using an adaptive Turbo code. EURASIP J. Image Video Process. **2007**(1), 047517 (2007). doi:10.1155/2007/47517
35. C-P Ho, C-J Tsai, Content-adaptive packetization and streaming of wavelet video over IP networks. EURASIP J. Image Video Process. **2007**(1), 045201 (2007). doi:10.1155/2007/45201
36. S Bezan, S Shirani, RD optimized, adaptive, error-resilient transmission of MJPEG2000-coded video over multiple time-varying channels. EURASIP J. Adv. Signal Process. **2006**(1), 079769 (2006). doi:10.1155/ASP/2006/79769
37. C Schwartz, F da Silva Marques, M da Silva Pinho, in *International Telecommunications Symposium (ITS)*. An unequal coding scheme for remote sensing systems based on CCSDS recommendations (IEEE, Sao Paulo, Brazil, 2014), pp. 1–5. doi:10.1109/ITS.2014.6947971
38. D Pascual Biosca, M Agueh, in *Mobile Multimedia Communications. Lecture Notes of the Institute for Computer Sciences, Social Informatics and Telecommunications Engineering*, ed. by L Atzori, J Delgado, and Giusto D. Optimal interleaving for robust wireless JPEG 2000 images and video transmission, vol. 79 (Springer, Berlin Heidelberg, 2012), pp. 217–226. doi:10.1007/978-3-642-30419-4-19
39. M Agueh, S Ataman, H Soude, in *Fourth International Conference on Communications and Networking in China (ChinaCOM 2009)*. A low time-consuming smart FEC rate allocation scheme for robust wireless JPEG 2000 images and video transmission (IEEE, Xi'an, China, 2009), pp. 1–5. doi:10.1109/CHINACOM.2009.5339854
40. J Abot, C Olivier, C Perrine, Y Pousset, A link adaptation scheme optimized for wireless JPEG 2000 transmission over realistic MIMO systems. Signal Process. Image Commun. **27**(10), 1066–1078 (2012). doi:10.1016/j.image.2012.08.003
41. F Fiorucci, G Baruffa, P Micanti, F Frescura, in *IEEE International Conference on Multimedia and Expo (ICME)*. A real-time, DSP-based JPWL implementation for wireless High Definition video transmission (IEEE, Barcelona, Spain, 2011), pp. 1–4. doi:10.1109/ICME.2011.6012054
42. MI Iqbal, H-J Zepernick, in *International Symposium on Communications and Information Technologies (ISCIT)*. Error protection for wireless imaging: providing a trade-off between performance and complexity (IEEE, Tokyo, Japan, 2010), pp. 249–254. doi:10.1109/ISCIT.2010.5664847
43. MI Iqbal, H-J Zepernick, A framework for error protection of region of interest coded images and videos. Signal Process. Image Commun. **26**(4–5), 236–249 (2011). doi:10.1016/j.image.2011.03.001
44. S Bahmani, IV Bajic, A HajShirmohammadi, Joint decoding of unequally protected JPEG2000 bitstreams and Reed-Solomon codes. IEEE Trans. Image Process. **19**(10), 2693–2704 (2010). doi:10.1109/TIP.2010.2049529
45. M Ouaret, F Dufaux, T Ebrahimi, Error-resilient scalable compression based on distributed video coding. Signal Process. Image Commun. **24**(6), 437–451 (2009). doi:10.1016/j.image.2009.02.011
46. E Baccaglini, T Tillo, G Olmo, Image and video transmission: a comparison study of using unequal loss protection and multiple description coding. Multimedia Tools Appl. **55**(2), 247–259 (2011). doi:10.1007/s11042-010-0574-3
47. Z Chen, M Xu, L Yin, J Lu, in *International Conference on Wireless Communications and Signal Processing (WCSP)*. Unequal error protected JPEG 2000 broadcast scheme with progressive fountain codes (IEEE, Nanjing, China, 2011), pp. 1–5. doi:10.1109/WCSP.2011.6096843

48. T Nakachi, Y Tonomura, T Fujii, in *7th International Conference on Signal Processing and Communication Systems (ICSPCS)*. A conceptual foundation of NSCW transport design using an MMT standard (IEEE, Gold Coast, Australia, 2013), pp. 1–6. doi:10.1109/ICSPCS.2013.6723976
49. G Baruffa, F Frescura, P Micanti, B Villarini, in *Proc. of 19th IEEE Int. Conf. on Image Process. (ICIP 2012)*. An optimal method for searching UEP profiles in wireless JPEG 2000 video transmission (IEEE, Orlando, FL, 2012), pp. 1645–1648. doi:10.1109/ICIP.2012.6467192
50. L Pu, MW Marcellin, B Vasic, A Bilgin, in *Proc. of IEEE Int. Conf. on Image Process. (ICIP 2005)*. Unequal error protection and progressive decoding for JPEG2000, vol. 3 (IEEE, Genoa, Italy, 2005), pp. 896–899. doi:10.1109/ICIP.2005.1530537
51. S Feldmann, M Radimirsch, in *IEEE Int. Symp. on Pers., Indoor and Mobile Radio Commun. (PIMRC 2002)*. A novel approximation method for error rate curves in radio communication systems, vol. 3 (IEEE, Lisboa, Portugal, 2002), pp. 1003–1007. doi:10.1109/PIMRC.2002.1045178
52. I Amonou, N Cammas, S Kervadec, S Pateux, Optimized rate-distortion extraction with quality layers in the scalable extension of H.264/AVC. IEEE Trans. Circuits Syst. Video Technol. **17**(9), 1186–1193 (2007). doi:10.1109/TCSVT.2007.906870
53. Xiph.org media. https://media.xiph.org/video/derf/. Accessed 1 March 2016
54. Kakadu v. 6.0. http://www.kakadusoftware.com. Accessed 1 March 2016
55. D Taubman, M Marcellin, *JPEG2000: Image Compression Fundamentals, Standards and Practice*. (Springer, New York, NY, 2002)
56. JG Proakis, M Salehi, *Digital Communications: Fifth Edition*. (McGraw-Hill Education, Singapore, 2007)
57. V Sanchez, MK Mandal, in *Proc. Int. Conf. on of Consumer Electronics (ICCE 2002)*. Robust transmission of JPEG 2000 images over noisy channels (IEEE, Los Angeles, CA, USA, 2002), pp. 80–81. doi:10.1109/ICCE.2002.1013935
58. BA Banister, B Belzer, TR Fischer, Robust image transmission using JPEG2000 and turbo-codes. IEEE Signal Process. Lett. **9**(4), 117–119 (2002). doi:10.1109/97.1001646
59. G Baruffa, F Frescura, Adaptive error protectioncoding for wireless transmission of motion JPEG 2000 Video. http://dante.diei.unipg.it/~baruffa/uep2015/. Accessed 1 March 2016
60. J Lai, NB Mandayam, Performance of Reed-Solomon codes for hybrid-ARQ over Rayleigh fading channels under imperfect interleaving. IEEE Trans. Commun. **48**(10), 1650–1659 (2000). doi:10.1109/26.871390

7

Directional global three-part image decomposition

D. H. Thai[1,2*] and C. Gottschlich[1]

Abstract

We consider the task of image decomposition, and we introduce a new model coined directional global three-part decomposition (DG3PD) for solving it. As key ingredients of the DG3PD model, we introduce a discrete multi-directional total variation norm and a discrete multi-directional G-norm. Using these novel norms, the proposed discrete DG3PD model can decompose an image into two or three parts. Existing models for image decomposition by Vese and Osher (J. Sci. Comput. 19(1–3):553–572, 2003), by Aujol and Chambolle (Int. J. Comput. Vis. 63(1):85–104, 2005), by Starck et al. (IEEE Trans. Image Process. 14(10):1570–1582, 2005), and by Thai and Gottschlich are included as special cases in the new model. Decomposition of an image by DG3PD results in a cartoon image, a texture image, and a residual image. Advantages of the DG3PD model over existing ones lie in the properties enforced on the cartoon and texture images. The geometric objects in the cartoon image have a very smooth surface and sharp edges. The texture image yields oscillating patterns on a defined scale which are both smooth and sparse. Moreover, the DG3PD method achieves the goal of perfect reconstruction by summation of all components better than the other considered methods. Relevant applications of DG3PD are a novel way of image compression as well as feature extraction for applications such as latent fingerprint processing and optical character recognition.

Keywords: Image decomposition, Variational calculus, Cartoon image, Texture image, Image compression, Feature extraction, Latent fingerprint image processing, Optical character recognition, Fingerprint recognition

1 Introduction

Feature extraction, denoising, and image compression are key issues in computer vision and image processing. We address these main tasks based on the paradigm that an image can be regarded as the addition or montage of several meaningful components. Image decomposition methods attempt to model these components by their properties and to recover the individual components using an algorithm. Relevant component images include geometrical objects which have piece-wise constant values or a smooth surface like the characters in Fig. 1b or components which are filled with an oscillating pattern like the fingerprint in Fig. 1c.

Based on these observations, we define the following goals:

Goal 1: the cartoon component $\mathbf{u}$ contains only geometrical objects with a very smooth surface, sharp boundaries, and no texture.

Goal 2: the texture component $\mathbf{v}$ contains only geometrical objects with oscillating patterns and $\mathbf{v}$ shall be both smooth and sparse.

Goal 2: three-part decomposition and reconstruction $\mathbf{f} = \mathbf{u} + \mathbf{v} + \boldsymbol{\epsilon}$.

How does achieving these goals serve the tasks of feature extraction, denoising, and compression?

Extremely efficient representations of the cartoon image $\mathbf{u}$ and texture image $\mathbf{v}$ exist. These two component images are highly compressible as discussed with full details in Section 7.3. Depending on the application, $\mathbf{u}$ or $\mathbf{v}$ or both can be considered as feature images. For the application to latent fingerprints, we are especially interested in the texture image $\mathbf{v}$ as a feature for fingerprint segmentation and all subsequent processing steps. Example results for the very challenging task of latent fingerprint segmentation are given in Section 7.1. In optical character recognition (OCR), pre-processing includes the

*Correspondence: dhthai@samsi.info
[1]Institute for Mathematical Stochastics, University of Goettingen, Goldschmidtstr. 7, 37077 Goettingen, Germany
Full list of author information is available at the end of the article

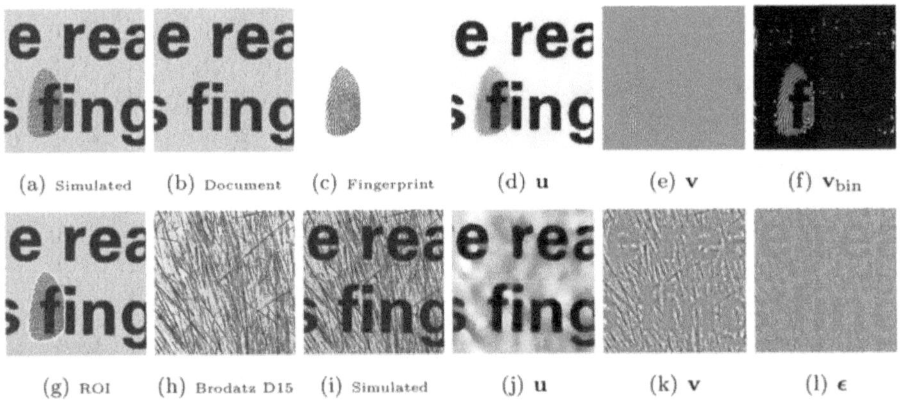

Fig. 1 The simulated latent fingerprint (**a**) is composed by adding the fingerprint (**c**) to image (**b**) which is a detail from a photo of a printed document. DG3PD decomposition of **a** obtains a smooth cartoon image u (**d**) and a smooth and sparse texture image v (**e**). All positive coefficients of v are visualized as white pixels in **f**. The region of interest shown in **g** is estimated from v_{bin} using morphological operations [57]. Image **i** is composed from **b** and **h** and decomposed by DG3PD into cartoon (**j**), texture (**k**), and residual (**l**)

removal of complex background and the isolation of characters. After three-part decomposition and depending on the scale of the characters, the cartoon image $\mathbf{u}$ contains the information of interest for OCR (see Fig. 1j), and the background is separated into $\mathbf{v}$ and $\boldsymbol{\epsilon}$ simultaneously in the minimization procedure. As a consequence of the requirements imposed on $\mathbf{u}$ and $\mathbf{v}$, noise and small scale objects are driven into the residual image $\boldsymbol{\epsilon}$ during the decomposition of $\mathbf{f}$. Therefore, the image $\mathbf{u} + \mathbf{v}$ can be regarded as a denoised version of $\mathbf{f}$ and the degree of denoising can be steered by the choice of parameters.

The paper is organized as follows. In Section 2, we begin by describing notation and preliminaries. After having established these prerequisites, we define the DG3PD model in Section 3 and in Section 4; we explain its relation to existing models in the literature for two-part and three-part decomposition. In Section 5, we describe an iterative, numerical algorithm which solves the DG3PD model for practical applications to discrete, two-dimensional images. In Section 6, we perform a detailed comparison of the DG3PD method to state-of-the-art decomposition approaches. Applications of DG3PD, especially feature extraction and image compression, are the topics in Section 7. Discussion and conclusions are given in Section 8. An overview of the algorithm and an additional comparison is given in the Appendix.

2 Notation and preliminaries

For simplification, we use a bold symbol to denote the coordinates of a two-dimensional signal, e.g., $\mathbf{x} = (x_1, x_2)$, $\mathbf{k} = [k_1, k_2]$, $\boldsymbol{\omega} = (\omega_1, \omega_2)$, and $e^{j\boldsymbol{\omega}} = \left[e^{j\omega_1}, e^{j\omega_2}\right]$. A two-dimensional image $f[\mathbf{k}] : \Omega \to \mathbb{R}_+$, the discretization of the continuous version $f(\mathbf{x})$ (i.e. $f[\mathbf{k}] = f(\mathbf{x})\,|_{\mathbf{x}=\mathbf{k}\in\Omega}$), is specified on the lattice:

$$\Omega = \left\{(k_1, k_2) \in [0, m-1] \times [0, n-1] \subset \mathbb{N}_0^2\right\}.$$

Let X be the Euclidean space whose dimension is given by the size of the lattice Ω, i.e., $X = \mathbb{R}^{|\Omega|}$. The 2D discrete Fourier transform $\mathcal{F}$ acting on $f[\mathbf{k}]$ is

$$f[\mathbf{k}] \overset{\mathcal{F}}{\longleftrightarrow} F(e^{j\boldsymbol{\omega}}) = \sum_{\mathbf{k}\in\Omega} f[\mathbf{k}] \cdot e^{-j\langle \mathbf{k}, \boldsymbol{\omega}\rangle_{\ell_2}},$$

where $\boldsymbol{\omega}$ is defined on the lattice:

$$\mathcal{I} = \left\{(\omega_1, \omega_2) = \left(\frac{2\pi n'}{n}, \frac{2\pi m'}{m}\right) \mid (n', m') \in \left[-\frac{n}{2}, \frac{n}{2}\right) \times \left[-\frac{m}{2}, \frac{m}{2}\right) \subset \mathbb{Z}^2\right\},$$

i.e., $\boldsymbol{\omega} \in [-\pi, \pi]^2$.

Forward and backward difference operators: Given the matrix

$$\mathbf{D_m} = \begin{pmatrix} -1 & 1 & 0 \ldots & 0 \\ 0 & -1 & 1 \ldots & 0 \\ \vdots & \vdots & \vdots \ddots & \vdots \\ 0 & 0 & 0 \ldots & 1 \\ 1 & 0 & 0 \ldots & -1 \end{pmatrix} \in \mathbb{R}^{m \times m},$$

the forward and backward difference operators with periodic boundary condition in convolution and matrix forms and their Fourier transform are explained in Table 1.

Discrete directional derivative: Let $\nabla^+ = \left[\partial_x^+, \partial_y^+\right]$ be the discrete forward gradient operator with ∂_x^+ and ∂_y^+ defined in Table 1. The discrete derivative operator

Table 1 Forward/backward difference with periodic boundary condition in convolution/matrix form and their Fourier transform

	Forward difference	Backward difference
Convolution	$\partial_x^+ f[\mathbf{k}] = f[k_1, k_2+1] - f[k_1, k_2]$	$\partial_x^- f[\mathbf{k}] = f[k_1, k_2] - f[k_1, k_2-1]$
form	$\partial_y^+ f[\mathbf{k}] = f[k_1+1, k_2] - f[k_1, k_2]$	$\partial_y^- f[\mathbf{k}] = f[k_1, k_2] - f[k_1-1, k_2]$
Matrix	$\left[\partial_x^+ f[\mathbf{k}]\right]_{\mathbf{k}\in\Omega} = \mathbf{f}\mathbf{D}_\mathbf{n}^{\mathrm{T}}$	$\left[\partial_x^- f[\mathbf{k}]\right]_{\mathbf{k}\in\Omega} = -\mathbf{f}\mathbf{D}_\mathbf{n}$
form	$\left[\partial_y^+ f[\mathbf{k}]\right]_{\mathbf{k}\in\Omega} = \mathbf{D}_\mathbf{m}\mathbf{f}$	$\left[\partial_y^- f[\mathbf{k}]\right]_{\mathbf{k}\in\Omega} = -\mathbf{D}_\mathbf{m}^{\mathrm{T}}\mathbf{f}$
Fourier	$\left(e^{j\omega_2} - 1\right) F\left(e^{j\boldsymbol{\omega}}\right)$	$-\left(e^{-j\omega_2} - 1\right) F\left(e^{j\boldsymbol{\omega}}\right)$
transform	$\left(e^{j\omega_1} - 1\right) F\left(e^{j\boldsymbol{\omega}}\right)$	$-\left(e^{-j\omega_1} - 1\right) F\left(e^{j\boldsymbol{\omega}}\right)$

$\mathbf{D}_\mathbf{n}^{\mathrm{T}}$ and $\mathbf{D}_\mathbf{m}^{\mathrm{T}}$ are the transposed matrices of $\mathbf{D}_\mathbf{n}$ and $\mathbf{D}_\mathbf{m}$, respectively

following the direction $\vec{d} = \left[\cos\frac{\pi l}{L}, \sin\frac{\pi l}{L}\right]^{\mathrm{T}}$ with $l = 0, \ldots, L-1$ is defined as

$$\partial_l^+ = \left\langle \vec{d}, \nabla^+ \right\rangle = \cos\left(\frac{\pi l}{L}\right)\partial_x^+ + \sin\left(\frac{\pi l}{L}\right)\partial_y^+ .$$

Thus, the discrete directional gradient operator is

$$\nabla_L^+ = \left[\partial_l^+\right]_{l\in[0,L-1]} . \tag{1}$$

2.1 Discrete directional TV norm

The continuous total variation norm has been defined in [1]. Due to the discrete nature of images, we define its discrete version with forward difference operators as

$$\left\|\nabla^+ \mathbf{u}\right\|_{\ell_1} = \sum_{\mathbf{k}\in\Omega} \sqrt{\left(\partial_x^+ u[\mathbf{k}]\right)^2 + \left(\partial_y^+ u[\mathbf{k}]\right)^2}.$$

We extend it into multi-direction L with the discrete directional gradient operator (1):

$$\left\|\nabla_L^+ \mathbf{u}\right\|_{\ell_1} = \sum_{\mathbf{k}\in\Omega} \sqrt{\sum_{l=0}^{L-1} \left(\partial_l^+ u[\mathbf{k}]\right)^2} = \left\| \sqrt{\sum_{l=0}^{L-1} \left(\cos\frac{\pi l}{L}\mathbf{u}\mathbf{D}_\mathbf{n}^{\mathrm{T}} + \sin\frac{\pi l}{L}\mathbf{D}_\mathbf{m}\mathbf{u}\right)^2} \right\|_{\ell_1} .$$

The discrete anisotropic total variation norm in a matrix form is

$$\left\|\nabla_L^+ \mathbf{u}\right\|_{\ell_1} = \sum_{l=0}^{L-1} \left\| \cos\frac{\pi l}{L}\mathbf{u}\mathbf{D}_\mathbf{n}^{\mathrm{T}} + \sin\frac{\pi l}{L}\mathbf{D}_\mathbf{m}\mathbf{u} \right\|_{\ell_1} . \tag{2}$$

2.2 Discrete directional G-norm

Discrete G-norm: Meyer [2] has proposed a space G of continuous functions to measure oscillating functions (texture and noise). The discrete version of the G-norm has been introduced by Aujol and Chambolle in [3]. We rewrite it with the matrix form of the forward difference operators as

$$\|\mathbf{v}\|_G = \inf\left\{ \left\| \sqrt{\mathbf{g}_1^2 + \mathbf{g}_2^2} \right\|_{\ell_\infty}, \mathbf{v} = \mathbf{g}_1\mathbf{D}_\mathbf{n}^{\mathrm{T}} + \mathbf{D}_\mathbf{m}\mathbf{g}_2, \left[\mathbf{g}_l\right]_{l\in[1,2]} \in X^2 \right\}. \tag{3}$$

Discrete directional G-norm: We extend (3) into multi-directions $S \in \mathbb{N}_+$ with the directional difference operator to obtain the discrete directional G-norm as

$$\|\mathbf{v}\|_{G_S} = \inf\left\{ \left\| \sqrt{\sum_{s=0}^{S-1} \mathbf{g}_s^2} \right\|_{\ell_\infty}, \underbrace{\mathbf{v} = \sum_{s=0}^{S-1}\left[\cos\frac{\pi s}{S}\mathbf{g}_s\mathbf{D}_\mathbf{n}^{\mathrm{T}} + \sin\frac{\pi s}{S}\mathbf{D}_\mathbf{m}\mathbf{g}_s\right]}_{\Leftrightarrow\, v[\mathbf{k}] = \sum_{s=0}^{S-1} \partial_s^+ g_s[\mathbf{k}], \forall \mathbf{k}\in\Omega}, \left[\mathbf{g}_s\right]_{s=0}^{S-1} \in X^S \right\}. \tag{4}$$

3 DG3PD

We define the DG3PD model for discrete directional three-part decomposition of an image into cartoon, texture, and residual parts as

$$\min_{(\mathbf{u},\mathbf{v},\boldsymbol{\epsilon})\in X^3} \left\{ \left\|\nabla_L^+ \mathbf{u}\right\|_{\ell_1} + \mu_1\|\mathbf{v}\|_{G_S} + \mu_2\|\mathbf{v}\|_{\ell_1} \quad \text{s.t.} \sup_{(i,l,k)\in\mathcal{K}} \left|\mathcal{C}_{i,l}\{\boldsymbol{\epsilon}\}[\mathbf{k}]\right| \le \delta,\ \mathbf{f} = \mathbf{u} + \mathbf{v} + \boldsymbol{\epsilon} \right\}, \tag{5}$$

where $\mathcal{C}$ is the curvelet transform [4, 5] with the index set $\mathcal{K}$. Please note that by setting the parameter $\delta = 0$ in (5), we obtain a two-part decomposition which can be considered as a special case of the DG3PD model. Next, we discuss how the DG3PD model relates to existing decomposition models.

4 Related work

In this section, we give an overview of prior work in two-part and three-part image decomposition in chronological order and we explain how it relates to the proposed DG3PD model.

Mumford and Shah: In 1989, Mumford and Shah [6] have proposed piecewise smooth (for image restoration) and piecewise constant (for image segmentation) models by minimizing the energy functional. This approach can be considered as a precursor for subsequent image decomposition models with texture components. However, due to the Hausdorff one-dimensional measure $\mathcal{H}^1$ in $\mathbb{R}^2$, it poses a challenging or even NP-hard problem in optimization to minimize the Mumford and Shah functional. Later, based on the Mumford and Shah model, Chan and Vese [7] have proposed an active contour for image segmentation which they solved by a level set method [8].

Rudin, Osher, and Fatemi: In 1992, Rudin et al. [1] pioneered image decomposition with a two-part model for denoising.

Meyer: The model defined by Meyer in 2001 [2] for a two-part decomposition in the continuous setting is comprised in the DG3PD model for the special case of $L = S = 2$ and $\mu_2 = \delta = 0$ in the discrete domain.

Vese and Osher: In 2003, Vese and Osher [9] solved Meyer's model for two-part decomposition in the continuous setting and they proposed to approximate the L_∞-norm in the G-norm by the L_1-norm and to apply the penalty method for reformulating the constraint. For practical application to images, they discretized their solution. This approach is extended in [10].

Aujol and Chambolle: Aujol and Chambolle in 2005 [3] adapted the work by Meyer for discrete two-part decomposition (see (5.49) in [3]), and they used the penalty method for the constraint. Their model is included in the DG3PD model with parameters $L = S = 2$, $\mu_1 = \mu_2 = 0$ and applying the supremum norm to the wavelet coefficients of the oscillating pattern, i.e., $\|\mathcal{W}\{\cdot\}\|_{\ell_\infty}$, instead of the curvelet coefficients $\|\mathcal{C}\{\cdot\}\|_{\ell_\infty}$ as in the DG3PD model. Moreover, Aujol and Chambolle proposed a model for discrete three-part decomposition (see (6.59) in [3]) which measures texture by the G-norm and noise by the supremum norm of wavelet coefficients and the penalty method for the constraint. Different from Vese and Osher as well as our approach explained later, they describe the G-norm for capturing texture using the indicator function defined on a convex set and they obtain the solution by Chambolle's projection onto this convex set. Their model is included in the DG3PD model for parameters $L = S = 2$, $\mu_2 = 0$ and using wavelets instead of curvelets for the residual as before.

Starck et al.: Starck et al. [11] introduced a model for two-part decomposition based on a dictionary approach. Their basic idea is to choose one appropriate dictionary for piecewise smooth objects (cartoon) and another suitable dictionary for capturing texture parts.

Aujol et al.: In 2006, Aujol et al. [12] proposed a two-part decomposition of an image into a structure component and a texture component using Gabor functions for the texture part.

Gilles: Gilles [13] proposed a three-part image decomposition method in 2007 which is similar to the Aujol-Chambolle model [3], but G-norm is used as a measurement of the residual (or noise) instead of Besov space $\dot{B}^{\infty}_{-1,\infty}$ with a local adaptability property. Their argument is that the more a function is oscillatory, the smaller is the G norm. Then, they propose a new "merged-algorithm" with a combination of a local adaptivity behavior and Besov space.

Maragos and Evangelopoulos: In 2007, Maragos and Evangelopoulos [14] have proposed a two-part decomposition model which relies on energy responses of a bank of 2D Gabor filters for measuring the texture component. They discuss the connection between Meyer's oscillating functions [2], Gabor filters [15], and AM-FM image modeling [16, 17].

Buades et al.: In 2010, Buades et al. [18] derived a nonlinear filter pair for two-part decomposition into cartoon and texture parts. Further models for two-part decomposition are listed in Table 1 of [18].

Maurel et al. and Chikkerur et al.: In 2011, Maurel et al. [19] proposed a decomposition approach which models the texture component by local Fourier atoms. For fingerprint textures, Chikkerur et al. [20] proposed in 2007 the application of local Fourier analysis (or short-time Fourier transform, STFT) for image enhancement. However, the usefulness of local Fourier analysis for capturing texture information depends on and is limited by the level of noise in the corresponding local window, see Figure 2c in [21] for an example in which STFT enhances some regions successfully and fails in other regions.

Ono et al.: In 2014, Ono et al. [22] proposed a cartoon-texture decomposition using the block nuclear norm (an generalized version of [23]) which interprets the texture component as the combination of overlapped and sheared blockwise low-rank matrices in different directions. The underlying assumption is that "texture, in general, is globally dissimilar but locally well patterned." Similar to our directional G-norm, the shear helps to handle patterns in non-horizontal or vertical. However, their cartoon component still contains some texture and their texture is not highly sparse, i.e., the non-zero coefficients are only due to texture component, see Fig. 4a and d, respectively. Moreover, they use ℓ_1 or ℓ_2 norm for the data-fidelity term. However, according to [2, 24], "oscillatory components do not have small norms in $L_2(\Omega)$ or $L_1(\Omega)$." In our case, we use the Banach $\|\mathcal{C}\{\cdot\}\|$ which is more suitable than the Banach space $E = B^{-1}_{\infty,\infty}$ in equation (1.3) [3] for measuring small-scale objects, e.g., noise.

G3PD: A model for discrete three-part decomposition of fingerprint images has recently been proposed by Thai and Gottschlich in 2015 [25] with the aim of obtaining a texture image $\mathbf{v}$ which serves as a useful feature for estimating the region of interest (ROI). The G3PD model is included in the DG3PD model by choosing $L = 2$ and replacing the directional G-norm in the DG3PD model by the ℓ_1-norm of curvelet coefficients (multi-scale and multi-orientation decomposition) to capture texture. However, a disadvantage of the ℓ_1-norm of curvelet coefficients is a tendency to generate the halo effect on the boundary of the texture region due to the scaling factor in curvelet decomposition (see Figure 3d in [25]), whereas the directional G-norm in the DG3PD model is capable to capture oscillating patterns (see [2]) without the halo effect.

Directional total variation and G-norm: In Section 2.1, we introduced the discrete directional total variation norm, and in Section 2.2, we introduced the discrete directional G-norm. Please note the aspect of summation over multiple directions in Eqs. (2) and (4).

The term "directional total variation" has previously been used by Bayram and Kamasak [26, 27] for defining and computing the TV norm in only one specific direction. They have treated the special case of images with one globally dominant direction and addressed those by two-part decomposition and for the purpose denoising. Zhang and Wang [28] proposed an extension of the work by Bayram and Kamasak for denoising images with more than one dominant direction.

5 Solution of the DG3PD model

Now, we present a numerical algorithm for obtaining the solution of the DG3PD model stated in (5). Given $\delta > 0$, denote $G^*\left(\frac{\epsilon}{\delta}\right)$ as the indicator function on the feasible convex set $A(\delta)$ of (5), i.e.,

$$A(\delta) = \left\{\boldsymbol{\epsilon} \in X \,:\, \|\mathcal{C}\{\boldsymbol{\epsilon}\}\|_{\ell_\infty} \le \delta\right\} \text{ and } G^*\left(\frac{\boldsymbol{\epsilon}}{\delta}\right) = \begin{cases} 0\,, & \boldsymbol{\epsilon} \in A(\delta) \\ +\infty\,, & \boldsymbol{\epsilon} \notin A(\delta) \end{cases}.$$

By analogy with the work of Vese and Osher, we consider the approximation of G-norm with ℓ_1 norm and the anisotropic version of directional total variation norm. The minimization problem in (5) is rewritten as

$$\left(\mathbf{u}^*, \mathbf{v}^*, \boldsymbol{\epsilon}^*, [\mathbf{g}_s^*]_{s=0}^{S-1}\right) = \underset{\left(\mathbf{u},\mathbf{v},\boldsymbol{\epsilon},[\mathbf{g}_s]_{s=0}^{S-1}\right)\in X^{S+3}}{\arg\min} \left\{ \sum_{l=0}^{L-1} \left\| \cos\left(\frac{\pi l}{L}\right)\mathbf{u}\mathbf{D}_\mathbf{n}^\mathrm{T} + \sin\left(\frac{\pi l}{L}\right)\mathbf{D}_\mathbf{m}\mathbf{u} \right\|_{\ell_1} + \mu_1 \sum_{s=0}^{S-1} \left\|\mathbf{g}_s\right\|_{\ell_1} + \mu_2\|\mathbf{v}\|_{\ell_1} + G^*\left(\frac{\boldsymbol{\epsilon}}{\delta}\right) \text{ s.t. } \begin{cases} \mathbf{f} = \mathbf{u} + \mathbf{v} + \boldsymbol{\epsilon} \\ \mathbf{v} = \sum_{s=0}^{S-1}\left[\cos\left(\frac{\pi s}{S}\right)\mathbf{g}_s\mathbf{D}_\mathbf{n}^\mathrm{T} + \sin\left(\frac{\pi s}{S}\right)\mathbf{D}_\mathbf{m}\mathbf{g}_s\right] \end{cases} \right\}. \tag{6}$$

To simplify the calculation, we introduce two new variables:

$$\begin{cases} \mathbf{r}_b = \cos\left(\frac{\pi b}{L}\right)\mathbf{u}\mathbf{D}_\mathbf{n}^\mathrm{T} + \sin\left(\frac{\pi b}{L}\right)\mathbf{D}_\mathbf{m}\mathbf{u}\,,\; b = 0, \dots, L-1, \\ \mathbf{w}_a = \mathbf{g}_a\,, a = 0, \dots, S-1\,. \end{cases}$$

Equation (6) is a constrained minimization problem. The augmented Lagrangian method (ALM) is applied to turn (6) into an unconstrained one as

$$\min_{\left(\mathbf{u},\mathbf{v},\boldsymbol{\epsilon},[\mathbf{r}_l]_{l=0}^{L-1},[\mathbf{w}_s]_{s=0}^{S-1},[\mathbf{g}_s]_{s=0}^{S-1}\right)\in X^{L+2S+3}} \mathcal{L}\left(\mathbf{u}, \mathbf{v}, \boldsymbol{\epsilon}, [\mathbf{r}_l]_{l=0}^{L-1}, [\mathbf{w}_s]_{s=0}^{S-1}, [\mathbf{g}_s]_{s=0}^{S-1}\,; [\boldsymbol{\lambda}_{1l}]_{l=0}^{L-1}, [\boldsymbol{\lambda}_{2s}]_{s=0}^{S-1}, \boldsymbol{\lambda}_3, \boldsymbol{\lambda}_4\right), \tag{7}$$

where the Lagrange function is

$$\begin{aligned} \mathcal{L}(\cdot\,;\cdot) = & \sum_{l=0}^{L-1} \|\mathbf{r}_l\|_{\ell_1} + \mu_1 \sum_{s=0}^{S-1} \|\mathbf{w}_s\|_{\ell_1} + \mu_2\|\mathbf{v}\|_{\ell_1} + G^*\left(\frac{\boldsymbol{\epsilon}}{\delta}\right) \\ & + \frac{\beta_1}{2} \sum_{l=0}^{L-1} \left\| \mathbf{r}_l - \cos\left(\frac{\pi l}{L}\right)\mathbf{u}\mathbf{D}_\mathbf{n}^\mathrm{T} - \sin\left(\frac{\pi l}{L}\right)\mathbf{D}_\mathbf{m}\mathbf{u} \frac{\boldsymbol{\lambda}_{1l}}{\beta_1} \right\|_{\ell_2}^2 \\ & + \frac{\beta_2}{2} \sum_{s=0}^{S-1} \left\| \mathbf{w}_s - \mathbf{g}_s + \frac{\boldsymbol{\lambda}_{2s}}{\beta_2} \right\|_{\ell_2}^2 \\ & + \frac{\beta_3}{2} \left\| \mathbf{v} - \sum_{s=0}^{S-1}\left[\cos\left(\frac{\pi s}{S}\right)\mathbf{g}_s\mathbf{D}_\mathbf{n}^\mathrm{T} + \sin\left(\frac{\pi s}{S}\right)\mathbf{D}_\mathbf{m}\mathbf{g}_s\right] + \frac{\boldsymbol{\lambda}_3}{\beta_3} \right\|_{\ell_2}^2 \\ & + \frac{\beta_4}{2} \left\| \mathbf{f} - \mathbf{u} - \mathbf{v} - \boldsymbol{\epsilon} + \frac{\boldsymbol{\lambda}_4}{\beta_4} \right\|_{\ell_2}^2. \end{aligned}$$

Due to the minimization problem with multi-variables, we apply the alternating directional method of multipliers to solve (7). Its minimizer is numerically computed through iterations $t = 1, 2, \dots$

$$\begin{aligned} & \left(\mathbf{u}^{(t)}, \mathbf{v}^{(t)}, \boldsymbol{\epsilon}^{(t)}, \left[\mathbf{r}_l^{(t)}\right]_{l=0}^{L-1}, \left[\mathbf{w}_s^{(t)}\right]_{s=0}^{S-1}, \left[\mathbf{g}_s^{(t)}\right]_{s=0}^{S-1}\right) = \\ & \arg\min \mathcal{L}\left(\mathbf{u}, \mathbf{v}, \boldsymbol{\epsilon}, [\mathbf{r}_l]_{l=0}^{L-1}, [\mathbf{w}_s]_{s=0}^{S-1}, [\mathbf{g}_s]_{s=0}^{S-1}\,; \left[\boldsymbol{\lambda}_{1l}^{(t-1)}\right]_{l=0}^{L-1}, \left[\boldsymbol{\lambda}_{2s}^{(t-1)}\right]_{s=0}^{S-1}, \boldsymbol{\lambda}_3^{(t-1)}, \boldsymbol{\lambda}_4^{(t-1)}\right) \end{aligned} \tag{8}$$

and the Lagrange multipliers are updated after every step t with a rate γ. We initialize $\mathbf{u}^{(0)} = \mathbf{f}, \mathbf{v}^{(0)} = \boldsymbol{\epsilon}^{(0)} = \left[\mathbf{r}_l^{(0)}\right]_{l=0}^{L-1} = \left[\mathbf{w}_s^{(0)}\right]_{s=0}^{S-1} = \left[\mathbf{g}_s^{(0)}\right]_{s=0}^{S-1} = \left[\boldsymbol{\lambda}_{1l}^{(0)}\right]_{l=0}^{L-1} =$

$\left[\boldsymbol{\lambda}_{2a}^{(0)}\right]_{a=0}^{S-1} = \boldsymbol{\lambda}_3^{(0)} = \boldsymbol{\lambda}_4^{(0)} = \mathbf{0}$. In each iteration, we first solve the following six subproblems in the listed order and then we update the four Lagrange multipliers:

The "$[\mathbf{r}_l]_{l=0}^{L-1}$-problem": Fix $\mathbf{u}, \mathbf{v}, \boldsymbol{\epsilon}, [\mathbf{w}_s]_{s=0}^{S-1}, [\mathbf{g}_s]_{s=0}^{S-1}$ and

$$\min_{[\mathbf{r}_l]_{l=0}^{L-1} \in X^L} \left\{ \sum_{l=0}^{L-1} \|\mathbf{r}_l\|_{\ell_1} + \frac{\beta_1}{2} \sum_{l=0}^{L-1} \left\| \mathbf{r}_l - \cos\left(\frac{\pi l}{L}\right) \mathbf{u} \mathbf{D}_{\mathbf{n}}^{\mathrm{T}} - \sin\left(\frac{\pi l}{L}\right) \mathbf{D}_{\mathbf{m}} \mathbf{u} + \frac{\boldsymbol{\lambda}_{1l}}{\beta_1} \right\|_{\ell_2}^2 \right\} \tag{9}$$

Due to its separability, we consider the problem at $b = 0, \ldots, L-1$. The solution of (9) is

$$\mathbf{r}_b^* = \operatorname{Shrink}\left(\cos\left(\frac{\pi b}{L}\right) \mathbf{u} \mathbf{D}_{\mathbf{n}}^{\mathrm{T}} + \sin\left(\frac{\pi b}{L}\right) \mathbf{D}_{\mathbf{m}} \mathbf{u} - \frac{\boldsymbol{\lambda}_{1b}}{\beta_1}, \frac{1}{\beta_1} \right).$$

The operator Shrink$(\cdot,\cdot)$ is defined in [25].

The "$[\mathbf{w}_s]_{s=0}^{S-1}$-problem": Fix $\mathbf{u}, \mathbf{v}, \boldsymbol{\epsilon}, [\mathbf{r}_l]_{l=0}^{L-1}, [\mathbf{g}_s]_{s=0}^{S-1}$ and

$$\min_{[\mathbf{w}_s]_{s=0}^{S-1} \in X^S} \left\{ \mu_1 \sum_{s=0}^{S-1} \|\mathbf{w}_s\|_{\ell_1} + \frac{\beta_2}{2} \sum_{s=0}^{S-1} \left\| \mathbf{w}_s - \mathbf{g}_s + \frac{\boldsymbol{\lambda}_{2s}}{\beta_2} \right\|_{\ell_2}^2 \right\} \tag{10}$$

Similarly, the solution of (10) for each separable problem $a = 0, \ldots, S-1$ is

$$\mathbf{w}_a^* = \operatorname{Shrink}\left(\underbrace{\mathbf{g}_a - \frac{\boldsymbol{\lambda}_{2a}}{\beta_2}}_{:= \mathbf{t}_{\mathbf{w}_a}}, \frac{\mu_1}{\beta_2} \right). \tag{11}$$

The "$[\mathbf{g}_s]_{s=0}^{S-1}$-problem": Fix $\mathbf{u}, \mathbf{v}, \boldsymbol{\epsilon}, [\mathbf{r}_l]_{l=0}^{L-1}, [\mathbf{w}_s]_{s=0}^{S-1}$ and

$$\min_{[\mathbf{g}_s]_{s=0}^{S-1} \in X^S} \left\{ \frac{\beta_2}{2} \sum_{s=0}^{S-1} \left\| \mathbf{w}_s - \mathbf{g}_s + \frac{\boldsymbol{\lambda}_{2s}}{\beta_2} \right\|_{\ell_2}^2 + \frac{\beta_3}{2} \left\| \mathbf{v} - \sum_{s=0}^{S-1} \left[\cos\left(\frac{\pi s}{S}\right) \mathbf{g}_s \mathbf{D}_{\mathbf{n}}^{\mathrm{T}} + \sin\left(\frac{\pi s}{S}\right) \mathbf{D}_{\mathbf{m}} \mathbf{g}_s \right] + \frac{\boldsymbol{\lambda}_3}{\beta_3} \right\|_{\ell_2}^2 \right\} \tag{12}$$

For the discrete finite frequency coordinates $\boldsymbol{\omega} = [\omega_1, \omega_2] \in \mathcal{I}$, let be $\mathbf{z} = [\mathbf{z}_1, \mathbf{z}_2] = \left[e^{j\omega_1}, e^{j\omega_2}\right]$. We denote by $W_a(\mathbf{z}), \Lambda_{2a}(\mathbf{z}), V(\mathbf{z}), G_s(\mathbf{z})$, and $\Lambda_3(\mathbf{z})$ the discrete Fourier transforms of $w_a[\boldsymbol{k}], \lambda_{2a}[\boldsymbol{k}], v[\boldsymbol{k}], g_s[\boldsymbol{k}]$ and $\lambda_3[\boldsymbol{k}]$, respectively. Due to the separability, the solution of (12) is obtained for $a = 0, \ldots, S-1$ as

$$\mathbf{g}_a^* = \operatorname{Re}\left[\mathcal{F}^{-1}\{\mathcal{A}(\mathbf{z}) \cdot \mathcal{B}(\mathbf{z})\}\right] \tag{13}$$

with

$$\mathcal{A}(\mathbf{z}) = \left[\beta_2 + \beta_3 \left[\sin\frac{\pi a}{S}\left(\mathbf{z}_1^{-1} - 1\right) + \cos\frac{\pi a}{S}\left(\mathbf{z}_2^{-1} - 1\right) \right] \left[\sin\frac{\pi a}{S}(\mathbf{z}_1 - 1) + \cos\frac{\pi a}{S}(\mathbf{z}_2 - 1) \right] \right]^{-1},$$

$$\mathcal{B}(\mathbf{z}) = \beta_2 \left[W_a(\mathbf{z}) + \frac{\Lambda_{2a}(\mathbf{z})}{\beta_2} \right] + \beta_3 \left[\sin\left(\frac{\pi a}{S}\right)\left(\mathbf{z}_1^{-1} - 1\right) + \cos\left(\frac{\pi a}{S}\right)\left(\mathbf{z}_2^{-1} - 1\right) \right] \times \left[V(\mathbf{z}) - \sum_{s=[0,S-1]\setminus\{a\}} \left[\cos\left(\frac{\pi s}{S}\right)(\mathbf{z}_2 - 1) + \sin\left(\frac{\pi s}{S}\right)(\mathbf{z}_1 - 1) \right] G_s(\mathbf{z}) + \frac{\Lambda_3(\mathbf{z})}{\beta_3} \right].$$

The "$\mathbf{v}$-problem": Fix $\mathbf{u}, \boldsymbol{\epsilon}, [\mathbf{r}_l]_{l=0}^{L-1}, [\mathbf{w}_s]_{s=0}^{S-1}, [\mathbf{g}_s]_{s=0}^{S-1}$ and

$$\min_{\mathbf{v} \in X} \left\{ \mu_2 \|\mathbf{v}\|_{\ell_1} + \frac{\beta_3}{2} \left\| \mathbf{v} - \left(\sum_{s=0}^{S-1} \left[\cos\left(\frac{\pi s}{S}\right) \mathbf{g}_s \mathbf{D}_{\mathbf{n}}^{\mathrm{T}} + \sin\left(\frac{\pi s}{S}\right) \mathbf{D}_{\mathbf{m}} \mathbf{g}_s \right] - \frac{\boldsymbol{\lambda}_3}{\beta_3} \right) \right\|_{\ell_2}^2 + \frac{\beta_4}{2} \left\| \mathbf{v} - \left(\mathbf{f} - \mathbf{u} - \boldsymbol{\epsilon} + \frac{\boldsymbol{\lambda}_4}{\beta_4} \right) \right\|_{\ell_2}^2 \right\} \tag{14}$$

The solution of (14) is defined as

$$\mathbf{v}^* = \operatorname{Shrink}\left(\mathbf{t}_{\mathbf{v}}, \frac{\mu_2}{\beta_3 + \beta_4} \right), \tag{15}$$

with

$$\mathbf{t}_{\mathbf{v}} := \frac{\beta_3}{\beta_3 + \beta_4} \left(\sum_{s=0}^{S-1} \left[\cos\left(\frac{\pi s}{S}\right) \mathbf{g}_s \mathbf{D}_{\mathbf{n}}^{\mathrm{T}} + \sin\left(\frac{\pi s}{S}\right) \mathbf{D}_{\mathbf{m}} \mathbf{g}_s \right] - \frac{\boldsymbol{\lambda}_3}{\beta_3} \right) + \frac{\beta_4}{\beta_3 + \beta_4} \left(\mathbf{f} - \mathbf{u} - \boldsymbol{\epsilon} + \frac{\boldsymbol{\lambda}_4}{\beta_4} \right). \tag{16}$$

The "$\mathbf{u}$-problem": Fix $\mathbf{v}, \boldsymbol{\epsilon}, [\mathbf{r}_l]_{l=0}^{L-1}, [\mathbf{w}_s]_{s=0}^{S-1}, [\mathbf{g}_s]_{s=0}^{S-1}$ and

$$\min_{\mathbf{u} \in X} \left\{ \frac{\beta_1}{2} \sum_{l=0}^{L-1} \left\| \mathbf{r}_l - \cos\left(\frac{\pi l}{L}\right) \mathbf{u} \mathbf{D}_{\mathbf{n}}^{\mathrm{T}} - \sin\left(\frac{\pi l}{L}\right) \mathbf{D}_{\mathbf{m}} \mathbf{u} + \frac{\boldsymbol{\lambda}_{1l}}{\beta_1} \right\|_{\ell_2}^2 + \frac{\beta_4}{2} \left\| \mathbf{f} - \mathbf{u} - \mathbf{v} - \boldsymbol{\epsilon} + \frac{\boldsymbol{\lambda}_4}{\beta_4} \right\|_{\ell_2}^2 \right\} \tag{17}$$

We denote $F(\mathbf{z}), \mathcal{E}(\mathbf{z}), \Lambda_4(\mathbf{z}), R_l(\mathbf{z})$, and $\Lambda_{1l}(\mathbf{z})$ as the discrete Fourier transforms of $f[\boldsymbol{k}], \epsilon[\boldsymbol{k}], \lambda_4[\boldsymbol{k}], r_l[\boldsymbol{k}]$, and $\lambda_{1l}[\boldsymbol{k}]$, respectively. This (17) is solved in the Fourier domain by

$$\mathbf{u}^* = \operatorname{Re}\left[\mathcal{F}^{-1}\{\mathcal{X}(\mathbf{z}) \cdot \mathcal{Y}(\mathbf{z})\}\right] \tag{18}$$

with

$$\mathcal{X}(\mathbf{z}) = \left[\beta_4 + \beta_1 \sum_{l=0}^{L-1} \left[\sin\left(\frac{\pi l}{L}\right)\left(\mathbf{z}_1^{-1} - 1\right) + \cos\left(\frac{\pi l}{L}\right)\left(\mathbf{z}_2^{-1} - 1\right)\right]\left[\sin\left(\frac{\pi l}{L}\right)(\mathbf{z}_1 - 1) + \cos\left(\frac{\pi l}{L}\right)(\mathbf{z}_2 - 1)\right]\right]^{-1},$$

$$\mathcal{Y}(\mathbf{z}) = \beta_4\left[F(\mathbf{z}) - V(\mathbf{z}) - \mathcal{E}(\mathbf{z}) + \frac{\Lambda_4(\mathbf{z})}{\beta_4}\right] + \beta_1 \sum_{l=0}^{L-1}\left[\sin\left(\frac{\pi l}{L}\right)\left(\mathbf{z}_1^{-1} - 1\right) + \cos\left(\frac{\pi l}{L}\right)\left(\mathbf{z}_2^{-1} - 1\right)\right]\left[R_l(\mathbf{z}) + \frac{\Lambda_{1l}(\mathbf{z})}{\beta_1}\right].$$

The "$\boldsymbol{\epsilon}$-problem": Fix $\mathbf{u}$, $\mathbf{v}$, $[\mathbf{r}_l]_{l=0}^{L-1}$, $[\mathbf{w}_s]_{s=0}^{S-1}$, $[\mathbf{g}_s]_{s=0}^{S-1}$, and

$$\min_{\boldsymbol{\epsilon}\in X}\left\{G^*\left(\frac{\boldsymbol{\epsilon}}{\delta}\right) + \frac{\beta_4}{2}\left\|\boldsymbol{\epsilon} - \left(\mathbf{f} - \mathbf{u} - \mathbf{v} + \frac{\boldsymbol{\lambda}_4}{\beta_4}\right)\right\|_{\ell_2}^2\right\} \quad (19)$$

Let $\mathcal{C}^*$ be the inverse curvelet transform [4]. The minimization of (19) is solved by (see [3])

$$\boldsymbol{\epsilon}^* = \left(\mathbf{f} - \mathbf{u} - \mathbf{v} + \frac{\boldsymbol{\lambda}_4}{\beta_4}\right) \underbrace{- \mathcal{C}^*\left\{\text{Shrink}\left(\mathcal{C}\left\{\mathbf{f} - \mathbf{u} - \mathbf{v} + \frac{\boldsymbol{\lambda}_4}{\beta_4}\right\}, \delta\right)\right\}}_{:=\,\text{CST}\left(\mathbf{f}-\mathbf{u}-\mathbf{v}+\frac{\boldsymbol{\lambda}_4}{\beta_4}, \delta\right)}$$

or by the projection method with the component-wise operators

$$\boldsymbol{\epsilon}^* = \mathcal{C}^*\left\{\frac{\delta\,\mathcal{C}\left\{\mathbf{f} - \mathbf{u} - \mathbf{v} + \frac{\boldsymbol{\lambda}_4}{\beta_4}\right\}}{\max\left(\delta, \left|\mathcal{C}\left\{\mathbf{f} - \mathbf{u} - \mathbf{v} + \frac{\boldsymbol{\lambda}_4}{\beta_4}\right\}\right|\right)}\right\}.$$

Update Lagrange multipliers $\left([\boldsymbol{\lambda}_{1l}]_{l=0}^{L-1}, [\boldsymbol{\lambda}_{2a}]_{a=0}^{S-1}, \boldsymbol{\lambda}_3, \boldsymbol{\lambda}_4\right) \in X^{L+S+2}$.

$$\begin{aligned}
\boldsymbol{\lambda}_{1b}^{(t)} &= \boldsymbol{\lambda}_{1b}^{(t-1)} + \gamma\beta_1\left(\mathbf{r}_b - \cos\left(\frac{\pi b}{L}\right)\mathbf{u}\mathbf{D}_\mathbf{n}^{\mathrm{T}} - \sin\left(\frac{\pi b}{L}\right)\mathbf{D}_\mathbf{m}\mathbf{u}\right), \quad b = 0, \ldots, L-1 \\
\boldsymbol{\lambda}_{2a}^{(t)} &= \boldsymbol{\lambda}_{2a}^{(t-1)} + \gamma\beta_2(\mathbf{w}_a - \mathbf{g}_a), \quad a = 0, \ldots, S-1 \\
\boldsymbol{\lambda}_3^{(t)} &= \boldsymbol{\lambda}_3^{(t-1)} + \gamma\beta_3\left(\mathbf{v} - \sum_{s=0}^{S-1}\left[\cos\left(\frac{\pi s}{S}\right)\mathbf{g}_s\mathbf{D}_\mathbf{n}^{\mathrm{T}} + \sin\left(\frac{\pi s}{S}\right)\mathbf{D}_\mathbf{m}\mathbf{g}_s\right]\right) \\
\boldsymbol{\lambda}_4^{(t)} &= \boldsymbol{\lambda}_4^{(t-1)} + \gamma\beta_4(\mathbf{f} - \mathbf{u} - \mathbf{v} - \boldsymbol{\epsilon})
\end{aligned}$$

Choice of parameters

Due to the ℓ_1-norms in the minimization problem (7) which corresponds to the shrinkage operator with parameters μ_1 and μ_2, these are defined as

$$\begin{aligned}
\mu_1 &= c_{\mu_1}\beta_2 \cdot \max_{\mathbf{k}\in\Omega}\left(\left|t_{\mathbf{w}_a}[\mathbf{k}]\right|\right) \quad \text{and} \\
\mu_2 &= c_{\mu_2}(\beta_3 + \beta_4) \cdot \max_{\mathbf{k}\in\Omega}\left(\left|t_\mathbf{v}[\mathbf{k}]\right|\right),
\end{aligned} \quad (20)$$

where $t_{\mathbf{w}_a}[\mathbf{k}]$ and $t_\mathbf{v}[\mathbf{k}]$ are defined in (11) and (16), respectively. Note that the choice of c_{μ_1} and c_{μ_2} is adapted to specific images.

In order to balance between the smoothing terms and the updated terms for the solutions of the **g**-problem in (13), the **v**-problem in (15), and the **u**-problem in (18), we choose

$$\beta_2 = c_2\beta_3, \beta_3 = \frac{\theta}{1-\theta}\beta_4, \theta \in (0,1) \text{ and } \beta_1 = c_1\beta_4.$$

The choice of δ mainly impacts the smoothness and sparsity of the texture **v**. The first row of Fig. 2 shows the effect of selecting the threshold $\delta = 0$ which corresponds to a two-part decomposition, i.e., the residual image $\boldsymbol{\epsilon} = \mathbf{0}$ in (d). This case also demonstrates the limitation of all two-part decomposition approaches: for this choice of δ, very small scale objects are assigned to the texture image **v** in Fig. 2b which is obvious in its binarization $\mathbf{v}_{\text{bin}}$ shown in (c). In order to remove these and to yield a smoother and sparser texture **v**, one can increase the value of δ, say, e.g., by choosing $\delta = 10$. The effect of this choice can be seen in the binarized version $\mathbf{v}_{\text{bin}}$ in Fig. 2g and small-scale objects are moved to the residual image $\boldsymbol{\epsilon}$ in (h). Therefore, the value of δ defines the level of the residual $\boldsymbol{\epsilon}$.

6 Comparison of DG3PD with prior art

As stated before, the main objective of the DG3PD model is to achieve the following three goals (see Section 1):

- Goal 1: **u** contains only geometrical objects with a very smooth surface, sharp boundaries, and no texture.
- Goal 2: **v** contains only objects with sparse oscillating patterns and **v** shall be both smooth and sparse.
- Goal 3: Perfect reconstruction of **f**, i.e., $\mathbf{f} = \mathbf{u} + \mathbf{v} + \boldsymbol{\epsilon}$.

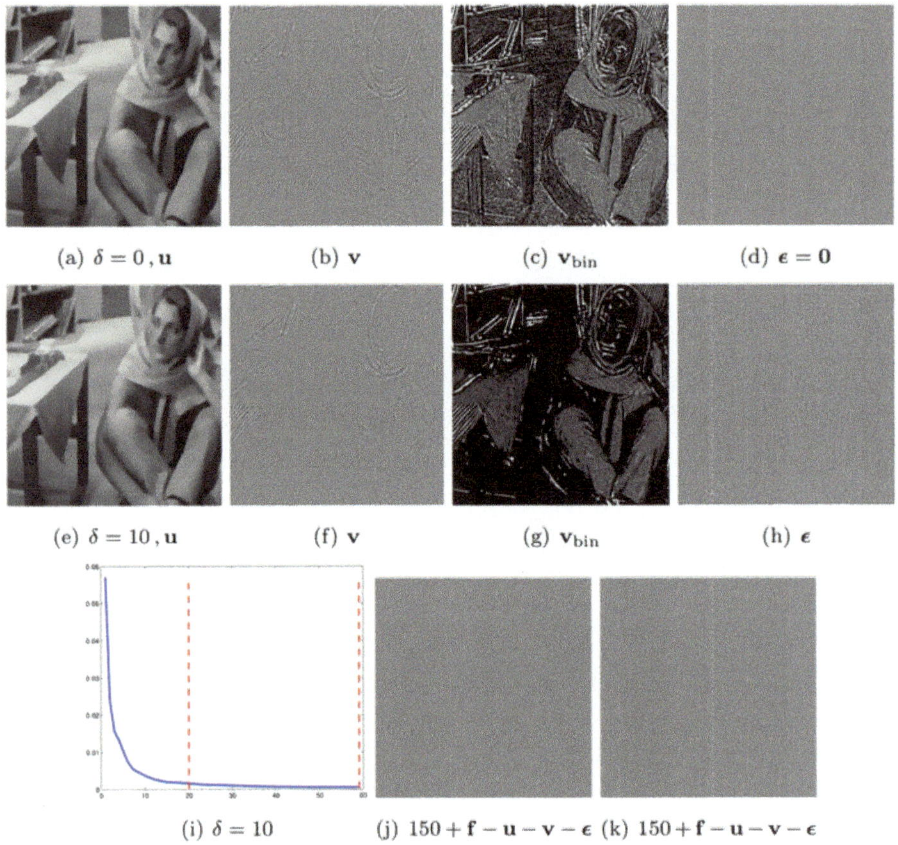

Fig. 2 Visualization of the decomposition results by DG3PD with $\delta = 0$ (**a–d**) and $\delta = 10$ (**e–h**) after 20 iterations. The convergence rates for these decompositions with $\delta = 0$ and $\delta = 10$ are depicted in Fig. 6m, **i**, respectively, which plots the relative error (*y*-axis) as defined in [25] versus the number of iterations (*x*-axis). The parameters are $\beta_4 = 0.04, \theta = 0.9, c_1 = 1, c_2 = 1.3, c_{\mu_1} = c_{\mu_2} = 0.03, \gamma = 1$, and $S = L = 9$. Error images are illustrated in **j** after 20 iterations and in **k** after 60 iterations with $\delta = 10$

Based on these criteria, we compare the proposed DG3PD model in this section with the state-of-the-art methods using the original Barbara image. We highlight selected regions for an improved conspicuousness of the differences between the considered methods; see Fig. 3:

- Images without noise: Rudin, Osher, and Fatemi (ROF) [1], Vese and Osher (VO) [9], Starck, Elad, and Donoho (SED) [11], and TV Gabor (TVG) by Aujol et al. [12] models.
- Images corrupted by i.i.d. Gaussian noise $\mathcal{N}(0, \sigma)$: the Aujol and Chambolle (AC) model [3].

For better visibility of differences between the various models, we show decomposition results for the ROF, VO, SED, TVG, and DG3PD models for two magnified parts of the original image (see Fig. 3b, c) in Figs. 4 and 5. We observe two main differences between the compared models and the proposed DG3PD model:

- Two-part decomposition instead of three-part decomposition.
- Quadratic penalty method (QPM) for solving the constrained minimization instead of ALM.

Goal 1: cartoon **u**. Regarding the cartoon **u**, we observe that the VO and SED models still contain texture on the scarf (see Fig. 4b, c, respectively). For the SED, the cartoon **u** is blurred and there are some small-scale objects under the table (see Fig. 5c). Comparing all five methods, VO and SED are furthest away from achieving the first goal, whereas ROF and TVG generate better cartoon images in terms of smoother surfaces without texture and sharper boundaries. Recently, Schaeffer and Osher [23] suggested a two-part decomposition approach. Note that the cartoon image by their decomposition which is shown in Figure 7a of [23] also contains texture and does not meet the first goal. The cartoon images produced by the DG3PD method come closest to the first goal, see Fig. 4i, m and Fig. 5e.

Fig. 3 The original Barbara image (**a**) and highlighted details (**b**, **c**)

Goal 2: texture **v**. Concerning the texture **v**, among the state-of-the-art methods, the decomposition by the SED model results in the sparsest texture (see Figs. 4g and 5h), while the texture images of the ROF, VO, and TVG have more coefficients different from zero. In addition, the texture component obtained by the ROF model also contains some geometry information which should have been assigned to the cartoon component, see Figs. 4e and 5f. The DG3PD model yields an even sparser texture than the SED model due to the $\|\mathbf{v}\|_{\ell_1}$ in the minimization (5), see the binarized versions with threshold "0" for visualization in Fig. 4o or Fig. 2g.

Goal 3: reconstruction by summation of all components. Figures 6 and 7 illustrate the effects of QPM and ALM. The decomposition by the ROF model results in a relatively large error $(\mathbf{f} - \mathbf{u})$ which contains geometry and texture information; see [9] and Fig. 6i. In the VO model, the error $(\mathbf{f} - \mathbf{u} - \mathbf{v})$ is reduced in comparison to the ROF model, but some information still remains in the error image; see Fig. 6n. In case of the DG3PD model with the ALM-based approach for solving the constrained minimization, the error $(\mathbf{f} - \mathbf{u} - \mathbf{v} - \boldsymbol{\epsilon})$ is significantly reduced and numerically the error tends to $\mathbf{0}$ as the number of iterations increases. For a comparison to ROF and VO using the same detail, see Fig. 6o for a visualization of the error after 20 iterations and (p) after 60 iterations. The error for the whole image after 20 and 60 iterations is displayed in Fig. 2j, k, respectively. To the best of our knowledge, this effect can be explained by using ALM for solving the constrained minimization instead of QPM. For more details about QPM and ALM, we refer the reader to [29, 30] and Chapter 3 in [31].

Comparison with Aujol and Chambolle. Figure 8 illustrates a situation in which the image is corrupted by i.i.d. Gaussian noise $\mathcal{N}(0, \sigma)$ with $\sigma = 20$ and compares DG3PD with the AC model [3] for three-part decomposition. It shows that under "heavy" noise, our DG3PD model still meets the criteria for cartoon **u** and texture **v**, i.e.,

- Our cartoon **u** contains smooth surfaces with sharp edges and no texture; see Fig. 8e. However, the cartoon **u** from the AC model is blurry with texture on the scarf; see Fig. 8a.
- Our texture **v** is sparse and smooth; see Fig. 8f and its binarization (h). However, the texture from the AC model is not sparse; see Fig. 8b.

However, there is a limitation for both methods: the noise image $\boldsymbol{\epsilon}$ contains some pieces of information due to the value of δ which defines the level of the noise. Similar to [3], we modify the classical threshold for curvelet coefficients, i.e., $\sigma\sqrt{2\log|\mathcal{K}|}$, with a weighting parameter η as follows $\delta = \eta\sigma\sqrt{2\log|\mathcal{K}|}$ and $|\mathcal{K}|$ is total number of curvelet coefficients.

Summary. We observe that the DG3PD method meets all three requirements much more closely than the other methods for images without noise, like the original Barbara image. And in particular, for images with additive noise, the DG3PD method still achieves all three goals as shown in the comparison with Aujol and Chambolle.

7 Applications

Here, we limit ourselves to consider three important applications of the DG3PD method: feature extraction, denoising, and image compression.

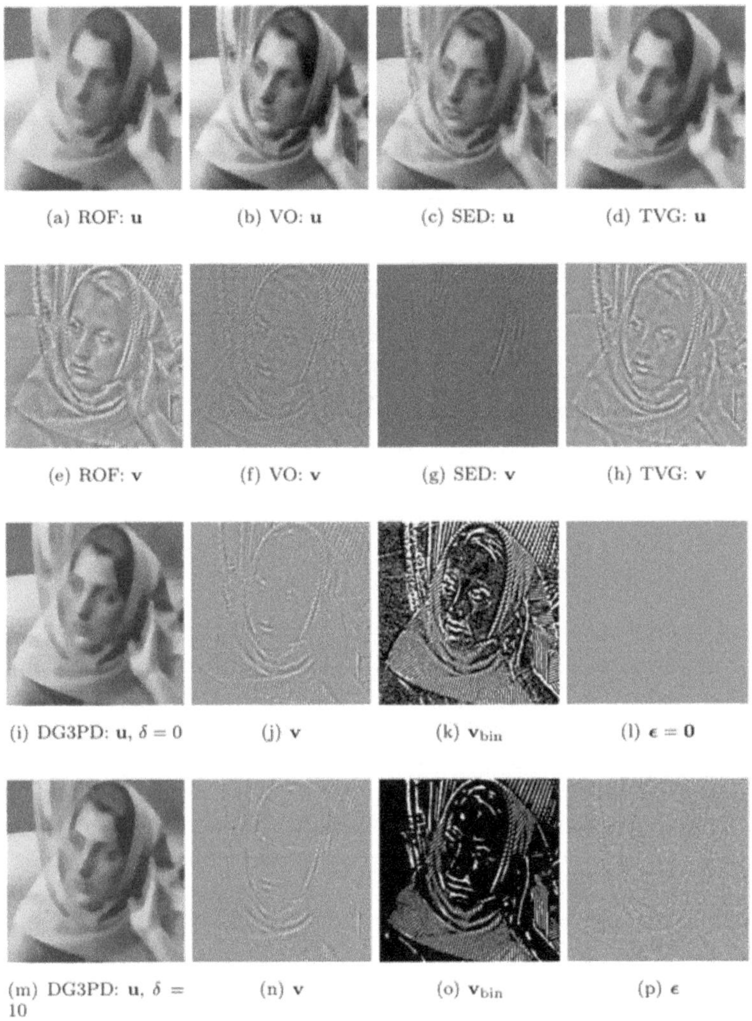

Fig. 4 a–p The *images* in the *first row* depict the comparison of the cartoon u for different methods in the literature on the original highlighted region (see Fig. 3c). The *images* in the *second row* show their corresponding texture v. Note that the ROF is reported in [12]. The *images* in the *third row* are obtained by the DG3PD model with $\delta = 0$ and $\delta = 10$ in the *fourth row* (see Fig. 2 for the whole image in these two cases)

7.1 Feature extraction

Depending on the specific field of application, the cartoon or texture, or both can be viewed as feature images. For the application of DG3PD to fingerprints, we are especially interested in the texture image **v** as a feature for subsequent processing steps like segmentation, orientation field estimation [32] and ridge frequency estimation [15], and fingerprint image enhancement [15, 21]. The first of these processing steps is to separate the foreground from the background [25, 57]. The foreground area (or region of interest) contains the relevant information for a fingerprint comparison. Segmentation is still a challenging problem for latent fingerprints [33] which are very low-quality fingerprints lifted from crime scenes. Both the foreground and background area can contain "noise" on all scales, from small objects or dirt on the surface to written or printed characters (on paper) and large-scale objects like an arc drawn by the forensic examiner. Standard fingerprint segmentation methods cannot cope with this variety of noise, whereas the texture image by decomposition with the DG3PD method crops out to be an excellent feature for estimating the region of interest. Figure 9 depicts a detailed example of the latent fingerprint segmentation by the DG3PD decomposition. In Fig. 10, we show further examples of segmentation results obtained using the texture image extracted by the DG3PD method and morphological postprocessing as described in [25, 57].

7.2 Denoising

The DG3PD model can be used for denoising images with texture because noise and small-scale objects are

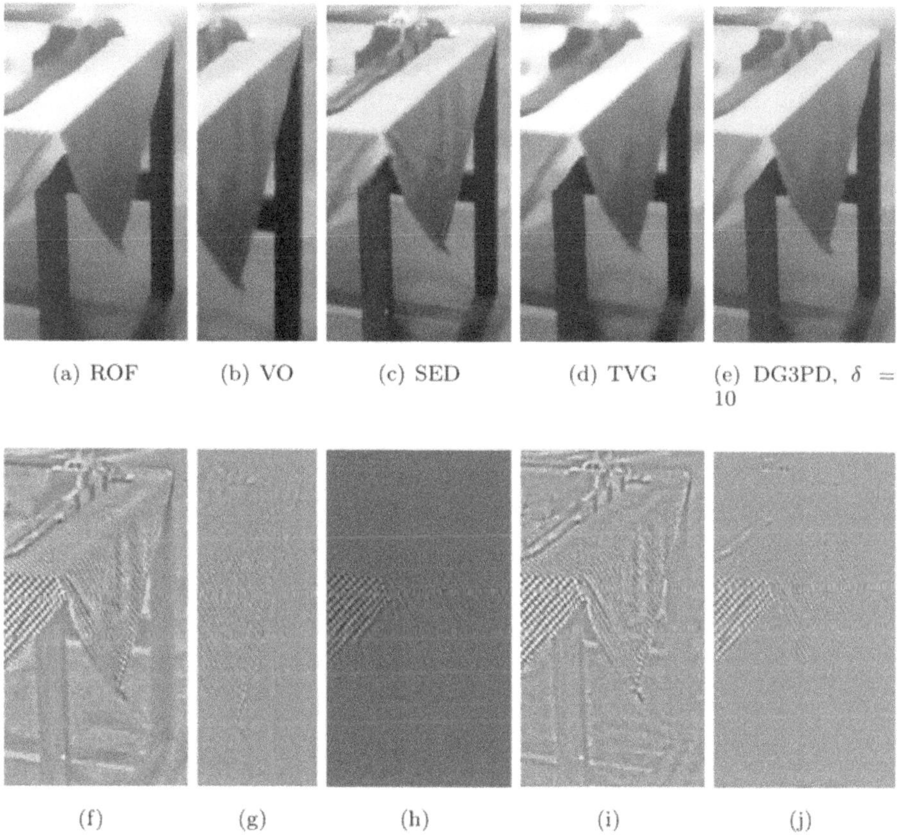

Fig. 5 a–j The *images* in the *first row* depict the comparison of the cartoon u for different methods on the original *highlighted region* (see Fig. 3(b)). The *images* in the *second row* show their corresponding texture v

moved into the residual image ϵ during the decomposition of $\mathbf{f}$ due to the supremum norm of the curvelet coefficients of ϵ. Therefore, the image $\mathbf{f}_{\text{denoised}} = \mathbf{u} + \mathbf{v}$ can be regarded as a denoised version of $\mathbf{f}$ and the degree of denoising can be steered by the choice of parameters, especially δ. For $\delta = 0$ which is equal to two-part decomposition, we obtain the original image again. As we increase δ, more noise is driven into ϵ and thereby removed from $\mathbf{f}_{\text{denoised}}$. Denoising images with texture, in particular with texture parts on different scales, is a relevant problem which we plan to address in future works.

Jung and Kang [34] proposed a variational minimization for vector-valued (color) image decomposition and restoration. Different from our directional total variation, their energy function involves a weighted second-order regularization for the cartoon component also to reduce the staircase effect and to provide image restoration of higher quality. Similar to the Vese-Osher model, they use the L_2 norm for measuring the residual which is different from our proposed $\|\mathcal{C}\{\cdot\}\|_\infty$. Moreover, their reconstructed texture is not sparse due to a lack of the assumption on its sparsity in their minimization model; see Figure 14 b in [34]. In a different view of convex minimization for image reconstruction, Tschumperle [35] proposed a curvature-preserving method for anisotropic smoothing of multi-valued images while preserving natural curvature constraints (or the edges). Then, line integral convolutions are applied for a numerical scheme to this tensor-driven diffusion PDE with two main advantages: namely, it preserves the orientation of thin image structure and the cost of computation is smaller in comparison to the classical explicit scheme. This two-part decomposition scheme is applied for denoising, inpainting, and resizing of vector-valued images.

7.3 Compression

Based on the DG3PD model, we propose a novel approach to image compression. The core idea is to perform image decomposition by the DG3PD method first and subsequently to compress the three component images by three different algorithms, each particularly suited for compressing the specific type of image.

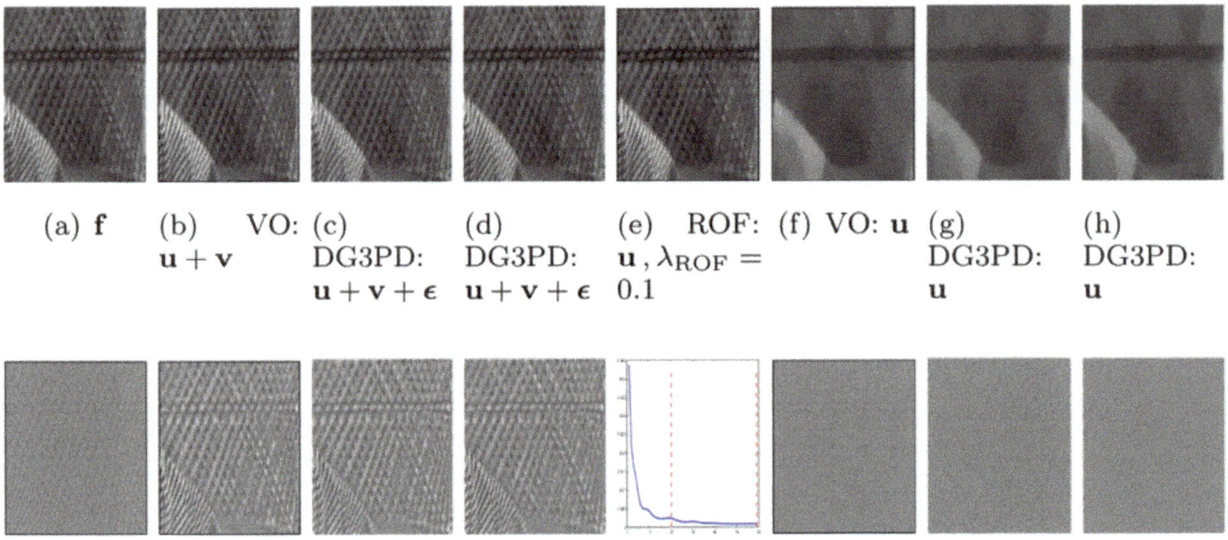

(a) $\mathbf{f}$ (b) VO: $\mathbf{u}+\mathbf{v}$ (c) DG3PD: $\mathbf{u}+\mathbf{v}+\boldsymbol{\epsilon}$ (d) DG3PD: $\mathbf{u}+\mathbf{v}+\boldsymbol{\epsilon}$ (e) ROF: $\mathbf{u}$, $\lambda_{\text{ROF}} = 0.1$ (f) VO: $\mathbf{u}$ (g) DG3PD: $\mathbf{u}$ (h) DG3PD: $\mathbf{u}$

(i) ROF: $150 + (\mathbf{f} - \mathbf{u})$ (j) VO: $150 + \mathbf{v}$ (k) DG3PD: $150 + \mathbf{v}$ (l) DG3PD: $150 + \mathbf{v}$ (m) DG3PD: $\delta = 0$ (n) $150 + (\mathbf{f} - \mathbf{u} - \mathbf{v})$ (o) $150 + (\mathbf{f} - \mathbf{u} - \mathbf{v} - \boldsymbol{\epsilon})$ (p) $150 + (\mathbf{f} - \mathbf{u} - \mathbf{v} - \boldsymbol{\epsilon})$

Fig. 6 a–l Comparison of QPM and ALM for the ROF, VO, and DG3PD models: The images for ROF and VO models are obtained from [9]. The parameters for the DG3PD model are $\delta = 0$ and 20 iterations for the *third* and 60 for the *fourth column*, and the other parameters are the same as in Fig. 2. The relative error (*y*-axis) versus the number of iterations (*x*-axis) is illustrated in **m**. The *first row* shows that the VO and the DG3PD models can achieve good reconstructed images, see **b**, **c**, **d**, in comparison with the original magnified image (**a**). As mentioned in [9], the error image from the VO model (**n**) contains much less geometry and texture than the one from the ROF model. However, the error image from our model is much further reduced in comparison to the VO model. After 20 iterations some pieces of information still remain in the error image, see (**o**). As the number of iterations increases, the error numerically tends to **0**; see **p** after 60 iterations

This scheme can be used for lossy as well as lossless compression.

7.3.1 Cartoon image compression

As stated in our definition of goals, the cartoon image consists of geometric objects with a very smooth or piecewise constant surface and sharp edges. This special kind of images is highly compressible, and a very effective approach is based on diffusion. Anisotropic diffusion [36] is useful for many purposes in image processing, e.g., fingerprint image enhancement by oriented diffusion filtering [21].

The basic idea of diffusion-based compression is store information for only a few sparse locations which encode the edges of the cartoon image. The surface areas are inpainted using a linear or non-linear diffusion process. Please note that the cartoon image obtained by the DG3PD method is much better suited for this type of compression due to the property of sharper edges between geometric objects in comparison to the cartoon images of the other decomposition approaches. Moreover, some difficulties and drawbacks of diffusion-based compression for arbitrary images do not apply to this special case. In general, it is a challenging question how and where to select locations for diffusion seed points. In our case, this task is easily solvable because of the sharp edges between homogeneous regions in the DG3PD cartoon image. This allows for an extremely sparse selection of locations on corners and edges.

Image compression with edge-enhancing anisotropic diffusion (EED) has been studied by Galic et al. [37] and has been improved by Schmaltz et al. [38]. Compression of cartoon-like images with homogeneous diffusion has been analyzed by Mainberger et al. [39].

A viable alternative to diffusion-based compression of cartoon images is a dictionary-based approach [40] in which the dictionary is optimized for cartoon images. Another very promising possibility to compress the DG3PD cartoon component is the usage of linear splines over adaptive triangulations which has been proposed in the work of Demaret et al. [41].

7.3.2 Texture image compression

Tailor-made solutions are available for texture image compression and especially for compressing oscillating patterns like fingerprints.

Larkin and Fletcher [17] achieved a compression rate of 1:239 for a fingerprint image using amplitude and

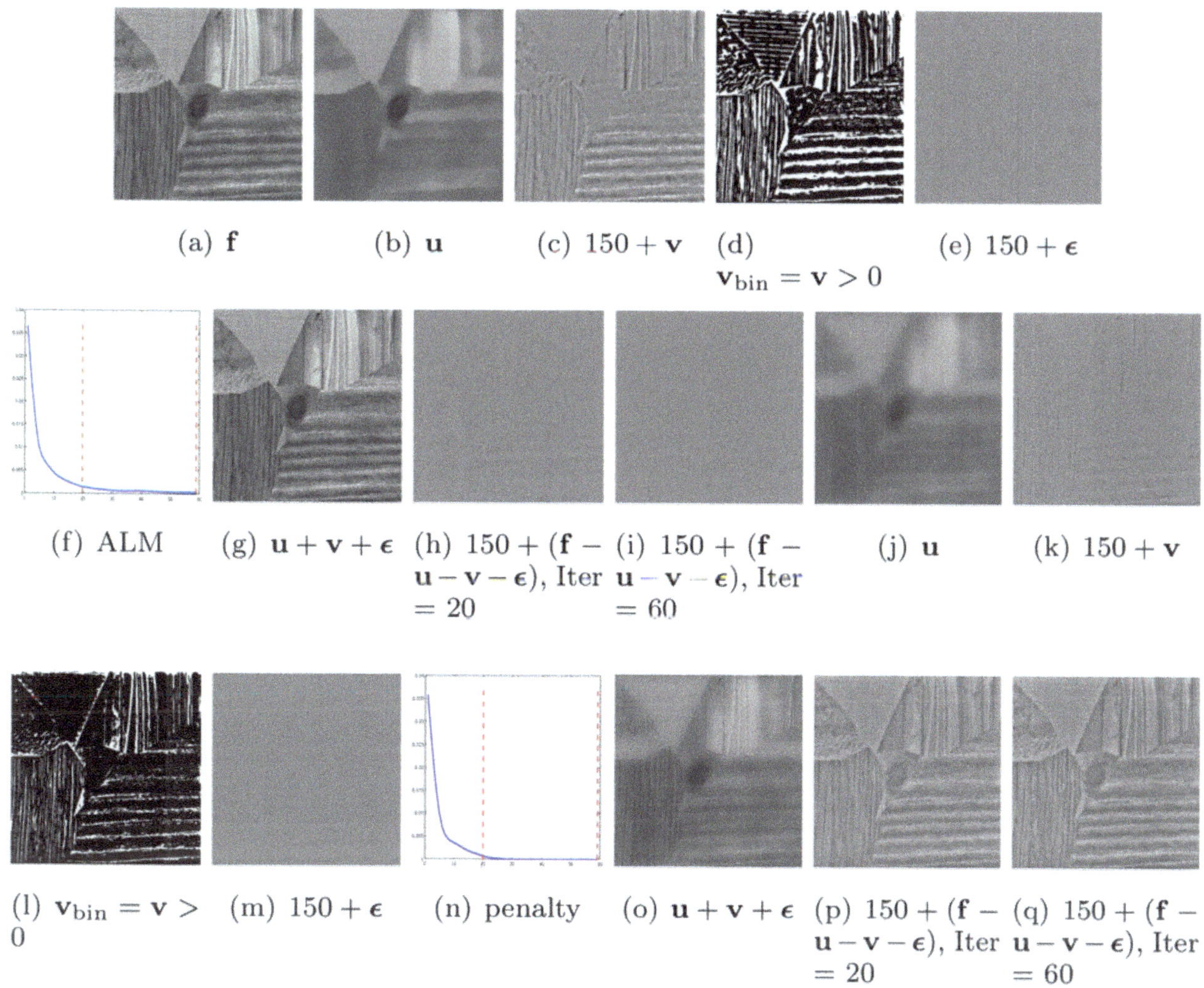

Fig. 7 The comparison of ALM (**a**–**i**) and QPM (**j**–**q**) for the DG3PD model with the same parameters and 20 iterations: $\beta_4 = 0.025, \theta = 0.9, c_1 = 10, c_2 = 1.3, c_{\mu_1} = c_{\mu_2} = 0.03, \gamma = 1, S = L = 32, \delta = 5$. We observe that the error image of QPM still contains geometry and texture information; see after 20 iterations in **p** and after 60 in **q**. Using QPM for constrained minimization is similar to decomposing the original image into four parts, namely cartoon (**j**), texture (**k**), residual (**m**), and error (**p**). The amount of information in the error image by QPM strongly depends on the choice of parameters $(\beta_1, \beta_2, \beta_3, \beta_4)$. However, for ALM, the updated Lagrange multipliers $(\boldsymbol{\lambda}_1, \boldsymbol{\lambda}_2, \boldsymbol{\lambda}_3, \boldsymbol{\lambda}_4)$ compensate for the choice of $(\beta_1, \beta_2, \beta_3, \beta_4)$. Thus, the error numerically tends to **0** as the number of iterations increases, see the error image after 20 iterations in **h** and after 60 iterations in **i**

frequency modulated (AM-FM) functions. They decompose a fingerprint image into four parts, and this idea can be applied to the texture image **v** obtained by the DG3PD method:

$$\mathbf{v} \approx \boldsymbol{a} + [\boldsymbol{b} \cdot \cos(\boldsymbol{\Psi}_C + \boldsymbol{\Psi}_S)]$$

Each of the four components is again highly compressible and can be stored with only a few bytes (see Figure 5 in [17]). This is remarkable and we would like to offer another perspective on the AM-FM model. Storing a minutiae template can be viewed as a lossy form of fingerprint compression. The minutiae of a fingerprint are locations where ridges (dark lines) end or bifurcate, and a template stores the locations and directions of these minutiae. Several algorithms have been proposed for reconstructing the orientation field (OF) from a minutia template [42]. The continuous phase $\boldsymbol{\Psi}_C$ can be derived from the unwrapped reconstructed OF and the spiral phase $\boldsymbol{\Psi}_S$ directly constructed from the minutiae template. Choosing appropriate values for $\boldsymbol{a}$ and $\boldsymbol{b}$ leads to a fingerprint image. A survey of further methods for reconstructing fingerprints from their minutiae template is given in [43]. An alternative way of lossy fingerprint image compression is wavelet scalar

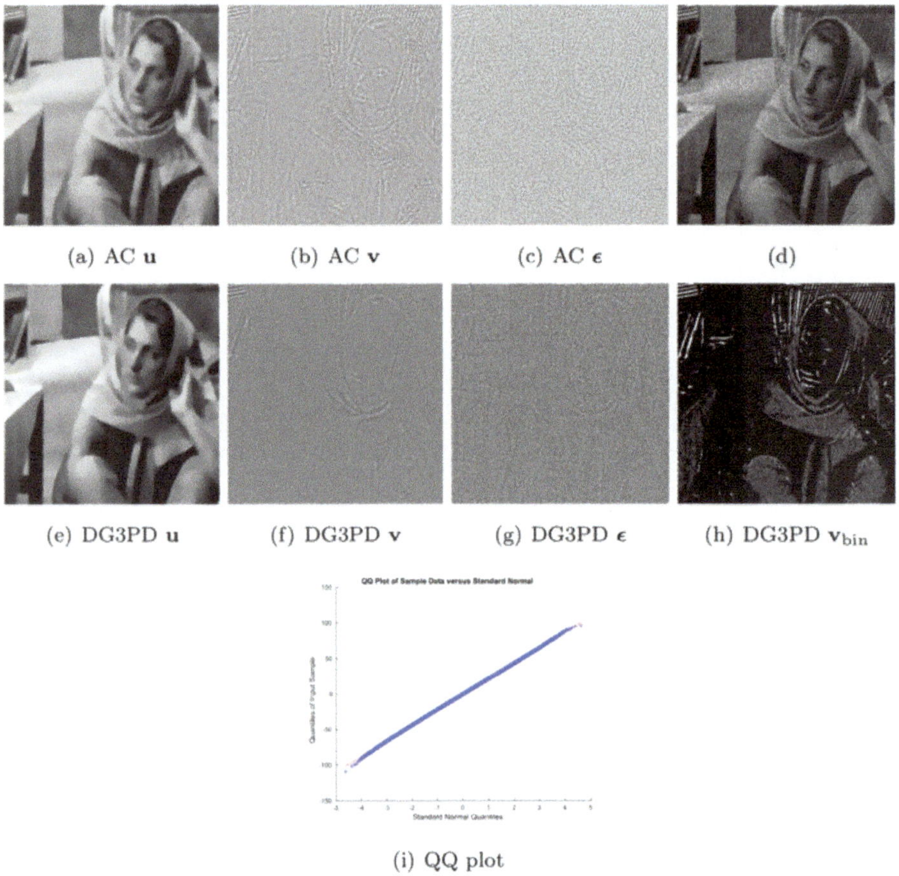

Fig. 8 The Barbara image (**d**) with additive Gaussian white noise ($\sigma = 20$) is decomposed by the Aujol and Chambolle model (**a**–**c**) and the DG3PD model (**e**–**h**) with $\delta = 16$ and the other parameters are the same as in Fig. 2. Comparing **a** and **e**, we observe that the cartoon image u obtained by the DG3PD model (**e**) has a smoother surface and sharper edges than **a**. Comparing the texture images v, we note that **f** is smoother and sparser than **b**. In order to highlight the sparseness of the DG3PD texture, all positive coefficients are visualized as white pixels in **h**. For visualization, we add 150 to the value of the residual ϵ in **g**. The residual in **g** still contains some texture, but mainly the Gaussian noise which is obviously shown in the QQ plot (**i**). There are some differences at the end of the tail in **i** probably due to the remaining texture and the numerical simulation of Gaussian noise

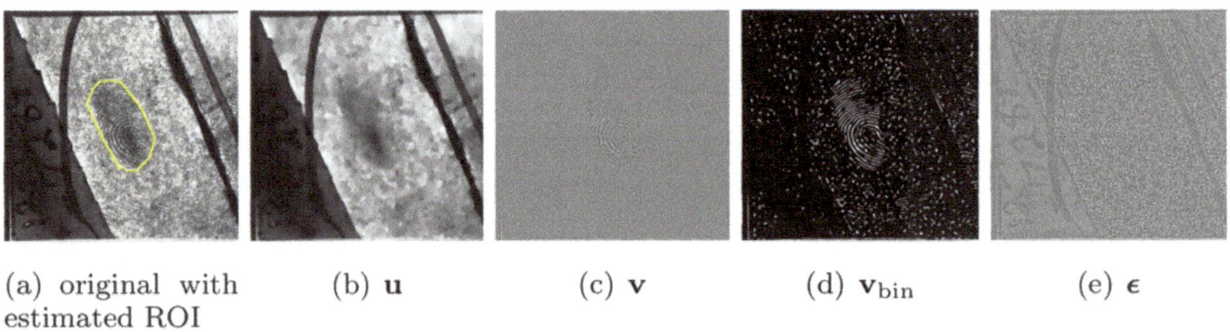

Fig. 9 DG3PD decomposition of a latent fingerprint for segmentation (**a**) with $\delta = 60$. **b**–**e** The ROI is obtained from the binarized texture v_{bin} and morphological postprocessing as described in [57]

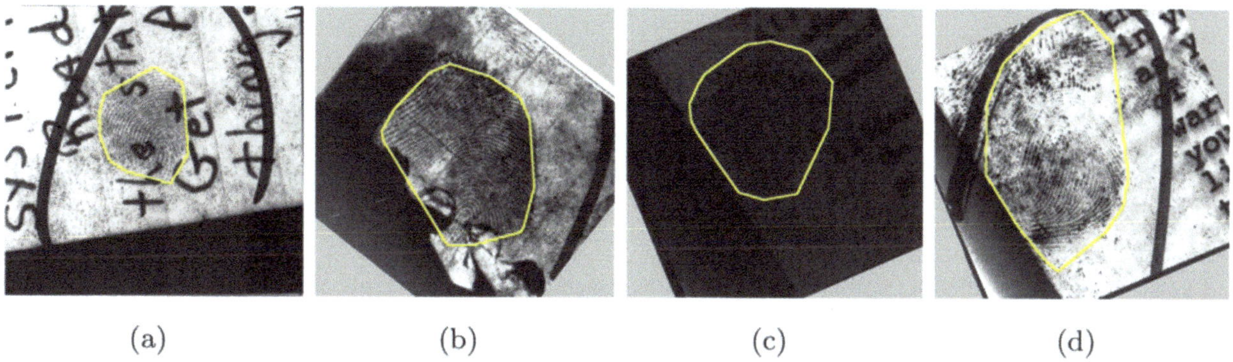

Fig. 10 Latent fingerprint images from NIST SD27. The boundary of the foreground estimated by the DG3PD method is drawn in *yellow* (**a**–**d**)

quantization (WSQ) [44] which has been a compression standard for fingerprints used by the Federal Bureau of Investigation in the USA. See Fig. 11f–h for application example of WSQ to the texture of the Barbara image.

A third, very good compression possibility is dictionary learning [40] with optimization of the dictionary for the texture component $\mathbf{v}$. For fingerprint images, this problem has recently been studied by Shao et al. [45].

7.3.3 Residual image compression

For image compression using DG3PD, we propose the following steps in this order: First, image decomposition $\mathbf{f} = \mathbf{u} + \mathbf{v} + \boldsymbol{\epsilon}$. Second, a tailor-made, lossy, high compression of the cartoon component $\mathbf{u}$ and the texture component $\mathbf{v}$. Third, decompressing $\mathbf{u}$ and $\mathbf{v}$ in order to compute the compression residual image $\mathbf{s} = \mathbf{f} - \mathbf{u}_d - \mathbf{v}_d$, where $\mathbf{u}_d$ is the cartoon image and $\mathbf{v}_d$ the texture image after decompression. Fourth, compression of $\mathbf{s}$.

In steps two and four, the term "compression" denotes the whole process including coefficient quantization and symbol encoding (see [46] for scalar quantization, Huffman coding, LZ77, LZW, and many other standard techniques).

Let be $\mathbf{e}_u = \mathbf{u} - \mathbf{u}_d$, the difference between the cartoon component before and after compression, i.e., the compression error, and $\mathbf{e}_v = \mathbf{v} - \mathbf{v}_d$, then we can rewrite

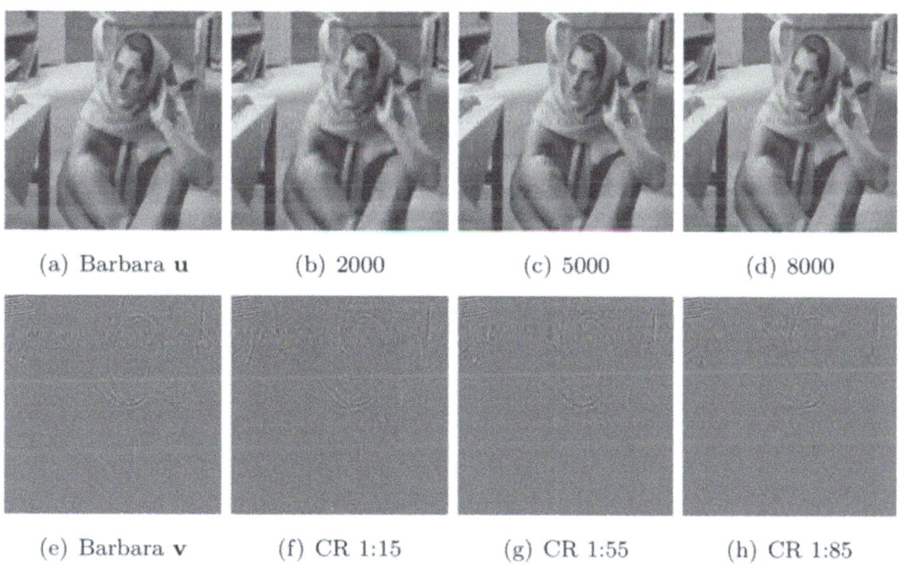

Fig. 11 The cartoon component u (**a**) and texture component v (**e**) of the Barbara image obtained by DG3PD. Images **b**–**d** display the reconstruction of u from the 2000, 5000, and 8000 largest wavelet coefficients (from a total of 262,144 coefficients). **f**–**h** show the decompressed images after compression of v by WSQ at compression rates between 1:15 and 1:85

$\mathbf{r} = \mathbf{f} - \mathbf{u}_d - \mathbf{v}_d = \mathbf{u} + \mathbf{v} + \boldsymbol{\epsilon} - \mathbf{u}_d - \mathbf{v}_d = \boldsymbol{\epsilon} + \mathbf{e}_u + \mathbf{e}_v$. Hence, the residual image $\mathbf{s}$ computed in step four contains the residual component $\boldsymbol{\epsilon}$ plus the compression errors of the other two components. Now, lossless compression can be achieved by lossless compression of $\mathbf{s}$. If the goal is lossy compression with a certain target quality or target compression rate, this can be achieved by adapting the lossy compression of $\mathbf{s}$ accordingly. See Fig. 11 for the effects of different compression rates on the decompressed cartoon $\mathbf{u}_d$ and the decompressed texture $\mathbf{v}_d$.

An additional advantage of decompression beginning with $\mathbf{u}_d$, followed by $\mathbf{v}_d$ and finally $\mathbf{s}_d$ is the fast generation of a preview image which mimics the effects of interlacing. In a scenario with limited bandwidth for data transmission, e.g., sending an image to a mobile phone, the user can be shown a preview based on the compressed, transmitted, and decompressed $\mathbf{u}$ image. During the transmission of the compressed $\mathbf{v}$ and $\mathbf{s}$, the user can decide whether to continue or abort the transmission.

8 Conclusions

The DG3PD model is a novel method for three-part image decomposition. We have shown that the DG3PD method achieves the goals defined in the introduction much better than other relevant image decomposition approaches. The DG3PD model lays the groundwork for applications such as image compression, denoising, and feature extraction for challenging tasks such as latent fingerprint processing. We follow in the footsteps of Aujol and Chambolle [3] who pioneered three-part decomposition and DG3PD generalizes their approach. We believe that three-part decomposition is the way forward to address many important problems in image processing and computer vision. Buades et al. [18] asked in 2010: "Can images be decomposed into the sum of a geometric part and a textural part?" Our answer to that question is no if an image contains other parts than cartoon and texture, i.e., noise or small-scale objects. Consider, e.g., the noisy Barbara image in Fig. 8d. If the sum of the cartoon and texture images shall reconstruct the input image $\mathbf{f}$, a two-part decomposition has to assign the noise parts either to the cartoon or to the texture component. In principle, not even the best two-part decomposition model can fully achieve both goals regarding the desired properties of the cartoon and texture component simultaneously. The solution is that noise and small-scale objects which do not belong to the cartoon or texture have to be allotted to a third component.

In our future work, we intend to optimize the DG3PD method for specific applications, especially image compression and latent fingerprint processing. Issues for improvement include the data-driven, automatic parameter selection, and the convergence rate (can the same decomposition be achieved in fewer iterations?) Furthermore, we plan to explore and evaluate specialized compression approaches for cartoon, texture, and residual images.

Additionally, a very interesting application for the residual component can be biometric liveness detection. Recently, Gragnaniello et al. [47] have concluded that high-pass filtering before computing local image descriptors improves the accuracy of their proposed iris liveness detection algorithm. The residual image obtained by DG3PD contains the high-frequency components of the input image. Applications of DG3PD to iris or fingerprint liveness detection [48, 49] are therefore very promising. A survey of local image descriptors for fingerprint, iris, and face liveness detection can be found in [50].

Optimal solutions of transportation problems are the key to compute the earth mover's distance (EMD) or Wasserstein distance [51]. Recently, Brauer and Lorenz [52] discussed an interesting connection between Meyer's G-norm and transportation problems. Solving transportation problems for images of dimension 512×512 pixels or larger can be a computationally extremely challenging problem (depending on the number of producers and consumers, and their distribution over the image domain) even for state-of-the-art methods such as the shortlist method [51]. However, in the special case of $p = 1$, the Kantorovich-Rubinstein duality provides a loophole which allows it to avoid solving the associated transportation problem. Based on this property, Brauer and Lorenz [52] proposed a three-part image decomposition with transport norms. Lellmann et al. [53] proposed the use of transport norms for image denosing and two-part image decomposition. We believe that the commonalities between the Vese-Osher model [9], Meyer's G-norm, and the recently proposed models using transport norms deserve further research.

Moreover, our experiments have shown that the curvelet transform as part of the DG3PD model is very suitable for capturing the residual component in three-part image decomposition in terms of i.i.d. (or weakly correlated) Gaussian (or non-Gaussian) noise. In principle, it is also conceivable to apply the shearlet transform [54], the contourlet transform [55], or the steerable wavelet transform [56] instead of the curvelet transform, and we deem it worth investigating if especially for specific types of images one of these transforms performs considerably better than the others—as part of a DG3PD model with directional total variational norms, directional G-norms, or transport norms.

Appendix

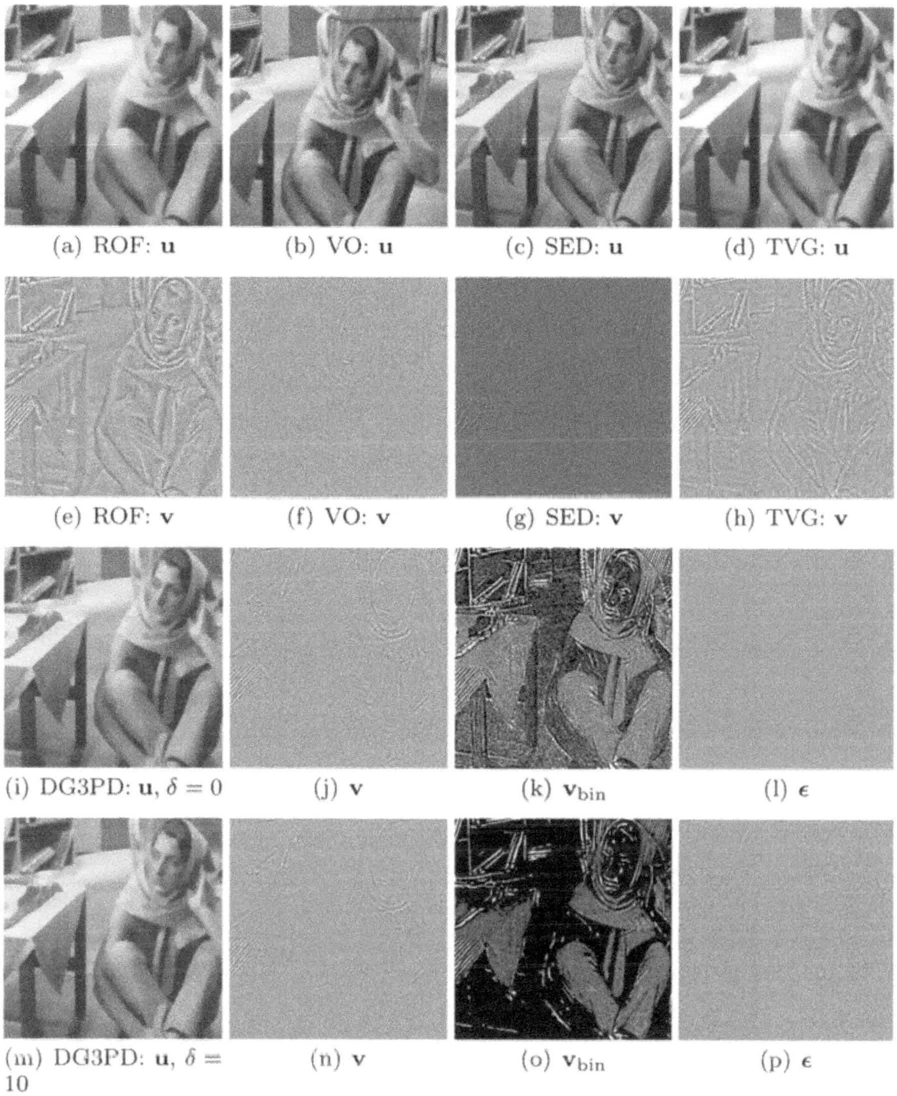

Fig. 12 a–f Comparison of decomposition results by the ROF [1], VO [9], SED [11], TVG [12], and DG3PD models. Parameters for DG3PD are $\beta_4 = 0.04, \theta = 0.9, c_1 = 1, c_2 = 1.3, c_{\mu_1} = c_{\mu_2} = 0.03, \gamma = 1, S = L = 9$

Algorithm 1 The Discrete DG3PD Model

Initialization: $\mathbf{u}^{(0)} = \mathbf{f}, \mathbf{v}^{(0)} = \boldsymbol{\epsilon}^{(0)} = \left[\mathbf{r}_l^{(0)}\right]_{l=0}^{L-1} = \left[\mathbf{w}_s^{(0)}\right]_{s=0}^{S-1} = \left[\mathbf{g}_s^{(0)}\right]_{s=0}^{S-1} = \left[\boldsymbol{\lambda}_{1l}^{(0)}\right]_{l=0}^{L-1} = \left[\boldsymbol{\lambda}_{2a}^{(0)}\right]_{a=0}^{S-1} = \boldsymbol{\lambda}_3^{(0)} = \boldsymbol{\lambda}_4^{(0)} = \mathbf{0}.$

for $t = 1, \dots, T$ **do**

1. Compute $\left(\left[\mathbf{r}_b^{(t)}\right]_{b=0}^{L-1}, \left[\mathbf{w}_a^{(t)}\right]_{a=0}^{S-1}, \left[\mathbf{g}_a^{(t)}\right]_{a=0}^{S-1}, \mathbf{v}^{(t)}, \mathbf{u}^{(t)}, \boldsymbol{\epsilon}^{(t)}\right) \in X^{L+2S+3}$:

$$\mathbf{r}_b^{(t)} = \text{Shrink}\left(\cos\left(\frac{\pi b}{L}\right)\mathbf{u}^{(t-1)}\mathbf{D}_\mathbf{n}^\mathrm{T} + \sin\left(\frac{\pi b}{L}\right)\mathbf{D}_\mathbf{m}\mathbf{u}^{(t-1)} - \frac{\boldsymbol{\lambda}_{1b}^{(t-1)}}{\beta_1}, \frac{1}{\beta_1}\right), \quad b = 0, \dots, L-1$$

$$\mathbf{w}_a^{(t)} = \text{Shrink}\left(\mathbf{t}_{\mathbf{w}_a} := \mathbf{g}_a^{(t-1)} - \frac{\boldsymbol{\lambda}_{2a}^{(t-1)}}{\beta_2}, \frac{\mu_1}{\beta_2}\right), \quad a = 0, \dots, S-1$$

$$\mathbf{g}_a^{(t)} = \text{Re}\left[\mathcal{F}^{-1}\left\{\mathcal{A}^{(t)}(\mathbf{z}) \cdot \mathcal{B}^{(t)}(\mathbf{z})\right\}\right], \ a = 0, \dots, S-1$$

$$\mathbf{v}^{(t)} = \text{Shrink}\Bigg(\mathbf{t}_\mathbf{v} := \frac{\beta_3}{\beta_3 + \beta_4}\left(\sum_{s=0}^{S-1}\left[\cos\left(\frac{\pi s}{S}\right)\mathbf{g}_s^{(t)}\mathbf{D}_\mathbf{n}^\mathrm{T} + \sin\left(\frac{\pi s}{S}\right)\mathbf{D}_\mathbf{m}\mathbf{g}_s^{(t)}\right] - \frac{\boldsymbol{\lambda}_3^{(t-1)}}{\beta_3}\right)$$
$$+ \frac{\beta_4}{\beta_3 + \beta_4}\left(\mathbf{f} - \mathbf{u}^{(t-1)} - \boldsymbol{\epsilon}^{(t-1)} + \frac{\boldsymbol{\lambda}_4^{(t-1)}}{\beta_4}\right), \frac{\mu_2}{\beta_3 + \beta_4}\Bigg)$$

$$\mathbf{u}^{(t)} = \text{Re}\left[\mathcal{F}^{-1}\left\{\mathcal{X}^{(t)}(\mathbf{z}) \cdot \mathcal{Y}^{(t)}(\mathbf{z})\right\}\right]$$

$$\boldsymbol{\epsilon}^{(t)} = \left(\mathbf{f} - \mathbf{u}^{(t)} - \mathbf{v}^{(t)} + \frac{\boldsymbol{\lambda}_4^{(t-1)}}{\beta_4}\right) - \text{CST}\left(\mathbf{f} - \mathbf{u}^{(t)} - \mathbf{v}^{(t)} + \frac{\boldsymbol{\lambda}_4^{(t-1)}}{\beta_4}, \delta\right)$$

2. Update $\left(\left[\boldsymbol{\lambda}_{1b}^{(t)}\right]_{b=0}^{L-1}, \left[\boldsymbol{\lambda}_{2a}^{(t)}\right]_{a=0}^{S-1}, \boldsymbol{\lambda}_3^{(t)}, \boldsymbol{\lambda}_4^{(t)}\right) \in X^{L+S+2}$:

$$\boldsymbol{\lambda}_{1b}^{(t)} = \boldsymbol{\lambda}_{1b}^{(t-1)} + \gamma\beta_1\left(\mathbf{r}_b^{(t)} - \cos\left(\frac{\pi b}{L}\right)\mathbf{u}^{(t)}\mathbf{D}_\mathbf{n}^\mathrm{T} - \sin\left(\frac{\pi b}{L}\right)\mathbf{D}_\mathbf{m}\mathbf{u}^{(t)}\right), \quad b = 0, \dots, L-1$$

$$\boldsymbol{\lambda}_{2a}^{(t)} = \boldsymbol{\lambda}_{2a}^{(t-1)} + \gamma\beta_2\left(\mathbf{w}_a^{(t)} - \mathbf{g}_a^{(t)}\right), \quad a = 0, \dots, S-1$$

$$\boldsymbol{\lambda}_3^{(t)} = \boldsymbol{\lambda}_3^{(t-1)} + \gamma\beta_3\left(\mathbf{v}^{(t)} - \sum_{s=0}^{S-1}\left[\cos\left(\frac{\pi s}{S}\right)\mathbf{g}_s^{(t)}\mathbf{D}_\mathbf{n}^\mathrm{T} + \sin\left(\frac{\pi s}{S}\right)\mathbf{D}_\mathbf{m}\mathbf{g}_s^{(t)}\right]\right)$$

$$\boldsymbol{\lambda}_4^{(t)} = \boldsymbol{\lambda}_4^{(t-1)} + \gamma\beta_4\left(\mathbf{f} - \mathbf{u}^{(t)} - \mathbf{v}^{(t)} - \boldsymbol{\epsilon}^{(t)}\right)$$

end for

$$\mathcal{A}(\mathbf{z}) = \left[\beta_2 \mathbf{1}_{\mathbf{mn}} + \beta_3\left[\sin\frac{\pi a}{S}(\mathbf{z}_1^{-1} - 1) + \cos\frac{\pi a}{S}(\mathbf{z}_2^{-1} - 1)\right]\left[\sin\frac{\pi a}{S}(\mathbf{z}_1 - 1) + \cos\frac{\pi a}{S}(\mathbf{z}_2 - 1)\right]\right]^{-1},$$

$$\mathcal{B}(\mathbf{z}) = \beta_2\left[W_a(\mathbf{z}) + \frac{\Lambda_{2a}(\mathbf{z})}{\beta_2}\right] + \beta_3\left[\sin\left(\frac{\pi a}{S}\right)(\mathbf{z}_1^{-1} - 1) + \cos\left(\frac{\pi a}{S}\right)(\mathbf{z}_2^{-1} - 1)\right] \times$$
$$\left[V(\mathbf{z}) - \sum_{s=[0,S-1]\backslash\{a\}}\left[\cos\left(\frac{\pi s}{S}\right)(\mathbf{z}_2 - 1) + \sin\left(\frac{\pi s}{S}\right)(\mathbf{z}_1 - 1)\right]G_s(\mathbf{z}) + \frac{\Lambda_3(\mathbf{z})}{\beta_3}\right],$$

$$\mathcal{X}(\mathbf{z}) = \left[\beta_4 \mathbf{1}_{\mathbf{mn}} + \beta_1\sum_{l=0}^{L-1}\left[\sin\left(\frac{\pi l}{L}\right)(\mathbf{z}_1^{-1} - 1) + \cos\left(\frac{\pi l}{L}\right)(\mathbf{z}_2^{-1} - 1)\right]\left[\sin\left(\frac{\pi l}{L}\right)(\mathbf{z}_1 - 1) + \cos\left(\frac{\pi l}{L}\right)(\mathbf{z}_2 - 1)\right]\right]^{-1},$$

$$\mathcal{Y}(\mathbf{z}) = \beta_4\left[F(\mathbf{z}) - V(\mathbf{z}) - \mathcal{E}(\mathbf{z}) + \frac{\Lambda_4(\mathbf{z})}{\beta_4}\right] + \beta_1\sum_{l=0}^{L-1}\left[\sin\left(\frac{\pi l}{L}\right)(\mathbf{z}_1^{-1} - 1) + \cos\left(\frac{\pi l}{L}\right)(\mathbf{z}_2^{-1} - 1)\right]\left[R_l(\mathbf{z}) + \frac{\Lambda_{1l}(\mathbf{z})}{\beta_1}\right].$$

Choice of Parameters

$$\mu_1 = c_{\mu_1}\beta_2 \cdot \max_{\mathbf{k}\in\Omega}\left(\left|t_{\mathbf{w}_a}[\mathbf{k}]\right|\right), \ \mu_2 = c_{\mu_2}(\beta_3 + \beta_4) \cdot \max_{\mathbf{k}\in\Omega}\left(\left|t_\mathbf{v}[\mathbf{k}]\right|\right) \text{ and } \beta_2 = c_2\beta_3, \beta_3 = \frac{\theta}{1-\theta}\beta_4, \beta_1 = c_1\beta_4.$$

Competing interests

The authors declare that they have no competing interests.

Acknowledgements

The authors would like to thank Luminita Vese for her permission to show the images in Figs. 4(b, f), 5(b, g), 6(b, f, j, n, e, i) and 12(a, b). The authors would like to thank Jean-Francois Aujol for his permission to depict the images in Figs. 4(d, h), 5(d, i), 8(a, b, c) and 12(d, h, a, e). The authors would like to thank Jean-Luc Starck for his permission to display the images in Figs. 4(c, g), 5(c, h), and 12(c, g). The authors gratefully acknowledge support by the German Research Foundation (DFG) and the Open Access Publication Funds of the University of Göttingen. D.H. Thai is supported by the National Science Foundation under Grant DMS-1127914 to the Statistical and Applied Mathematical Sciences Institute. C. Gottschlich gratefully acknowledges support by the Felix-Bernstein-Institute for Mathematical Statistics in the Biosciences and the Niedersachsen Vorab of the Volkswagen Foundation.

Author details

[1]Institute for Mathematical Stochastics, University of Goettingen, Goldschmidtstr. 7, 37077 Goettingen, Germany. [2]Statistical and Applied Mathematical Science Institute (SAMSI), 19 T. W. Alexander Drive, 27709-4006 NC, USA.

References

1. L Rudin, S Osher, E Fatemi, Nonlinear total variation based noise removal algorithms. Physica D. **60**(1–4), 259–268 (1992)
2. Y Meyer, *Oscillating Patterns in Image Processing and Nonlinear Evolution Equations*. (American Mathematical Society, Boston, MA, USA, 2001)
3. J-F Aujol, A Chambolle, Dual norms and image decomposition models. Int. J. Comput. Vis. **63**(1), 85–104 (2005)
4. E Candès, L Demanet, D Donoho, L Ying, Fast discrete curvelet transforms. Multiscale Model. Simul. **5**(3), 861–899 (2006)
5. J Ma, G Plonka, The curvelet transform. IEEE Signal Process. Mag. **27**(2), 118–133 (2010)
6. D Mumford, J Shah, Optimal approximations by piecewise smooth functions and associated variational problems. Commun. Pure Appl. Math. **42**(5), 577–685 (1989)
7. TF Chan, LA Vese, Active contours without edges. IEEE Trans. Image Process. **10**(2), 266–277 (2001)
8. S Osher, JA Sethian, Fronts propagating with curvature-dependent speed: algorithms based on Hamilton-Jacobi formulations. J. Comput. Phys. **79**(1), 12–49 (1988)
9. LA Vese, S Osher, Modeling textures with total variation minimization and oscillatory patterns in image processing. J. Sci. Comput. **19**(1–3), 553–572 (2003)
10. LA Vese, S Osher, Image denoising and decomposition with total variation minimization and oscillatory functions. J. Math. Imaging Vis. **20**(1–2), 7–18 (2004)
11. J-L Starck, M Elad, DL Donoho, Image decomposition via the combination of sparse representations and a variational approach. IEEE Trans. Image Process. **14**(10), 1570–1582 (2005)
12. J-F Aujol, G Gilboa, T Chan, S Osher, Structure-texture image decomposition—modeling, algorithms, and parameter selection. Int. J. Comput. Vis. **67**(1), 111–136 (2006)
13. J Gilles, Noisy image decomposition: a new structure, texture and noise model based on local adaptivity. J. Math. Imaging Vis. **28**(3), 285–295 (2007)
14. P Maragos, G Evangelopoulos, in *Proc. Int. Symp. Mathematical Morphology*. Leveling cartoons, texture energy markers, and image decomposition, (Rio de Janeiro, Brazil, 2007), pp. 125–138
15. C Gottschlich, Curved-region-based ridge frequency estimation and curved Gabor filters for fingerprint image enhancement. IEEE Trans. Image Process. **21**(4), 2220–2227 (2012). http://dx.doi.org/10.1109/TIP.2011.2170696
16. JP Havlicek, DS Harding, AC Bovik, Multidimensional quasi-eigenfunction approximations and multicomponent AM-FM models. IEEE Trans. Image Process. **9**(2), 227–242 (2000)
17. KG Larkin, PA Fletcher, A coherent framework for fingerprint analysis: are fingerprints holograms? Optics Express. **15**(14), 8667–8677 (2007)
18. A Buades, TM Le, J-M Morel, LA Vese, Fast cartoon + texture image filters. IEEE Trans. Image Process. **19**(8), 1978–1986 (2010)
19. P Maurel, J-F Aujol, G Peyre, Locally parallel texture modeling. SIAM J. Imaging Sci. **4**(1), 413–447 (2011)
20. S Chikkerur, A Cartwright, V Govindaraju, Fingerprint image enhancement using STFT analysis. Pattern Recognit. **40**(1), 198–211 (2007)
21. C Gottschlich, C-B Schönlieb, Oriented diffusion filtering for enhancing low-quality fingerprint images. IET Biometrics. **1**(2), 105–113 (2012)
22. S Ono, T Miyata, I Yamada, Cartoon-texture image decomposition using blockwise low-rank texture characterization. IEEE Trans. Image Process. **23**(3), 1128–1142 (2014)
23. H Schaeffer, S Osher, A low patch-rank interpretation of texture. SIAM J. Imaging Sci. **6**(1), 226–262 (2013)
24. JB Garnett, PT Jones, TM Le, LA Vese, Modeling oscillatory components with the homogeneous spaces $B\dot{M}O^{-\alpha}$ and $\dot{W}^{-\alpha,p}$. Pure Appl. Math. Q. **7**(2), 275–318 (2011)
25. DH Thai, C Gottschlich, Global variational method for fingerprint segmentation by three-part decomposition. IET Biometrics (to appear, doi:10.1049/iet-bmt.2015.0010)
26. I Bayram, ME Kamasak, in *Proc. EUSIPCO*. A directional total variation, (Bucharest, Romania, 2012), pp. 265–269. http://digital-library.theiet.org/content/journals/10.1049/iet-bmt.2015.0010
27. I Bayram, ME Kamasak, Directional total variation. IEEE Signal Process. Lett. **19**(12), 781–784 (2012). http://dx.doi.org/10.1109/LSP.2012.2220349
28. H Zhang, Y Wang, Edge adaptive directional total variation. J Eng, 1–2 (2013). http://digital-library.theiet.org/content/journals/10.1049/joe.2013.0116
29. R Courant, Variational methods for the solution of problems of equilibrium and vibrations. Bull. Amer. Math. Soc. **49**(1), 1–23 (1943)
30. S Boyd, N Parikh, E Chu, B Peleato, J Eckstein, Distributed optimization and statistical learning via the alternating direction method of multipliers. Foundations Trends Mach. Learn. **3**(1), 1–122 (2011)
31. M Fortin, R Glowinski, *Augmented Lagrangian Methods. Applications to the Numerical Solution of Boundary-value Problems*. (North-Holland Pub., Amsterdam, Netherlands, 1983)
32. C Gottschlich, P Mihăilescu, A Munk, Robust orientation field estimation and extrapolation using semilocal line sensors. IEEE Trans. Inform. Forensics Secur. **4**(4), 802–811 (2009)
33. A Sankaran, M Vatsa, R Singh, Latent fingerprint matching: a survey. IEEE Access. **2**, 982–1004 (2014)
34. M Jung, S Kang, Simultaneous cartoon and texture image restoration with higher-order regularization. SIAM J. Imaging Sci. **8**(1), 721–756 (2015)
35. D Tschumperlé, Fast anisotropic smoothing of multi-valued images using curvature-preserving PDE's. Int. J. Comput. Vis. **68**(1), 65–82 (2006)
36. J Weickert, *Anisotropic Diffusion in Image Processing*. (Teubner, Stuttgart, Germany, 1998)
37. I Galic, J Weickert, M Welk, A Bruhn, A Belyaev, H-P Seidel, Image compression with anisotropic diffusion. J. Math. Imaging Vis. **31**(2–3), 255–269 (2008)
38. C Schmaltz, P Peter, M Mainberger, F Ebel, J Weickert, A Bruhn, Understanding, optimising, and extending data compression with anisotropic diffusion. Int. J. Comput. Vis. **108**(3), 222–240 (2014)
39. M Mainberger, A Bruhn, J Weickert, S Forchhammer, Edge-based compression of cartoon-like images with homogeneous diffusion. Pattern Recognit. **44**(9), 1859–1873 (2011)
40. M Elad, M Aharon, Image denoising via sparse and redundant representations over learned dictionaries. IEEE Trans. Image Process. **15**(12), 3736–3745 (2006)
41. L Demaret, N Dyn, A Iske, Image compression by linear splines over adaptive triangulations. Signal Process. **86**(7), 1604–1616 (2006)
42. L Oehlmann, S Huckemann, C Gottschlich, in *Proc. IWBF*. Performance evaluation of fingerprint orientation field reconstruction methods, (Gjovik, Norway, 2015), pp. 1–6
43. C Gottschlich, S Huckemann, Separating the real from the synthetic: minutiae histograms as fingerprints of fingerprints. IET Biometrics. **3**(4), 291–301 (2014). http://dx.doi.org/10.1049/iet-bmt.2013.0065
44. T Hopper, C Brislawn, J Bradley, WSQ gray-scale fingerprint image compression specification. Technical report, Federal Bureau of Investigation (February 1993). http://www.nist.gov/itl/iad/ig/wsq.cfm

45. G Shao, Y Wu, A Yong, X Liu, T Guo, Fingerprint compression based on sparse representation. IEEE Trans. Image Process. **23**(2), 489–501 (2014)
46. D Salomon, *Data Compression*, Fourth edition. (Springer, London, UK, 2007)
47. D Gragnaniello, C Sansone, L Verdoliva, Iris liveness detection for mobile devices based on local descriptors. Pattern Recognit. Lett. **57**, 81–87 (2015)
48. C Gottschlich, Convolution comparison pattern: an efficient local image descriptor for fingerprint liveness detection. PLoS ONE. **11**(2), 0148552 (2016)
49. C Gottschlich, E Marasco, AY Yang, B Cukic, in *Proc. IJCB*. Fingerprint liveness detection based on histograms of invariant gradients (Clearwater, FL, USA, 2014), pp. 1–7
50. D Gragnaniello, G Poggi, C Sansone, L Verdoliva, An investigation of local descriptors for biometric spoofing detection. IEEE Trans. Inform. Forensics Secur. **10**(4), 849–863 (2015)
51. C Gottschlich, D Schuhmacher, The shortlist method for fast computation of the earth mover's distance and finding optimal solutions to transportation problems. PLoS ONE. **9**(10), 110214 (2014)
52. C Brauer, D Lorenz, in *Proc. SSVM*. Cartoon-texture-noise decomposition with transport norms (Lege-Cap Ferret, France, 2015), pp. 142–153
53. J Lellmann, D Lorenz, C-B Schönlieb, T Valkonen, Imaging with Kantorovich–Rubinstein discrepancy. SIAM J. Imaging Sci. **7**(4), 2833–2859 (2014)
54. (G Kutyniok, D Labate, eds.), *Shearlets. Multiscale Analysis for Multivariate Data*. (Birkhäuser, Boston, MA, USA, 2012)
55. MN Do, M Vetterli, The contourlet transform: an efficient directional multiresolution image representation. IEEE Trans. Image Process. **14**(12), 2091–2106 (2005)
56. M Unser, DVD Ville, Wavelet steerability and the higher-order Riesz transform. IEEE Trans. Image Process. **19**(3), 636–652 (2010)
57. DH Thai, S Huckemann, C Gottschlich, Filter design and performance evaluation for fingerprint image segmentation (2015). arXiv:1501.02113 [cs.CV]

8

Spatio-temporal consistent depth-image-based rendering using layered depth image and inpainting

Suryanarayana M. Muddala, Roger Olsson and Mårten Sjöström*

Abstract

Depth-image-based rendering (DIBR) is a commonly used method for synthesizing additional views using video-plus-depth (V+D) format. A critical issue with DIBR-based view synthesis is the lack of information behind foreground objects. This lack is manifested as disocclusions, holes, next to the foreground objects in rendered virtual views as a consequence of the virtual camera "seeing" behind the foreground object. The disocclusions are larger in the extrapolation case, i.e. the single camera case. Texture synthesis methods (inpainting methods) aim to fill these disocclusions by producing plausible texture content. However, virtual views inevitably exhibit both spatial and temporal inconsistencies at the filled disocclusion areas, depending on the scene content. In this paper, we propose a layered depth image (LDI) approach that improves the spatio-temporal consistency. In the process of LDI generation, depth information is used to classify the foreground and background in order to form a static scene sprite from a set of neighboring frames. Occlusions in the LDI are then identified and filled using inpainting, such that no disocclusions appear when the LDI data is rendered to a virtual view. In addition to the depth information, optical flow is computed to extract the stationary parts of the scene and to classify the occlusions in the inpainting process. Experimental results demonstrate that spatio-temporal inconsistencies are significantly reduced using the proposed method. Furthermore, subjective and objective qualities are improved compared to state-of-the-art reference methods.

Keywords: View synthesis, Depth-image-based rendering, Image inpainting, Texture synthesis, Hole filling, Disocclusions, Layered depth image, Temporal consistency

1 Introduction

Three-dimensional television (3DTV) and free viewpoint television (FTV) technologies are actively pursued in both research and compression standardization, which is supported by the development of enabling display technologies [1]. The 3DTV application aims to provide realistic depth impressions by presenting several views of a scene simultaneously from slightly different viewpoints. In comparison, the FTV application offers users the ability to look around the scene by generating new views (in 2D or 3D) from any arbitrary position [2]. Hence, both applications require the display of a large number of views, all from different viewpoints, in order to provide a smooth transition between the views and thereby an immersive visual experience.

However, capturing and transmission of the required number of views is not a feasible solution from a bandwidth point-of-view. Therefore, view synthesis is employed as an efficient solution to produce the necessary number of views at the receiver side using a smaller set of captured and transmitted views. Depth-image-based rendering (DIBR) methods generate these views from formats such as video-plus-depth (V+D) and multiview video-plus-depth (MVD) [3]. In this work, we define the input views that a DIBR method operates on as original views and output views it produces as the virtual (i.e. rendered) views. A virtual view is produced using DIBR by utilizing texture (the 2D color information) and a corresponding depth image. Using geometrical 3D warping, which relies on knowledge about scene depth and camera parameters,

*Correspondence: marten.sjostrom@miun.se
Department of Information and Communication Systems, Mid Sweden University, Holmgatan 10, 85170 Sundsvall, Sweden

pixels from the original view are projected into the virtual view's image plane. An illustration of the view rendering is shown in Fig. 1, where the V+D format is adopted.

Warped images, which are rendered using DIBR, exhibit several artifacts due to the inherent properties of warping as shown in Fig. 2a. DIBR artifacts are ghosting and uncovered areas (i.e. holes). Ghosting artifacts are blended texture values caused by objects that are neighboring in the original view but becomes overlapped in the virtual view due to the depth and texture misalignments. A typical cause of this misalignment is the noise introduced in the depth map around object borders, frequently occurring in depth from stereo matching. Ghosting artifacts also cause inconsistent hole filling. Holes are defined as pixels in the warped image that receives no texture values from the forward warping process (see Fig. 2a). Holes can be classified into three groups: *cracks, disocclusions* and *out-of-field areas.* Cracks are holes of size one to two pixel wide caused by rounding the floating point position of a warped pixel to make it fit the pixel grid of the virtual view image plane. Disocclusions and out-of-field areas are holes of larger size caused by the depth discontinuities between objects within the scene and the limited field of view of the camera. The size of the holes increases as the distance between the original and warped camera view position increases. Dealing with disocclusions and out-of-field areas in a visually plausible manner is a much greater challenge. In the next section, we will elaborate on how we address these problems.

In order to solve the problems caused by the holes, we use the following terminology. We define foreground as the part of the scene closer to the camera that occludes other objects and background to be the part of the scene farther from the camera partially occluded by the foreground. Disocclusions are occluded regions in the original view that are revealed in the virtual view as a result of warping. Disocclusions occur at the borders of the object depending on the relative position translation between the original and virtual view camera. Out-of-field areas are caused by the limited field of view in the original view and they occur at the image boundaries of the virtual view. Moreover, out-of-field areas usually occur in the linear camera set-ups, whereas in converging camera set-up may not. The size of the holes (disocclusions and out-of-field areas) increases when the size of the of the inter-camera distance (i.e. the baseline) increases.

1.1 Related work

The literature proposes several pre-processing and post-processing methods to overcome DIBR artifacts. Pre-processing methods smooth the depth map prior to warping using, e.g. symmetric or asymmetric Gaussian filter [4, 5]. However, pre-processing of the depth map may imply geometrical distortions in the warped

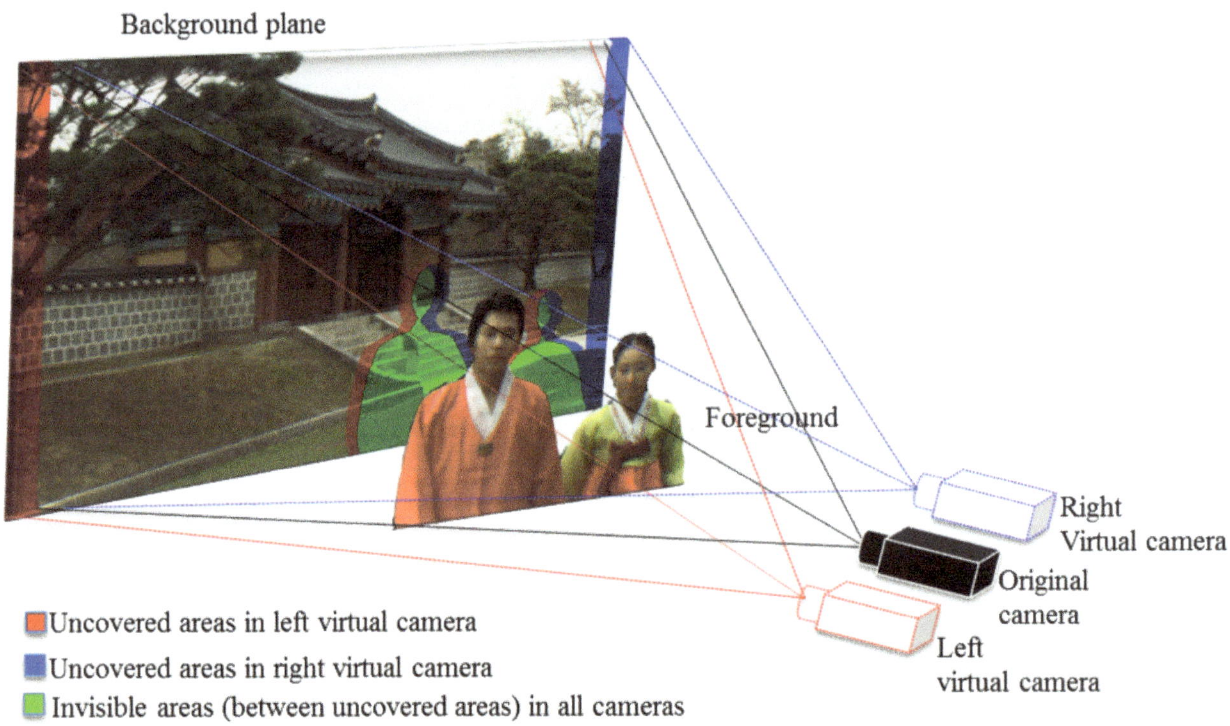

Fig. 1 Illustration of DIBR view extrapolation from V+D. Invisible uncovered areas in original camera are visible in the virtual cameras, so they need to be filled

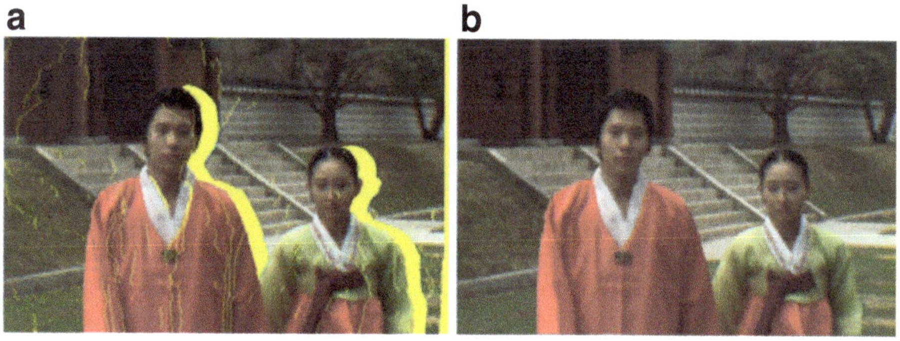

Fig. 2 View synthesis results. **a** Synthesized image (holes in *yellow color*). **b** Proposed synthesized image

image. However, post-processing techniques add complicated steps after warping in the warped image [6–9]. Image-domain warping methods are another way of reconstructing the virtual views using sparse disparity information [10, 11]. A number of methods combine both pre-processing and post-processing techniques to address rendering problems [8].

Several hole-filling methods have been proposed in the literature. In the context of virtual view synthesis from a single V+D, the hole-filling methods can be generally classified into two groups. In the first group, the data required to fill the hole is synthesized from the image itself. We term this group spatial domain hole-filling method. In the second group, the data required to fill the hole is recovered from both the spatial and temporal domain, so we term them temporal domain hole-filling methods.

Spatial domain hole filling: Spatial domain hole-filling methods can be classified into interpolation, extrapolation and image inpainting. Interpolation and extrapolation methods use the texture of neighbor pixels around the holes to fill missing areas in the warped images [12]. However, for large holes, the results yield annoying artifacts and unnatural textures. In contrast, inpainting methods use structural and textural properties of the image in the neighborhood of the hole to fill missing areas [13, 14]. The view synthesis reference software (VSRS) is used in the standardization activities of 3DTV and FTV and combines various techniques to produce virtual views [15]. For example, VSRS applies a diffusion-based inpainting method to fill any holes. In general, diffusion methods suffer from the introduction of blurred results in the virtual image, more so when the hole is large. In addition, VSRS suffers from limitations in how foreground-background is identified during inpainting, which results in holes being partially filled with foreground textures, and not only background as appropriate. To reduce this problem, Oh et al. [16] replaced the foreground pixels on the boundary with background values.

To overcome the limitations of these diffusion methods, exemplar methods combine both structural and textural properties when filling in missing areas [17]. In exemplar methods, a filling order is first computed on the hole boundary using the number of known pixels in a patch (confidence term) and structure in the neighborhood (data term). Next, a highest priority patch (target patch) is selected to be filled using the best patch (source patch) from the neighborhood using patch matching. Although the exemplar method has the advantage of recovering the structure, this priority order is not suitable for hole filling in the view synthesis context, since holes should be filled only from the background texture. Several methods have been proposed by improving filling priorities and adding various foreground-background classification techniques in the exemplar methods including the methods in [18–23]. Despite improvements, these methods still have spatial texture inconsistencies and moreover, they severely suffer from temporal inconsistencies (flickering).

Another approach to filling missing areas is to generate the hole information in the original view before warping, instead of filling the holes after DIBR. This approach is similar to methods working with the layered depth video (LDV) format. LDV contains occlusion information along with V+D and share many properties with the layered depth image (LDI) [24]. The method in [25] has been presented to generate this occlusion data from a stereo pair. Several ways to produce the necessary occlusion data from different camera set-ups are presented in [26]. However, an explicit way to produce LDI data from a single V+D is not presented. Recently, a method has been presented in [27] to generate LDI from V+D. Although this method has improved spatial consistency, it still suffers from flickering in the temporal domain.

Temporal domain hole filling: Video inpainting has been an active research field in computer vision for the past decade with increasing video editing applications. It makes sense to use the temporal true texture instead of the synthesized texture from the inpainting. Chen et al. proposed a method to recover the disocclusion information from temporal frames by using the motion vectors of H.264/AVC bit stream [28]. However, since motion vectors are sparse and do not correspond to the real motion, quality gains are limited. In other attempts, background sprites are generated by using the temporal frame information, after which disocclusions are filled and updated using the true texture and finally, inpainting is applied to fill holes in virtual views [29–32]. The method provided by Schmeing et al. requires a spatial hole filling for disocclusions caused by static foregrounds [30]. Regardless, reusing the synthesized background from the sprite and inpainting the remaining holes at the moving foreground causes spatial inconsistencies. Another sprite-based hole-filling method presented in [33–35] extracts the static background information using depth classification by probability analysis, a Gaussian mixed model and structural similarity index. Yet, these methods are prone to exhibit spatio-temporal inconsistencies for moving foreground objects. Hsu et al. proposed a method to construct the disocclusions using global optimization [36]. However, the results show spatial inconsistencies at the structured background due to the inpainting method used. Choi et al. used a method where the disocclusions are filled by finding the best patches in the past frames along with current frames [37]. This method yields spatially acceptable results but suffers from flickering artifacts particularly when it comes to the static foreground objects.

Although the above-mentioned temporal hole-filling methods show improvements and new advances, they still suffer from spatial and temporal inconsistencies. These inconsistencies consequently deteriorate the visual quality of the virtual view.

In this paper, we propose a spatio-temporal consistent view synthesis method for extrapolating virtual views. The proposed method produces a layered depth image by identifying the occlusions and inpainting them in the original view using temporal information and motion estimation. When the layered depth image is rendered to the virtual view, no disocclusions appear and views are consistent in the spatial and temporal domain.

This paper is organized as follows: Section 2 explains the proposed spatio-temporal consistent view synthesis with inpainting. The experimental set-up and the evaluation criteria are described in Section 3. Results and analysis of the proposed method are presented in Section 4. Finally, in Section 5, we discuss challenges and conclude the work.

2 Proposed view synthesis method

Temporal inconsistencies are the result of filling holes in a frame irrespective of temporal information. To reduce the temporal inconsistencies, we construct a static scene sprite over time and fill the holes by using a depth-based inpainting method. The sprite is formed such that the occlusion information in the original view is generated in a spatially and temporally consistent manner. Using the occlusion-free sprite, a LDI is generated such that no disocclusions will appear after warping.

In the sprite construction process, we have used a set of future frames, motion estimation, occlusion region, occlusion depth threshold and foreground-background classification. Future frames not only fetch the true information of the holes but also reduce temporal inconsistencies by adding a smoothing constraint. At each time instant t, the sprite makes use of information in all previous frames as well as future frames up to time instant $t + B_l$. The buffer length B_l depends on the applications and must be selected accordingly. Initially, the sprite is constructed by considering the information in the first B_l frames by employing the process described below in this section. Then, the sprite is updated with new information of frame $(t + B_l)$ at each iteration. That is, the sprite always contains information from the B_l future frames. After filling the sprite with this information, there are still holes to be filled. These holes are filled by employing the inpainting procedure described in Section 2.4. The inpainting procedure is rather time-consuming and is therefore carried out every B_l frame. Both for information in a new frame and for inpainted data, the process gradually transforms the existing information with new information such that it is fully replaced at time $t + B_l$. The identified occlusion, occlusion depth threshold and foreground-background classification help in finding the background information to fill the holes from neighboring frames and guide the inpainting process to fill from background. As the global foreground-background classification is not consistent over time, motion estimation helps to construct the static scene sprite. Next, the holes in the sprite are labeled using motion estimation such that holes that share information in the temporal domain are filled with previous inpainted information and the remaining holes are inpainted over a set of frames. This step not only guarantees the temporal consistency but also improves the spatial consistency at moving foregrounds.

The proposed view synthesis method using LDI is shown in Fig. 3. The process of the method is as follows: first, a spatio-temporal consistent LDI is generated. Next, the LDI is warped and combined with the warped image. The final step consists of hole-filling post-processing. The details of each part are presented in the following sections. The LDI generates the occlusion layers in the original view, where the occlusions and depth thresholds from V+D are identified using the work of [27].

2.1 Identification of occlusion and depth threshold

The LDI generation consists of identifying occlusion layers and inpainting them in the original view before warping. Given the properties of DIBR, occlusions belong to the background and are identified using the depth values and the projected locations of pixels in the virtual view.

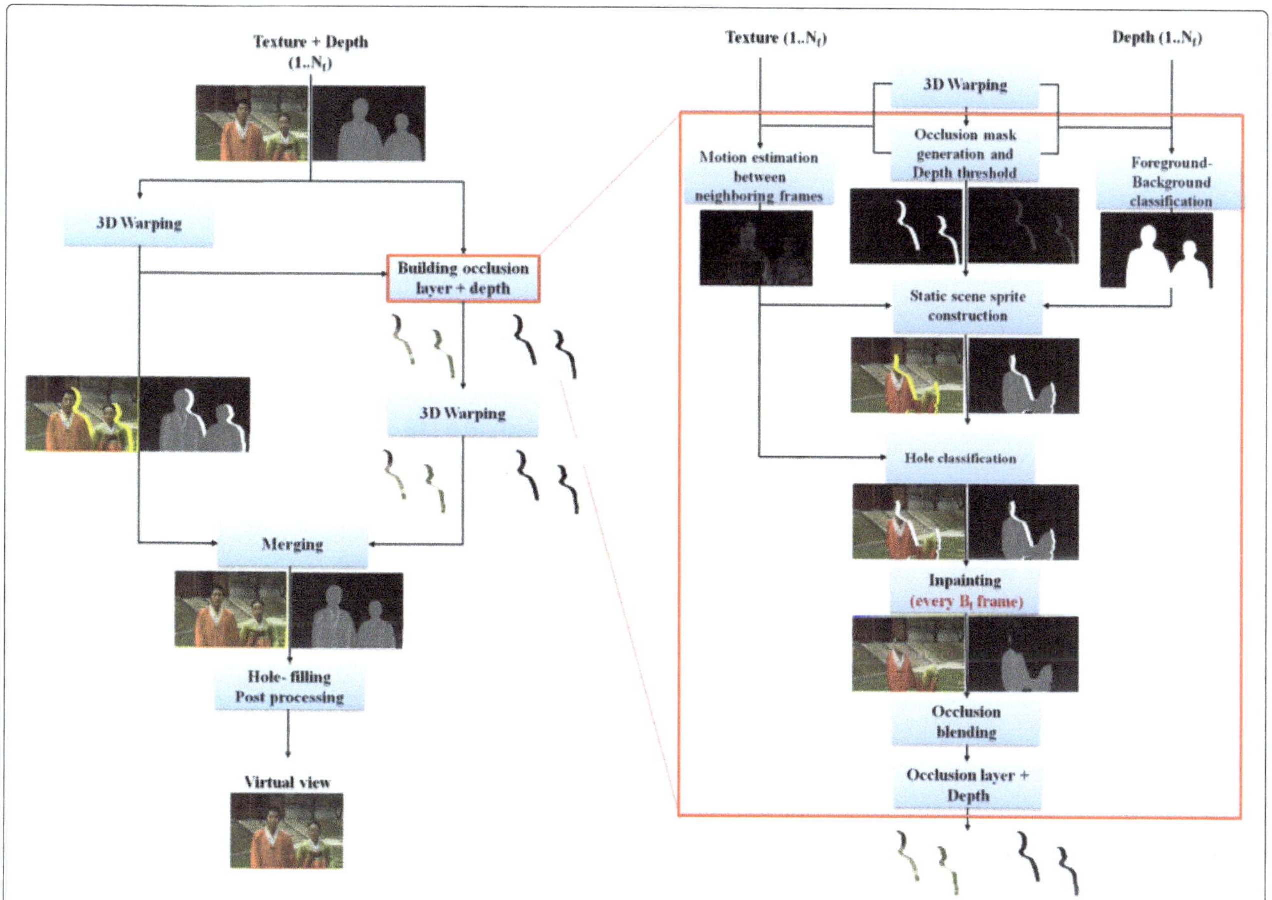

Fig. 3 Proposed view synthesis method. Spatio-temporal consistent view snythesis using LDI and inpainting

The occlusions and the corresponding depth thresholds are identified as follows:

2.1.1 Depth discontinuity pixel identification

Disocclusions occur between pixels that are neighboring in the original image, have a distinct difference in depth, and that become separated in the warped image. For example, the occlusion behind a foreground object in the original image becomes visible as a disocclusion in the virtual image. A disocclusion appears to the right of the foreground object (see Fig. 4), when the virtual image is rendered from a camera positioned to the right of the original camera.

Let Λ be the warping operation that projects the original view image I_o to the virtual view image I_v. We define a depth discontinuity pixel (DDP) to be a neighboring pixel pair $\{\mathbf{f}_o, \mathbf{p}_o\}$ positioned in the original view as shown in Fig. 4 that satisfy the following conditions:

$$\text{DDP} = \{\mathbf{f}_o, \mathbf{p}_o\}\,;\, \|\mathbf{d}_v\|_2 > \eta, \tag{1}$$

where η is an occlusion threshold, and $\mathbf{d}_v$ is a displacement vector given by

$$\mathbf{d}_v = \begin{pmatrix} d_x \\ d_y \end{pmatrix} = \mathbf{f}_v - \mathbf{p}_v, \tag{2}$$

where $\mathbf{f}_v, \mathbf{p}_v$ are projected pixels in I_v:

$$\begin{aligned} \Lambda &: \mathbf{f}_o \rightarrow \mathbf{f}_v, \\ \Lambda &: \mathbf{p}_o \rightarrow \mathbf{p}_v. \end{aligned} \tag{3}$$

The above definition of DDP is presented within the context of a right warped view. It is straightforward to change Eq. (1) for a left warped image. The summary of the notation used in this paper is presented in Table 1.

In the next step, the pixels in the DDP are labeled foreground and background using their respective depth values. When there are multiple depth layers, DDPs might overlap and produce inconsistent foreground-background labeling, such that a pixel-labeled background in one DDP might be labeled foreground in another DDP. To circumvent inconsistent foreground-background labeling, we apply a simple labeling refinement step that selects either background or foreground based on the position difference of the projected pixels, that is, the pixel in DDP

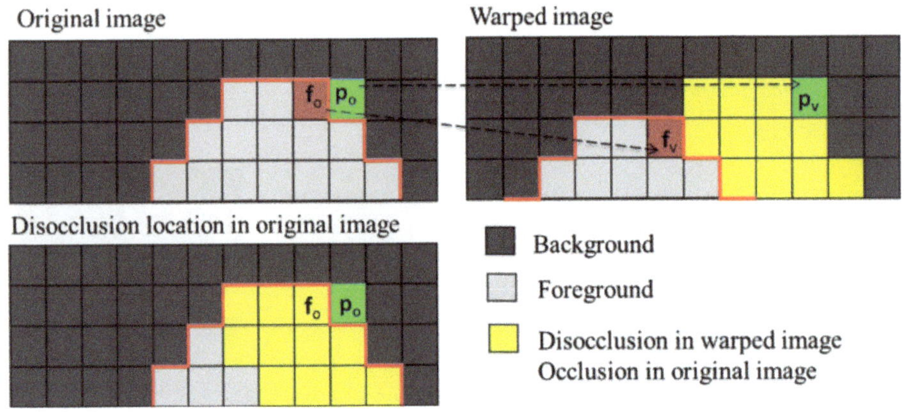

Fig. 4 Identification of occlusion location

is labeled as background when its projected pixels difference is less than its neighbor pixel in the overlapped DDP.

2.1.2 *Occlusion identification*

After identifying and labelling the DDPs, occlusions are located in the original image by horizontal and vertical displacements and their corresponding depth values. The occlusions at the DDP $\{\mathbf{p}_o, \mathbf{f}_o\}$ are identified as the following set:

$$\Omega_{O(DDP)} = \{\mathbf{p}; \mathbf{p} \in l \wedge (Z(\mathbf{p}) > \xi)\} \quad (4)$$

where l is a line joining pixels $\mathbf{p}_o$ and $\mathbf{r}_o = \mathbf{p}_o - \mathbf{d}_v$, and ξ is an occlusion depth threshold, given by the arithmetic mean over the DDP depths:

$$\xi = \frac{Z(\mathbf{f}_o) + Z(\mathbf{p}_o)}{2}, \quad (5)$$

The full occlusion in the image consists of all such pixels:

$$\Omega_O = \bigcup_{\text{all DDP}} \Omega_{O(DDP)} \quad (6)$$

An example of an occlusion is shown in Fig. 5d.

2.1.3 *Occlusion depth threshold*

Identified occlusions belong to the background and should be filled with consistent background information.

Table 1 Notation

Variable	Description
f,p,q	Pixels in an image
I	Texture image
Z	Depth map
Ω	Hole region
$\delta\Omega$	Hole boundary
Φ	Source region $(I - \Omega)$
Ψ	2D patch

Hence, two depth threshold images are created, Z_A and Z_B, which contain the average depth, and background depth values of the DDPs at occlusion areas, respectively:

$$\left.\begin{aligned} Z_A(\mathbf{q}) &= \xi, \\ Z_B(\mathbf{q}) &= Z(\mathbf{p}_o), \end{aligned}\right\} \quad \forall \mathbf{q} \in l, \quad (7)$$

where l is a line defined in Eq. (4). Note that pixels on different lines l are calculated from different DDPs. Examples of Z_A and Z_B are shown in Fig. 5e, f.

2.2 Static scene sprite construction

The static scene sprite construction step builds a global sprite during the view synthesis process by collecting texture and depth information of the static scene from temporal frames. The global sprite consists of both the texture I_S and depth Z_S. The construction process continues for a set of frames. The static information from the set of frames is extracted and stored in the constructed sprites I_S and Z_S. Example of texture sprite at different time instants is shown in Fig. 6. The remaining holes in the sprites are filled using the inpainting process.

The global sprite formation is guided by the following parameter specific for the sprite: The occlusions Ω_{OS} indicate the holes, the occlusion threshold Z_{AS} indicates allowable depth values to fill the holes, the global foreground-background classification Ω_{FS} identifies the absolute foreground-background and finally the motion pixels Ω_{MS} indicate the moving objects.

We use a simple depth map segmentation on the whole image to identify the background in order to form the sprite with the background information [38]. The Otsu method selects an optimal threshold by maximizing the class variance and it works well when the entire distribution is segmented into two classes. We denote the identified foreground pixels as Ω_F.

We want all static information to be part of the sprite. Therefore, motion is estimated to detect the moving

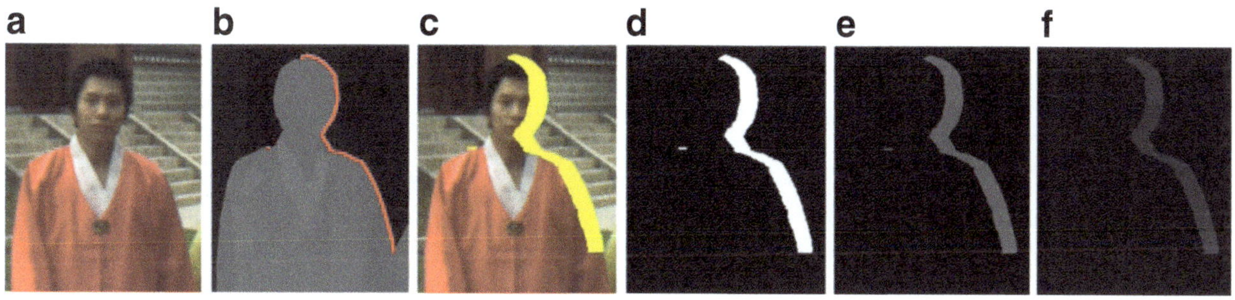

Fig. 5 Occlusion layer identification and depth threshold generation. **a** Original image. **b** Depth discontinuity pixels (DDPs) (in *red color*). **c** Occlusion in original image (in *yellow color*). **d** Occlusion region. **e** Average depth threshold image. **f** Background depth threshold image. Note that the depth values in the threshold images are scaled for illustration purpose

objects in the scene. The optical flow describes the apparent motion of objects, surfaces and edges between images, so we use the Lukas-Kanade algorithm, a simple and efficient method [39]. The motion pixels are given by

$$\Omega_M - \{\mathbf{p}; |\mathrm{V}(\mathbf{p})| > \rho\} \tag{8}$$

where $|\mathrm{V}(\mathbf{p})|$ is the absolute optical flow value at $\mathbf{p}$. ρ is the optical flow threshold, which allows to change the noise sensitivity of the optical flow calculations.

The static sprite is formed in three steps: (i) Hole regions in the sprite are filled with the future frame using the occlusion depth threshold Z_A. (ii) The foreground and moving information in the sprite are replaced by background, static foreground and motion information. (iii) Finally, the holes in the upcoming frame are located in the sprite using the depth threshold Z_A. In the sprite construction process of the static scene, the sprite is initialized with information from the first frame ($t = 0$).

$$\begin{aligned} I_S &:= I^0, \\ Z_S &:= Z^0, \\ Z_{AS} &:= Z_A^0, \\ \Omega_{OS} &:= \Omega_O^0, \\ \Omega_{MS} &:= \Omega_M^0, \\ \Omega_{FS} &:= \Omega_F^0. \end{aligned} \tag{9}$$

Once the sprite is initialized, the occlusions should be filled with new texture as it becomes available. Firstly,

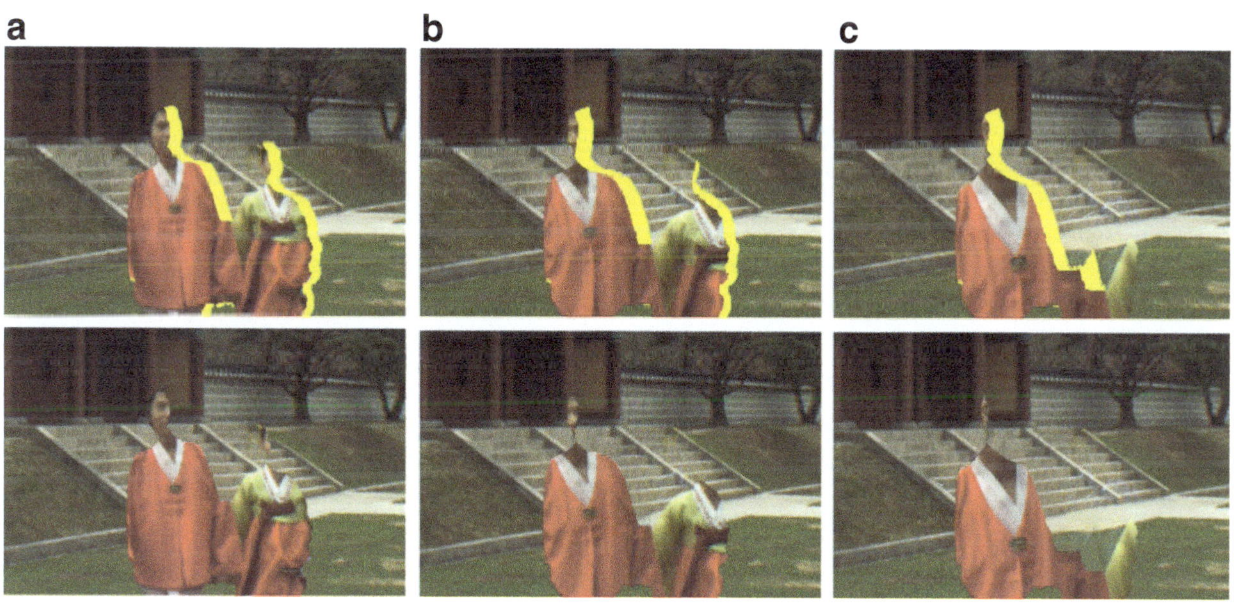

Fig. 6 Static scene sprites at different time instants. A static scene sprite is initialized with the first frame and updated with background information from subsequent frames at holes and foreground regions. Sprites after locating holes are in first row and sprites after inpainting are in second row; **a** at frame 6; **b** at frame 46; **c** at frame 96

the holes are filled with static background from the new frame:

$$\left.\begin{array}{l} I_S(\mathbf{p}) := I^t(\mathbf{p}) \\ Z_S(\mathbf{p}) := Z^t(\mathbf{p}) \end{array}\right\} \forall \mathbf{p} \in \Omega_S \backslash \Omega_O^t; Z^t(\mathbf{p}) < Z_{AS}(\mathbf{p}) \tag{10}$$

The set of occlusion pixels in the sprite Ω_{OS} are then updated by removing the pixels considered in Eq. (10). Next, the moving object pixels and foreground pixels in the sprite are replaced by information in the new frame:

$$\left.\begin{array}{l} I_S(\mathbf{p}) := I^t(\mathbf{p}) \\ Z_S(\mathbf{p}) := Z^t(\mathbf{p}) \end{array}\right\} \begin{array}{l} \forall \mathbf{p} \in \Omega_{MS} \cup \Omega_{FS}; \\ Z^t(\mathbf{p}) < Z_S(\mathbf{p}) - \vartheta, \end{array} \tag{11}$$

where ϑ is depth tolerance, which is set according to the variations of the scene depth. The set of motion and foreground pixels in the sprite, Ω_{MS} and Ω_{FS}, are then updated with the corresponding pixel information in the new frame.

Finally, the average depth threshold in the sprite Z_{AS} is updated with the corresponding information of holes in the new frame Ω^t:

$$Z_{AS}(\mathbf{p}) = \frac{Z_S(\mathbf{p}) + Z_A^t(\mathbf{p})}{2}, \forall \mathbf{p} \in \Omega_O^t; Z_A^t(\mathbf{p}) > Z_S(\mathbf{p}), \tag{12}$$

whereby the set of occlusion pixels in the sprite, Ω_{OS} is updated with the corresponding pixel information in the new frame.

An example of texture sprite at different time instants is shown in Fig. 6. The remaining holes in the sprites are filled using the inpainting process.

2.3 Hole classification

Holes in each frame are classified as static and dynamic by using the motion of the foreground objects (see Fig. 7e). Holes which appear at the moving foreground objects are labeled dynamic holes. These holes are filled with either true texture from consequent frames or by the inpainting process. Holes which appear at static foreground objects are labeled static holes. Using this classification, static holes are filled only once in the entire view synthesis process and are placed in the sprite to be reused when needed. In contrast, dynamic holes are inpainted over a set of frames in order for the inpainted texture to be spatially and temporally consistent with the background. As not all motion pixels belong to the foreground object that causes holes, only the pixels that belong to the dynamic holes are given by

$$\Omega_D = \left\{\mathbf{p}; \mathbf{p} \in \Omega_{O(n)} \wedge (\nu_n > \rho)\right\} \tag{13}$$

$$\nu_n = \frac{1}{\Gamma} \sum_{\mathbf{p} \in E_n} \max\left(\left|V_{\mathbf{p}}\right|\right), \tag{14}$$

where $\Omega_{O(n)}$ is nth hole in a occlusion region. E_n is a set that contains the DDPs corresponding to the nth hole. ν_n is the average of maximum optical values in the patch centered at DDPs on the Ω_n in a frame (see Fig. 7b). Γ is the number of DDPs in E_n.

Once the hole classification for the frames in the buffer are obtained, then the dynamic holes in the sprites are given by $\Omega_{DS} = \Omega_{OS} \cap \Omega_D$.

2.4 Inpainting

The holes remaining in the sprite, after compensating the true texture from the temporal information, are divided into small and large holes. This is because small holes will have no inpainting priority due to the foreground presence within a patch. As a result, small holes will not be filled until after the large holes have been filled. This implies inconsistent selection and filling of large holes when they lie within a search region of a given hole to find a best match. Therefore, small holes with a size smaller than a patch area are filled using the interpolation method. Then inpainting is applied on the large holes, where the inpainting method is adopted from [27], which is derived from [17]. Note that the inpainting method is applied on the sprite instead of each frame. The steps of the inpainting process are as follows:

1. Identify the background boundary of the holes using the hole depth threshold Z_{AS}.

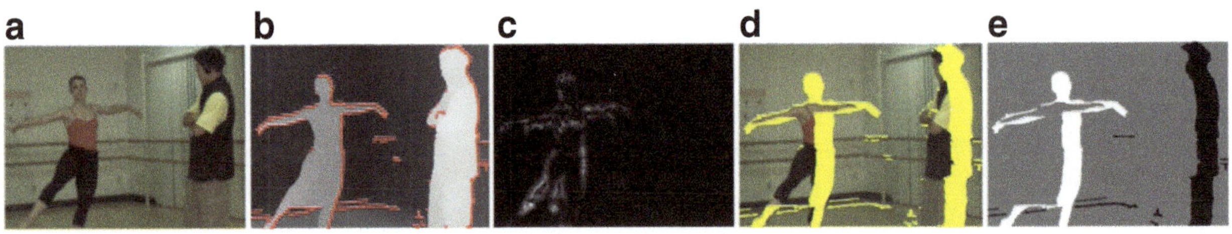

Fig. 7 Hole classification using optical flow. **a** Texture. **b** Depth image DDPs (in *red color*). **c** Optical flow image (*brighter region* indicates motion and *black regions* indicate no motion). **d** Occlusions in texture (in *yellow color*). **e** Hole classification region (*white areas* indicate dynamic holes and *black* indicates static holes)

2. Compute the priority on the background boundary using the curvature data term and depth-based confidence term and select a patch with the highest priority.
3. Find a good match to the target patch in both texture and depth. Fill holes in the target patch with a weighted combination of the best patches.
4. Update the terms and repeat the above steps until the holes in the sprite are filled.

In the next sections, we will give further details on the inpainting steps.

2.4.1 Foreground-background boundary extraction

The background boundary of the hole is identified by classifying the hole boundary into the foreground and background using the depth threshold image Z_{AS}. The classification proceeds as follows: Initially, the hole boundary $\delta\Omega$ is obtained by applying a Laplacian convolution on the hole region. Next, the hole boundary is classified into the foreground boundary $\delta\Omega_F$ and the background boundary $\delta\Omega_B$, i.e. $\delta\Omega = \delta\Omega_F \cup \delta\Omega_B$.

Let us begin by defining the maximal value of the occlusion threshold on the 2D patch at point $\mathbf{p}$, $\Psi_\mathbf{p}$, coinciding with the source region Φ:

$$\mu = \max\left(Z_{AS}(\mathbf{q})\,|_{\mathbf{q}\in(\Psi_\mathbf{p}\cap\Omega)}\right) \tag{15}$$

$$\delta\Omega_F = \delta\Omega \;\; \forall\mathbf{p} : Z_S(\mathbf{p}) > \mu, \tag{16}$$

$$\delta\Omega_B = \delta\Omega - \delta\Omega_F, \tag{17}$$

2.4.2 Priority

The priority term on the background boundary is the product of confidence and data terms:

$$P(\mathbf{p}) = C_z(\mathbf{p}) \cdot D_k(\mathbf{p}), \tag{18}$$

where $C_z(\mathbf{p})$ is the depth-based confidence term, and $D_k(\mathbf{p})$ is the curvature data term at $\mathbf{p}$.

Depth-based confidence term computes the percentage of background pixels in a patch.

$$C_z(\mathbf{p}) = \frac{1}{|\Psi_\mathbf{p}|}\sum_{\mathbf{q}\in\Psi_\mathbf{p}\cap\Phi} C(\mathbf{q}), \tag{19}$$

$$C(\mathbf{q}) = \begin{cases} 1, & \mathbf{q}\in\Phi_B; \\ 0, & \text{otherwise.}\end{cases} \tag{20}$$

where Φ_B is the background source region, which is given by

$$\Phi_B = \{\mathbf{p}; \mathbf{p}\in\Phi \wedge (Z_S(\mathbf{p}) < \mu)\} \tag{21}$$

Curvature data term finds the structure details in the neighborhood of the patch using the curvature, where it consists of both geometry and strength of the isophote. The data term gives the priority to the patches that consists of linear structures, which is given as

$$D_k(\mathbf{p}) = 1 - |k_\mathbf{p}|. \tag{22}$$

$$k_\mathbf{p} = \nabla\cdot\left(\frac{\nabla I_\mathbf{p}}{|\nabla I_\mathbf{p}|}\right), \tag{23}$$

where $k_\mathbf{p}$ is the curvature of the isophote through a pixel $\mathbf{p}$, $\nabla\cdot$ is the divergence at $\mathbf{p}$ that gives a total isophote flow from $\mathbf{p}$.

2.4.3 Patch matching and filling

To avoid the patch selection from the foreground during the inpainting process, the source region is classified into the foreground-background locally. The threshold for the classification is selected from the previously generated threshold image Z_{AS}. Similar to [23], the block matching is performed with only background pixels in the target patch by removing foreground pixels in order to avoid artifacts at foreground objects.

$$\Phi_F = \Phi - \Phi_B, \tag{24}$$

where Φ is the source region and subscripts F and B denote the foreground and background, respectively. Φ_F is the foreground source region, whose depth values are larger than the depth threshold μ. Thus, the block matching equation becomes

$$\Psi_\mathbf{q} = \arg\min_{\Psi_\mathbf{q}\in\Phi_B}\left\{\left\|\Psi_{\hat{\mathbf{p}}} - \Psi_\mathbf{q}\right\|_2^2 + \beta\cdot\left\|Z_{\hat{\mathbf{p}}} - Z_\mathbf{q}\right\|_2^2\right)\right\}, \tag{25}$$

where $\Psi_{\hat{\mathbf{p}}}$ and $Z_{\hat{\mathbf{p}}}$ are the target texture and depth patches, and $\Psi_{\hat{\mathbf{q}}}$ is the source patch. $\left\|\Psi_{\hat{\mathbf{p}}} - \Psi_\mathbf{q}\right\|_2^2$ is the sum of squared difference between known pixels in two patches and β is a weighting coefficient to equalize the effect of the depth and texture. Similar to [22], holes in the depth map are filled by copying the depth values from the best patch location. Holes in the texture image are filled with a weighted average of N patches of texture values.

2.5 Moving weighted averaging and blending

Although the static scene construction process recovers the true texture from the temporal information of the set of future frames, the scene and holes could change for each buffer. Thus, sudden changes could occur in the construction of the sprite. Sudden changes (flickering) appear in the texture content in the following cases: (i) available true texture in the current frame for the holes that are inpainted in the previous frame; (ii) holes in the current frame and previous frame are inpainted separately. To avoid flickering that occurs at the occlusion areas, a moving weighted average blending is introduced. This creates a smooth transition over frames at hole regions. The blending of textures is done as follows:

$$\mathrm{I}_{WS}^{t+i}(\mathbf{p}) = (1-\omega_i)\mathrm{I}_S^t(\mathbf{p}) + \omega_i \mathrm{I}^{t+B_l}(\mathbf{p}), \forall \mathbf{p} \in \Omega_{OS}^t \cap \Phi^{t+B_l} \quad (26)$$

$$\omega_i = \frac{t \mod B_l}{B_l} \quad (27)$$

where, $\mathrm{I}_S^t(\mathbf{p})$, $\mathrm{I}^{t+B_l}(\mathbf{p})$ are inpainted texture in the sprite at frame t and true texture in the frame $t + B_l$ at pixel $\mathbf{p}$. $\mathrm{I}_{WS}^t(\mathbf{p})$ is a blended texture image. Φ^{t+B_l} is the source region in frame I^{t+B_l} and Ω_{OS}^t is the hole region before filling the sprite at frame t. ω_i is the weight that linearly combines future frames, and i is an index number within the set.

By using the sprite, a spatially and temporally consistent layered depth image is generated, which is later used to recover the disocclusions after warping.

2.6 Warping layered depth image and merging

To avoid artifacts after projection, layers are warped using background depth values of occlusion areas, which are stored in Z_B. Then, the warped occlusion layers are merged with the warped image using Z-buffering (the pixel closest to the camera is selected when two pixels are projected at the same location).

2.7 Crack removal and hole-filling post-processing

By using the LDI from the sprite, spatial and temporal consistency is achieved at the disocclusions in the virtual view. However, certain types of holes could still be present after a virtual view is produced by warping, making hole-filling post-processing necessary. These holes are a result of the forward warping and are manifested as cracks and out-of-field areas. Typically, depth maps have significantly varying depth values surrounding the occlusion area, which will exhibit minor depth discontinuities in the filled area. These discontinuities are also manifested as disocclusions with an extent that is slightly larger than cracks but significantly smaller than conventional disocclusions. These artifacts are removed by using hole-filling post-processing.

Similar to the hole classification used in the sprite filling, small holes are filled by using the interpolation method. The remaining holes are filled by using the proposed inpainting method, operating without depth classification. This is because the threshold for classification is not identified, since the foreground is not available during the occlusion identification step. Out-of-field areas occur at image borders and are static and so they are filled only once and then reused for the subsequent frames in such a way that temporal consistency is achieved.

3 Experimental set-up

We use four MVD test sequences to evaluate the performance of the proposed view synthesis method. The test sequences are "Ballet" [40], "Book Arrival" [41], "Lovebird1" [42] and "Poznan Street" [43], which cover various texture and depth characteristics (see Table 2) to validate our method in different conditions. The hole size in the table is computed as follows: first, the percentage of hole pixels is calculated for each warped image; then the average of the calculated percentage is computed over the sequence. Note that only the static camera test sequence is used, since the proposed method is limited to static scenes. MVD sequences are used in the evaluation, for the purpose of measuring the quality of the virtual view using full reference objective metrics. We rendered extrapolated virtual views using V+D at the available real camera views, so that rendered views match the ground truth textures in the MVD, as a pre-requisite for measuring full reference quality evaluations.

In the quality evaluations of our proposed method, we use both objective measurements and visual inspection. As the proposed method is aimed at the spatial and temporal consistency of the virtual views, we divide the quality evaluation set-up into two parts; spatial consistency evaluation and temporal consistency evaluation.

1. Spatial consistency evaluation
2. Temporal consistency evaluation

Table 2 Test input data characteristics

Sequence name	Resolution	Frames	Camera arrangement	Texture background	Depth properties	Hole size (Average)
Ballet	1024×768	1–100	8 cameras with 30-cm spacing, 1D arc arrangement	Low structured	Large depth discontinuities and many layers	13.5 %
Book Arrival	1024×768	1–100	16 cameras with 6.4-cm spacing, 1D arrangement	Low structured	Small depth discontinuities and many layers	6.4 %
Lovebird1	1024×768	50–200	12 cameras with 3.5-cm spacing, 1D arrangement	Structured	Large depth discontinuities and few layers	4.3 %
Poznan Street	1920×1088	1–100	9 cameras with 13.5-cm spacing, 1D arrangement	Structured	Small depth discontinuities and many layers	5 %

3.1 Spatial consistency evaluation

Inpainting is an ill-posed problem, and so investigating the inpainting quality is as complex as the disocclusion problem. Hence, we used the commonly used evaluation methodology in the view synthesis context. We use full reference objective metrics peak signal-to-noise ratio of the luminance component (YPSNR) and mean structural similarity index (MSSIM) to evaluate the spatial consistency of virtual views. PSNR measures the absolute difference between images, whereas MSSIM measures structural similarities between images and closely corresponds to the perceptual visual quality [44]. In line with commonly used evaluation methodology of view synthesis, both metrics are applied on the full image.

3.2 Temporal consistency evaluation

Temporal consistency of the virtual view is evaluated by measuring the amount of flickering. We adopt the method in [45] where the amount of flickering is computed as

$$\tau_D^t = \frac{1}{\left|\Omega^t\right|} \sum_{\mathbf{p}\in\Omega^t} \left|\mathrm{I}^t(\mathbf{p}) - \mathrm{I}^{t-1}(\mathbf{p})\right|, \forall t = 2 \ldots N_f \quad (28)$$

where τ_D^t is the inter-frame average absolute difference at holes in frame t, N_f is the number of frames, $\left|\Omega^t\right|$ is the number of hole pixels in the frame t and the flicker measured between the frames is given as

$$\text{F-score} = \left|\tau_{\mathrm{Dg}}^t - \tau_{\mathrm{Dv}}^t\right| \quad (29)$$

where τ_{Dg}^t and τ_{Dv}^t represent the inter-frame average absolute difference at holes of the ground truth and the virtual view at frame t, respectively. A lower F-score indicates less flickering. Note that flickering is only measured at the inpainted holes.

In addition to the objective spatial and temporal consistency measurements, synthesized views are also presented to allow qualitative visual comparison.

3.3 Complexity evaluation

The proposed method is implemented in MATLAB without any optimization, which is not the case for reference methods which are implemented in different platforms. However, few insights about the processing times are given. Note that the main objective is to improve the quality of the rendered view.

We use five of the state-of-the-art methods for comparison purposes. The comparison is carried out with various view synthesis methods that use diffusion and exemplar inpainting to fill holes in spatial and temporal domains. MPEG VSRS version 3.5 (M1) is a commonly known view synthesis method, employing several tools to handle artifacts, used to render virtual views [15]. Diffusion-based inpainting is one of the tools used for rendering. Other recent view synthesis methods presented by Ahn et al. [21] and Choi et al. [37] are based on exemplar inpainting. We denote these methods as M2 and M3, respectively. Method M2 applies inpainting by using only spatial information, whereas M3 applies inpainting by using both spatial and temporal information. Moreover, methods M2 and M3 are similar to our inpainting, since M2 and M3 uses exemplar-based inpainting. The remaining two methods are VSRS-alpha-Gist and VSRS-apha-Etri, which include different hole-filling methods and these methods are developed for view extrapolation scenarios [46, 47]. The methods VSRS-alpha-Gist and VSRS-apha-Etri are denoted as M4 and M5, respectively. All the state-of-the art methods are used in the traditional scenario, where hole filling is performed after warping to the virtual view. To make a fair comparison, and since we only have access to the low-resolution results of a few datasets from M3, we downscale the "Ballet" and "Poznan Street" sequence results during comparison.

In the notation used in the article, e.g. v4→v3 indicates that the original camera view 4 is warped to the virtual camera view 3. Parameter values used in the algorithm are empirically selected as follows: optical flow threshold $\rho = 0.05$ and depth tolerance $\vartheta = 10$ is used in the sprite construction step. The number of future frames $B_l = 5$ is used. Similar to [27], the following parameters are used in the inpainting process: the occlusion threshold $\eta = 5$ for a small baseline, a search region of 120×120 pixels, a patch

Table 3 Objective spatial quality evaluation

Sequence	YPSNR						SSIM					
	Proposed	M1	M2	M3	M4	M5	Proposed	M1	M2	M3	M4	M5
Poznan Street v3→v1	29.56	29.49	*29.60*	25.89	29.09	29.36	*0.8921*	0.8885	0.8892	0.8225	0.8858	0.8892
Ballet v4→v3	*30.77*	26.05	29.87	27.40	27.68	25.38	*0.9363*	0.9097	0.9268	0.8572	0.9205	0.9089
Ballet v3→v4	*29.32*	27.18	28.78	-	27.34	26.37	0.8499	0.8534	0.8456	-	*0.8590*	0.8511
Lovebird1 v6→v8	*27.63*	27.56	27.29	-	27.30	27.52	0.8693	*0.8747*	0.8653	-	0.8722	0.8733
Lovebird1 v8→v6	*28.56*	28.29	27.92	-	27.86	28.25	*0.8707*	0.8702	0.8651	-	0.8673	0.8693
Book Arrival v8→v10	*25.29*	25.00	25.10	-	18.75	25.51	0.8636	*0.8662*	0.8580	-	0.8472	0.8635

Values in italics indicates the highest score values

Table 4 Objective temporal quality evaluation: average F-score

Sequence	F-score					
	Proposed	M1	M2	M3	M4	M5
Poznan Street v3→v1	*0.33*	1.32	7.78	1.90	2.17	1.11
Ballet v4→v3	*0.65*	2.19	3.70	1.56	1.74	1.76
Ballet v3→v4	*0.54*	2.37	4.74	-	1.07	2.13
Lovebird1 v6→v8	*0.32*	2.72	6.47	-	2.05	2.93
Lovebird1 v8→v6	*0.50*	2.57	7.58	-	2.17	2.75
Book Arrival v8→v10	0.62	*0.33*	13.67	-	0.69	1.32

Values in italics indicates the highest score values

size of 11×11 pixels, number of best patches $N = 5$, and $\beta = 3$ in (25). The number of occlusion layers identified and inpainted is one.

4 Results and analysis

4.1 Spatial consistency evaluation

The objective spatial consistency evaluation results are presented in Table 3. Objective measurement results demonstrate that the proposed method performs better than the state-of-the art reference methods. In addition to the objective results, a set of images is shown in Figs. 9, 10, 11 and 12 for visual inspection. As method M1 fills the holes using the structured based inpainting without boundary classification, results exhibit foreground propagation and blurring artifacts (see row 2 in Figs. 9, 10, 11 and 12). Although method M2 used the background classification while inpainting the holes, holes are still filled with inconsistent background textures due to foreground-background classification (see row 3 in Fig. 9 at the woman's legs, in Fig. 11 at stairs and in Fig. 12 at wall clock). Method M3 uses the temporal information to fill the holes, however, there are still artifacts: such as the artifacts at the holes near moving objects (see row 4 in Fig. 9 at holes near the woman), the foreground propagation into the holes near moving objects and inconsistent filling at static objects (see row 4 in Fig. 10 at holes near the woman and the pole in front of the car). Methods M4 and M5 are aimed at filling holes in view extrapolation; however, holes are filled with smoothed textures and structure discontinuity (see rows 5 and 6 in Fig. 9, Fig. 10 rows 4 and 5 in Fig. 11 and Fig. 12).

The proposed method results in Figs. 9, 10, 11 and 12 show that holes are filled with spatial consistent background textures when the holes belong to the absolute background. However, the proposed method is limited when it comes to filling the holes when they occur between relative foreground and background. In this case, holes are filled with absolute background textures since the proposed method does not classify the absolute and

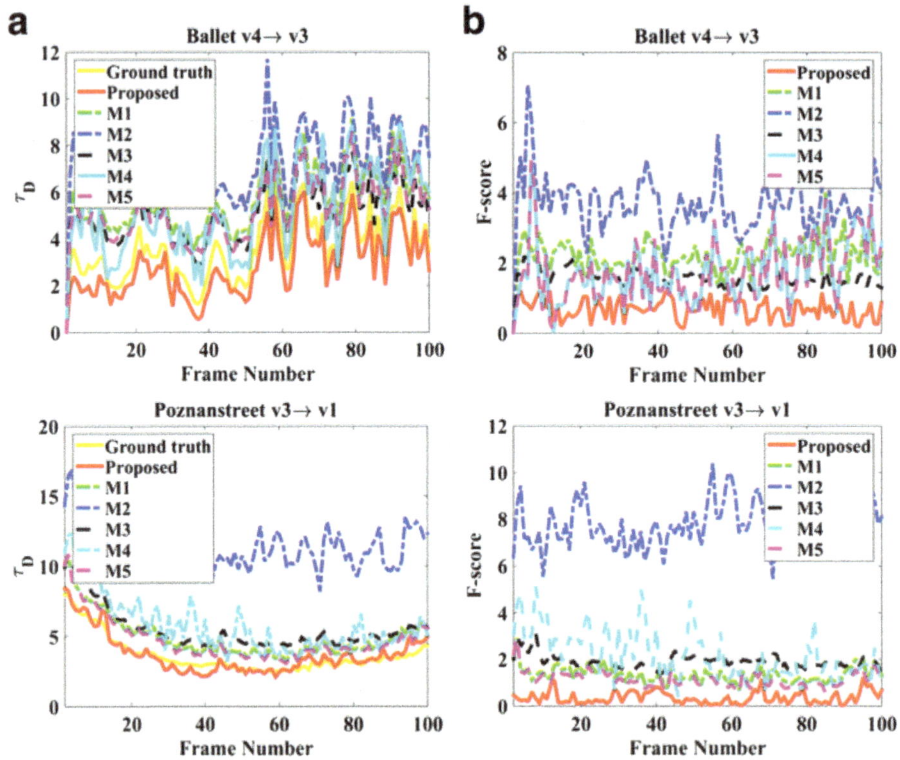

Fig. 8 Temporal objective quality evaluation. **a** Inter-frame average absolute difference at holes (τ_D). **b** Flickering score (F-score)

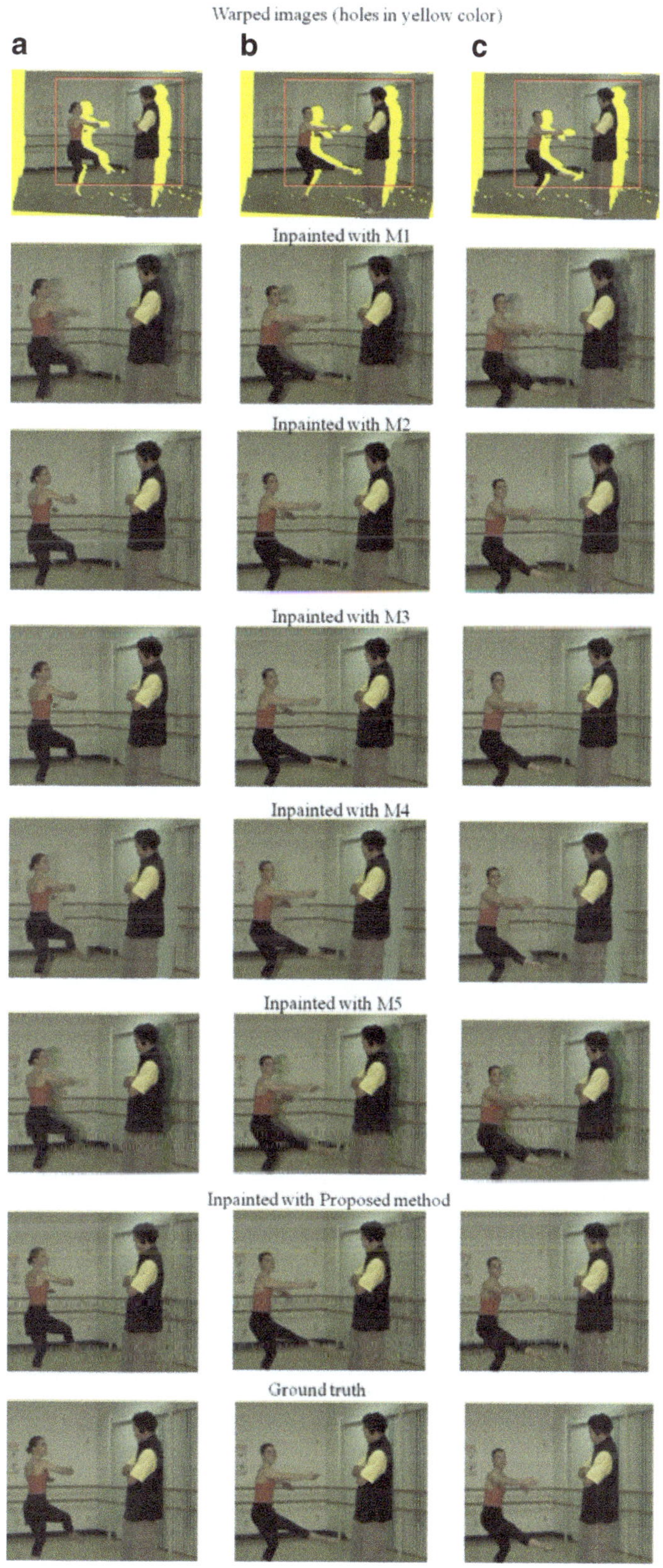

Fig. 9 Visual quality results of "Ballet" v4→v3 sequential frames. **a** Frame 66. **b** Frame 67. **c** Frame 68

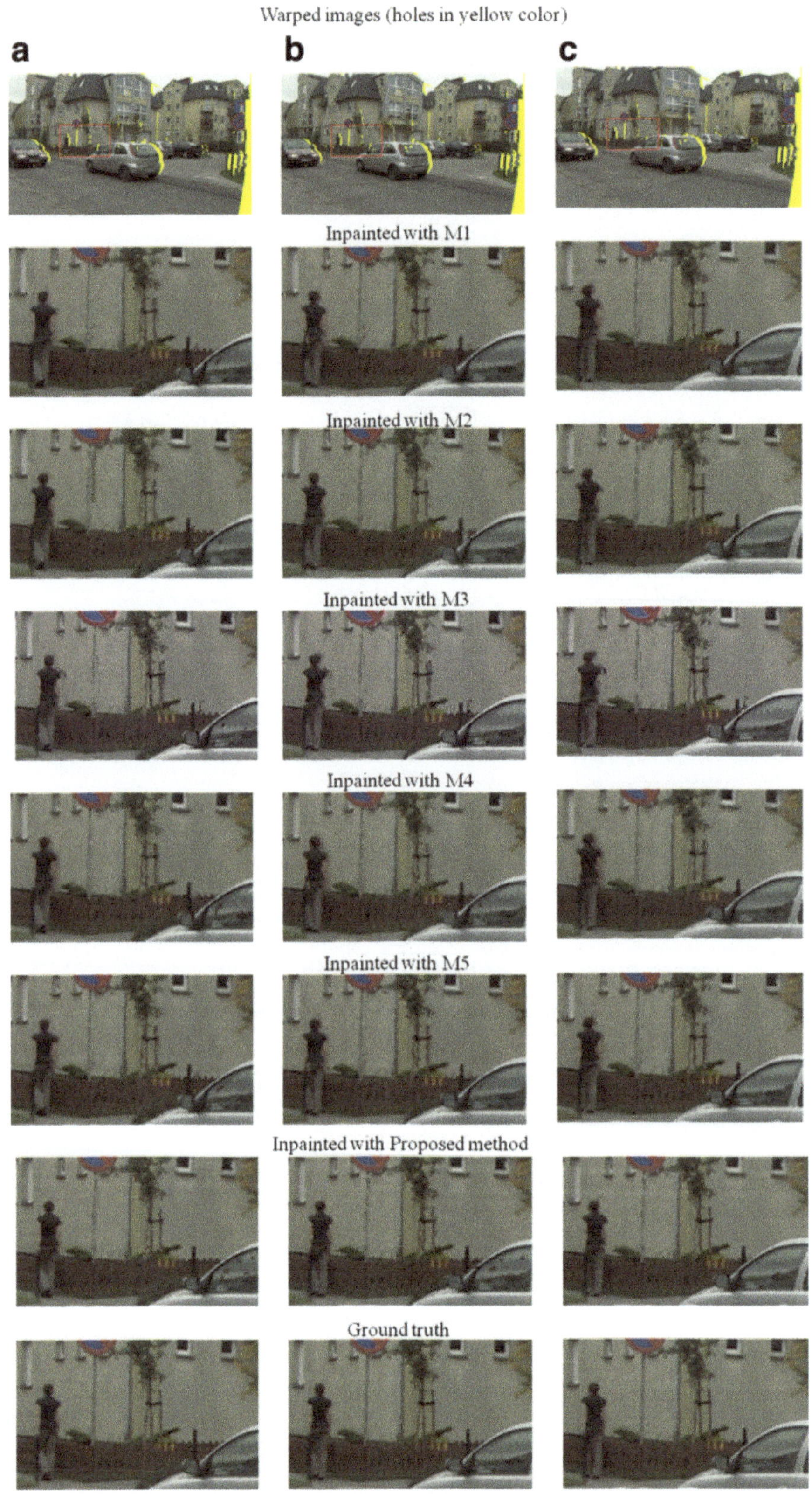

Fig. 10 Visual quality results of "Poznan Street" v3→v1 sequential frames. **a** Frame 18. **b** Frame 19. **c** Frame 20

Fig. 11 Visual quality results of "Lovebird1" v6→v8 sequential frames. **a** Frame 111. **b** Frame 112. **c** Frame 113

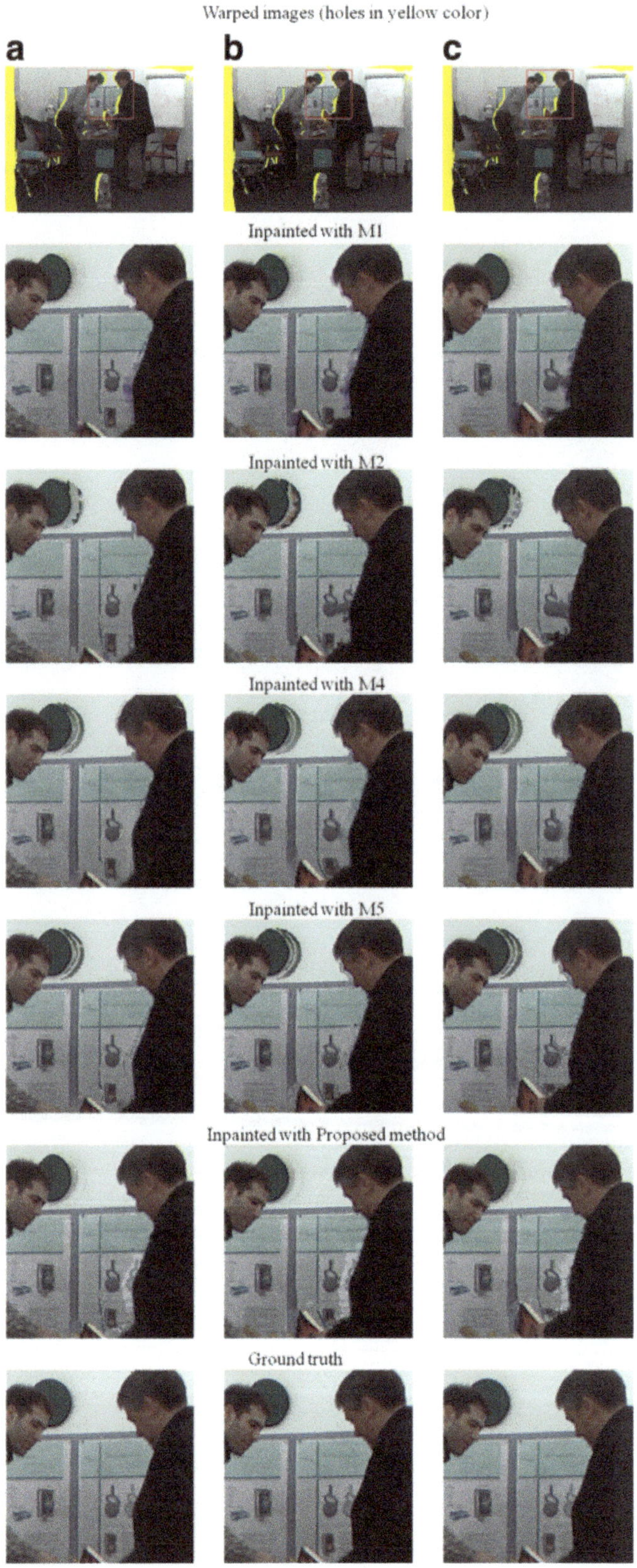

Fig. 12 Visual quality results of "Book Arrival" v8→v10 sequential frames. **a** Frame 57. **b** Frame 58. **c** Frame 59

relative background. We partially addressed this problem in our previous work by identifying the holes between relative foreground and background but only in the spatial domain [27].

4.2 Temporal consistency evaluation

Results from the objective temporal consistency measurements are presented in Table 4. In addition to the average F-score values, the inter-frame absolute difference at the holes and F-score of few frames are presented to show the similarity with the original signal (see Fig. 8). The average flickering score demonstrates that the proposed method performs better than the state-of-the art reference methods (consistent with the results from the spatial domain). Figure 8a shows that the result from the proposed method is similar to the ground truth data. In addition to objective measurements, the visual comparison clearly shows that the holes are temporally stable at the static holes (see row 7 in Fig. 9 at the holes near the man and row 7 in Fig. 10 at the holes near poles and the woman). Holes at the moving objects are also filled smoothly over frames (see row 6 in Fig. 11 at the holes near the woman, the edge of the stairs and in Fig. 12 at the holes near the man). Although flickering artifacts are significantly reduced at both static and moving objects' holes, apparent flickering is noticeable in the foreground of "Poznan Street" and less apparent in "Book Arrival" sequences due to erroneous depth. However, these are not well reflected by the F-score in the case of "Poznan Street" due to the flickering areas being significantly smaller than other disocclusions in the scène.

4.3 Complexity evaluation

Using the proposed method, it takes about 11 min to render a single frame of "Ballet" v4→v3, whereas using the reference method M1 takes 0.4 min, M2 takes 35 min, M3 takes 5 min, M4 takes 0.12 min and M5 takes 0.05 min, respectively. Note that M1, M3, M4 and M5 are implemented in C++, whereas M2 is implemented in MATLAB. Compared to the spatial hole-filling method M2, the computation time for the proposed method is reduced. This is because the holes at the static objects are inpainted once and the inpainted information is reused until they move, which means that there is no need to inpaint every frame for the holes at the dynamic objects. Because of the LDI generation and motion estimation and iterative inpainting, the proposed method is relatively slower than the reference methods.

5 Conclusions

We proposed a new view synthesis method with a spatial and temporal consistent layered depth image and depth-based inpainting, in order to improve the rendered view quality. The proposed method handles disocclusions in the virtual view in a spatial and temporal consistent manner. Furthermore, the proposed hole classification using the optical flow ensures the temporal consistency at static object holes by reusing the previously inpainted information. Spatio-temporal consistency is also ensured at moving objects' holes by using the true texture from temporal frames and inpainting. The computational time was reduced compared to reference methods by using the hole classification and reusing the inpainted textures to fill the holes at static objects. Moving weighted average blending between the textures within the hole regions reduced flickering and created a smooth transition. The objective and subjective test results consistently demonstrated that the proposed view synthesis method improved virtual view quality of the static background scenes. The results highlight the importance of the hole classification in improving the spatio-temporal consistency of extrapolated views.

The proposed method is aimed at static backgrounds and fills the holes with absolute background only, which means that the method is limited when it comes to handling the occlusion between relative foreground and background objects. Investigating possible improvements by identifying the relative foreground and background and building layered sprite is a matter for the future work. Furthermore, if the scene has illumination variations, reusing the inpainted information might cause artifacts. Investigations and finding a way to handle these problems is also a matter for future work. Identifying moving objects using other optimized and dense optical flow algorithms and finding the foreground objects locations in relation to the temporal domain is still challenging and is left for the future work. Optimizing the computation time is another issue to be dealt with in future investigations.

Additional file

Supplementary videos can be found at the paper web page www.miun.se/stc/Realistic3D/Muddala-2015.

Abbreviations

2D, two-dimensional; 3D, three-dimensional; 3DTV, three-dimensional television; FTV, free viewpoint television; DIBR, depth-image-based rendering; LDI, layered depth image; PSNR, peak signal-to-noise ratio; MSSIM, mean structural similarity index.

Competing interests

The authors declare that they have no competing interests.

Acknowledgements

This work has partly been supported by grant 20120328 of the KK Foundation, Sweden. We would like to thank Sunghwan Choi for providing their test results. We would like to thank Mitra Damghanian and Martin Köppel for their help. We would also like to acknowledge the reviewers for their valuable and helpful suggestions.

References

1. O Schreer, P Kauff, T Sikora, *3D Video Communication Algorithms, Concepts and Real-time Systems in Human Centered Communication*. (John Wiley & Sons Ltd., Chichester, 2005)
2. M Tanimoto, in *Computer Vision and Pattern Recognition Workshop, 2006. CVPRW '06. Conference On*. FTV (free viewpoint television) for 3D scene reproduction and creation, (2006), pp. 172–172. doi:10.1109/CVPRW.2006.84
3. C Fehn, in *Proc. SPIE Stereoscopic Displays and Virtual Reality Systems XI*. Depth-image-based rendering (DIBR), compression, and transmission for a new approach on 3D-TV, (2004), pp. 93–104. doi:10.1117/12.524762
4. L Zhang, WJ Tam, in *IEEE Transactions on Broadcasting*. Stereoscopic image generation based on depth images for 3D TV, (2005), pp. 191–199. doi:10.1109/TBC.2005.846190
5. M Köppel, MB Makhlouf, M Muller, P Ndjiki-Nya, in *Visual Communications and Image Processing (VCIP), 2013*. Temporally consistent adaptive depth map preprocessing for view synthesis, (2013), pp. 1–6. doi:10.1109/VCIP.2013.6706436
6. K Muller, A Smolic, K Dix, P Merkle, P Kauff, T Wiegand, View synthesis for advanced 3D video systems. EURASIP J. Image Video Process. (2008). doi:10.1155/2008/438148
7. Y Mori, N Fukushima, T Yendo, T Fujii, M Tanimoto, View generation with 3D warping using depth information for FTV. Signal Process. Image Commun. **24**(1–2), 65–72 (2009). doi:10.1016/j.image.2008.10.013
8. D Tian, PL Lai, P Lopez, C Gomila, View synthesis techniques for 3D video. Proc. SPIE. **7443**, 74430–7443011 (2009). doi:10.1117/12.829372
9. SM Muddala, M Sjöström, R Olsson, S Tourancheau, Edge-aided virtual view rendering for multiview video plus depth. Proc. SPIE. **8650**, 86500–865007 (2013). doi:10.1117/12.2004116
10. N Plath, S Knorr, L Goldmann, T Sikora, Adaptive image warping for hole prevention in 3D view synthesis. Image Process. IEEE Trans. **22**(9), 3420–3432 (2013). doi:10.1109/TIP.2013.2268940
11. N Stefanoski, O Wang, M Lang, P Greisen, S Heinzle, A Smolic, Automatic view synthesis by image-domain-warping. Image Process. IEEE Trans. **22**(9), 3329–3341 (2013). doi:10.1109/TIP.2013.2264817
12. SM Muddala, M Sjöström, R Olsson, in *International Conference on 3D Imaging 2012 (IC3D)*. Edge-preserving depth-image-based rendering method, (2012). doi:10.1109/IC3D.2012.6615113
13. M Bertalmio, G Sapiro, V Caselles, C Ballester, in *Proceedings of ACM Conf. Comp. Graphics (SIGGRAPH)*. Image inpainting, (2000), pp. 417–424. doi:10.1145/344779.344972
14. A Telea, An image inpainting technique based on the fast marching method. J. Graphics GPU Game Tools. **9**(1), 23–34 (2004). doi:10.1145/344779.344972
15. Report on experimental framework for 3D video coding (2010). ISO/IEC JTC1/SC29/WG11 MPEG2010/N11631. Guangzhou, China
16. KJ Oh, S Yea, YS Ho, in *Proceedings of the 27th Conference on Picture Coding Symposium, Chicago, IL, USA*. Hole filling method using depth based in-painting for view synthesis in free viewpoint television (FTV) and 3D video, (2009), pp. 233–236
17. A Criminisi, P Pérez, K Toyama, Region filling and object removal by exemplar-based image inpainting. IEEE Trans. Image Process. **13**, 1200–1212 (2004). doi:10.1109/TIP.2004.833105
18. I Daribo, H Saito, A novel inpainting-based layered depth video for 3DTV. Broadcasting IEEE Trans. **57**(2), 533–541 (2011). doi:10.1109/TBC.2011.2125110
19. V Jantet, C Guillemot, L Morin, in *3D Research*. Joint projection filling method for occlusion handling in depth-image-based rendering, (2011). doi:10.1007/3DRes.04(2011)4
20. H Lim, YS Kim, S Lee, O C, JDK Kim, C Kim, in *Image Processing (ICIP), 2011 18th IEEE International Conference On*. Bi-layer inpainting for novel view synthesis, (2011), pp. 1089–1092. doi:10.1109/ICIP.2011.6115615
21. I Ahn, C Kim, A novel depth-based virtual view synthesis method for free viewpoint video. Broadcasting, IEEE Trans. **59**(4), 614–626 (2013). doi:10.1109/TBC.2013.2281658
22. SM Muddala, R Olsson, M Sjöström, Depth-included curvature inpainting for disocclusion filling in view synthesis. Int. J. Adv. Telecommun. **6**(3 & 4), 132–142 (2013)
23. SM Muddala, M Sjöström, R Olsson, in *3DTV-Conference: The True Vision - Capture, Transmission and Display of 3D Video (3DTV-CON), 2014*. Depth-based inpainting for disocclusion filling, (2014), pp. 1–4. doi:10.1109/TBC.2005.846190
24. JW Shade, SJ Gortler, L-W He, R Szelisk, in *Proceedings of the 25th Annual Conference on Computer Graphics and Interactive Techniques. SIGGRAPH '98*. Layered depth images, (1998), pp. 231–242. doi:10.1145/280814.280882
25. B Barenbrug, R-PM Berretty, RK Gunnewiek, Robust image, depth, and occlusion generation from uncalibrated stereo. Proc. SPIE 6803, 68031–680318 (2008). doi:10.1117/12.765508
26. B Bartczak, P Vandewalle, O Grau, G Briand, J Fournier, P Kerbiriou, M Murdoch, M Müller, R Goris, R Koch, R van der Vleuten, Display-independent 3D-TV production and delivery using the layered depth video format. Broadcasting IEEE Trans. **57**(2), 477–490 (2011). doi:10.1109/TBC.2011.2120790
27. SM Muddala, Free view rendering for 3D video – Edge-aided rendering and depth-based image inpainting (2015). Doctoral Thesis No. 226, Mid Sweden University, Sundsvall, Sweden
28. K-Y Chen, P-K Tsung, P-C Lin, H-J Yang, L-G Chen, in *3DTV-Conference: The True Vision - Capture, Transmission and Display of 3D Video (3DTV-CON), 2010*. Hybrid motion/depth-oriented inpainting for virtual view synthesis in multiview applications, (2010), pp. 1–4. doi:10.1109/3DTV.2010.5506309
29. RK Gunnewiek, R-PM Berretty, B Barenbrug, JPM aes, Coherent spatial and temporal occlusion generation. Proc. SPIE. **7237**, 723713–72371310 (2009)
30. M Schmeing, X Jiang, in *3DTV-Conference: The True Vision - Capture, Transmission and Display of 3D Video (3DTV-CON), 2010*. Depth image based rendering: a faithful approach for the disocclusion problem, (2010), pp. 1–4. doi:10.1109/3DTV.2010.5506596
31. P Ndjiki-Nya, M Köppel, D Doshkov, H Lakshman, P Merkle, K Muller, T Wiegand, Depth image-based rendering with advanced texture synthesis for 3-D video. Multimedia, IEEE Trans. **13**(3), 453–465 (2011). doi:10.1109/TMM.2011.2128862
32. M Köppel, X Wang, D Doshkov, T Wiegand, P Ndjiki-Nya, in *19th IEEE International Conference on Image Processing, ICIP, 2012*. Depth image-based rendering with spatio-temporally consistent texture synthesis for 3-D video with global motion, (2012), pp. 2713–2716. doi:10.1109/ICIP.2012.6467459
33. W Sun, OC Au, L Xu, Y Li, W Hu, in *Image Processing (ICIP), 2012 19th IEEE International Conference On*. Novel temporal domain hole filling based on background modeling for view synthesis, (2012), pp. 2721–2724. doi:10.1109/ICIP.2012.6467461
34. C Yao, T Tillo, Y Zhao, J Xiao, H Bai, C Lin, Depth map driven hole filling algorithm exploiting temporal correlation information. Broadcasting, IEEE Trans. **60**(2), 394–404 (2014). doi:10.1109/TBC.2014.2321671
35. M Xi, L-H Wang, Q-Q Yang, D Li, M Zhang, Depth-image-based rendering with spatial and temporal texture synthesis for 3DTV. EURASIP J. Image Video Process. **2013**, 1–18 (2013). doi:10.1186/1687-5281-2013-9
36. H-A Hsu, C-K Chiang, S-H Lai, Spatio-temporally consistent view synthesis from video-plus-depth data with global optimization. Circuits Syst. Video Technol. IEEE Trans. **24**(1), 74–84 (2014). doi:10.1109/TCSVT.2013.2276699
37. S Choi, B Ham, K Sohn, Space-time hole filling with random walks in view extrapolation for 3D video. Image Process. IEEE Trans. **22**(6), 2429–2441 (2013). doi:10.1109/TIP.2013.2251646
38. N Otsu, A threshold selection method from gray level histograms. IEEE Trans. Systems, Man Cybernetics. **9**, 62–66 (1979). doi:10.1109/TSMC.1979.4310076
39. BD Lucas, T Kanade, in *Proceedings of the 7th International Joint Conference on Artificial Intelligence - Volume 2. IJCAI'81, Vancouver, B. C., Canada*. An iterative image registration technique with an application to stereo vision, (1981), pp. 674–679
40. CL Zitnick, SB Kang, M Uyttendaele, S Winder, R Szeliski, High-quality video view interpolation using a layered representation. ACM Trans. Graph. **23**(3), 600–608 (2004)
41. I Feldmann, M Mueller, F Zilly, R Tanger, K Mueller, A Smolic, P Kauff, T Wiegand, HHI test material for 3D video (2008). ISO/IEC JTC1/SC29/WG11 MPEG 2008/M15413
42. GM Um, G Bang, N Hur, J Kim, YS Ho, 3D video test material of outdoor scene (2008). ISO/IEC JTC1/SC29/WG11/M15371
43. M Domanski, T Grajek, K Klimaszewski, M Kurc, O Stankiewicz, J Stankowski, K Wegner, Poznan multiview video test sequences and camera parameters (2009). ISO/IEC JTC1/SC29/WG11 MPEG 2009/M17050. Xian, China
44. Z Wang, AC Bovik, HR Sheikh, EP Simoncelli, Image quality assessment: from error visibility to structural similarity. IEEE Trans. Image Process. **13**(4), 600–612 (2004). doi:10.1109/TIP.2003.819861

45. M Schmeing, J Xiaoyi, in *3DTV Conference: The True Vision - Capture, Transmission and Display of 3D Video (3DTV-CON), 2011*. Time-consistency of disocclusion filling algorithms in depth image based rendering, (2011), pp. 1–4. doi:10.1109/3DTV.2011.5877201
46. G Bang, MS Ko, J Yoo, WS Cheong, G Um, N Hur, Boundary noise removal and common hole filling method for VSRS 3.5 (2011). ISO/IEC JTC1/SC29/WG11 Doc. m19356
47. C Lee, YS Ho, Results of view synthesis using modified hole filling methods (2010). ISO/IEC JTC1/SC29/WG11 Doc. m19281. Guangzhou, China

9

From D-RGB-based reconstruction toward a mesh deformation model for monocular reconstruction of isometric surfaces

S. Jafar Hosseini* and Helder Araujo

Abstract

In this paper, we address 3D reconstruction of surfaces deforming isometrically. Given that an isometric surface is represented by means of a triangular mesh and that feature/point correspondences on an image are available, the goal is to estimate the 3D positions of the mesh vertices. To perform such monocular reconstruction, a common practice is to adopt *linear deformation model*. We also integrate this model into a least-squares optimization. However, this model is obtained through a learning process requiring an adequate data set of possible mesh deformations. Providing this prior data is the primary goal of this work and therefore a novel reconstruction technique is proposed for a mesh overlaid across a typical isometric surface. This technique consists in the use of a range camera accompanied by a conventional camera and implements the path from the depth of the feature points to the 3D positions of the vertices through convex programming. The idea is to use the high-resolution images from the RGB camera in combination with the low-resolution depth map to enhance mesh deformation estimation. With this approach, multiple deformations of the mesh are recovered with the possibility that the resulting deformation model is simply extended to any other isometric surfaces for monocular reconstruction. Experimental results show that the proposed approach is robust to noise and generates accurate reconstructions.

Keywords: Isometric surface, 3D reconstruction, Range camera, Convex programming

1 Introduction

The reconstruction of objects from a single image is under-constrained, meaning that the recovery of 3D shape is an inherently ambiguous problem. The case of non-rigid objects is even more complex and difficult [1–3]. Given a specific configuration of points on the image plane, different 3D non-rigid shapes and camera motions can be found that fit the measurements. The approaches proposed over the past years can be categorized in two major types: those involving physics-based models [4–6] and those relying on non-rigid structure-from-motion (NRSfM) approaches [7–13]. In most cases, the former type ends up designing a complex objective function to be minimized over the solution space. The latter, on the other hand, takes advantage of prior knowledge on the shape and motion, to constrain the solution so that the inherent ambiguity can be tackled, and it performs effectively provided that the 2D point tracks are accurate and reliable. For example, Aanaes et al. [14] impose the prior knowledge that the reconstructed shape does not vary much from frame to frame while Del Bue et al. [15] impose the constraint that some of the points on the object are rigid. The priors can be divided in two main categories: the statistical and physical priors. For instance, the methods relying on the low-rank factorization paradigm [14, 15] can be classified as statistical approaches. Learning approaches such as [16–21] also belong to the statistical approaches. Physical constraints include spatial and temporal priors on the surface to be reconstructed [22, 23]. Monocular reconstruction of deformable surface reconstruction has been extensively studied in the last few years [24, 25]. Strictly speaking, isometric reconstruction from perspective camera views has attracted much of the attention. A physical prior of particular interest in this case is the hypothesis of having an inextensible (i.e., isometric) surface [26–29]. In this paper, we consider this type

*Correspondence: jafar@isr.uc.pt
Institute of Systems and Robotics, Department of Electrical and Computer Engineering, University of Coimbra, Coimbra, Portugal

of surface. This hypothesis means that the length of the geodesics between every two points on the surface should not change across time, which makes sense for many types of material such as paper and some types of fabric.

The reconstruction of 3D deformable surfaces is becoming increasingly important and this can be visible considering its practical applications. Physics has inspired early approaches. These approaches amount to a minimization based on the physical behavior of the surface [4, 30, 31]. Although it makes sense that we integrate physical laws into our algorithms, the final framework will be affected by two major shortcomings:

- The material parameters, which are typically unknown, have to be determined.
- In order to estimate the parameters accurately, in the presence of large deformations, we need to build a complex cost functional (which is hard to optimize).

Methods that learn models from training data were introduced to overcome these limitations. In that case, surface deformations are expressed as linear combinations of deformation modes which are obtained from training data. NRSfM methods built on this principle recover simultaneously the shape and the modes from image sequences [11, 32, 33]. Although this is a very attractive idea, practical implementations are not easy since they require points to be tracked across the entire sequence. Moreover, they are only effective for relatively small deformations. There have also been a number of attempts at performing 3D surface reconstruction without using a deformation model. One approach is to use lighting information in addition to texture clues to constrain the reconstruction process [34], which has only been demonstrated under very restrictive assumptions on lighting conditions and is therefore not generally applicable. The algorithms for reconstructing deformable surfaces can be classified by the type of the surface model (or representation) used: Point-wise methods only reconstruct the 3D position of a relatively small number of feature points, resulting in a sparse reconstruction of the 3D surface [27]. Physics-based models such as superquadrics [35], triangular meshes [28], or thin-plate splines (TPS) [27] have been also utilized in other algorithms. In TPS, the 3D surface is represented as a parametric 2D-3D map between the template image space and the 3D space. Then, a parametric model is fit to a sparse set of reconstructed 3D points in order to obtain a smooth surface which is not actually used in the 3D reconstruction process.

Having an isometric surface means that the length of the geodesics between pairs of points remains unchanged when the surface deforms and the deformed surface can be obtained by applying an isometric transformation (map) to a template surface. In many cases, computation of the geodesics is not trivial and involves the application of differential geometry. Instead, the Euclidean distance, which is much easier to estimate, has been regarded as a good approximation to the geodesic distance, on condition that it does not drop too much below the geodesics. Euclidean approximation is better when there are a large number of points. Although it can work well in some cases, it gives poor results when creases appear in the 3D surface. In this case, the Euclidean distance between two points on the surface can shrink. For this reason, the "upper bound approach" has been proposed which relies on the fact that the Euclidean distance between two random points on a plane is necessarily less than (or equal to) the corresponding geodesics, which is known as inextensibility constraint. As a result, early approaches relax the non-convex isometric constraints to inextensibility with the so-called maximum depth heuristic [24, 27]. The idea is to maximize point depths so that the Euclidean distance between every pair of points is upper bounded by its geodesic distance, computed in the template [18, 28]. In these papers, a convex cost function combining the depth of the reconstructed points and the negative of the reprojection error is maximized while enforcing the inequality constraints arising from the surface inextensibility. The resulting formulation can be easily turned into a SOCP problem. This problem is convex and gives accurate reconstructions. A similar approach is explored in [26]. The approach of [27] is a point-wise method. The approaches of [18, 26, 28] use a triangular mesh as surface model, and the inextensibility constraints are applied to the vertices of the mesh. Recently, analytical solutions for isometric and conformal deformations have been provided by posing them as a system of partial differential equations [36, 37].

1.1 Problem formulation

In this paper, we aim at the reconstruction of surfaces that undergo isometric deformations. Assuming that a triangular mesh is used to represent an isometric surface and that a set of feature/point correspondences on an image of the surface have been provided, the objective is to determine the 3D positions of the mesh vertices. To carry out this monocular reconstruction, we formulate a non-linear least-squares optimization that integrates the linear deformation model, deformation-based constraints which we call *isometric constraints*, and the projection equations in order to solve for 3D positions of the mesh vertices.

Main contribution: Several reconstruction methods have previously relied on the linear deformation model as a crucial element that can reduce the ambiguity of infinite solutions. This model is specially useful when using the mesh representation. It is typically obtained from prior training data that corresponds to various possible deformations of the mesh. As a result, it is required to reconstruct these mesh deformations beforehand, which is challenging without some sort of supporting 3D information. Furthermore, the precision of the training data is important and must be ensured. For this purpose, we propose an innovative technique to acquiring such data with high accuracy. This technique aims to estimate a regular 3D mesh overlaid across a generic isometric surface and is used to recover several different deformations of the mesh in a way that makes it possible to extend the computed deformation model to other isometric surfaces for monocular reconstruction. In developing this approach, we use a conventional RGB camera aided by a range camera. Our emphasis is, in fact, on the use of a time-of-flight (ToF) camera in conjunction with the RGB camera. Most RGB cameras provide high-resolution images. With these cameras, one can use efficient algorithms to calculate the depth of the scene, recover object shape or reveal structure, but at a high computational cost. ToF cameras deliver a depth map of the scene in real time but with insufficient resolution for some applications. So, a combination of a conventional camera and a ToF camera can exploit the capabilities of both. We assume that the fields of view of both cameras mostly overlap. From the depth map, the depth of the feature points can be extracted by adopting a registration technique for the camera combination. This allows the depth of the mesh vertices to be subsequently computed using either a linear system of equations or a linear programming problem. Given the mesh depth data, the complete 3D positions of the vertices can be recovered through a second-order cone programming (SOCP) problem. Applying the approach just described to a variety of mesh deformations leads to the required data, thereby computing the deformation model.

1.2 Outline of the paper

This paper is organized as follows: Section 2 discusses the background of our work, including the notation used, mesh representation, and the linear deformation model. Section 3 describes the monocular reconstruction. Section 4 is assigned to a detailed explanation of our D-RGB-based reconstruction. Section 5 presents experimental results and quantitative evaluations, demonstrating the efficiency of our reconstruction schemes. In Section 6, we discuss conclusions.

2 Background

2.1 Notation

Matrices are represented as bold capital letters ($\mathbf{A} \in \mathbb{R}^{n\times m}$, n rows and m columns). Vectors are represented as bold small letters ($\mathbf{a} \in \mathbb{R}^{n}$, n elements). By default, a vector is considered a column. Small letters (a) represent one-dimensional elements. By default, the jth column vector of $\mathbf{A}$ is specified as $\mathbf{a}_j$. The jth element of a vector $\mathbf{a}$ is written as a_j. The element of $\mathbf{A}$ in the row i and column j is represented as $\mathbf{A}_{i,j}$. $\mathbf{A}^{(1:2)}$ and $\mathbf{a}^{(1:2)}$ indicate the first two rows of $\mathbf{A}$ and $\mathbf{a}$. $\mathbf{A}^{(3)}$ and $\mathbf{a}^{(3)}$ denote the third row of $\mathbf{A}$ and $\mathbf{a}$, respectively. Regular capital letters (A) indicate one-dimensional constants. We use $\mathbb{R}$ after a vector or matrix to denote that it is represented up to a scale factor.

2.2 Mesh representation

Assume that a set of 3D feature points $\mathbf{p}^{\text{ref}} = \{\mathbf{p}_1^{\text{ref}}, \cdots, \mathbf{p}_N^{\text{ref}}\}$ on a template with a known shape (usually a flat surface), and a set of 2D image points $\mathbf{q} = \{\mathbf{q}_1, \cdots, \mathbf{q}_N\}$ tracked on the RGB input image of the same surface, but with a different and unknown deformation are given. As already stated, we represent the surface as a triangular 3D mesh with n_v vertices $\mathbf{v}_i$ (and n_{tr} triangles) concatenated in a vector $\mathbf{s} = \left[\mathbf{v}_1^T, \cdots, \mathbf{v}_{n_v}^T\right]^T$, and denote by $\mathbf{s}^{\text{ref}}$ the template mesh, and $\mathbf{s}$ the mesh we seek to estimate—see Fig. 1. Let $\mathbf{p}_i$ be a feature point on the mesh $\mathbf{s}$ corresponding to the point $\mathbf{p}_i^{\text{ref}}$ in the template. We can express $\mathbf{p}_i$ in terms of the barycentric coordinates of the triangle it belongs to:

$$\mathbf{p}_i = \sum_{j=1}^{3} a_{ij}\mathbf{v}_j^{[i]} \quad (1)$$

where the a_{ij} are the barycentric coordinates and $\mathbf{v}_j^{[i]}$ are the vertices of the triangle containing the point $\mathbf{p}_i$. Mesh representation has the advantage of simplifying reconstructions in view of the fact that the isometric type of deformation imposes the constraint that the length of the edges of a mesh with a dense distribution of vertices stay nearly the same, as the surface deforms. As a result, we may treat the mesh triangles as rigid, allowing us to consider that the barycentric coordinates remain constant for each point. These coordinates are easily computed from points $\mathbf{p}_i^{\text{ref}}$ and the mesh $\mathbf{s}^{\text{ref}}$. Let us denote by $\mathbf{A} = \{\mathbf{a}_1, \cdots, \mathbf{a}_N\}$ the set of barycentric coordinates associated with the feature points, where $\mathbf{a}_i = \left[a_{i1}, a_{i2}, a_{i3}\right]$.

2.3 Linear deformation model

The space of possible deformed shapes of the surface is constrained by applying a deformation model. This model adequately fills in the missing information while being flexible enough to allow reconstruction of complex deformations [17]. A mesh deformation is thus modeled as a linear combination of a mean shape $\mathbf{s}_0$ and n_m basis shapes (deformation modes) $\mathbf{S} = \left[\mathbf{s}_1, \ldots, \mathbf{s}_{n_m}\right]$:

$$\mathbf{s} = \mathbf{s}_0 + \sum_{k=1}^{n_m} w_k\mathbf{s}_k = \mathbf{s}_0 + \mathbf{S}\mathbf{w} \quad (2)$$

These modes can be obtained by applying principal component analysis (PCA) to a plenary set of training deformations. In our work, this training data is acquired

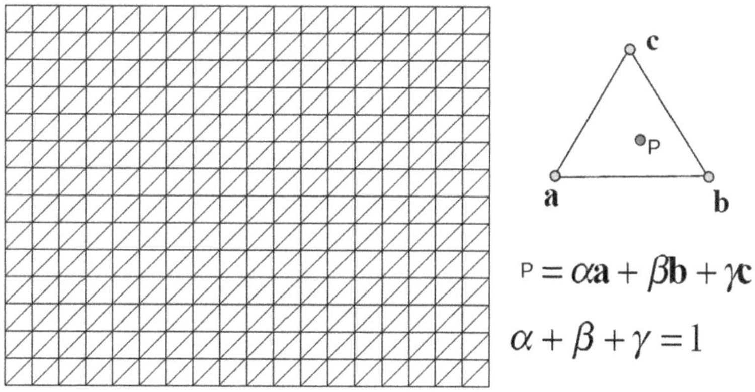

Fig. 1 A regular triangular mesh and barycentric coordinates

using a high-resolution image combined with the knowledge of the depth of a set of feature points.

3 Monocular reconstruction from a single view

Given that the linear deformation model has been computed, the objective is to proceed with an efficient algorithm which is intended to demonstrate the use of the linear deformation model in monocular reconstruction of mesh deformations. For this purpose, we introduce an algorithm that falls within a particular class of methods which follow the same basic principle, namely, mesh representation along with linear deformation model [17, 18, 24, 28]. Our algorithm is slightly different, which is composed of two non-linear constraints. It is capable of performing such reconstruction that the shape of any isometrically deformed surface is estimated by using only a conventional camera.

Isometric constraint: This constraint is the difference between the observed and the predicted length of an edge. We formulate a non-linear constraint as

$$e_{iso} = \sum_{i=1}^{n_e} \left(L_i - \left\| \mathbf{s}_1^{[i]} - \mathbf{s}_2^{[i]} \right\| \right)^2 \tag{3}$$

where L_i is the length of the edge i, computed on the template. $\mathbf{s}_1^{[i]}$ and $\mathbf{s}_2^{[i]}$ denote the two entries of the mesh that account for the ending vertices of the edge i.

Reprojection error: In addition, there are also reprojection errors, that is, errors on the image position of the feature points. We should thus account for the reprojection error by adding a term to the function to be optimized. By combining Eqs. 1 and 2, we will have

$$\mathbf{p}_i = \sum_{j=1}^{3} a_{ij} \mathbf{s}_j^{[i]} \tag{4}$$

where $\mathbf{s}_{0j}^{[i]}$ and $\mathbf{S}_j^{[i]}$ are the subvector of $\mathbf{s}_0$ and the submatrix of $\mathbf{S}$ (respectively), corresponding to the vertex j of the triangle in which the feature point i resides. The term corresponding to the reprojection error can be obtained as indicated below.

$$e_{re} = \sum_{i=1}^{N} \left\| \lambda_i \begin{bmatrix} \mathbf{q}_i \\ 1 \end{bmatrix} - \left[\mathbf{K}^{\circ}_{rgb} \cdot \mathbf{p}_i \right] \right\|^2 \tag{5}$$

where N is the number of feature points. The λ_i add extra unknowns to the optimization problem. Therefore, it is advantageous to reformulate the above equations so that the λ_i can be eliminated. Consider the equation below:

$$\lambda_i \begin{bmatrix} \mathbf{q}_i \\ 1 \end{bmatrix} = \mathbf{K}^{\circ}_{rgb} \left[\sum_{j=1}^{3} a_{ij} \mathbf{s}_j^{[i]} \right] \tag{6}$$

After some simple algebraic manipulation, we obtain

$$\begin{bmatrix} a_{i1}\mathbf{A}_i & a_{i2}\mathbf{A}_i & a_{i3}\mathbf{A}_i \end{bmatrix}_{2\times 9} \begin{bmatrix} \mathbf{s}_1^{[i]} \\ \mathbf{s}_2^{[i]} \\ \mathbf{s}_3^{[i]} \end{bmatrix}_{9\times 1} =$$

$$\begin{bmatrix} e_{1,i} \\ e_{2,i} \end{bmatrix}_{2\times 1} = 0 \quad \text{where } \mathbf{A}_i = \mathbf{K}^{\circ(1:2)}_{rgb} - \mathbf{q}_i \mathbf{K}^{\circ(3)}_{rgb} \tag{7}$$

This equation provides 2 linear constraints as: $e_{1,i} = 0$ and $e_{2,i} = 0$. Thus, the modified e_{re} takes a form where the λ_i does not exist, as follows:

$$e_{mre} = \sum_{i=1}^{N} \left((e_{1,i})^2 + (e_{2,i})^2 \right) \tag{8}$$

where e_{mre} denotes the modified e_{re}.

Objective function: We have now derived two constraints, described as two separate non-linear expressions. However, we intend to integrate both constraints into a single objective function so that they are taken into account at one time, while estimating all the parameters. To do so, we minimize the weighted summation of them in such a way that the reprojection error term is assigned a weight m that accounts for its relative influence within the combined objective function.

$$\min_{\mathbf{w}} e_{\text{tot}} = (e_{\text{iso}} + m.e_{\text{mre}}) \quad (9)$$

The above optimization scheme is a non-linear least-squares minimization problem, typically solved using an iterative algorithm such as Levenberg-Marquardt.

4 Reconstruction using a D-RGB camera setup

In order to build an adequate data set of mesh deformations for learning the deformation model, we propose a reconstruction approach for a typical surface based on a D-RGB camera setup. The completed deformation model can be then extended for monocular reconstruction of any other surfaces that undergo isometric deformations. Using the result of the registration described below, we can obtain an estimate for the depth of the feature points. The idea behind our D-RGB-based reconstruction is to determine the 3D positions of the mesh vertices, given this depth data. This is done in two steps: first the depth of the vertices is estimated and then their xy-coordinates.

Registration between depth and RGB images: The resolutions of the depth and RGB images are different. A major issue that directly arises from the difference in resolution is that a pixel-to-pixel correspondence between the two images cannot be established even if the FOVs fully overlap. Therefore, the two images have to be registered so that a mapping between the pixels in the depth image and in the RGB image can be established. The depth images provided by the ToF camera are sparse and affected by errors. Several methods can be used to improve their resolution [38–41], allowing the estimation of dense depth images. However, to estimate depth for all the pixels of the RGB image, based on the depth map given by the ToF camera, simple linear procedures are used as follows:

We use a pinhole camera model for both cameras and assume that they are calibrated internally and that also the relative pose between both cameras, specified by the rotation matrix $\mathbf{R}'$ and translation vector $\mathbf{t}'$ has been estimated. Let $\mathbf{p}_{\text{tof}}$ and $\mathbf{p}_{\text{rgb}}$ represent the 3D positions of a point in the coordinate system of the ToF and the RGB cameras, respectively. $\mathbf{p}_{\text{tof}}$ is obtained directly from the calibrated ToF camera. Thus, $\mathbf{p}_{\text{rgb}}$ can be easily calculated by $\mathbf{p}_{\text{rgb}} = \mathbf{R}'\mathbf{p}_{\text{tof}} + \mathbf{t}'$—see Fig. 2. For each point of the RGB image, we select the four closest neighbors whose depth was obtained from the depth image. Then, a bilinear interpolation is performed. Another possibility could be to select the three closest neighboring points (therefore, defining a triangle) and assume that the corresponding 3D points define a plane. An estimate for the depth of the point could be then obtained by intersecting its projecting ray with the 3D plane defined by the three 3D points. As a result, the depth of the N feature points is computed accurately and we indicate by $p_{z,k}$ the depth of the feature point k.

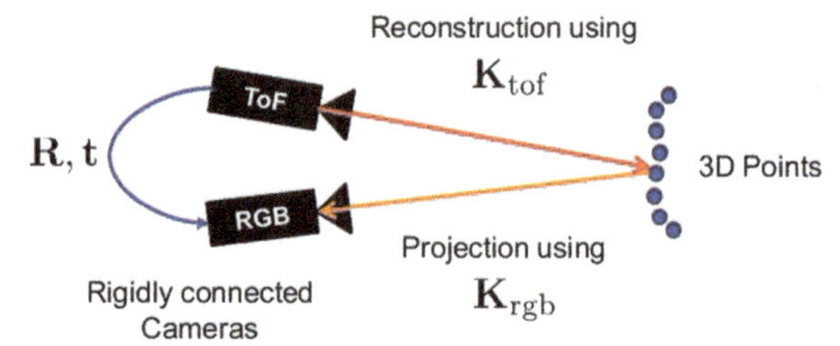

Fig. 2 ToF/RGB camera setup

4.1 Step 1: recovery of the depth of the vertices

Given $p_{z,k}$ for all ks, the goal is to estimate the depth of the vertices. Let z_i and rz_j denote the depth of the vertex i and the relative depth of the edge j, respectively. The vertices are numbered and sorted according to a particular ordering. The same goes for the set of all relative depths. In addition, a relative depth needs to conform to either of the two directions along its edge, i.e., $rz_{25} = z_{16} - z_7$ or vice versa. So, a predefined set of selected directions is applied to all edges. As a matter of fact, the rigidity of a closed triangle enforces the fact that the sum of the depth differences between every two vertices concatenated around the triangle, be zero. This can be expressed with relative depths and gives us n_{tr} equations which, in conjunction with the equations associating the relative depths with the depth of vertices, add up to $n_{\text{tr}} + n_e$ (the number of triangles + the number of edges) linear equations. We augment this linear system with the depth of the feature points. From Eq. 1, we can derive $p_{z,i} = a_{i1}z_1^{[i]} + a_{i2}z_2^{[i]} + a_{i3}z_3^{[i]}$. Having this equation for every feature point results in N linear independent equations. Putting together all the equations available, we end up with $n_{\text{tot}} = n_{\text{tr}} + n_e + N$ linear equations where the only unknowns are the depth of vertices and of the edges (i.e., $n_v + n_e$ unknowns), which means that the resulting linear system is overdetermined. We denote this linear system as $\mathbf{Mx} = \begin{bmatrix} \mathbf{0} \\ \mathbf{p}_z \end{bmatrix}$. We now propose two algorithms for determining the depth of the mesh vertices below.

Algorithm 1: solving a linear system of equations: The linear system above has $n_e + N$ independent equations out of n_{tot} and this is not yet enough to find the right single solution because there are still an infinitude of solutions that satisfy this linear system. One possible alternative to handle this is to fit an initial mesh using polynomial interpolation, to the data. This fitting consists in xy-coordinates of the feature points on the template as input and their z-coordinates on the input deformation as output. Once the parameters of the interpolant have been found, we can obtain initial estimates for depth of the vertices, with their xy-coordinates on the template as input. Let z_i' be the interpolated depth of the vertex i. By adding this result as an extra equation to the linear system described earlier, we obtain the modified linear system $\mathbf{M}_{\text{mod}}\mathbf{x} = \mathbf{b}$, which has most likely full column rank. So, the number of independent equations out of $n_{\text{tot}} + n_v$ will be $n_e + n_v$. Since the number of independent equations is equal to that of unknowns, there must be a unique solution which can be computed via the normal equations. In general, the use of least-squares minimization leads to better results.

Algorithm 2: a linear programming problem: An LP can be also defined to estimate the depth of the vertices. The linear system $\mathbf{M}\mathbf{x} = \begin{bmatrix} \mathbf{0} \\ \mathbf{p}_z \end{bmatrix}$ is used as a set of constraints in this LP. However, it is essentially useful to have additional estimates for the depth of the mesh vertices in order to ensure accurate results. For this purpose, we use again the output of the polynomial interpolation. From this, additional constraints on the depth of the vertices can be defined as $z_i^{'} - \sigma \leq z_i \leq z_i^{'} + \sigma$. σ is set to a small value (e.g., 0.5 cm) depending on the object's deformations. Apart from these constraints, we need to define an objective function that is best suited to our particular problem. This objective function is defined as summation over all relative depths, which is equal to a linear expression g in terms of the depth of the vertices (with coefficients -1, 0, or $+1$), depending on the direction of the edges. For example, using our conventional directions for a 9×9 mesh, we would have $\sum_{j=1}^{n_e} rz_j = g = z_{73} - z_9$. As a result, the error e_z to be minimized will be:

$$e_z = \sum_{j=1}^{n_e} rz_j - g \tag{10}$$

However, this error must be close to zero but strictly positive. Therefore, we need to specify $e_z \geq 0$. Finally, the depth of all vertices can be estimated via the linear program expressed as

$$\min_z e_z \quad \text{such that}$$

$$e_z = \sum_{j=1}^{n_e} rz_j - g, \quad e_z \geq 0, \quad z_i^{'} - \sigma \leq z_i \leq z_i^{'} + \sigma$$

$$\mathbf{M}\mathbf{x} = \begin{bmatrix} \mathbf{0} \\ \mathbf{p}_z \end{bmatrix}, \quad z_i \geq 0, \quad \forall i \in \{1, \cdots, n_v\} \tag{11}$$

where $\mathbf{M}$ is a $(n_{tr} + n_e + N) \times (n_e + n_v)$ matrix containing the coefficients of the linear system, $\mathbf{x}$ represents the vector comprising z_i and rz_j for all is and js and $\mathbf{p}_z$ indicates the set of all $p_{z,i}$s. This LP problem provides accurate estimates, as will be shown in the experimental results.

4.2 Step 2: estimation of the *xy*-coordinates of the vertices

Assuming that $\mathbf{K}^{\circ}_{\text{rgb}}$ is the calibration matrix of the RGB camera, an optimization procedure is formulated to estimate the variables $\mathbf{q}^{\circ}_{v,i}$ and $\mathbf{q}^{\circ}_{v,i}$ $\left(\mathbf{q}^{\circ}_{v,i} = z_i\mathbf{q}_{v,i} = \begin{bmatrix} u_i^{\circ} & v_i^{\circ} \end{bmatrix}^T\right)$ of vertex i. We call these variables *unnormalized image coordinates*. Such estimation is based on what we call *unnormalized projected lengths* and is performed by means of second-order cone programming (SOCP), consequently determining the full 3D positions of the vertices. This SOCP includes a linear objective function and a set of linear and conic constraints.

Unnormalized projected length: Let us represent $\mathbf{v}_1$ and $\mathbf{v}_2$ as $\mathbf{v}_1 = \begin{bmatrix} x_1 & y_1 & z_1 \end{bmatrix}^T$ and $\mathbf{v}_2 = \begin{bmatrix} x_2 & y_2 & z_2 \end{bmatrix}^T$, respectively . We can derive the difference between the corresponding unnormalized image points $\left(\mathbf{q}^{\circ}_{v,1} = \begin{bmatrix} z_1u_1 & z_1v_1 \end{bmatrix}^T \text{ and } \mathbf{q}^{\circ}_{v,2} = \begin{bmatrix} z_2u_2 & z_2v_2 \end{bmatrix}^T\right)$ as follows: $z_1u_1 - z_2u_2 = f(x_1 - x_2), \quad z_1v_1 - z_2v_2 = f(y_1 - y_2)$. By squaring and subsequently computing the sum of these two equations, we obtain this:

$$(z_1u_1 - z_2u_2)^2 + (z_1v_1 - z_2v_2)^2 = f^2\left[(x_1 - x_2)^2 + (y_1 - y_2)^2\right] \tag{12}$$

Note that the 3D length of an edge can be expressed as

$$L^2 = (x_1 - x_2)^2 + (y_1 - y_2)^2 + (z_1 - z_2)^2 \tag{13}$$

For isometric deformations, the geodesics between any two points on the surface is constrained to a constant value. The Euclidean distance between these two points can be assumed to equal the corresponding geodesics when the edge connecting them is generally short length and the deformations do not cause sharp creases along this edge. Therefore, let us assume that L does not change and can be pre-computed from the template. Note that z_1 and z_2 have been already determined. With these results, Eq. 13 can be rewritten as $L^2 - (z_1 - z_2)^2 = (x_1 - x_2)^2 + (y_1 - y_2)^2$. Thus, the right hand side of Eq. 12 can be easily calculated with the equation above. We define as the *unnormalized projected length* the square root of the left-hand side of Eq. 12

$$l = \sqrt{(z_1u_1 - z_2u_2)^2 + (z_1v_1 - z_2v_2)^2} \tag{14}$$

SOCP optimization: Equation 14 introduces a quadratic constraint. Such a constraint may not be satisfied if folds between mesh vertices occur. To deal with that issue, we replace the above constraint by a variation that allows the vertices to move closer. So, it can be relaxed into a conic constraint as

$$\sqrt{(u^{\circ}_{i,1} - u^{\circ}_{i,2})^2 + (v^{\circ}_{i,1} - v^{\circ}_{i,2})^2} \leq l \tag{15}$$

where $i \in \{1, \cdots, n_e\}$. The above conic constraint is applied to each edge of the mesh. According to Eq. 1, the unnormalized image coordinates of feature point k (i.e., $\mathbf{q}_k^\circ = p_{z,k}\mathbf{q}_k = \left[\, u_{f,k}^\circ \; v_{f,k}^\circ \,\right]^T$) can be represented as

$$\begin{aligned} u_{f,k}^\circ &= a_{k1}u_1^\circ + a_{k2}u_2^\circ + a_{k3}u_3^\circ, \\ v_{f,k}^\circ &= a_{k1}v_1^\circ + a_{k2}v_2^\circ + a_{k3}v_3^\circ, \end{aligned} \tag{16}$$

where $k \in \{1, \cdots, N\}$. The linear equations above hold for all the feature points. In these equations, the left-hand side represents the observed unnormalized image coordinates of the feature points, while the right-hand side represents the estimated coordinates. The cost function being minimized is the geometric distance between these two terms. However, in formulating our optimization as a SOCP, this error is not used as the objective function but as a conic constraint:

$$\left(\sum_{k=1}^{N} \left\| \begin{bmatrix} u_{f,k}^\circ \\ v_{f,k}^\circ \end{bmatrix} - \sum_{j=1}^{3} a_{kj} \begin{bmatrix} u_j^\circ \\ v_j^\circ \end{bmatrix}^{[k]} \right\|^2 \right)^{\frac{1}{2}} \leq \sigma_{uv} \tag{17}$$

Finally, the appropriate SOCP is formulated like this: $\min_{\mathbf{q}_v^\circ} \sigma_{uv}$ such that Eqs. 15 and 17 are satisfied. When applied to a number of different mesh deformations of a generic isometric surface, the approach detailed in this section results in the training data set required to reconstruct other isometric surfaces by the use of only a normal camera and from a single view, as discussed in the previous section.

5 Experiments and results

5.1 Synthetic data

With synthetic data that exactly simulates and conforms to various deformations of a 9 × 9 mesh, we have evaluated both reconstruction schemes proposed, in order to validate their efficiency. The evaluation comprised a number of experiments, conducted with a set of feature points ($N = 60$) well distributed over the mesh triangles. From the planar template mesh, the barycentric coordinates can be computed—see Fig. 3. The virtual RGB camera model is defined such that the focal length is $f = 268$ pixels. With this model, point correspondences across the simulated deformations were projected onto the virtual RGB image plane, assuming that the simulated mesh is placed 50 cm in front of the camera (along the optical axis). To perform the quantitative evaluation, it is necessary to define some numerical metrics as follows:

- To evaluate the results from mesh depth recovery, obtained by linear programming, the following criterion is adopted:
  $\text{DepthAccuracy} = \frac{1}{n_v}\sum_{i=1}^{n_v}\left[\left\|z_{v,i} - \hat{z_{v,i}}\right\|^2 / \left\|\hat{z_{v,i}}\right\|^2\right]$
  Mesh depth estimates are strongly affected by errors mainly due to the errors on the depth estimates of the feature points—see Fig. 4.
- Point reconstruction error (PRE): the normalized Euclidean distance between the observed ($\hat{\mathbf{p}}_i$) and estimated ($\mathbf{p}_i$) feature points:
  $\text{PRE} = \frac{1}{N}\sum_{i=1}^{N}\left[\left\|\mathbf{p}_i - \hat{\mathbf{p}}_i\right\|^2 / \left\|\hat{\mathbf{p}}_i\right\|^2\right]$

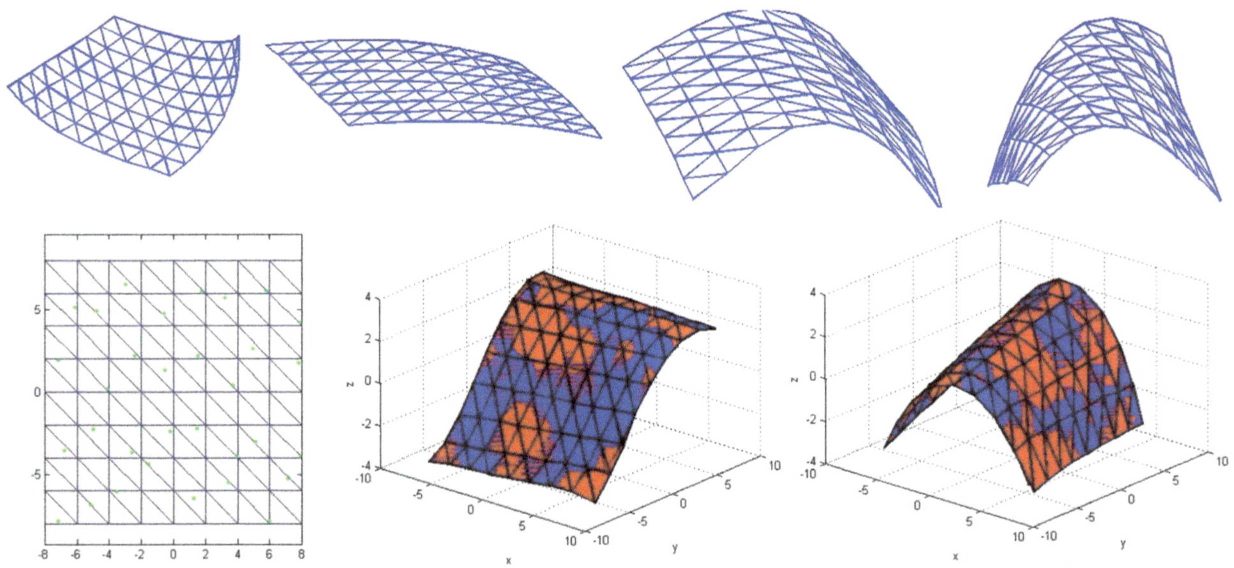

Fig. 3 *Top*—some simulated deformations. *Bottom—left*: A 9 × 9 template mesh with feature points—radius = 20 cm. *Right*: metric coordinates in centimeters—overlap between the ground-truth shape (*blue*) and the estimated shape (*red*)

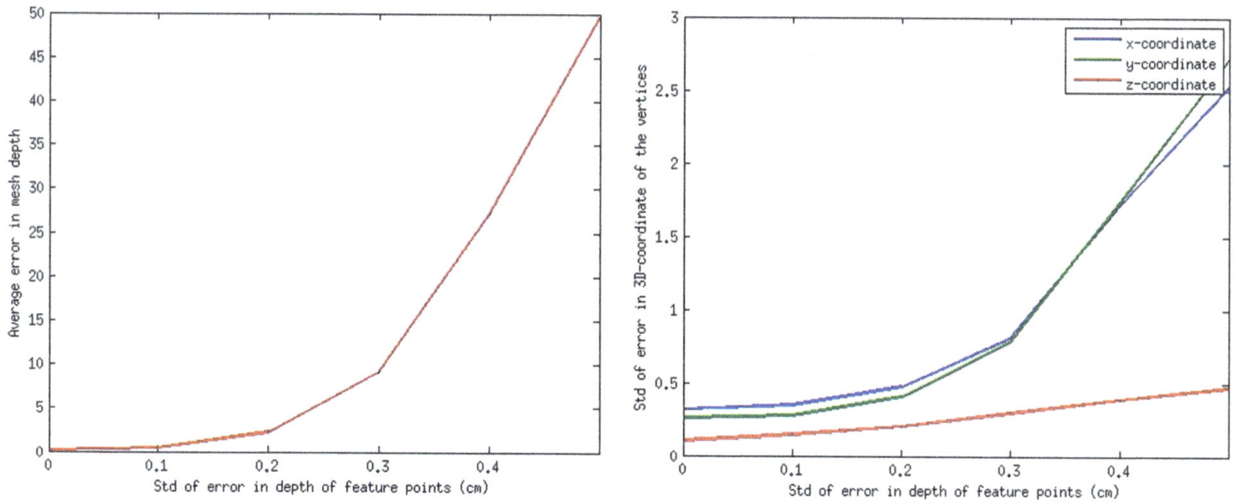

Fig. 4 D-RGB-based reconstruction. *Left*: the error on the depth estimates of the mesh vertices, computed via LP. *Right*: The std of the global error on the estimates of the positions of the mesh vertices

- Mesh reconstruction error (MRE): the normalized Euclidean distance between the observed ($\hat{\mathbf{v}}_i$) and estimated ($\mathbf{v}_i$) 3D vertices of the mesh, computed as $\mathrm{MRE} = \frac{1}{n_v}\sum_{i=1}^{n_v}\left[\left\|\mathbf{v}_i - \hat{\mathbf{v}}_i\right\|^2 / \left\|\hat{\mathbf{v}}_i\right\|^2\right]$
- The re-projection error of the feature points is also another measure of precision: $\mathrm{ReprErr} = \frac{1}{N}\sum_{i=1}^{N}\left[\left\|q_i - \hat{q}_i\right\|^2 / \left\|\hat{q}_i\right\|^2\right]$
- The standard deviation of the errors on the estimates of 3D positions of the mesh vertices: the standard deviation of the global error in each coordinate of the mesh vertices estimated with the monocular optimization algorithm (calculated separately for each coordinate).

Note that all quantitative results represent an average obtained from five deformations randomly selected. By performing 500 trials for each deformation, each average value was acquired from 2500 trials.

Experiments on D-RGB-based reconstruction: We obtained results in a set of experiments where Gaussian random noise with five different standard deviation values added to the depth of the synthetic feature points. Noise levels with standard deviations greater than 0.3 cm prevent the LP from giving good results, as shown in Fig. 4. Since image points are also used in the 3D reconstruction, the effect of the noise in the image points was evaluated. To do so, Gaussian noise was also added to the image points (with standard deviation in increments of 0.5 pixels). Figure 5 shows how the reconstruction accuracy behaves as a function of noise level. In the left-hand plot, a zero-mean Gaussian noise with 0.1-cm std in the depth estimates of the feature points was also considered in all the relevant tests. In the right-hand plot, on the other hand, a zero-mean Gaussian noise with 1-pixel standard deviation in image points was also considered in all the relevant tests. Two of the recovered deformations and their equivalent ground truth are illustrated in Fig. 3.

Experiments on monocular reconstruction: To perform such experiments, a deformation model has to be estimated. This model was directly obtained by applying PCA to a comprehensive set of synthetic mesh deformations. In these experiments, Gaussian noise was also added to the image points (with standard deviation in increments of 0.5 pixels). Figure 6 shows how the reconstruction accuracy behaves toward noise.

Comparative evaluation: In the literature, there are several approaches for 3D reconstruction of deformable surfaces. To compare against the approaches described in this paper, we chose the approach presented in [17]. The main reason for selecting this work is because the authors have used both the linear deformation model and the mesh representation and this enables us to make reliable comparisons. They also propose a SOCP problem for the reconstruction and their approach is known to be robust and efficient, in which a linear local deformation model is used to combine local patches into a global surface. We obtain results from this approach and the proposed reconstruction schemes in the absence of noise. The linear deformation model for the approach being compared and our monocular reconstruction was computed from the results of the D-RGB-based reconstruction of the synthetic mesh deformations. Two different cases have been examined: (1) simple deformations with small, moderate creases; (2) complex deformations with large, sharp creases. Accordingly, the comparative results are divided in two cases, as shown in Fig. 7. The charts reveal that the monocular reconstruction outperforms the other two

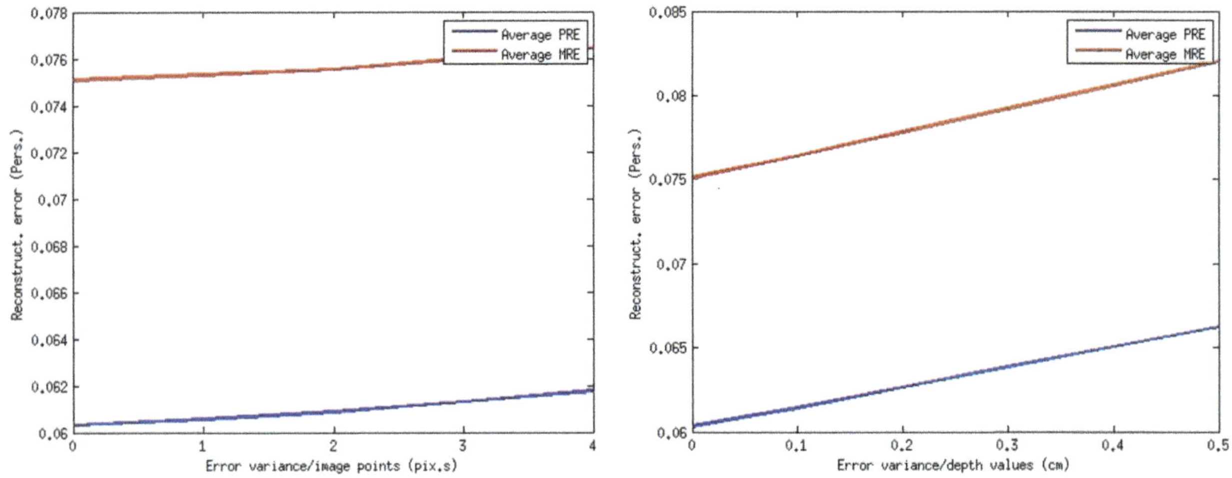

Fig. 5 D-RGB-based reconstruction. *Left*: average PRE and average MRE with respect to the increasing noise in image points. *Right*: Average PRE and average MRE with respect to the increasing noise in depth data

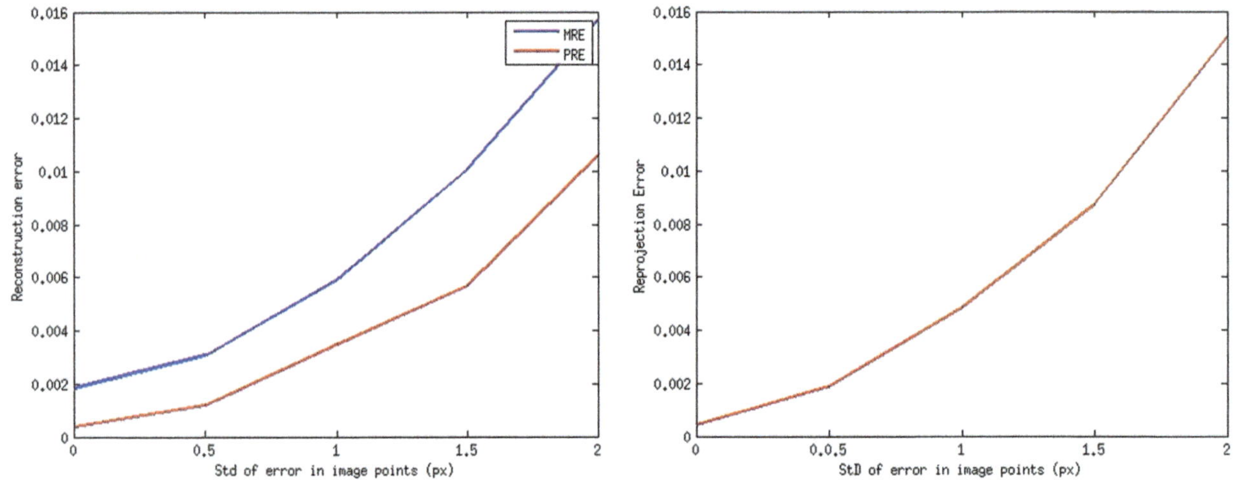

Fig. 6 Monocular reconstruction. Metrics as a function of the noise in image points. *Left*: average PRE and average MRE. *Right*: average reprojection error

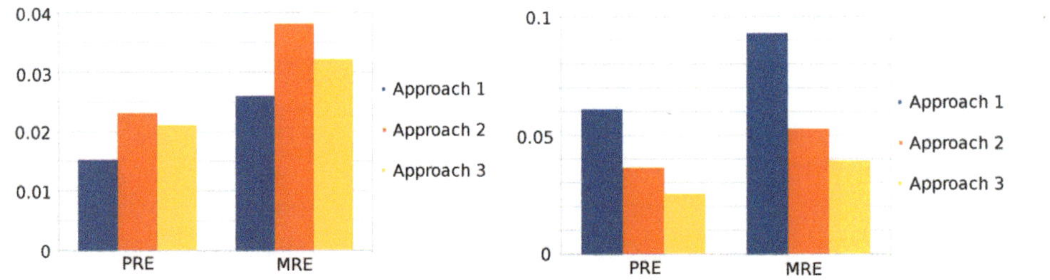

Fig. 7 Left-hand chart: the case of simple deformations; right-hand chart: the case of complex deformations. Approaches 1, 2, and 3 refer to our monocular reconstruction, the approach presented in [17], and the D-RGB-based reconstruction, respectively

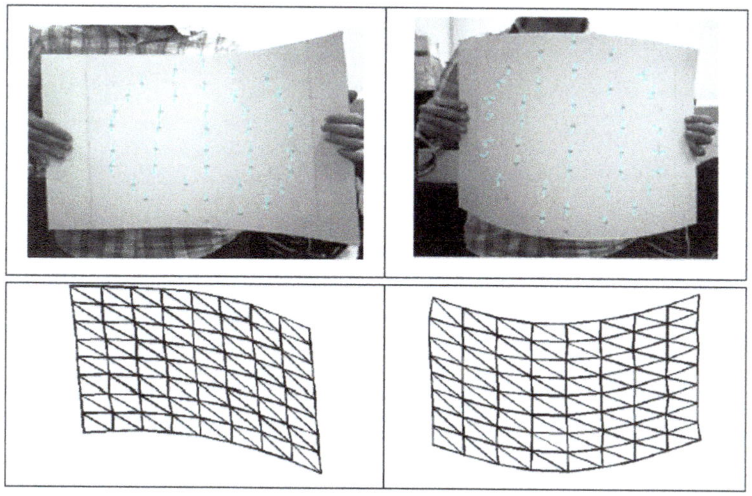

Fig. 8 Real deformations. A 20 × 20-cm square was selected from the intermediate part of the cardboard and then reconstructed

approaches in the case of simple deformations but its performance declines significantly in the case of complex deformations, while the D-RGB-based reconstruction maintains satisfactorily stable performance under different situations (i.e., the results do not vary dramatically).

5.2 Real data

For qualitative assessment, the reconstruction schemes have been tested with real data. A camera setup made up of a high-quality ToF camera and a high-resolution RGB camera was prepared for D-RGB-based reconstruction. The two cameras were calibrated both internally and externally. In the experiments, we used a piece of cardboard flexible enough to allow creating as many different deformations as possible so that the deformation model learned from the reconstruction results could be generalized to other surfaces of different material. The camera setup was located 60 cm in front of the surface being reconstructed, guaranteeing that the FOV of the ToF camera was completely covered by the RGB camera. A regular 9 × 9 mesh was again used to represent the surface, with the positions of the feature points available in relation to the positions of the vertices on the planar template. Such positioning data enabled the calculation of the barycentric coordinates for the feature points. Correspondence of these points across the image sequence was established with respect to the template. The depth of the points, given by the ToF image, was registered with respect to the RGB image. The rest of the implementation was just the same as in the experiments with synthetic data. After having applied the D-RGB-based reconstruction to multiple deformations of the cardboard, the training data necessary to estimate the deformation model was acquired, as shown in Fig. 8. In the case of monocular reconstruction, this deformation model was then employed to reconstruct the same mesh overlaid across such isometric surfaces as those given in Fig. 9, by using only the RGB camera. The qualitative results have shown that this reconstruction scheme yields good results (although no quantitative assessment was possible because of lack of the ground truth). It is worth mentioning that the left and right surfaces in Fig. 9 resemble the cardboard in Fig. 8 in terms of flexibility in particular, whereas the middle surface was made with a different material. However, we reached the conclusion that while the results for the left and right surfaces appeared slightly better than those for the middle

Fig. 9 Isometric surfaces. Real deformations. Courtesy of [18]

surface, a readily deformable cardboard is a proper choice for deriving the linear deformation model.

6 Conclusions

In this paper, we dealt with reconstruction of isometric surfaces. To perform such monocular reconstruction, an algorithm based on the linear deformation model and consisting of a non-linear least-squares optimization was proposed. To find the proper deformation model, prior training data should be used. We therefore provided this prior data by proposing a novel approach for the reconstruction of a typical surface so that the computed deformation model can be also extended to other isometric surfaces. This approach was founded on a range camera along with a conventional camera and its goal is to estimate the 3D positions of the mesh vertices from the depth of the feature points. By applying this approach to multiple mesh deformations, we acquired the training data required. Experimental results showed that both the proposed reconstruction schemes are efficient and result in accurate reconstructions.

Competing interests

The authors declare that they have no competing interests.

References

1. S Srivastava, A Saxena, C Theobalt, S Thrun, *Rapid interactive 3D reconstruction from a single image.* (VMV, 2009), pp. 9–28
2. M Paladini, AD Bue, M Stosic, M Dodig, J Xavier, L Agapito, Factorization for non-rigid and articulated structure using metric projections. Proc. IEEE Conf. Comput. Vis. Pattern Recognit, 2898–2905 (2009)
3. Y Dai, H Li, M He, in *CVPR.* A simple prior-free method for non-rigid structure-from-motion factorization, (2012), pp. 2018–2025
4. D Terzopoulos, J Platt, A Barr, K Fleicher. Elastically deformable models, vol. 21 (ACM SIGGRAPH, 1987), pp. 205–214
5. C Nastar, N Ayache, Frequency-based nonrigid motion analysis. PAMI, 1067–1079 (1996)
6. A Pentland, S Sclaroff, Closed-form solutions for physically based shape modeling and recognition. PAMI, 715–729 (1991)
7. X Llado, AD Bue, L Agapito, *Non-rigid 3D factorization for projective reconstruction.* (BMVC, 2005)
8. I Akhter, Y Sheikh, S Khan, *In defense of orthonormality constraints for nonrigid structure from motion.* (CVPR, 2009), pp. 1534–1541
9. J Xiao, JX Chai, T Kanade, *A closed-form solution to non-rigid shape and motion recovery.* (ECCV, 2004), pp. 573–587
10. H Zhou, X Li, AH Sadka, *Nonrigid structure-from-motion from 2-D images using Markov chain Monte Carlo.* (MultMed, 2012), pp. 168–177
11. J Xiao, T Kanade, in *ICCV.* Uncalibrated perspective reconstruction of deformable structures, vol. 2 (ICCV, 2005), pp. 1075–1082
12. M Brand, *Morphable 3D models from video.* (CVPR, 2001), pp. 456–463
13. A Bartoli, S Olsen, *A Batch Algorithm For Implicit Non-Rigid Shape and Motion Recovery: In ICCV Workshop on Dynamical Vision,* (2005)
14. H Aanæs, F Kahl, *Estimation of deformable structure and motion. Workshop on Vision and Modelling of Dynamic Scenes, ECCV,* (Copenhagen, Denmark, 2002)
15. A Del-Bue, X Llad, L Agapito, *Non-rigid metric shape and motion recovery from uncalibrated images using priors. IEEE Conference on Computer Vision and Pattern Recognition,* vol. 1, (New York, 2006), pp. 1191–1198
16. V Gay-Bellile, M Perriollat, A Bartoli, P Sayd, Image registration by combining thin-plate splines with a 3D morphable model. Int. Conf. Image Process, 1069–1072 (2006)
17. M Salzmann, R Hartley, P Fua, Convex optimization for deformable surface 3D tracking. IEEE Int. Conf. Comput. Vis, 1–8 (2007)
18. M Salzmann, P Fua, Reconstructing sharply folding surfaces: a convex formulation. IEEE Conf. Comput. Vis. Pattern Recognit, 1054–1061 (2007)
19. L Li, S Jiang, Q Huang, Learning hierarchical semantic description via mixed-norm regularization for image understanding. IEEE Trans. Multimed. **14**(5), 1401–1413 (2012)
20. Y Zhang, J Xu, et al., Efficient parallel framework for HEVC motion estimation on many-core processors. IEEE Trans. Circ. Syst. Video Technol. **24**, 2077–2089 (2014)
21. C Yan, Y Zhang, et al., A highly parallel framework for HEVC coding unit partitioning tree decision on many-core processors. IEEE Signal Proc. Lett. **21**, 573–576 (2014)
22. N Gumerov, A Zandifar, A Duraiswami, LS Davis, *Structure of applicable surfaces from single views*, vol. 3023, (Heidelberg, 2004), pp. 482–496
23. M Prasad, A Zisserman, AW Fitzgibbon, Single view reconstruction of curved surfaces. IEEE Conf. Comput. Vis. Pattern Recognit. **2**, 1345–1354 (2006)
24. M Salzmann, R Urtasun, P Fua, Local deformation models for monocular 3D shape recovery. IEEE Conf. Comput. Vis. Pattern Recognit, 1–8 (2008)
25. M Perriollat, R Hartley, A Bartoli, *Monocular template-based reconstruction of inextensible surfaces.* (BMVC, 2008)
26. S Shen, W Shi, Y Liu, Monocular 3-D tracking of inextensible deformable surfaces under L2-norm. IEEE Trans. Image Process. **19**, 512–521 (2010)
27. M Perriollat, R Hartley, A Bartoli, *Monocular template-based reconstruction of inextensible surfaces*, (2010)
28. M Salzmann, F Moreno-Noguer, V Lepetit, P Fua, Closed-form solution to non-rigid 3D surface registration. Eur. Conf. Comput. Vis, 581–594 (2008)
29. M Perriollat, A Bartoli, in *CVPR BenCos Workshop.* A quasi-minimal model for paper-like surfaces, (2007)
30. L Cohen, I Cohen, Finite-element methods for active contour models and balloons for 2-d and 3-d images. PAMI, 1131–1147 (1993)
31. D Metaxas, D Terzopoulos, Constrained deformable superquadrics and nonrigid motion tracking. PAMI, 580–591 (1993)
32. X Llado, ADel Bue, L Agapito. Non-rigid 3D factorization for projective reconstruction, vol. 28 (BMVC, 2005)
33. L Torresani, A Hertzmann, C Bregler, Nonrigid structure-from-motion: estimating shape and motion with hierarchical priors. PAMI, 878–892 (2008)
34. R White, D Forsyth. Combining cues: shape from shading and texture (CVPR, 2006), pp. 1809–1816
35. D Metaxas, D Terzopoulos, Constrained deformable superquadrics and nonrigid motion tracking. PAMI. **15**, 580–591 (1993)
36. F Brunet, R Hartley, A Bartoli, N Navab, R Malgouyres. Monocular Template-based Reconstruction of Smooth and Inextensible Surfaces. Tenth Asian Conference on Computer Vision (ACCV), (Queenstown (New Zealand), 2010)
37. A Bartoli, Y Grard, F Chadebecq, T Collins, in *CVPR.* On template-based reconstruction from a single view: Analytical solutions and proofs of well-posedness for developable, isometric and conformal surfaces, (2012), pp. 2026–2033
38. J Diebel, S Thrun, in *Proc. NIPS.* An application of Markov random fields to range sensing, (2005), pp. 291–298
39. R Yang, J Davis, D Nister, *Spatial-Depth Super Resolution for Range Images. Computer Vision and Pattern Recognition, CVPR '07.* (IEEE Conference, Minneapolis, MN, 2007), pp. 1–8
40. H Kim, YW Tai, MS Brown, *High Quality Depth Map Upsampling for 3D-TOF Cameras. Inso Kweon Computer Vision (ICCV).* (IEEE International Conference, Barcelona, 2011), pp. 1623–1630
41. YM Kim, C Theobalt, J Diebel, J Kosecka, B Miscusik, S Thrun, *Multi-view image and, ToF sensor fusion for dense 3D reconstruction.* (Computer Vision Workshops (ICCV Workshops), 2009), pp. 1542–1549

10

Unified piecewise epipolar resampling method for pushbroom satellite images

Jin-Woo Koh[1,2*] and Hyun-Seung Yang[1]

Abstract

Computational stereo is in the fields of computer vision and photogrammetry. In the computational stereo and surface reconstruction paradigms, it is very important to achieve appropriate epipolar constraints during the camera-modeling step of the stereo image processing. It has been shown that the epipolar geometry of linear pushbroom imagery has a hyperbola-like shape because of the non-coplanarity of the line of sight vectors. Several studies have been conducted to generate resampled epipolar image pairs from linear pushbroom satellites images; however, the currently prevailing methods are limited by their pixel scales, skewed axis angles, or disproportionality between x-parallax disparities and height. In this paper, a practical and unified piecewise epipolar resampling method is proposed to generate stereo image pairs with zero y-parallax, a square pixel scale, and proportionality between x-parallax disparity and height. Furthermore, four criteria are suggested for performance evaluations of the prevailing methods, and experimental results of the method are presented based on the suggested criteria. The proposed method is shown to be equal to or an improvement upon the prevailing methods.

Keywords: Pushbroom high-resolution satellite imagery, Piecewise epipolar resampling, Stereo image pair

1 Introduction

The paradigm of computational stereo [1] is that it proceeds following six major steps: image acquisition, camera modeling, feature acquisition, image matching, distance or depth determination, and interpolation. It is very important to determine appropriate epipolar constraints as a preprocessing step for image matching after the camera or sensor modeling because many stereo image processing algorithms use these constraints to extract three-dimensional information such as positioning and surface data. Among the conditions determined using epipolar constraints are restrictions on the search space for stereo matching, identification and elimination of blunders or outliers from the matching results, and others.

It is well known that stereo images acquired by frame camera(s) have epipolar geometry, that is, there exist epipolar lines and the epipolar pairs on each line throughout the space. However, it has been shown [2] that the epipolar geometry of linear pushbroom camera images are not lines but hyperbola-like non-linear curves because of the non-coplanarity of line of sight vectors. Therefore, an epipolar curve for the entire image can only be approximated incrementally in linear segments, and epipolar pairs exist only locally.

Several studies have been conducted to establish the epipolar geometry of the pushbroom camera to achieve effective epipolar image resampling. For example, Oh et al. [3] proposed a piecewise approach to this problem with rational polynomial coefficient (RPC); this is an image space-based approach and was shown to achieve almost zero y-parallax, as well as a linear relationship between the x-parallax and the ground height. Alternatively, Wang et al. [4] suggest a method that implements a new epipolarity model based on the projection reference plane (PRP); this is an object space-based approach, and experimental results showed that the vertical parallaxes all attained sub-pixel levels in along-track and cross-track stereo images from actual equipment, including Earth observation satellites SPOT-5, IKONOS, IRS-P5, and QuickBird.

Because zero y-parallax is one of the most important features of epipolar constraints in stereo image processing applications, zero y-parallax should be used as an

* Correspondence: xinukoh@kaist.ac.kr
[1]Korea Advanced Institute of Science and Technology (KAIST), School of Computing, 291 Daehak-ro, Yuseong-gu, Daejeon 34141, Republic of Korea
[2]Agency for Defense Development (ADD), P.O. Box 35, Yuseong, Daejeon 34188, Republic of Korea

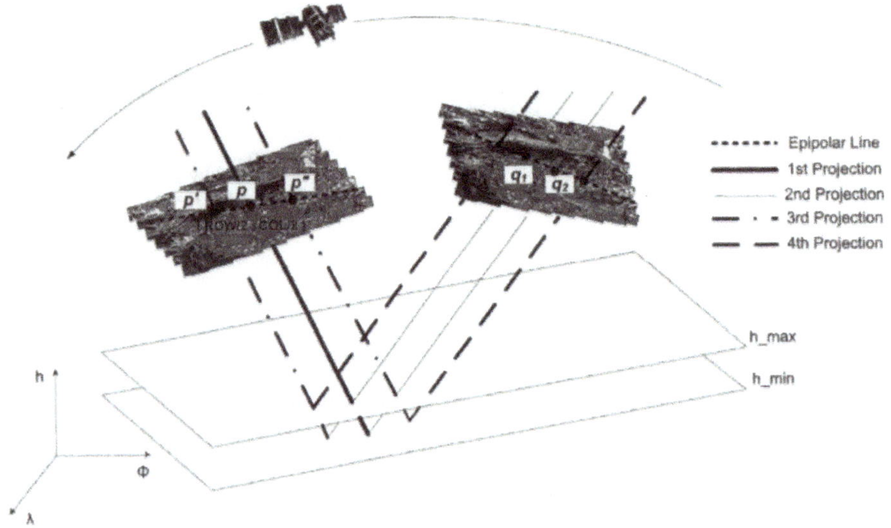

Fig. 1 Piecewise epipolar curve pair generation [3]

evaluation criterion. Therefore, the experimental results were presented to show whether the proposed algorithms satisfied this criterion and could be considered valid for use in applications.

In this paper, we suggest a modified piecewise epipolar resampling method to obtain an epipolar geometry that can be effectively applied to construct stereo image pairs from linear pushbroom sensors. The modified algorithm is based on the piecewise approach [3, 4]; however, in the proposed method, the distance in the epipolar image space is determined by the distance in the object space, as described in detail in Section 3.

Furthermore, we suggest four performance evaluation criteria for epipolar resampling algorithms: (1) geopositioning accuracy of the resampled epipolar images relative to the original sensor model, (2) square pixel scale and perpendicularity of axis in object space, (3) near zero y-parallax in epipolar image space, and (4) proportionality between the disparity in the x-parallax and the ground height. Using these criteria, performance evaluations of the previously suggested algorithms and the algorithm proposed in this paper are presented based on experimental data from satellite images, including optical images such as those from IKONOS, Pleiades, and WorldView, as well as synthetic aperture radar (SAR) images as those from KOMPSAT-5 and TerraSAR-X.

2 Previous studies

2.1 Image space-based method

Oh et al. [3] suggested a piecewise approach to epipolar resampling of pushbroom satellite images with RPC [5, 6]. It was based on the researches presented by Habib et al. [7] and Kim [2]. These previous works showed that epipolar curves of linear pushbroom images are not lines but hyperbola-like non-linear curves and that an epipolar curve for the entire image can only be approximated by piecewise linear segments, with epipolar curve pairs

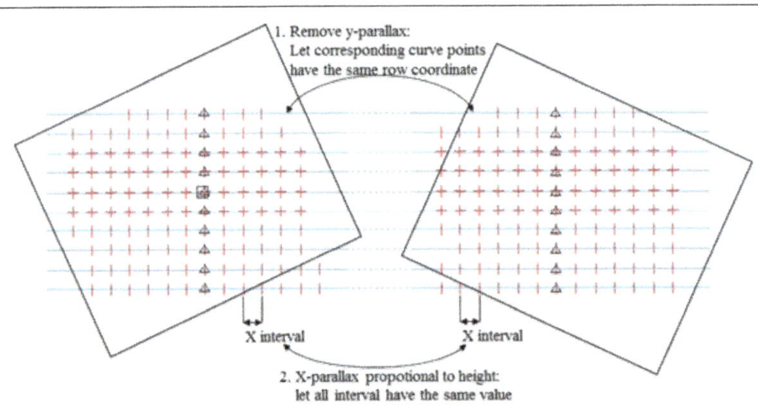

Fig. 2 Extracted points on the piecewise epipolar curves [3]

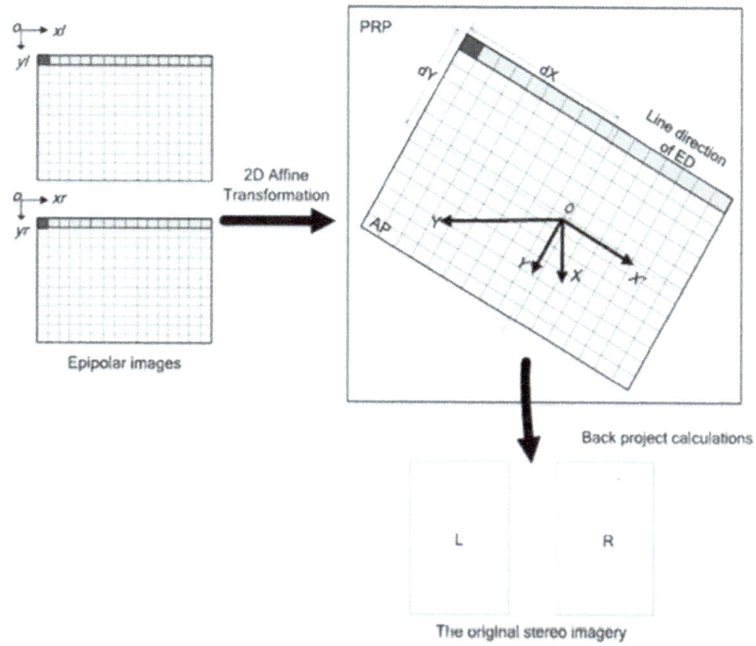

Fig. 3 Coordinate transformation relation of epipolar and the original images [4]

existing only locally. Based on the previous results, they showed locally existing epipolar pairs for the entire space represented by the epipolar curve.

Figure 1 shows the essential elements of this method. For a selected point in the left image, *p*, the corresponding control point in the right image for *p*at the maximum ground height, h_max, is $q1$, and $q2$ is the corresponding control point for *p* at the minimum ground height, h_min. The curve between $q1$ and $q2$ in the right image can be approximated as a line segment. Likewise, the piecewise epipolar curve in the left image for $q1$, now a reference point in the right image, can be established in the same way, from control points p' and p'' This process is continued until the entire spaces covered by the left and/or right images are delineated.

Figure 2 shows all of the extracted points on the piecewise epipolar curves for the left and right images. Within this epipolar image space, the control points on each approximately horizontal line should be relocated to a truly horizontal line, and the intervals between points should

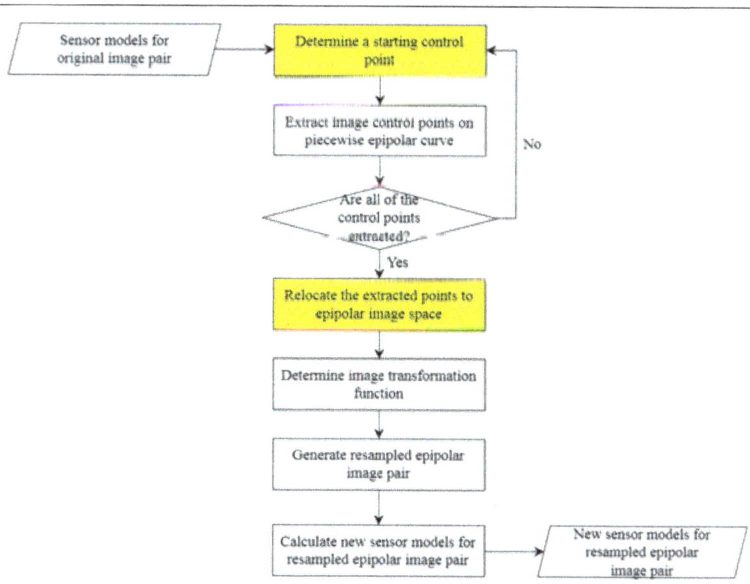

Fig. 4 Basic steps of the proposed algorithm for piecewise epipolar resampling

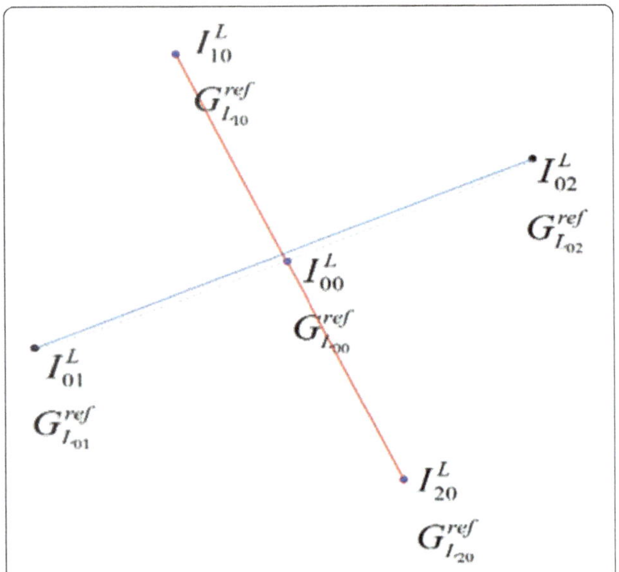
Fig. 5 Determination of two starting control points

produce unexpected results, such as a non-square pixel scale or a non-orthogonal axis in object space, discussed more detail in subsequent sections.

all have the same pixel distance, guaranteeing (1) almost zero y-parallax, (2) x-parallax proportional to height.

This method efficiently generates piecewise resampled epipolar image pairs because most of the steps are performed in original or epipolar image space, except when calculating ground coordinates. They also described that the starting point in each epipolar curve, which is marked with triangle in Fig. 2, can be established along the orthogonal direction to the trajectory, i.e., orthogonal to the fitted line using the first epipolar curve generated, and a predefined interval. This process may

2.2 Object space-based method

Wang et al. [4] suggested a new epipolarity model based on the PRP in object space, which was first introduced using a virtual horizontal plane (VHP) [8]. Like the image space-based method, the PRP method also extracts control points from the original image space but assigns the control points' coordinates to new coordinates on a PRP in object space instead of image space. This method does not require any ground control points, which is essential to the method suggested by Morgan et al. [9]. And this method may not cause a non-orthogonal axis in the object space.

The generated epipolar curve image points are first mapped on the PRP, which is defined such that points on the PRP have the same elevation in the local vertical coordinate system. After determining the direction of the approximated epipolar curve, only an affine transformation is needed to generate resampled epipolar images. Figure 3 shows the coordinate transformation between the epipolar images and the original stereo imagery.

Wang et al. described experimental results demonstrating feasibility of this method, with residuals from the fitted straight lines of less than 1 mm, and a root mean square of the vertical parallax of less than half of a pixel. Unlike the image space-based

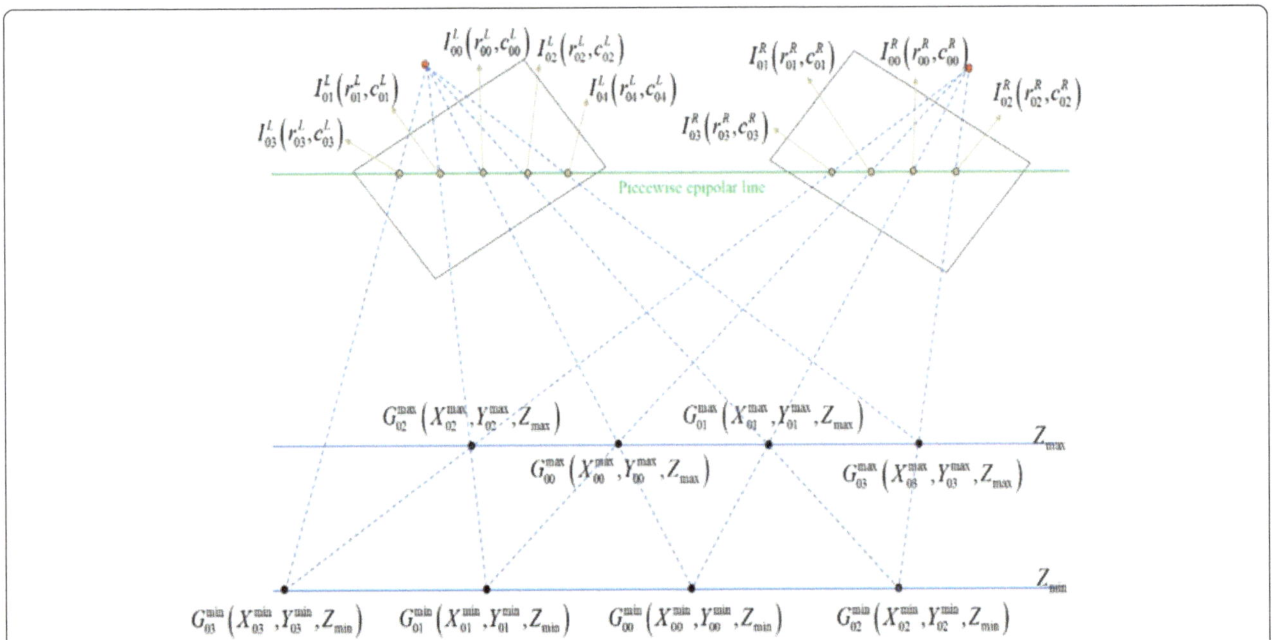

Fig. 6 Piecewise epipolar curve generation using the proposed algorithm

method, the PRP method uses the distance on the PRP as the pixel distance between the neighboring control points in epipolar image space; as a result, the horizontal parallax may not be proportional to the ground height.

3 Proposed method

This section describes a modified algorithm to perform piecewise epipolar resampling that ensures (1) preservation of the geo-positioning accuracy of the original sensor model in the resampled epipolar images, (2) scales with equidistant pixels and perpendicularity of the axis, (3) near zero y-parallax, and (4) proportionality between the disparity in the x-parallax and the ground height. The proposed algorithm is a unified one because it can be applied to both along-track and cross-track stereo images and can also be applied to SAR stereo images for radargrammetry, as shown by experimental results reported in Section 5. The prevailing methods can also be applied to all of these types of stereo images, but the proposed one uses only the CSM [10] interface, stated in Section 3.2, so that it can be applied to all of stereo image pairs regardless of the sensor types, and furthermore, it shows better performance in terms of the four performance evaluation criteria.

The basic steps of the proposed algorithm are shown in Fig. 4. The most significant innovation of the proposed algorithm is implemented in the first and the third step. In the first step, a starting control point on each horizontal line is determined such that it is located on the orthogonal line in the object space. In the third step, the generated epipolar curve image points are relocated to epipolar image space with the distances in the object space such that the two axes are perpendicular to each other and the image points satisfy the epipolar constraint.

3.1 Determine a starting control point

The first step is to determine a starting control point in the reference image to generate the control points on each epipolar curve pair. For example, the starting control point is p in Fig.1. The first starting control point can be determined as the one in the center of the reference image. The image control point I_{00}^{L}in Fig. 5 is a starting control point, and I_{01}^{L} and I_{02}^{L} are the neighbor image control points of I_{00}^{L} on the same epipolar curve. To fit $\overline{I_{01}^{L}I_{00}^{L}}$ and $\overline{I_{00}^{L}I_{02}^{L}}$to a horizontal line$l$in epipolar image space, the next step is to determine two vertical points, i.e., two starting control points, on a vertical line v passing through the control point I_{00}^{L}, where $l \cdot v = 0$. This vertical line can be approximately computed as shown in Fig. 5. The line l passes I_{01}^{L} and I_{02}^{L}, and the line v is perpendicular to l. The two vertical points, I_{10}^{L} and I_{20}^{L}, are selected among the points on the line v, and the two intervals, $\overline{I_{00}^{L}I_{10}^{L}}$ and $\overline{I_{00}^{L}I_{20}^{L}}$, can be determined to be the same as the interval between I_{01}^{L} and I_{00}^{L}. The interval determination is not important in this step because all the distances between points, horizontal and vertical, are here calculated in the local reference plane prior to their relocation to epipolar image space, as detailed in the next section.

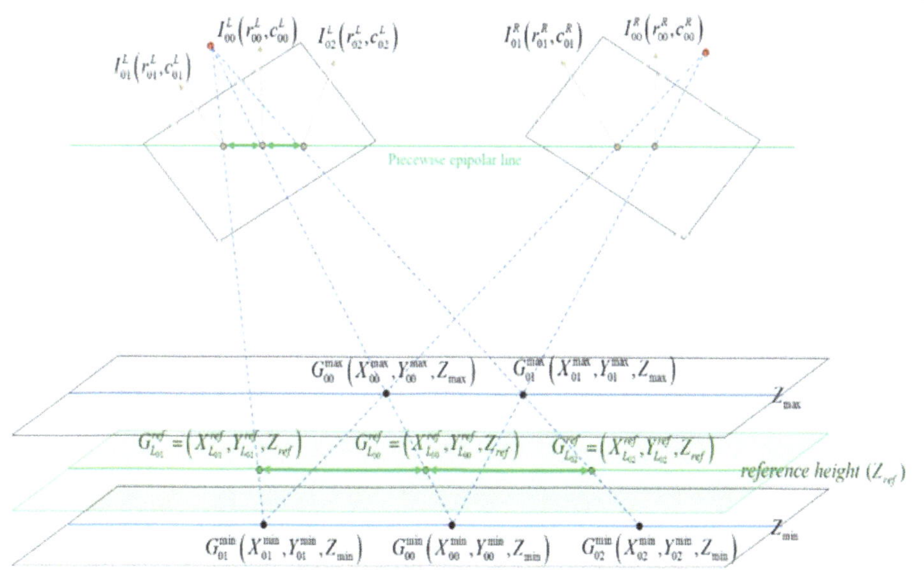

Fig. 7 Determination of epipolar image coordinates using ground space organized as a local reference plane

Table 1 Features of stereo image pairs from different satellites for experimental evaluation

Stereo imagery	Acquisition date	Image size (pixels)	Nominal GSD (m)
IKONOS-2 (I)	2002-02-07	11004 × 11004	1.0
	2002-02-07	11004 × 11004	1.0
WV-2 (W)	2010-04-23	19484 × 33990	0.5
	2010-04-23	19484 × 35180	0.5
Pleiades (P)	2013-04-22	18827 × 17411	0.5
	2013-04-22	19047 × 17208	0.5
Kompsat-5 (K)	2014-03-20	6194 × 7534	1
	2014-03-17	3544 × 8125	1
TerraSAR-X (T)	2008-11-14	16480 × 28559	1
	2008-11-20	17408 × 29861	1

3.2 Extract image control points on piecewise epipolar curve

The second step is to generate piecewise epipolar curve image points from original stereo images. As with the previously reported methods, this step is based on the fact that the hyperbola-like epipolar curve for a reference image point exists between two control points that correspond to the minimum and maximum ground height of the reference image point.

As shown in Fig. 6, a point $I_{00}^{L}(r_{00}^{L}, c_{00}^{L})$ in the reference image (left or right) is selected as the reference image point for determination of a piecewise epipolar curve with $G_{00}^{\min}(X_{00}^{\min}, Y_{00}^{\min}, Z^{\min})$ and $G_{00}^{\max}(X_{00}^{\max}, Y_{00}^{\max}, Z^{\max})$ describing the reference image point at minimum and maximum ground height $Z^{\min}$ and $Z^{\max}$, respectively. This can be done using one of the community sensor model (CSM) [10] application program interface (API) commands, "image2-Ground()," or any other type of rigorous sensor model or RPC [6].

The corresponding control points, $I_{00}^{R}(r_{00}^{R}, c_{00}^{R})$ and $I_{01}^{L}(r_{01}^{L}, c_{01}^{L})$, in the stereo image pair (right or left) for $G_{00}^{\min}$ and $G_{00}^{\max}$ are computed by direct coordinate transformation, using the command "ground2Image()" in CSM APIs. Subsequently, $I_{01}^{L}(r_{01}^{L}, c_{01}^{L})$ and $I_{02}^{L}(r_{02}^{L}, c_{02}^{L})$ are computed for $I_{00}^{R}(r_{00}^{R}, c_{00}^{R})$ and $I_{00}^{R}(r_{00}^{R}, c_{00}^{R})$ using the ground points $G_{01}^{\min}(X_{01}^{\min}, Y_{01}^{\min}, Z^{\min})$ and $G_{01}^{\max}(X_{01}^{\max}, Y_{01}^{\max}, Z^{\max})$, respectively. Successive iterations of this process continue until one of the image coordinates extends beyond the boundary of the reference image space.

The coordinate transformation between the object space and the original image space is denoted by the following equations, in which i is the i th line, j is the jth sample in i th line, and I_{ij}^{L} and I_{ij}^{R} are the image coordinates in the left and right image spaces, respectively.

$$
\begin{aligned}
I_{ij}^{L} &= \begin{cases} \text{ground2Image}\left(G_{ij}^{\min}\right), & \text{if j is odd} \\ \text{ground2Image}\left(G_{i(j-1)}^{\max}\right), & \text{if j is even} \end{cases} \quad i \ge 0,\ j > 0 \\
I_{ij}^{R} &= \begin{cases} \text{ground2Image}\left(G_{ij}^{\min}\right), & \text{if j is even} \\ \text{ground2Image}\left(G_{i(j-1)}^{\max}\right), & \text{if j is odd} \end{cases} \quad i \ge 0,\ j > 0 \\
G_{ij}^{\min} &= \begin{cases} \text{image2Ground}\left(I_{ij}^{R}, Z^{\min}\right), & \text{if j is odd} \\ \text{image2Ground}\left(I_{ij}^{L}, Z^{\min}\right), & \text{if j is even} \end{cases} \quad i \ge 0,\ j \ge 0 \\
G_{ij}^{\max} &= \begin{cases} \text{image2Ground}\left(I_{i(j-1)}^{R}, Z^{\max}\right), & \text{if j is odd} \\ \text{image2Ground}\left(I_{i(j-1)}^{L}, Z^{\max}\right), & \text{if j is even} \end{cases} \quad i \ge 0,\ j \ge 0
\end{aligned}
\tag{1}
$$

3.3 Relocate the extracted points to epipolar image space

The next step is to relocate the control points to epipolar image space. The point I_{00}^{L} and its conjugate point I_{00}^{R} are relocated to the origin in epipolar image space. Note

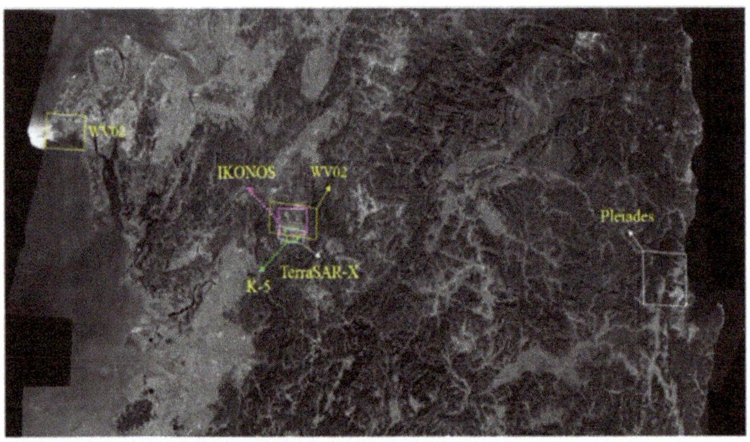

Fig. 8 Areas covered stereo image pairs from different satellites for experimental evaluation

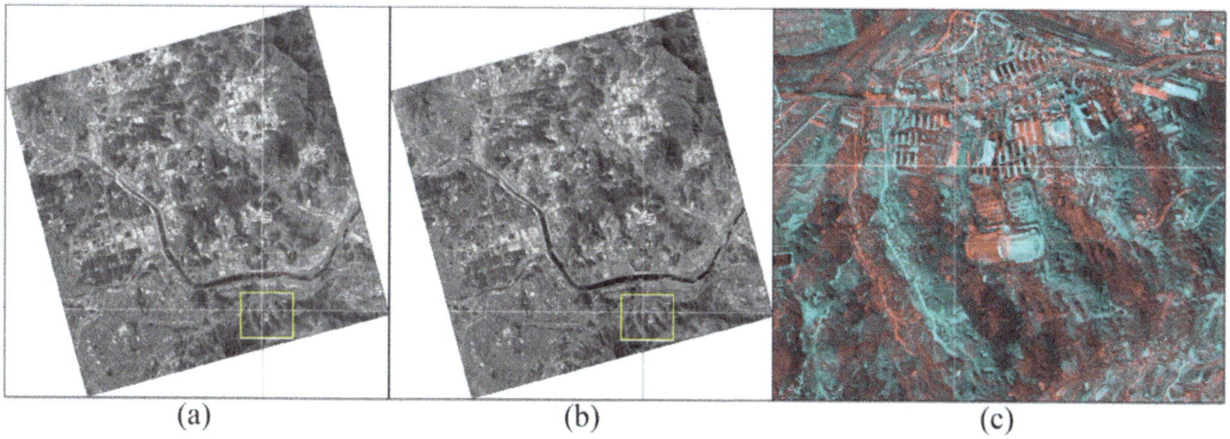

Fig. 9 Epipolar resampled stereo image pair generated by the proposed method (IKONOS-2). **a** Left image. **b** Right image. **c** Segment of anaglyph stereo image pair

that I_{00}^{R} represents the conjugate point for I_{00}^{L} at the minimum ground height; however, it can be any point between I_{00}^{R} and I_{01}^{R}.

To relocate all of the control points to epipolar image space requires the determination of all of the intervals between neighboring control points on the horizontal lines approximated by each piecewise epipolar curve. In addition, the intervals between all of the neighboring control points on the vertical lines have to be computed in the same way.

A local reference plane is used to compute the intervals between control points. As shown in Fig. 7, the distance between two neighboring control points is computed on a local reference plane with ground height Z^{ref}. For example, $G_{L_{0j}}^{\text{ref}}$ is the ground point where the control point I_{0j}^{L} is projected to a local plane with ground height of Z^{ref}. The distance between $G_{L_{00}}^{\text{ref}}$ and $G_{L_{01}}^{\text{ref}}$ can be calculated, and this distance is used to determine the pixel distance between I_{00}^{L} and I_{01}^{L} in epipolar image space. All of the pixel distances are determined in this way.

3.4 Determine the image transformation function

In the previous step, all of the original control points were relocated to epipolar image space, where they

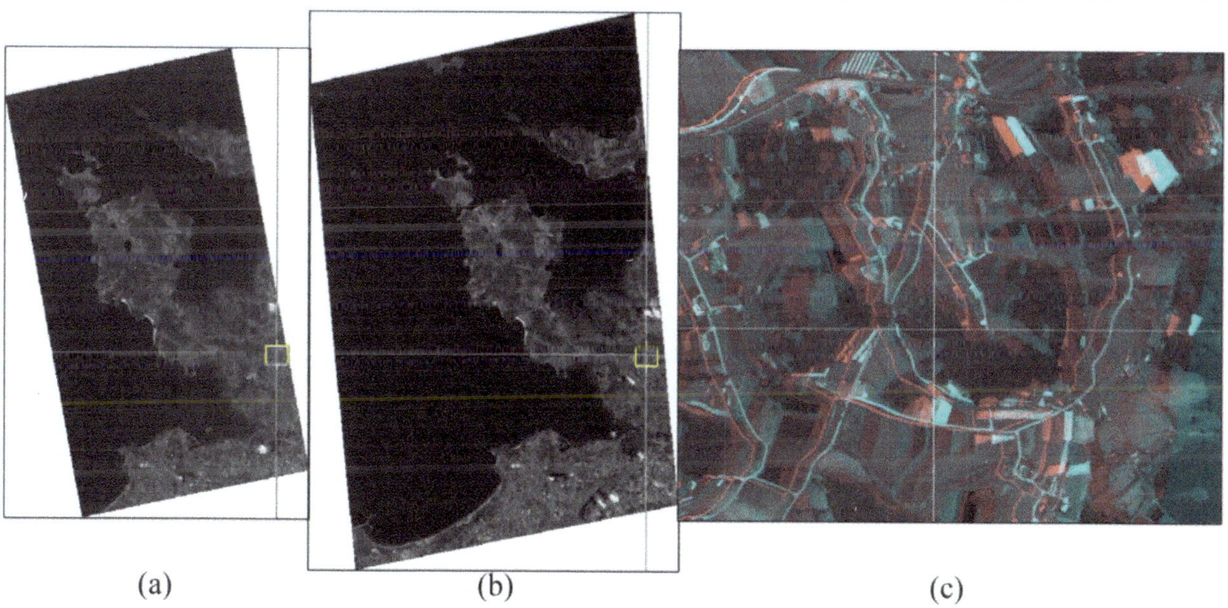

Fig. 10 Epipolar resampled stereo image pair generated by the proposed method (WorldView-2). **a** Left image. **b** Right image. **c** Segment of anaglyph stereo image pair

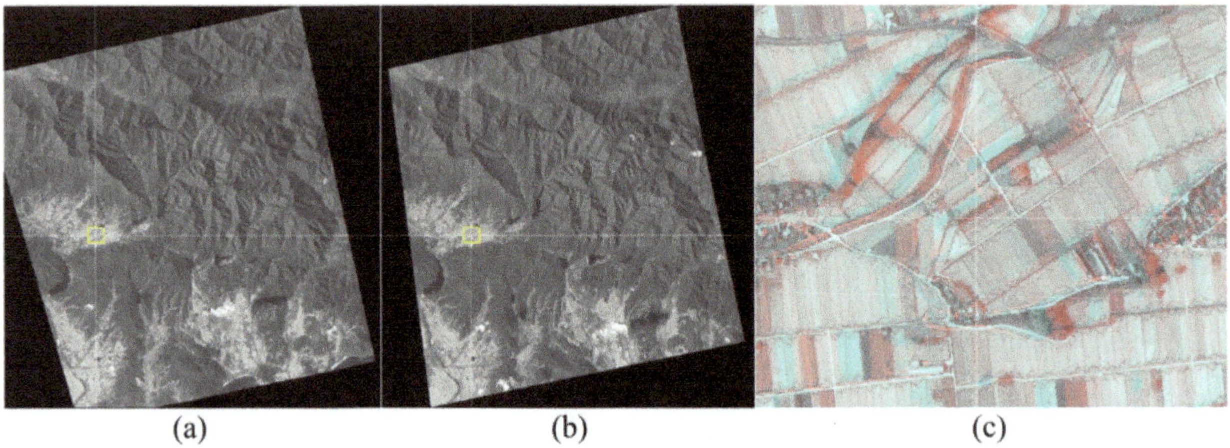

Fig. 11 Epipolar resampled stereo image pair generated by the proposed method (Pleiades). **a** Left image. **b** Right image. **c** Segment of anaglyph stereo image pair

become epipolar image points, expressed as $I_{ij}^{L'}\left(r_{ij}^{L'}, c_{ij}^{L'}\right)$ and $I_{ij}^{R'}\left(r_{ij}^{R'}, c_{ij}^{R'}\right)$. Subsequently, the polynomial image transformation function, as well as the reverse transformation function, can be established as follows.

$$\begin{aligned} r'_{ij} &= f(a) = a_1 + a_2 r_{ij} + a_3 c_{ij} + a_4 r_{ij} c_{ij} + \cdots \\ &\quad + a_{\frac{(n+1)(n+2)}{2}-1} r_{ij}^n + a_{\frac{(n+1)(n+2)}{2}} c_{ij}^n \\ c'_{ij} &= g(b) = b_1 + b_2 r_{ij} + b_3 c_{ij} + b_4 r_{ij} c_{ij} + \cdots \\ &\quad + b_{\frac{(n+1)(n+2)}{2}-1} r_{ij}^n + b_{\frac{(n+1)(n+2)}{2}} c_{ij}^n \end{aligned} \tag{2}$$

The coefficients of each polynomial function can be computed using the least squares method. The order of the polynomial function can be adapted to achieve the desired precision for the image transformation. We used the fifth order polynomial function for the performance evaluation of the algorithms.

3.5 Calculate new sensor models for resampled epipolar image pair

It is required to calculate or generate new sensor models for each resampled epipolar image; for example, new RPCs can be easily generated from the equation proposed by Tao and Hu [5]. First, the ground coordinates in a cubic grid are generated within the effective ground space of the original stereo image pair. Subsequently, epipolar image coordinates for the pair are computed using a "ground2Image()" for each. It is possible to compute image coordinates in epipolar image space using the image transformation function established in the previous section. It has been shown that both the replacement sensor model (RSM) [11] and the

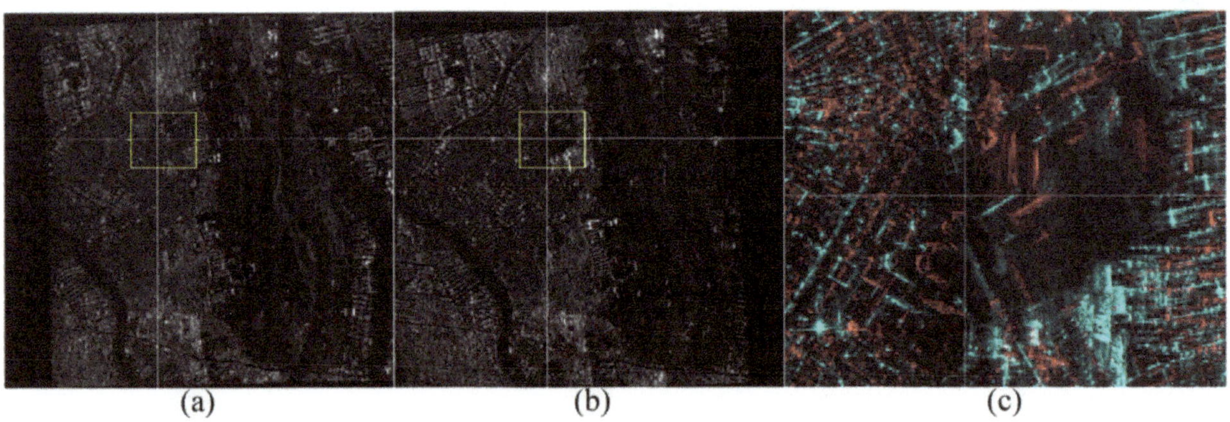

Fig. 12 Epipolar resampled stereo image pair generated by the proposed method (KOMPSAT-5). **a** Left image. **b** Right image. **c** Segment of anaglyph stereo image pair

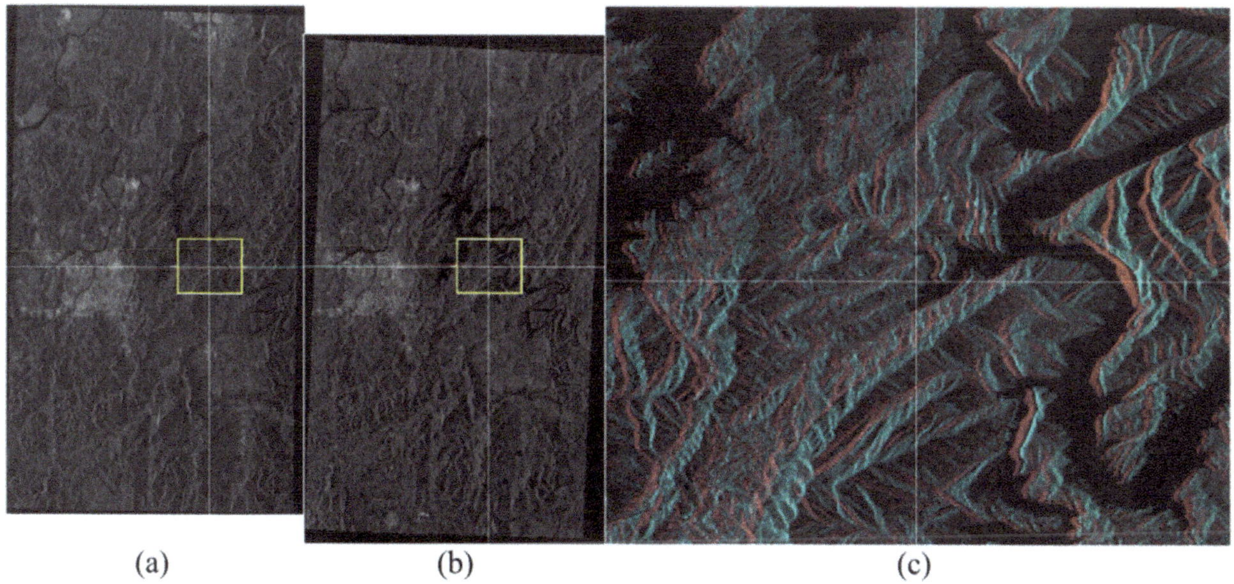

Fig. 13 Epipolar resampled stereo image pair generated by the proposed method (TerraSAR-X). **a** Left image. **b** Right image. **c** Segment of anaglyph stereo image pair

RPC [12] can express the relationship between ground and image space for most of sensor types.

A strict requirement of the process of determining the image transformation function is that the function must preserve geo-positioning accuracy with respect to the original sensor model. Therefore, the newly generated RPCs should maintain the relationships expressed by the forward and reverse coordinate transformation functions.

4 Performance evaluation criteria and analysis procedure

As stated in the introduction, we suggest four performance evaluation criteria for piecewise epipolar resampling algorithms: geo-positioning accuracy, pixel scale and perpendicularity of axis in object space, near zero y-parallax in epipolar image space, and proportionality between x-parallax disparity and ground height. The third and the fourth criteria have

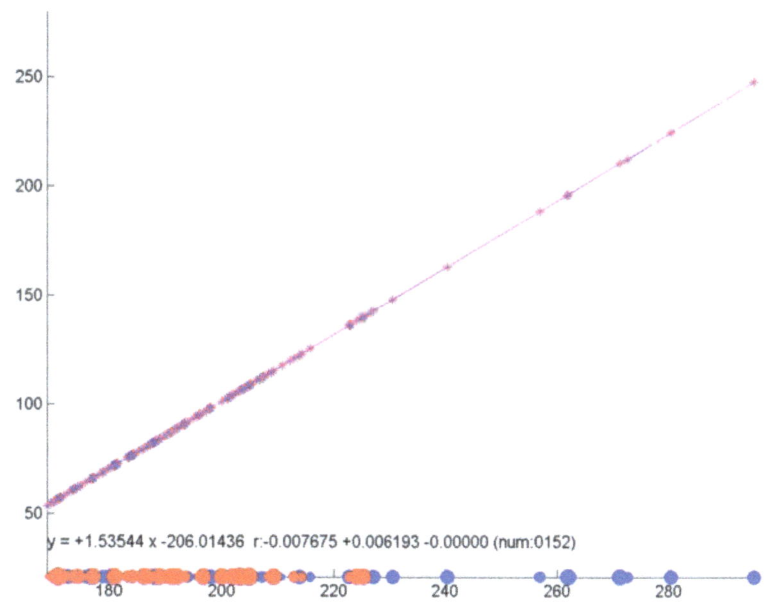

Fig. 14 Proportionality between x-parallax disparity and ground height

Fig. 15 An image matching result between the left and right epipolar image pair from IKONOS-2. **a**, **b** the left and the right image, respectively. **c** The left image which shows the reversed matching result

already been mentioned in several previous studies [3, 4, 7, 8].

4.1 Geo-positioning accuracy

The geo-positioning accuracy of the epipolar resampling algorithm can be measured by comparing the original and generated sensor models' forward coordinate transformation results for epipolar image point pairs. Ground coordinates, *G*, extracted from the cubic grid in the effective ground space are used to compute image coordinates from each of the original reference image points, and the image coordinates are relocated into epipolar image space using the image transformation polynomial function. New

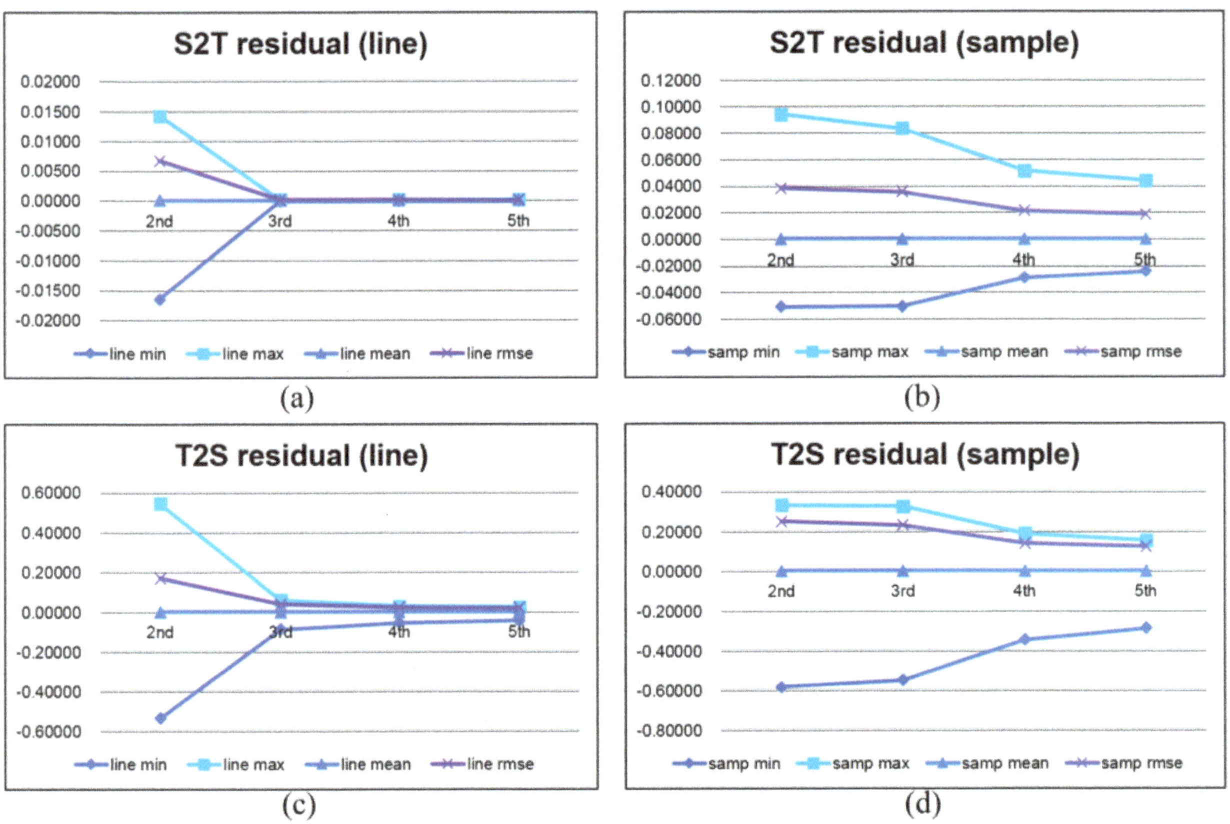

Fig. 16 Accuracy of image transformation function according to the order

Table 2 Geo-positioning accuracy of each method (unit: meter)

		Image space-based			Object space-based			Proposed		
		Horizontal		Vertical	Horizontal		Vertical	Horizontal		Vertical
		Easting	Northing		Easting	Northing		Easting	Northing	
I	μ	0.000	−0.000	−0.000	−0.000	0.000	0.000	−0.000	0.000	0.000
	σ	0.001	0.003	0.001	0.001	0.000	0.001	0.000	0.000	0.000
W	μ	0.000	−0.000	−0.000	0.000	−0.000	0.000	0.000	−0.000	0.000
	σ	0.002	0.001	0.002	0.002	0.000	0.001	0.002	0.002	0.003
P	μ	0.000	−0.000	0.000	−0.000	0.001	−0.000	−0.000	0.000	−0.000
	σ	0.002	0.001	0.000	0.003	0.002	0.000	0.003	0.001	0.001
K	μ	0.000	0.000	0.000	−0.000	0.000	−0.000	0.000	0.000	0.000
	σ	0.000	0.000	0.000	0.010	0.002	0.010	0.000	0.000	0.000
T	μ	0.002	0.000	0.001	0.001	0.000	0.000	−0.001	−0.000	−0.000
	σ	0.019	0.003	0.011	0.032	0.005	0.024	0.009	0.001	0.007

ground coordinates, G', can be calculated from these epipolar image coordinates using the newly generated RPCs. The differences between the coordinates, G-G', represent the geo-positioning accuracy of the resampled epipolar image pair.

4.2 Pixel scale and perpendicularity of axis in object space

Certain satellite images, especially SAR images, have different pixel scales, and their pixels are not always square, which inhibits manual stereo plotting. To confirm the degree to which the generated pixel scale is square, the pixel scale in meters of the grid points extracted within the effective ground space is computed along the x- and y-axis of the epipolar image space using the generated RPCs. To determine the perpendicularity of the axis in object space, the angle between the x- and y-axis in epipolar image space can be converted into different ground spaces including local tangential coordinate space and others.

4.3 Near zero y-parallax in epipolar image space

y-parallax between the resampled epipolar image pair can be measured by a stereo matching algorithm. All of the reliable matched image points from epipolar image pair are used to compute y-parallax. The points, which are used in stereo matching, are extracted using the Harris corner detector and the bi-directional area-based stereo matching algorithm with sub-pixel accuracy is used to get more reliable result.

4.4 Proportionality between disparity and height

Combined with the criteria for a square pixel scale, proportionality between the x-parallax disparity and ground height provides more favorable condition for stereo image processing. For example, in the field of computer vision technology, many dense stereo matching

Table 3 Pixel scale of each method (unit: meter)

		Source image		Resampled epipolar image					
				Image space-based		Object space-based		Proposed	
		Left	Right	Left	Right	Left	Right	Left	Right
I	V	1.000	1.000	0.995	0.995	1.000	1.000	1.000	1.000
	H	1.000	1.000	0.994	0.994	1.000	1.000	1.000	1.000
W	V	0.502	0.525	0.521	0.521	0.500	0.500	0.500	0.500
	H	0.726	0.587	0.495	0.495	0.500	0.500	0.500	0.500
P	V	0.544	0.557	0.573	0.573	0.500	0.500	0.500	0.500
	H	0.570	0.606	0.534	0.534	0.500	0.500	0.500	0.500
K	V	0.677	0.628	0.672	0.672	1.000	1.000	1.000	1.000
	H	0.882	1.519	0.878	0.878	1.000	1.000	1.000	1.000
T	V	1.976	1.894	1.945	1.945	2.000	2.000	2.000	2.000
	H	1.952	1.890	1.937	1.937	2.002	2.007	2.001	2.001

Table 4 Axis angle (unit: degree) in object space in each method

	Source image		Resampled epipolar image					
			Image space-based		Object space-based		Proposed	
	Left	Right	Left	Right	Left	Right	Left	Right
I	90.004	90.001	90.003	90.003	90.003	90.000	90.003	89.998
W	91.228	96.378	89.851	89.889	90.000	90.007	90.001	90.040
P	84.694	94.588	96.529	96.534	90.001	90.000	90.003	90.003
K	89.844	89.895	91.771	91.748	89.998	89.991	90.000	89.990
T	89.844	89.895	89.973	89.918	89.994	90.002	90.048	90.599

algorithms [13] are implemented to compute depth values from x-parallax disparity.

The proportionality of the x-parallax disparity and ground height can also be validated by area-based stereo matching algorithm described in the previous section. The x-parallax disparity between the conjugate image points from resampled epipolar image pair and the ground heights of the corresponding ground points, which can be calculated by the newly generated RPCs, are investigated the proportionality between them. The difference between the fitted lines is used as a performance metric for these criteria.

5 Performance evaluation of previous and proposed methods

In this section, we describe the experimental results from three different algorithms: the image space-based method proposed by Oh et al., the object space-based method proposed by Wang et al., and the method proposed in this paper. As detailed in the preceding section, the image space-based method uses pixel distance to re-allocate reference image coordinates to epipolar image coordinates, and the object space-based method uses distance on the PRP. The suggested algorithm uses ground distance on the local plane.

5.1 Data set

For experimental verification, we used stereo image pairs from five different satellites: IKONOS-2(I), WV-2(W), Pleiades(P), Kompsat-5(K), and TerraSAR-X(T). These satellites cover different areas in South Korea and have been acquired between the years 2002 and 2014. The acquisition dates, image sizes, and nominal ground sample distance (GSD) for these images from these satellites are shown in Table 1, and the areas they cover are shown in Fig. 8.

5.2 Results by the proposed method

Figures 9, 10, 11, 12, and 13 show a segment of an IKONOS-2 anaglyph image generated by the proposed method, in which the horizontal line indicates that near zero y-parallax is achieved using the proposed method.

Figure 14 illustrates the line fitting result showing proportionality between the x-parallax disparity and ground height of the resampled epipolar IKONOS image pair. In Fig. 14, the x-axis represents the disparity or the pixel distance between the matched points, and the y-axis represents the ground height. As is apparent from the figure, the slope of the line indicated by the 152 matched points is 1.53544 and negligible differences between actual ground heights and those computed from the x-parallax disparity.

Figure 15 shows an image matching result between the left and right images of a resampled epipolar IKONOS-2; (a) and (b) are the left and the right image, respectively, and (c) is again the left image which shows the reversed matching result. The image coordinates and the normalized correlation coefficient are also shown in the figure.

5.3 Geo-positioning accuracy

Geo-positioning accuracy can be evaluated by comparing the ground coordinates from the source image pair with those from the epipolar image pair using the generated RPCs. The differences between these ground coordinates for all three methods are shown in which it can be

Table 5 Epipolarity in terms of y-parallax for each method

		Image space-based	Object space-based	Proposed
I	min	−0.726	−0.772	−0.659
	max	0.583	0.463	0.677
	mean	−0.069	−0.078	0.106
	RMSE	0.221	0.208	0.235
W	min	0.012	−0.813	−0.278
	max	0.430	0.877	0.478
	mean	0.235	0.105	0.083
	RMSE	0.120	0.340	0.180
P	min	−0.611	−0.399	−0.410
	max	0.485	0.267	0.309
	mean	−0.037	−0.040	−0.054
	RMSE	0.122	0.139	0.119

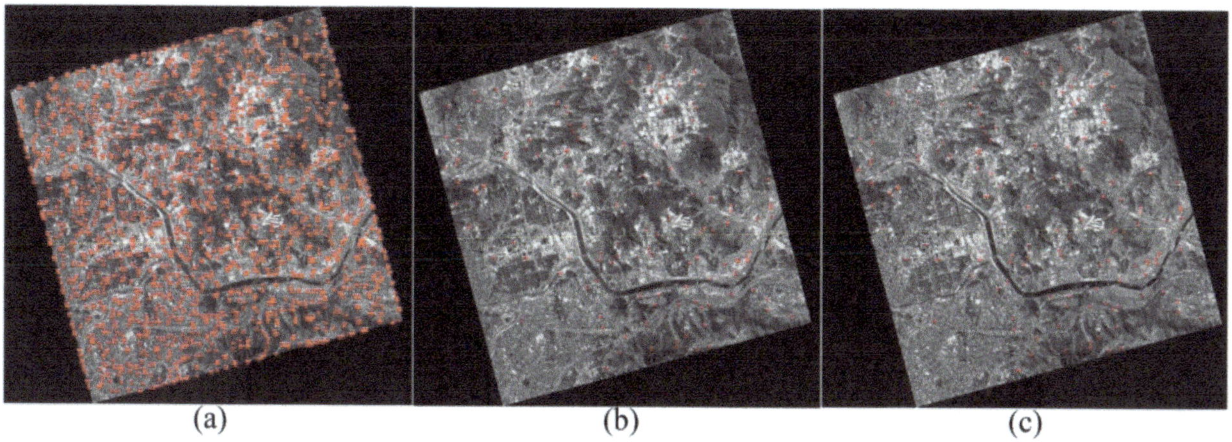

Fig. 17 Image matching result to evaluate epipolarity and y-parallax (IKONOS-2). **a** Harris corner points. **b**, **c** Matching results after outlier pruned, left and right, respectively

concluded that all of the three methods preserve the original geometric quality.

The geo-positioning accuracy is affected by the accuracy of image transformation and RPC fitting results. As shown in Fig. 16, the third or higher order polynomial function gives more accurate results than the second or lower order polynomial function. The fifth order polynomial function is used to transform original images to epipolar resampled images for all experiments.

As shown in Table 2, the geo-positioning accuracy can be preserved, that is, the difference of geo-positioning results from before and after the resampling process is negligibly small.

5.4 Pixel scale and perpendicularity of axis in object space

The pixel scales and the perpendicularity of the axes in object space for all three methods are shown in Tables 3 and 4. Regularly spaced image points are selected to compute pixel scale from epipolar resampled stereo image pair. The distance between the selected image point and the upper/right image point can be computed in meters. And, at each image point, axis angle between the two axes can be computed in ground space.

Because the image space-based method uses image space to relocate image control points, the pixel scale and the axis angle of the epipolar image pair depend on the source images. In contrast, the results show that the object space-based and the proposed methods can generate epipolar image pairs with almost square pixel scale and perpendicular axes.

5.5 Epipolarity and y-parallax

The y-parallax for each method was computed for each test data set by sub-pixel stereo matching, and the results are shown in Table 5. In this regard, all of the three methods show favorable performance for the generation of stereo image pair with epipolar constraints.

As explained in Section 4.3, bi-directional area-based stereo matching is used to determine conjugate points. Initially, epipolar image space is divided into 30 × 30 blocks, and three Harris corner points are extracted from each block (Fig. 17). Nine hundred corner points are used to extract matching points. The matched result whose normalized cross correlation value is below threshold, 0.8~0.9, is pruned. For example, in Fig. 15, 0.92 in (b) and 0.93 in (c) represents normalized cross correlation value. The unexpected matching results which caused by some changes between stereo images were manually discarded. For example, the matched results from the feature point extracted on the corner of floatages on the surface are eliminated.

Table 6 Proportionality between x-parallax disparity and ground height for each method

		Image space-based	Object space-based	Proposed
I	min	−0.003	−1.522	−0.007
	max	0.009	3.032	0.006
	mean	0.000	0.000	0.000
W	min	−0.018	−2.539	−0.011
	max	0.013	4.925	0.011
	mean	0.000	0.000	0.000
P	min	−0.001	−3.104	−0.006
	max	0.001	6.895	0.004
	mean	0.000	0.000	0.000

Table 7 Summary of the performance evaluation results

Criterion	Image-spaced method	Object-spaced method	Proposed method
Geo-positioning accuracy	Good	Good	Good
Pixel scale and perpendicularity of axis in object space	May not be square	Square	Square
	May not be perpendicular	Perpendicular	Perpendicular
Near zero y-parallax in epipolar image space	Good	Good	Good
Proportionality between disparity and height	Proportional	May not be proportional	Proportional

5.6 Proportionality between x-parallax disparity and ground height

Table 6 shows the proportionality between x-parallax disparity and ground height for each method. The image space-based and the proposed methods generate resampled epipolar image pairs in which x-parallax disparities are proportional to the ground height. In contrast, the object space-based method uses a local tangential plane with a single origin for the entire space; its x-parallax disparity is not always proportional to the ground height.

6 Conclusions

In this paper, we have suggested a practical and unified piecewise epipolar resampling method and have described the experimental results based on the suggested performance criteria. The proposed method showed performance equivalent to other methods in geo-positioning accuracy and near zero y-parallax. In addition, it showed an improved result in its square pixel scale, axis angle, and proportionality between x-parallax disparity and height. Table 7 shows the summary of the performance evaluation results.

The proposed method will be practical and useful in generating stereo image pairs for the many applications that employ dense matching algorithms [13], because the proposed method efficiently yields constraints on the search space, with near zero y-parallax and the x-parallax bounded by the minimum and maximum heights using any type of sensor models.

Competing interests

The authors declare that they have no competing interests.

Authors' contributions

JK carried out the analysis of the prevailing epipolar resampling method and suggested the core idea for improving the performance of the algorithm. Also, JK carried out the experiments and performed the analysis based on the performance criteria suggested in this paper. HY conceived the study, participated in its design and coordination, and drafted the manuscript. Both authors read and approved the final manuscript.

Authors' information

Jin-Woo Koh is a principal researcher at the Agency for Defense Development, Korea. He received his B.S. and M.S. degrees from the Department of Computer Science, Sogang University, Korea, and he received his Ph. D. degree from the School of Computing at KAIST, Korea. His research interests include image processing and computer vision, especially in military surveillance and reconnaissance field.

Hyun-Seung Yang is currently a Dean of Research and a Professor in the Department of Computer Science at KAIST. He received his B.S. degree from the Department of Electronics Engineering at Seoul National University, Korea and both M. S. and Ph. D. degrees from the School of Electrical Engineering at Purdue University, West Lafayette, Indiana, USA. His research activities include artificial Intelligence and vision, human robotics, virtual reality, ubiquitous and wearable Computing, media art, and technology.

Acknowledgements

The authors would like to acknowledge the support of the Target Information Directorate, ISR R&D Institute, Agency for Defense Development, Korea, for providing the data used in this paper.

References

1. T. Poggio, V. Torre, C Koch, Computational vision and regularization theory. Nature. **317**, 314–319 (1985).
2. T. Kim, A study on the epipolarity of linear pushbroom images. Photogramm. Eng. Remote. Sens. **66**(8), 961–966 (2000).
3. J. Oh, W.H. Lee, C.K. Toth, D.A. Grejner-Brzezinska, C. Lee, A piecewise approach to epipolar resampling of pushbroom satellite images based on RPC. Photogramm. Eng. Remote. Sens. **76**(12), 1353–1363 (2010).
4. M. Wang, F. Hu, J. Li, Epipolar resampling of linear pushbroom satellite imagery by a new epipolarity model. ISPRS J. Photogramm. Remote Sens. **66**, 347–355 (2011).
5. C.V. Tao, Y. Hu, A comprehensive study of the rational function model for photogrammetric processing. Photogramm. Eng. Remote. Sens. **67**(12), 1347–1357 (2001).
6. J. Grodecki, G. Dial, J. Lutes, in *Mathematical model for 3D feature extraction from multiple satellite images described by RPCs*, Proceeding of the ASPRS 2004 Annual Conference (Denver, 2004). http://libra.msra.cn/Publication/10420993/mathematical-model-for-3d-feature-extraction-from-multiplesatellite-images-described-by-rpcs
7. A.F. Habib, M. Morgan, S. Jeong, K. Kim, Analysis of epipolar geometry in linear array scanner scenes. Photogramm. Rec. **20**, 27–47 (2005).
8. F. Hu, M. Wang, D. Li, in *A Novel Epipolarity model of satellite stereo-imagery based on virtual horizontal plane of object-space*, Proceeding SPIE 7285, International Conference on Earth Observation Data Processing and Analysis (ICEODPA), (2008). http://spie.org/Publications/Proceedings/Paper/10.1117/12.815932
9. M. Morgan, K. Kim, S. Jeong, A. Habib, Epipolar resampling of space-borne linear array scanner scenes using parallel projection. Photogramm. Eng. Remote. Sens. **72**(11), 1255–1263 (2006).
10. NGA, *Community sensor model (CSM) technical requirements document (TRD), Version 3.0*, NGA Standardization Document (2010). http://www.gwg.nga.mil/documents/csmwg/documents/CSM_TRD_Version_3.0__15_November_2010.pdf
11. C.R. Taylor, J.T. Dolloff, M.M. Iiyama, *Replacement sensor model (RSM) performance for triangulation and geopositioning*, ASPRS 2008 Annual Conference, Portland, Oregon, April 28 - May 2 (2008).
12. J. Grodecki, G. Dial, Block adjustment of high-resolution satellite images described by rational polynomials. Photogramm. Eng. Remote Sens. 69(1), 59-68 (2003). http://www.asprs.org/a/publications/pers/2003journal/january/2003_jan_59-68.pdf
13. D. Scharstein, R. Szeliski, A taxonomy and evaluation of dense two-frame stereo correspondence algorithms. Int'l Jr. of Computer Vision **47**, 7–42 (2002).

11

High-capacity watermarking of high dynamic range images

Emanuele Maiorana* and Patrizio Campisi

Abstract

High dynamic range (HDR) imaging techniques address the need to capture the full range of color and light that the human eyes can perceive in the real world. HDR technology is becoming more and more pervasive. In fact, most of the cameras and smartphones available on the market are capable of capturing HDR images. Among the challenges posed by the spread of this new technology there is the increasing need to design proper techniques to protect the intellectual property of HDR digital media. In this paper, we speculate about the use of watermarking techniques to cope with the peculiarities of HDR media to prevent the misappropriation of HDR images.

Keywords: High dynamic range, Data hiding, Watermarking

1 Introduction

High dynamic range (HDR) imaging technologies provide a step forward in representing real scenes as they are perceived by the human eye. With respect to traditional imaging techniques capable of acquiring low dynamic range (LDR) images or video, HDR methodologies [1, 2] are capable of generating and rendering images and videos with a ratio between the luminance of the lightest and darkest areas far greater than the one provided by standard imaging technologies. It is evident that HDR media are richer in terms of content than their LDR counterparts and therefore much more valuable. It is therefore crucial to provide, since the early stages of the development of such technology, proper tools for protecting the intellectual property of digital HDR media. Data hiding, and more specifically digital watermarking [3–5], has emerged in the last decade as an enabling technology for copyright protection among other possible applications. Roughly speaking, data hiding is the general process by which a discrete information stream, the mark, is merged within media content by imposing imperceptible changes on the original host signal, while allowing the message to be detected or extracted even in the presence of either malevolent or non-intentional attacks. Other requirements, other than the transparency one, can be needed according to the specific application that is taken into account. Although LDR image or video digital watermarking is a very mature area of research, watermarking in an HDR scenario is still in its infancy and needs to be significantly developed. In this paper, we crystallize the state of the art on HDR image watermarking techniques and the watermarking embedding method proposed by the authors in [6] is more extensively tested using a wider set of images as well as up to date metrics for performance assessment.

A brief overview on digital watermarking is given in Section 2. An extensive and comparative review of the state of the art on HDR image watermarking is given is Section 3. The watermarking method is detailed in Section 4. An extensive set of experimental results is given in Section 5, and conclusions are drawn in Section 6.

2 Multimedia digital watermarking

In the past decades, there has been an explosion in the use and distribution of digital multimedia data, essentially driven by the diffusion of the Internet. In this scenario, data hiding and more specifically digital watermarking techniques [3, 4] have been proposed to address the ever-growing need to protect the intellectual property of multimedia content (digital still images, 2D and 3D video sequences, text, or audio) in the World Wide Web. Although copyright protection was the very first application of watermarking, different uses have been

*Correspondence: emanuele.maiorana@uniroma3.it
Roma Tre University, Department of Engineering, Section of Applied Electronics, Via Vito Volterra 62, 00146 Rome, Italy

proposed in the literature. Fingerprinting, copy control, broadcast monitoring, data authentication, multimedia indexing, content-based retrieval applications, medical imaging applications, error concealment, quality assessment, and improved data compression are only a few of the applications where watermarking can be usefully employed. When these techniques are used to preserve the copyright ownership with the purpose of avoiding unauthorized data duplications, the embedded watermark should be detectable. This is required even if malicious attacks or non-malevolent signal processing (i.e., filtering and compression) are applied on the multimedia data. This requirement is known as *watermark security*. On the other hand, when the watermark is required to be resistant only to non-malicious manipulations, the watermarking techniques are referred to as *robust*. For some applications, when the robustness requirement is severely required, each attempt of removing the mark should result in irreversible data quality degradation. In some applications, the watermarked host data are intended to undergo a limited number of signal processing operations. Therefore, in the aforementioned scenario, the watermark needs to be robust only to a limited number of set of manipulations, in which case the technique is known as *semi-fragile watermarking*. On the contrary, when unwanted modifications of the watermarked data affect even the extracted watermark, the embedding scheme is known as *fragile*. Fragile watermarking can be used to obtain information about the tampering process. In fact, it indicates whether or not the data has been altered and supplies localization information as to where the data was altered. *Capacity* is another watermarking requirement, referring to the number of bits of information that can be embedded in the original data, which needs to be fulfilled, depending on the specific application. Robustness, imperceptibility, and capacity are requirements hindering each other; therefore, a trade-off driven by the application needs to be considered.

3 HDR image watermarking: state of the art

Despite digital watermarking of low dynamic range images having been deeply studied for more than two decades, watermarking of HDR media, both images and video, has not been extensively explored. In fact, a direct transposition of the techniques developed for LDR media is not straightforward since HDR media possess some peculiarities with respect to LDR ones which do not allow a direct transposition of the plethora of watermarking approaches already developed for LDR images and videos to the HDR case. In fact, the specific characteristics of the human visual system (HVS) related to the fruition of HDR media need to be exploited when designing HDR-tailored embedding techniques. Moreover, HDR-specific watermarking methods need to be robust against intentional attacks or signal processing manipulations like the use of tone-mapping operators (TMOs). The use of TMOs is necessary when HDR media need to be experienced using conventional displays in order to generate LDR data retaining as much information as possible from the original objects, while reducing the overall contrast. Therefore, it is highly desirable to design HDR watermarking schemes also robust against TMOs, thus allowing to either recover or detect the embedded mark not only from the marked HDR image but also from its tone-mapped version. In the following, an overview on the state of the art for HDR image watermarking is given. Specifically, in Section 3.1, the 1-bit embedding approaches are detailed, whereas the multi-bit approaches are summarized in Section 3.2.

3.1 HDR image watermarking: 1-bit embedding

Watermarking techniques aimed at embedding 1 bit of information imply the detection of the embedded watermark at the receiving side. In [7], two different watermarking methods, both of them splitting the cover HDR image into a host image where to embed the mark and a residual part, are investigated. The first approach employs the μ-law to characterize a generic TMO and applies it to the original HDR image to derive an LDR representation where the watermark can be embedded. The residual part is given by the ratio between the HDR and the LDR images. However, since the range of HDR images may be orders of magnitude greater than the one of LDR images, their ratio will be high in areas with high luminance, where the watermark will be therefore very strong. The second approach decomposes the HDR image into a detail and a coarse component by applying bilateral filtering. The watermark embedding is then performed in the detail component that preserves the images' edges. In fact, it is assumed that a generic TMO does not affect the image details and the color appearance, while addressing the problem of strong contrast reduction. Both methods proposed in [7] project the luminance of the detail component in the discrete wavelet domain. Then, the embedding is performed using spread-spectrum (SS) techniques. The experimental tests, conducted using five HDR images and four local TMOs, show that the visual quality of images marked with the μ-law-based approach, evaluated by means of the peak signal-to-noise ratio (PSNR) computed over tone-mapped images, is better than the one associated with the method relying on bilateral filtering. Specifically, the PSNR of the images marked with the first approach is within the 50–70-dB range, while the second one results in a PSNR of about 30–60 dB. However, bilateral filtering guarantees better performance in terms of robustness to the application of TMOs.

In [8], a general tone-mapping, represented by the logarithm function, is considered and applied to the luminance of the given HDR image, thus obtaining the LogLuv domain. The watermark embedding is performed by applying a quantization-index-modulation (QIM) approach [9] to the approximation subband of the discrete wavelet transform of the LogLuv component. Specifically, the image is divided into different blocks of random shapes, and the coefficients of each block are modified in order to make the block kurtosis equal to a non-uniformly quantized value determined by the bit to be embedded into the block. Mark imperceptibility is obtained by applying a local perceptual mask based on luminance, texture, and contrast that provides the maximum amount of distortion that each coefficient in the embedding domain can withstand without resulting in visible artifacts. Experimental tests conducted on 15 HDR images include an objective analysis on mark imperceptibility evaluated through the HDR-Visual Difference Predictor (HDR-VDP) metric [10] and on the robustness of watermark detection against seven different TMOs as well as to the addition of Gaussian noise in the HDR domain.

In [11], a non-linear hybrid embedding approach operating in the detail subbands of a one-level wavelet decomposition is used to watermark an HDR image. Specifically, the approach combines both additive and multiplicative watermark embedding and is based on a square-root embedding equation operating in the wavelet domain. The embedding approach is applied to 12 images, and its robustness is tested against the application of seven TMOs. The quality assessment is carried out using both the HDR-VDP metric and a pool of three observers on a native HDR display.

In [12], a bracketing approach mimicking the HDR image generation process, consisting of merging several single-exposure LDR images, is exploited for HDR image watermarking. In details, an HDR image is split into multiple LDR images, which are watermarked before being fused again into a single HDR image representation. The mark embedded in the HDR image thus generated results to be detectable in its LDR counterparts obtained by either applying a tone-mapping process or extracting and displaying only a specific range of interest of intensities. The HDR-VDP-2 metric [13] is employed to assess the performance. An equal error rate (EER) lower than 10^{-8} % is achieved when detecting a watermark from LDR images obtained with five different TMOS. EERs lower than 10^{-2} % are estimated when analyzing LDR images obtained by isolating, from the HDR data, a dynamic range different from the one considered during embedding.

A summary of the HDR image 1-bit watermarking approaches is given in in Table 1.

3.2 HDR image watermarking: multi-bit embedding

In [14], binary information is inserted into an HDR image encoded in RGBe format by substituting the least significant bits (LSBs) of each pixel with bits taken from the secret message. In more detail, the HDR image is divided into flat and boundary areas by comparing the exponents associated to neighboring pixels. The number of bits which can be embedded into each color channel is then adaptively determined on the basis of local contrast and depending on the considered area, embedding more bits in high contrast and dark regions than in smooth and bright ones. The method capacity, tested on seven HDR images, is around 10 bits per pixel (bpp). Its imperceptibility is evaluated by converting the HDR images into their LDR counterparts by means of a TMO and measuring the peak signal-to-noise ratio (PSNR) between the tone-mapped cover image and the tone-mapped watermarked image, obtaining values around 30 dB.

Images in uncompressed LogLuv TIFF format are considered in [15]. The LSB of each channel's mantissa is modified to insert a binary message, while the exponent is selected in order to minimize the difference between the final value and the original one. The visual quality of the resulting images is evaluated over seven HDR images by applying tone-mapping and computing the PSNR and the HDR-VDP metrics.

The RGBe format is considered in [16], where its property of expressing a given color according to different possible choices is employed to hide a message into an HDR image: the final color representation is selected among the equivalent ones on the basis of the bits of the secret message. Such an approach produces distortion-free watermarked images, being different with the original cover image due only to the conversion employed to determine the floating point value of the pixels. A capacity of about 0.12 bpp is estimated over five HDR images.

The blue component of a detail layer obtained exploiting bilateral filtering as in [7] is employed in [17] to embed into an HDR image binary messages with as many bits as the image's pixels. The mark is inserted additively yet proportionally to the pixels' luminance. The mark imperceptibility is measured only through the PSNR, evaluated over five HDR images, achieving values around 60 dB. No objective measurements of mark extraction capability are provided, and the robustness against TMO is not evaluated. It is worth noting that no intentional or unintentional attack is taken into account in any of the aforementioned approaches performing multi-bit message embedding.

This kind of analysis is performed for the first time in [18]. The preprocessing stage proposed in [7], consisting of splitting an HDR image into its LDR counterpart and a residual component, is employed also here. Four TMOs are taken into account for both embedding the watermark

Table 1 HDR image 1-bit watermarking: state of the art

Paper	Embedding domain	Embedding method	# of images database	Imperceptibility	Robusteness
					4 TMOs
Xue et al. [7]	μ-law and wavelet	Multiplicative	5[a]	PSNR on LDR $\sim$ 55 dB	Score > threshold at $P_{fa} = 10^{-8\%}$
	Bilateral and wavelet	Multiplicative		PSNR on LDR $\sim$ 45 dB	Score > threshold at $P_{fa} = 10^{-8\%}$
					7 TMOs
Guerrini et al. [8]	LogLuv and wavelet	QIM	15[a,b]	HDR-VDP$_{75}$ $\sim$ 0.46 %	$P_{\text{miss}} = 10^{-2}$ % at $P_{fa} = 10^{-6\%}$
				HDR-VDP$_{95}$ $\sim$ 0.21 %	Add. noise
					$P_{\text{miss}} = 10^{-12}$ % at $P_{fa} = 10^{-6\%}$
Autrusseau et al. [11]	Wavelet	Hybrid	5, 7[a,b]	HDR-VDP	8 TMOs
				Subjective (3 subjects)	qualitative analysis
Solachidis et al. [12]	Wavelet	Additive	6[a]	HDR-VDP-2$_{95}$ $\sim$ 5.00 %	6 TMOs

[a]http://www.anyhere.com/gward/hdrenc/pages/originals.html
[b]http://www.cis.rit.edu/fairchild/HDRPS/HDRthumbs.html

and attacking the marked HDR image when extracting the hidden data. Differently from [7], a multi-bit mark is blindly embedded into an image by associating each bit to a specific 8×8 luminance block and by manipulating two of its middle-frequency discrete cosine transform (DCT) coefficients to make their difference dependent on the bit to be embedded. Tests performed on four HDR images provide a PSNR for tone-mapped watermarked images of about 40 dB, while a bit-error-rate (BER) of about 5 % is estimated between the mark embedded in the HDR image and the one extracted from the tone-mapped watermarked image. Cropping, Gaussian blur, and Gaussian noise addition are considered as attacks to the LDR marked images, resulting in BERs of about 12, 23, and 15 %, respectively.

In [19], a blind multi-bit watermarking method for HDR images is proposed. The employed embedding domain is obtained by applying the wavelet transform to the just-noticeable difference (JND)-scaled representation of the original HDR image. Such domain is exploited in order to process values directly associated with specific characteristics of the HVS. A visual mask based on bilateral filtering is used for the embedding region selection while the watermark intensity in each wavelet decomposition subband is modulated according to a contrast sensitivity function. An extensive set of experimental tests is conducted considering three HDR images and seven TMOs. The imperceptibility in the HDR domain is evaluated through the HDR-VDP-2 metric, while an EER lower than $10^{-3\%}$ with a BER around 5 % is estimated over tone-mapped LDR images.

A summary of the HDR image multi-bit watermarking approaches is given in Table 2.

4 HDR image watermarking in the Radon DCT domain

The multi-bit embedding method designed by the authors in [6] is further explored in the following, and a more extensive set of experimental results is here presented.

Let us consider a message $\mathbf{m}$ of M bits to be embedded in an HDR image. With the goal to design a watermarking approach robust against non-malicious TMO application, we resort to perform mark embedding in the image edge region, since TMOs typically preserve the edges of the original image, whereas compressing its dynamic range and adapting it for rendering HDR images on devices with low contrast capabilities [20, 21]. In order to achieve the desired goal, the properties of the Radon DCT (RDCT) domain are here exploited. As detailed in [22], the RDCT is a two-stage transform that consists in performing, as a first step, the Radon transform which maps each line of the host image into a single point. The DCT follows, therefore, the energy of the Radon projection concentrated in the first coefficients of the resulting RDCT representation, which can be eventually used as the host signal where to embed the watermark. The proposed watermarking scheme is given in Fig. 1 and details on the watermark embedding and extraction are given in the following sections.

4.1 HDR embedding domain

Without any loss of generality, we assume that the original HDR image is represented in the *RGB* color space. The embedding is performed in the luminance component $\mathbf{Y}$. Actually, most of TMOs perform alterations in the luminance component $\mathbf{Y}$ while compressing the original dynamic range, often resulting in unpredictable alterations of the chrominance contributions. Therefore, the chrominance components are not suitable to host the watermark. More specifically, in our approach, we consider the LogLuv domain [8], derived from the application of the logarithm to the luminance $\mathbf{Y}$. The logarithmic operation simulates a specific property of the human visual system (HVS), that is, its higher sensitivity to contrast in shadow areas, whose dynamic is expanded by the logarithm, when compared to the contribution of highlights, which is on the contrary compressed. The resulting representation is then linearly normalized for generating an image $\mathbf{N}$ whose pixel values belong to the interval $[0, 1]$.

A discrete wavelet transform (DWT) decomposition, mimicking the processing performed by the HVS [23], of the image $\mathbf{N}$ is then performed, and one of the resulting subbands $\mathbf{N}_\gamma$, with $\gamma \in \{LH, HL, HH\}$, is selected.

The considered subband is then decomposed into B square blocks $\mathbf{f}^{(b)}$ with size $P \times P$, $b = 1, \ldots, B$, in order to perform a local analysis of the image properties. A message $\mathbf{m}^{(b)} = [m_1^{(b)}, \ldots, m_C^{(b)}]$ derived taking C bits from $\mathbf{m}$ is then embedded into each block $\mathbf{f}^{(b)}$ according to the processing illustrated in Fig. 2 and described in the following. Having indicated with $S_1 \times S_2$ the size of the HDR image, the number of generated blocks is $B = \lfloor \frac{S_1}{2P} \rfloor \cdot \lfloor \frac{S_2}{2P} \rfloor$, while the length of the embedded message $\mathbf{m}$ is therefore equal to $M = B \cdot C$ bits.

The RDCT transform is applied to each block $\mathbf{f}^{(b)}$ extracted from $\mathbf{N}_\gamma$, with $b = 1, \ldots, B$. The embedding of C bits is then applied in the transformed domain as in [22]. Specifically, the employed implementation is based on the invertible and non-redundant finite Radon transform (FRAT) presented in [24]. Given a real-valued function $\mathbf{f}^{(b)}$ defined over a finite square grid Z_P^2, being $Z_P = \{0, 1, \cdots, P-1\}$, its FRAT $\mathbf{r}^{(b)}$ is defined as

$$\mathbf{r}^{(b)}[k, l] = \frac{1}{\sqrt{P}} \sum_{(i,j) \in L_{k,l}} \mathbf{f}^{(b)}[i, j], \quad (1)$$

where $L_{k,l}$ defines the set of points that forms a line on Z_P^2:

Table 2 HDR image multi-bit watermarking: state of the art

Paper	Embedding domain	Embedding method	# of images database	Capacity	Imperceptibility	Robusteness
Cheng et al. [14]	RGBe	LSB	7 N/A	$\sim$ 10 bpp	PSNR on LDR (1 TMO) $\sim$ 30 dB	N/A
Li et al. [15]	LogLuv	LSB	10 N/A	$\sim$ 26 bpp	PSNR on LDR (1 TMO) $\sim$ 35 dB $\text{HDR-VDP}_{75} \sim 0.42\,\%$ $\text{HDR-VDP}_{95} \sim 0.04\,\%$	N/A
Yu et al. [16]	RGBe	Pixel exponent	125 N/A	$\sim$ 0.12 bpp	N/A	N/A
Rattanacharuchinda et al. [17]	Blue comp. of detail layer	Multiplicative	5[a]	1 bpp	PSNR $\sim$ 60 db	N/A
Wu et al. [18]	DCT	Modify adjacent DCT coeff.	4[a]	0.015 bpp	PSNR on LDR (4 TMO) [40, 70] dB	4 TMO BER $\sim$ 5 % HDR crop BER $\sim$ 12 % HDR blur BER $\sim$ 22 % HDR noise BER $\sim$ 15 %
Solachidis et al. [19]	JND wavelet	Multiplicative	3[b]	128	HDR-VDP-2 $\sim$ 1 % (1 TMO) $\sim$ 30 dB	7 TMOs EER$<10^{-3}$% BER $\sim$ 5 %

[a]http://www.anyhere.com/gward/hdrenc/pages/originals.html
[b]http://www.mpi-inf.mpg.de/resources/hdr/gallery.html

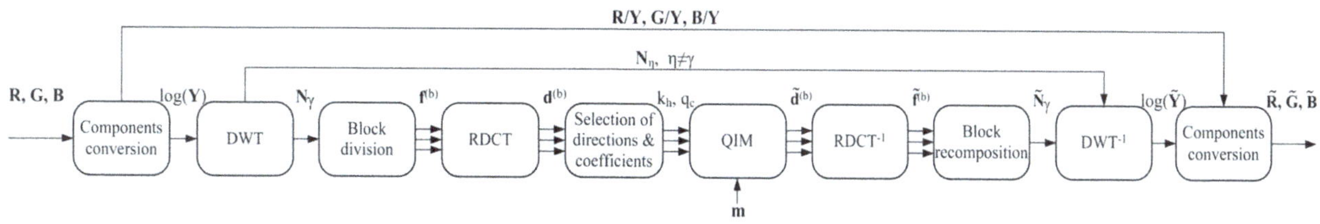

Fig. 1 Proposed embedding scheme

$$\begin{cases} L_{k,l} = \{(i,j) : j = k \cdot i + l \,(\text{mod}\, P), i \in Z_P\}, \\ L_{P,l} = \{(l,j) : j \in Z_P\} \end{cases} \tag{2}$$

being $k \in Z_{P+1}$ the line direction and $l \in Z_P$ its intercept.

The Radon projections, obtained as in (1), are further processed in order to concentrate their energies in few coefficients. This goal is achieved by applying the DCT to the Radon projections, thus obtaining the RDCT sequences as

$$\mathbf{d}^{(b)}[k,q] = \omega[l] \sum_{l=0}^{P-1} \mathbf{r}^{(b)}[k,l] \cos\left[\frac{\pi(2l+1)q}{2P}\right] \tag{3}$$

with $q \in Z_P$, $\omega[0] = \sqrt{1/P}$, and $\omega[l] = \sqrt{2/P}$, $l \neq 0$. The energy concentration provided by the RDCT is similar to that performed by the ridgelet transform [24], yet with a better behavior and performance for watermarking purposes provided by the RDCT, as shown in [22].

4.2 Host coefficients selection and QIM embedding

The host coefficients where to embed the mark are selected as those related to the presence of edges. This task is accomplished by considering the RDCT projections of a given block $\mathbf{f}^{(b)}$ and by sorting the $P+1$ directions available in $\mathbf{d}^{(b)}$ according to their energies in decreasing order and finally selecting H directions for mark embedding. Specifically, the directions from the $h_{\max}$-th to the $h_{\min}$-th most energetic ones, with $H = h_{\min} - h_{\max} + 1$, are selected to embed the C bits of $\mathbf{m}^{(b)}$ in each of them.

Having indicated with $\mathbf{d}^{(b)}[k_h, q]$ the RDCT projection with the h-th most energetic contribution, with $h \in \{h_{\max}, \ldots, h_{\min}\}$, C coefficients are taken from it and modified through QIM [9] to embed the desired message. It is worth specifying that, before doing this, the values in the considered RDCT projection are

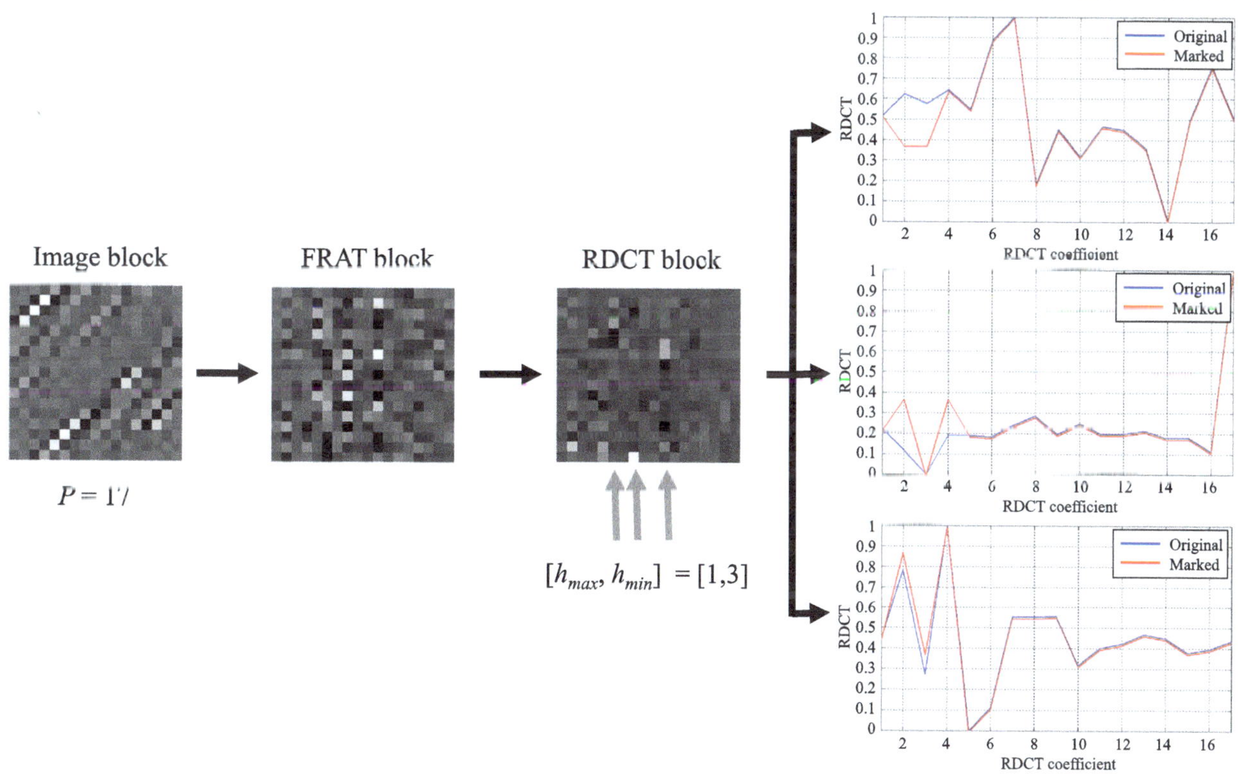

Fig. 2 Radon DCT pipeline: an example

linearly normalized between 0 and 1, thus producing the projections $\mathbf{d}_N^{(b)}[k_h, q]$. The coefficients having extreme values (0 and 1 after normalization) are not modified to not alter the range of the original values. Also, the RDCT coefficient $\mathbf{d}_N^{(b)}[k_h, 0]$, representing the DC component of the Radon projection, cannot be modified being connected with the mean value of all the projections computed from $\mathbf{f}^{(b)}$, as pointed in [22]. Therefore, the C coefficients to be marked are selected as the first ones of the available RDCT sequence, with the exception of the aforementioned elements, namely the DC coefficient and the extreme values.

The c-th bit of $\mathbf{m}^{(b)}$, with $c \in \{1, \ldots, C\}$, is therefore considered to alter the original coefficient $\mathbf{d}_N^{(b)}[k_h, q_c]$ according to a QIM approach (see Fig. 3). Specifically, keeping in mind that the considered coefficients lie in the interval $[0, 1]$, two code books $\mathcal{U}_0$ and $\mathcal{U}_1$ containing quantized values associated, respectively, to $m_c^{(b)} = 0$ and $m_c^{(b)} = 1$ can be defined as

$$\begin{cases} \mathcal{U}_0 = \left\{ \frac{1+4z}{2^{\Delta+1}}, z \in \{0, \ldots, 2^{\Delta-1} - 1\} \right\}, \\ \mathcal{U}_1 = \left\{ \frac{3+4z}{2^{\Delta+1}}, z \in \{0, \ldots, 2^{\Delta-1} - 1\} \right\} \end{cases} \quad (4)$$

being Δ the natural integer determining the quantization step $\frac{1}{2^{\Delta}}$ between admissible quantized values.

Watermark embedding is achieved by applying to $\mathbf{d}_N^{(b)}[k_h, q_c]$ either the quantizer associated to $\mathcal{U}_0$ or the one associated to $\mathcal{U}_1$:

$$\begin{cases} \mathcal{Q}_0\left(\mathbf{d}_N^{(b)}[k_h, q_c]\right) = \arg \min\limits_{u_{0,z} \in \mathcal{U}_0} |u_{0,z} - \mathbf{d}_N^{(b)}[k_h, q_c]|, \\ \mathcal{Q}_1\left(\mathbf{d}_N^{(b)}[k_h, q_c]\right) = \arg \min\limits_{u_{1,z} \in \mathcal{U}_1} |u_{1,z} - \mathbf{d}_N^{(b)}[k_h, q_c]| \end{cases} \quad (5)$$

where $u_{0,z}$ and $u_{1,z}$ are elements of $\mathcal{U}_0$ and $\mathcal{U}_1$, respectively. The marked coefficient $\tilde{\mathbf{d}}_N^{(b)}[k_h, q_c]$ can be expressed as

$$\tilde{\mathbf{d}}_N^{(b)}[k_h, q_c] = \begin{cases} \mathcal{Q}_0\left(\mathbf{d}_N^{(b)}[k_h, q_c]\right), & \text{if } m_c^{(b)} = 0, \\ \mathcal{Q}_1\left(\mathbf{d}_N^{(b)}[k_h, q_c]\right), & \text{if } m_c^{(b)} = 1. \end{cases} \quad (6)$$

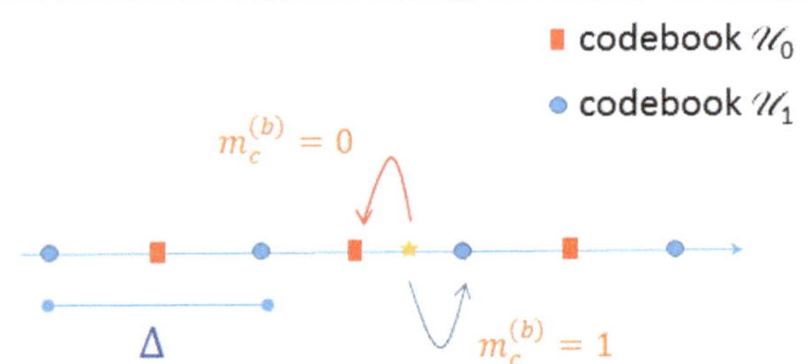

Fig. 3 QIM embedding procedure

The effects on the achievable performance, in terms of mark imperceptibility and robustness, of the employed parameters, namely the choice of the wavelet subband γ, the block size P, the embedding directions and coefficients, and the quantization parameter Δ, are discussed in Section 5. As for the capacity achievable with the proposed approach, it is easy to express it in a closed form as $\frac{C}{4 \cdot P^2}$ bpp. Practical values of the lengths of messages which can be embedded into HDR images are given in Section 5.

4.3 HDR image reconstruction

As described in the previous section, the block $\tilde{\mathbf{d}}_N^{(b)}$ is obtained from $\mathbf{d}_N^{(b)}$ by marking C coefficients of each direction k_h, with $h \in \{h_{\max}, \ldots, h_{\min}\}$. The RDCT representation $\tilde{\mathbf{d}}^{(b)}$ is derived by reversing the normalization process performed during the embedding. The watermarked block $\tilde{\mathbf{f}}^{(b)}$ is then determined through an inverse RDCT. The marked wavelet subband $\tilde{\mathbf{N}}_\gamma$ is reconstructed by recombining all the B-marked blocks. The marked normalized luminance $\tilde{\mathbf{N}}$ is then generated by combining $\tilde{\mathbf{N}}_\gamma$ with the other unaltered subbands $\tilde{\mathbf{N}}_\eta$, $\eta \neq \gamma$, and performing the inverse wavelet transform. The normalization process done during the embedding is reversed by applying an exponential operator thus obtaining the marked luminance $\tilde{\mathbf{Y}}$ which, as a result, is characterized by the same dynamic range of the original luminance $\mathbf{Y}$. The final watermarked image is generated by defining its trichromatic components as $\tilde{\mathbf{R}} = \frac{\mathbf{R}}{\mathbf{Y}} \cdot \tilde{\mathbf{Y}}$, $\tilde{\mathbf{G}} = \frac{\mathbf{G}}{\mathbf{Y}} \cdot \tilde{\mathbf{Y}}$ and $\tilde{\mathbf{B}} = \frac{\mathbf{B}}{\mathbf{Y}} \cdot \tilde{\mathbf{Y}}$. Therefore, the original chrominance contributions are kept unaltered, while the luminance of the resulting image allows to blindly recover the embedded information.

4.4 Mark extraction

In the watermark extraction stage, the marked HDR image undergoes the same operations described in Section 4.1 thus obtaining the RDCT projections $\bar{\mathbf{d}}^{(b)}[k_h, q]$ with $h \in \{h_{\max}, \ldots, h_{\min}\}$. Each projection is first normalized between 0 and 1, thus obtaining $\bar{\mathbf{d}}_N^{(b)}$, and then, the embedded bit estimation $\hat{m}_{h,c}^{(b)}$, $c = 1, \ldots, C$ is performed using a minimum distance decoder:

$$\hat{m}_{h,c}^{(b)} = \arg \min_{x \in \{0,1\}} \min_{u_{x,z} \in \mathcal{U}_x} |u_{x,z} - \bar{\mathbf{d}}_N^{(b)}[k_h, q_c]|, \quad z \in \{0, \ldots, 2^{\Delta-1} - 1\}. \quad (7)$$

The final estimations $\hat{m}_c^{(b)}$, $c = 1, \ldots, C$, of the C bits embedded into the block $\bar{\mathbf{d}}^{(b)}$ are made according to a majority-voting decision evaluated over the H estimates $\hat{m}_{h,c}^{(b)}$, $h \in \{h_{\max}, \ldots, h_{\min}\}$, available for each embedded bit.

It is worth specifying that the proposed approach takes into account more than a single direction for embedding C bits in each block to properly manage errors during the extraction process. Specifically, a relevant source of errors comes from the possibility of selecting for a block, during the mark extraction process, projections which differ from those employed during the embedding. In fact, if the energies of the RDCT projections of a given block are modified by some processing applied to the marked HDR image, the order with which the projections are selected to extract the mark may be altered, making it hard to recover the bits inserted into the considered block. The employed majority-voting decision strategy allows mitigating the effects of such events.

The considered approach allows also retrieving the embedded information from a tone-mapped LDR version of the watermarked HDR image. However, in this scenario, the mark is extracted directly from the luminance of the considered LDR image and not from its logarithmic counterpart. As in [8], it is in fact assumed that the application of a TMO produces a perceptual processing similar to the one performed by the logarithm operator.

5 Experimental results

An extensive set of experimental tests has been carried out to evaluate the performance of the proposed blind watermarking approach for HDR images. Specifically, to this aim, we have exploited 15 HDR images collected from http://www.mpi-inf.mpg.de/resources/hdr/gallery.html and http://www.anyhere.com/gward/hdrenc/pages/originals.html. The main characteristics of the employed images are listed in Table 3, while Fig. 4 shows all of them through their LDR representations, obtained with the ICAM06 TMO [25]. As can be seen, in the performed tests, we have considered images with different dimensions, dynamic ranges, and amount of edges and textured areas, in order to investigate the performance of the proposed method under a significant variability of possible practical conditions. Moreover, all the employed images have been selected to represent the depicted scenarios through real luminance values expressed in cd/m^2, being thus possible to exploit the HDR-VDP-2 metric [13] to objectively evaluate the imperceptibility of the embedded watermarks. In more details, the HDR-VDP-2 metric provides, for each pixel of a modified image, the probability of being detected as different from the original representation, that is, the percentage of people recognizing the pixel being modified. The latest HDR-VDP-2 version currently available (ver. 2.2.1) is employed in the performed tests.

The embedding capacities practically achievable for the considered images are shown in Table 4, for different values of P and using $C = 3$. As can be seen, the size of the embedded messages can grow very large, depending on the dimensions of the selected HDR image.

Table 4 also reports the processing time required for embedding messages into an HDR image according to the proposed method, when using Matlab [26] on a computer equipped with an Intel(R) Core(TM) i7-4500U at 1.8 GHz and 8.0 GB RAM. Given an HDR image, the required processing time is mainly dependent on the adopted block size P, and only the dependency on this parameter is therefore shown in Table 4, which provides values obtained with $\Delta = 2$, $[h_{\max}, h_{\min}] = [1, 5]$, and $C = 3$. As can be seen, large values of P are preferable for speeding up the marking process.

An analysis of the imperceptibility performance guaranteed by the proposed approach is then provided in Table 5, where the dependency on the employed parameters is investigated. Specifically, the table reports the maximum probability of detecting differences between the original and the marked data in at least the 5 % of the image pixels, for different choices of P, Δ, $[h_{\max}, h_{\min}]$ and C. It is worth specifying that the reported results are always referred to the mark embedding in the $\gamma = HH$ subband, since it provides far better imperceptibility performance with respect to the other first-level DWT subbands LH and HL. Actually, the evaluated probability of detecting a modification in at least 5 % of the image is 1.96 % when $\gamma = HH$, averaged over all the considered selections for the directions to be marked and the number of modified coefficients, with $P = 17$ and $\Delta = 2$. On the other hand, the same probability is 41.35 % when using the horizontal detail subband $\gamma = HL$ and reaches 53.13 % for the vertical detail subband $\gamma = LH$.

As can be seen from the results reported in Table 5, although decreasing the block dimension P would imply the possibility of embedding more bits in a given HDR

Table 3 Employed HDR images

Image ID	Image name	Dynamic range	Size ($S_1 \times S_2$)
1	AtriumMorning	$1.99 \cdot 2^{14}$	1016 × 760
2	AtriumNight	$1.62 \cdot 2^{28}$	1016 × 760
3	mpi_atrium_1	$1.48 \cdot 2^{14}$	676 × 1024
4	nancy_cathedral_1	$1.60 \cdot 2^{14}$	2048 × 1536
5	nancy_cathedral_2	$1.94 \cdot 2^{14}$	2048 × 1536
6	nave	$1.64 \cdot 2^{23}$	480 × 720
7	archway	$1.75 \cdot 2^{7}$	1080 × 721
8	medway	$1.99 \cdot 2^{8}$	1080 × 723
9	kitchenwindow	$1.77 \cdot 2^{11}$	1080 × 1650
10	memorial	$1.29 \cdot 2^{18}$	768 × 512
11	rosette	$1.83 \cdot 2^{17}$	480 × 720
12	snow	$1.02 \cdot 2^{10}$	1536 × 2048
13	rend07	$1.41 \cdot 2^{28}$	575 × 575
14	rend10	$1.20 \cdot 2^{26}$	1024 × 1024
15	Iwate	$1.08 \cdot 2^{19}$	1396 × 3720

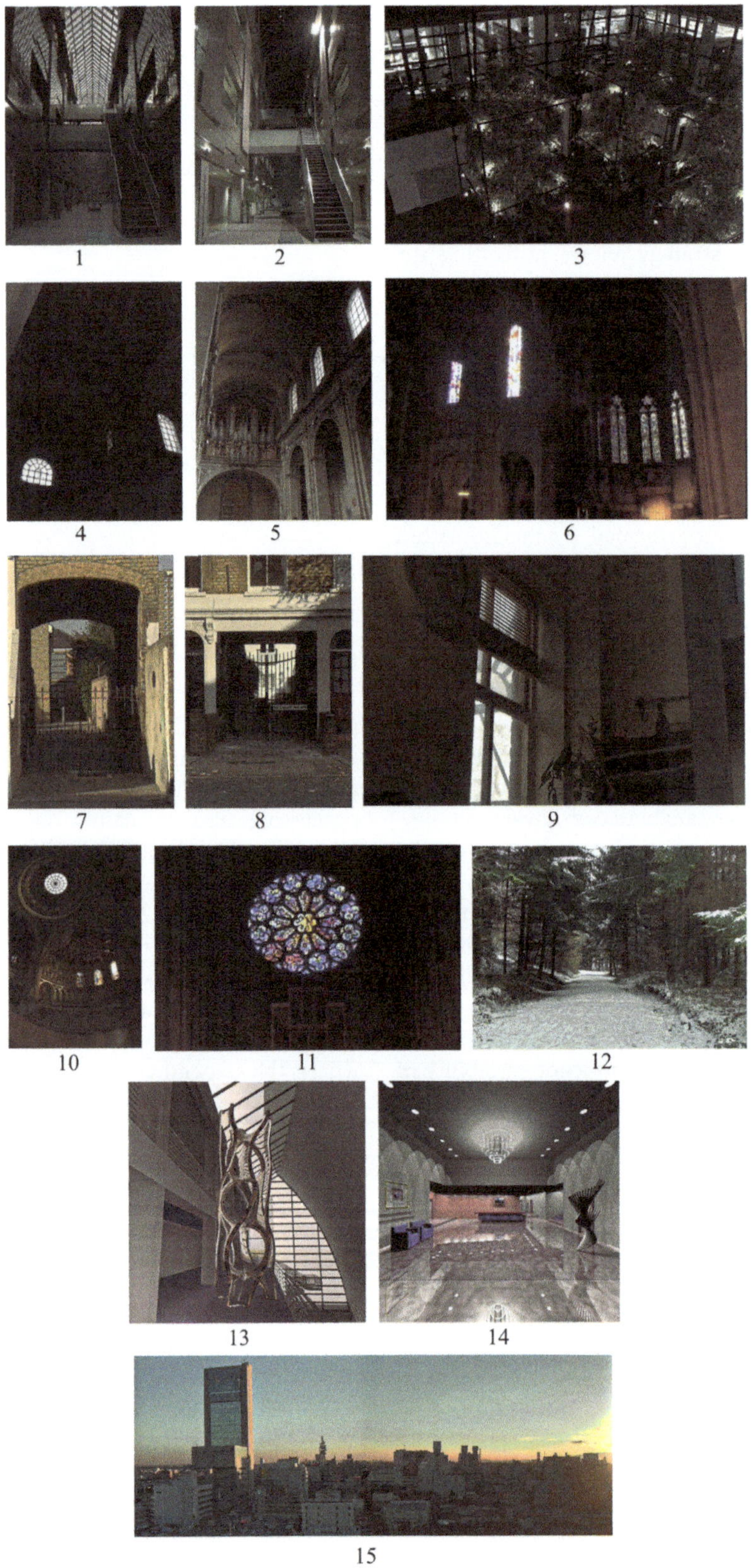

Fig. 4 Employed HDR images, rendered through the ICAM06 TMO [25]

Table 4 Capacity, in bpp and embeddable bits, of the proposed method, for $C = 3$. Also, the required embedding processing time is provided

	P	Image 1	2	3	4	5	6	7	8	9	10	11	12	13	14	15
Capacity (bpp)	7								0.0153							
	11								0.0062							
	17								0.0026							
Capacity (M)	7	11664	11664	10512	47742	47742	5202	11781	11781	27027	5832	5202	47742	5043	15987	78705
	11	4692	4692	4140	19251	19251	2016	4704	4704	11025	2346	2016	19251	2028	6348	31941
	17	1914	1914	1710	8100	8100	882	1953	1953	2346	990	882	8100	768	2700	13407
Processing time (s)	7	33.8	33.8	29.9	130.3	130.3	12.7	33.9	33.9	73.3	16.7	12.7	130.3	12.5	41.4	220.3
	11	14.5	14.5	12.6	57.4	57.4	5.3	14.6	14.6	32.9	7.2	5.3	57.4	5.2	18.1	97.4
	17	6.6	6.6	5.7	26.9	26.9	2.6	6.6	6.6	14.7	3.3	2.6	26.9	2.4	8.7	45.1

image as shown in Table 4, it would also result in an increased probability of detecting a modification in the marked HDR image. The obtained results also show that the embedded mark is more imperceptible as the parameter Δ is set to higher values, since a smaller quantization step $\frac{1}{2\Delta}$ is thus obtained. Moreover, selecting the $H = 3$ first most energetic directions to embed the mark in each image block guarantees a lower perceptual impact than using $H = 5$ directions. Within this regard, it can be also noticed that mark imperceptibility can be improved by discarding the most energetic projection for the embedding, as it is shown by the results achieved with $[h_{\max}, h_{\min}] = [2, 4]$ and $[h_{\max}, h_{\min}] = [2, 6]$. Eventually, the increase of the number of the marked coefficients C increases the number of embedded bits at the cost of worsening the imperceptibility performance.

A visual example on the imperceptibility of the embedded mark is given in Fig. 5, where the original image 1 is shown together with its processed version with $\gamma = HH$, $P = 17$, $[h_{\max}, h_{\min}] = [1, 5]$, $C = 3$ and $\Delta = 2$. The reported LDR representations are obtained by means of the ICAM06 TMO [25]. From Table 5, it can be seen that such choice of system parameters allows to guarantee, on average, a probability of detecting a modification in at least 5 % of the image which is equal to 6.4 %. In order to further demonstrate the effectiveness of the proposed approach, the achieved performance is compared with the one provided by the method in [18], the only one so far proposed that allows to embed large messages (>1000 bits) in an HDR image, while being able to recover them from tone-mapped marked images, as reported in Table 2. It has been noticed that, when applying the method in [18] to the 15 images considered for our experiments, the average probability of detecting a modification in at least 5 % of the marked images, measured with the HDR-VDP-2 metric, is equal to 27.6 %, being therefore

Table 5 Imperceptibility performance expressed through the HDR-VDP-2 metric (in %). Results averaged over the considered images

		$P = 7$		$P = 11$		$P = 17$	
$[h_{\max}, h_{\min}]$	C	$\Delta = 1$	$\Delta = 2$	$\Delta = 1$	$\Delta = 2$	$\Delta = 1$	$\Delta = 2$
[1, 3]	1	16.38	2.51	11.77	1.32	6.04	0.44
	2	28.18	5.79	16.32	3.23	9.96	1.21
	3	36.49	8.42	30.06	5.50	10.12	2.33
[2, 4]	1	15.69	1.48	7.51	0.87	6.20	0.50
	2	23.59	3.27	11.79	1.59	9.53	1.31
	3	29.51	8.41	24.57	4.48	15.44	2.10
[1, 5]	1	20.30	3.78	14.80	2.43	7.55	1.03
	2	35.00	8.55	26.85	4.54	16.51	1.79
	3	48.79	16.13	33.59	9.42	19.17	6.40
[2, 6]	1	16.92	1.70	13.30	1.78	8.09	0.69
	2	31.48	6.97	16.29	4.59	12.08	2.35
	3	36.68	15.19	26.50	8.38	18.82	3.40

a b

Fig. 5 Comparison between tone-mapped version of **a**: original image 1; **b**: marked image 1 (AtriumMorning)

much larger than the one obtained by the approach we propose. It is worth specifying that the parameters of the work in [18] have been selected in order to guarantee as much imperceptibility as possible, while still achieving admissible performance regarding data extraction from marked HDR images. The method in [18] in fact relies on the modification of a DCT coefficient in each 8×8 block of an LDR version of the considered HDR content. Nonetheless, large alterations are needed to achieve low BER values. The Durand TMO [20] is employed to generate the LDR images where the mark is embedded, as described in [18].

The parameter configuration with $\gamma = HH$, $P = 17$, $[h_{\max}, h_{\min}] = [1, 5]$, $C = 3$, and $\Delta = 2$ is taken into account to also evaluate the robustness of the proposed approach. Within this regard, Table 6 reports the BERs obtained when retrieving the embedded marks from the marked HDR images, as well as from their LDR counterparts obtained with the six TMOs listed in the table. From the reported results, it is evident that some errors may be encountered even when analyzing the marked HDR images, mainly due to the modification of the directions energies when performing the embedding and to the conversions performed when saving the marked images into an HDR format. Moreover, it can be seen that the application of distinct TMOs may represent a very different attack for the extraction of the mark from LDR representations. For instance, with the selected configuration, the best performance is achieved when considering the Reinhard TMO [27], while significantly worse results are obtained with the tonemap operator provided in Matlab [26]. It can be argued that the Reinhard TMO provides the best performance since it is based on bilateral filtering, therefore trying to accurately maintain the edges of an HDR image when converting it into its LDR version. Being the proposed approach based on data embedding close to the edges, a TMO preserving these components is also able to preserve the information there hidden.

The comparison between the robustness performance of the proposed approach, when still maintaining the aforementioned system parameters, and the method in [18], is given in Table 7. It can be seen that the method in [18] performs quite similarly for all the considered TMOs, yet it can never provide results as good as those achieved with our approach. Mark extraction is hard from HDR images, too, when an admissible imperceptibility has to be guaranteed with the method in [18].

A more detailed analysis on the capability of extracting a mark from tone-mapped versions of the marked HDR images is obtained by evaluating the BER achievable for different system configurations. As for the imperceptibility analysis, Table 8 reports the performance obtained for different choices of P, Δ, $[h_{\max}, h_{\min}]$, and C, while embedding the marks in the $\gamma = HH$ first-level decomposition subband.

Imperceptibility and robustness of the embedded marks are often conflicting requirements. This is experimentally confirmed by observing that the embedded marks can be extracted with less errors from LDR images obtained when using a lower value of Δ, since the resulting quantization step is much larger, even if the imperceptibility is negatively affected. It can be also seen that selecting a large number of directions for embedding purposes improves

Table 6 BER performance (in %) for data retrieval from marked HDR and LDR images, when considering $\gamma = HH$, $P = 17$, $\Delta = 2$, $[h_{max}, h_{min}] = [1, 5]$, and $C = 3$

Image	HDR	TMO					
		iCAM06	tonemap	Reinhard	Drago	Tumblin	Durand
ID		[25]	[26]	[27]	[28]	[29]	[20]
1	0.16	14.84	14.00	7.21	6.95	23.98	23.77
2	0.57	9.87	21.00	4.18	3.45	9.56	10.03
3	1.23	12.98	16.61	4.56	4.27	14.68	19.71
4	6.69	31.26	27.65	18.68	25.26	19.38	15.33
5	12.11	23.42	31.07	20.42	26.11	23.02	23.86
6	29.12	36.26	36.26	33.08	33.88	33.99	36.58
7	4.86	16.79	24.99	9.68	30.21	18.38	28.57
8	1.64	13.36	22.38	8.14	17.05	15.92	19.41
9	23.19	35.51	42.76	31.54	36.63	36.47	31.92
10	0.03	15.15	15.66	5.86	8.48	8.28	7.17
11	19.59	36.26	28.21	30.36	28.89	32.29	32.06
12	0.17	5.42	17.89	6.96	10.38	14.79	19.23
13	22.26	33.98	42.32	28.39	30.73	30.08	25.91
14	22.22	32.81	33.52	24.19	30.44	30.22	27.70
15	0.16	5.91	37.98	4.97	35.46	6.86	21.17

the mark extraction capabilities, although slightly affecting the imperceptibility of the performed modifications. Moreover, excluding the most energetic direction during the embedding worsens the achievable BERs, even if it is beneficial from the imperceptibility point of view, for both $H = 3$ and $H = 5$. As for the number of coefficients C marked for each direction, no meaningful difference in terms of BER can be noticed, thus suggesting to use $C = 3$, even if at the cost of a higher visibility of the modifications. It has also to be noticed that, although increasing the block dimension P significantly improves mark imperceptibility, it does not affect negatively the mark extraction capability, making thus preferable the selection of large block values such as $P = 17$. With such a selection, and exploiting a system configuration with $\Delta = 2$, $[h_{max}, h_{min}] = [1, 5]$, and $C = 3$, it is possible to guarantee a probability of detecting a modification in at least the 5 % of the image of about 6 %, while simultaneously achieving a BER for the mark extracted from LDR images at about 19 %, with marks in the order of at least a thousand of bits for the most of considerable images, as shown in Table 4.

It is also worth mentioning that even the selection of the DWT subband to be marked affects the robustness of the marking procedure: considering $P = 17$ and $\Delta = 2$, the BER obtained for $\gamma = HH$ is equal to 25.66 %, averaged over all the considered possibilities for marked directions and number of modified coefficients, while a better behavior is obtained when using the horizontal details ($\gamma = HL$) subband achieving a BER $= 15.31$ % and the vertical detail ($\gamma = LH$) subband achieving a BER $= 14.56$ %. The trade-off between imperceptibility and robustness is confirmed by this analysis too.

Eventually, besides analyzing the robustness of the proposed approach against non-malicious attacks such as those performed by means of TMOs, we have also evaluated the behavior of our watermarking schemes against malicious signal processing attacks carried out on the marked HDR images, still considering a system

Table 7 Comparison between our method and the one in [18] for the BER performance (in %) regarding data retrieval from marked HDR and LDR images, averaged over all the 15 considered images

	HDR	TMO					
		iCAM06	tonemap	Reinhard	Drago	Tumblin	Durand
		[25]	[26]	[27]	[28]	[29]	[20]
Proposed method	6.09	18.21	26.19	12.55	19.58	18.45	20.51
Wu et al. [18]	30.05	29.81	30.54	30.10	30.28	31.16	30.75

Table 8 Robustness performance in terms of BER (in %). Results averaged over the considered images and TMOs

		$P = 7$		$P = 11$		$P = 17$	
$[h_{max}, h_{min}]$	C	$\Delta = 1$	$\Delta = 2$	$\Delta = 1$	$\Delta = 2$	$\Delta = 1$	$\Delta = 2$
[1, 3]	1	17.78	27.67	15.62	24.78	16.34	23.37
	2	17.41	27.69	15.12	24.19	13.96	22.11
	3	17.59	28.07	14.90	24.07	13.64	21.99
[2, 4]	1	21.94	32.08	20.25	28.57	19.69	27.05
	2	22.68	31.76	20.35	29.03	18.53	26.34
	3	23.31	32.30	20.92	28.62	18.94	26.47
[1, 5]	1	15.82	26.28	13.73	22.58	13.21	20.36
	2	16.00	26.74	12.86	22.18	11.01	19.02
	3	16.25	26.91	13.23	22.16	11.29	19.52
[2, 6]	1	17.69	29.21	14.45	24.06	13.24	21.39
	2	18.22	29.11	14.74	23.96	12.57	20.65
	3	18.67	29.47	15.15	24.36	12.58	20.82

configuration with $P = 17$, $\Delta = 2$, $[h_{max}, h_{min}] = [1, 5]$, and $C = 3$. Specifically, Table 9 presents the BER performance, averaged over all the considered 15 images, achieved when extracting the embedded messages from HDR images attacked with additive Gaussian noise and with Gaussian blurring. In more detail, Gaussian noise with a standard deviation equal to 3 has been added to the employed images, while the employed Gaussian blurring filter has been defined with a [3, 3] window and a standard deviation set to 0.3. We have also investigated the possibility of extracting the embedded information from tone-mapped versions of the attacked HDR images. As the reported results testify, the proposed approach is able to provide proper resilience even against malicious signal processing attacks. Our method performs significantly better than the one in [18] when considering Gaussian filtering as an attack, while similar behaviors have been observed when attacking the marked HDR images with additive Gaussian noise.

6 Conclusions

HDR images represent valuable content, whose intellectual property needs to be protected by means of proper digital techniques. To this aim, a blind multi-bit watermarking method for HDR images has been presented in this paper. The proposed approach hides information in areas closely associated to image contours, thanks to the properties of the considered RDCT transform. The selected embedding domain guarantees proper imperceptibility of the inserted marks, while providing the means for blindly extracting the embedded information without the need of the original image. An extensive set of experimental tests, conducted on 15 different HDR images, has been conducted to estimate the performance of the proposed method. The HDR-VDP-2 metric is employed to evaluate the perceptibility of the performed modifications, while six different TMOs have been employed to evaluate the possibility of retrieving the embedded data from LDR images generated from the marked HDR ones. The effects of signal processing attacks such as Gaussian additive noise and filtering have been also evaluated. The obtained results testify the effectiveness of the proposed approach, also showing the trade-off relationships between imperceptibility, robustness, and capacity when varying the parameters determining the employed system configurations.

Table 9 Robustness against malicious signal processing attacks, expressed in terms of BER (in %). Results averaged over the considered images

			TMO					
		HDR	iCAM06 [25]	tonemap [26]	Reinhard [27]	Drago [28]	Tumblin [29]	Durand [20]
Gaussian noise	Proposed method	28.43	30.59	34.78	30.13	34.43	32.24	35.42
	Wu et al. [18]	31.07	31.10	30.69	30.79	30.88	31.02	31.08
Gaussian blur	Proposed method	11.59	22.31	29.25	17.82	23.22	22.28	23.92
	Wu et al. [18]	32.97	32.73	33.24	33.12	33.29	33.83	33.56

Competing interests

The authors declare that they have no competing interests.

References

1. F Banterle, A Artusi, K Debattista, A Chalmers, *Advanced High Dynamic Range Imaging: Theory and Practice*. (AK Peters (CRC Press), Natick, MA, USA, 2011)
2. JJ McCann, A Rizzi, *The Art and Science of HDR Imaging*. (Wiley, USA, 2011)
3. F Barni, F Bartolini, *Watermarking Systems Engineering: Enabling Digital Assets Security and Other Applications*. (CRC Press, USA, 2004)
4. IJ Cox, M Miller, JA Bloom, J Fridrich, T Kalker, *Digital Watermarking and Steganography*. (Morgan Kaufmann, USA, 2008)
5. J Fridrich, *Steganography in Digital Media: Principles, Algorithms, and Applications*. (Cambridge University Press, USA, 2009)
6. E Maiorana, V Solachidis, P Campisi, in *IEEE International Symposium on Image and Signal Processing and Analysis*. Robust multi-bit watermarking for HDR images in the Radon-DCT domain (IEEE, Trieste, Italy, 2013)
7. X Xue, T Jinno, J Xin, M Okuda, Watermarking for HDR image robust to tone mapping. IEICE Trans. Fundam. Electron. Commun. Comput. Sci. **94**(11), 2334–2341 (2011)
8. F Guerrini, M Okuda, N Adami, R Leonardi, High dynamic range image watermarking robust against tone-mapping operators. IEEE Trans. Inf. Forensics Sec. **6**(2), 283–295 (2011)
9. B Chen, GW Wornell, Quantization index modulation: a class of provably good methods for digital watermarking and information embedding. IEEE Trans. Inf. Theory. **47**(4), 1423–1443 (2001)
10. R Mantiuk, K Myszkowski, H-P Seidel, in *IEEE Internation Conference on Systems, Man and Cybernetics*. Visible difference predictor for high dynamic range images (IEEE, The Hague, Netherlands, 2004)
11. F Autrusseau, D Goudia, in *20th IEEE International Conference on Image Processing (ICIP)*. Non linear hybrid watermarking for high dynamic range images (IEEE, Melbourne, Australia, 2013)
12. V Solachidis, E Maiorana, P Campisi, F Banterle, in *IEEE International Conference on Digital Signal Processing (DSP)*. HDR image watermarking based on bracketing decomposition (IEEE, Santorini, Greece, 2013)
13. R Mantiuk, KJ Kim, AG Rempel, W Heidrich, HDR-VDP-2: a calibrated visual metric for visibility and quality predictions in all luminance conditions. ACM Trans. Graph. **30**(4), 1–14 (2011)
14. Y-M Cheng, C-M Wang, A novel approach to steganography in high-dynamic-range images. IEEE Multimedia. **16**(3), 70–80 (2009)
15. MT Li, NC Huang, CM Wang, A data hiding scheme for high dynamic range images. International Journal of Innovative Computing Information and Control (IJICIC). **7**(5A), 2021–2035 (2011)
16. C-M Yu, K-C Wu, C-M Wang, A distortion-free data hiding scheme for high dynamic range images. Displays. **32**(5), 225–236 (2011)
17. C Rattanacharuchinda, T Amornraksa, in *International Conference on Electrical Engineering/Electronics, Computer, Telecommunications and Information Technology (ECTI-CON)*. Image watermarking for high dynamic range images (IEEE, Hua Hin, Thailand, 2012)
18. J-L Wu, in *Watermarking - Volume 2*. Robust watermarking framework for high dynamic range images against tone-mapping attacks (InTech, USA, 2012), pp. 229–242
19. V Solachidis, E Maiorana, P Campisi, in *Proc. SPIE 8655, Image Processing: Algorithms and Systems XI, 865505*. HDR image multi-bit watermarking using bilateral-filtering-based masking (SPIE, San Francisco, USA, 2013)
20. F Durand, J Dorsey, Fast bilateral filtering for the display of high-dynamic-range images. ACM Trans. Graph. **21**(3), 257–266 (2002)
21. Z Farbman, R Fattal, D Lischinski, R Szeliski, Edge-preserving decompositions for multi-scale tone and detail manipulation. ACM Trans. Graph. **27**(3), 1–10 (2008)
22. E Maiorana, P Campisi, A Neri, in *IEEE Biometrics Symposium*. Biometric signature authentication using Radon transform-based watermarking techniques (IEEE, Baltimora, USA, 2007)
23. S Mallat, *A Wavelet Tour of Signal Processing: The Sparse Way*. (Accademic press, Waltham, MA, USA, 2008)
24. MN Do, M Vetterli, The finite ridgelet transform for image representation. IEEE Trans. Image Process. **12**(1), 16–28 (2003)
25. J Kuang, GM Johnson, MD Fairchild, iCAM06: a refined image appearance model for HDR image rendering. J. Vis. Commun. Image Represent. **18**(5), 406–414 (2007)
26. F Drago, K Myszkowski, T Annen, N Chiba, Adaptive logarithmic mapping for displaying high contrast scenes. EUROGRAPHICS. **22**(3), 419–426 (2003)
27. E Reinhard, M Stark, P Shirley, J Ferwerda, Photographic tone reproduction for digital images. ACM Trans. Graph. **21**(3), 267–276 (2002)
28. F Drago, K Myszkowski, T Annen, N Chiba, Adaptive logarithmic mapping for displaying high contrast scenes. **22**(3), 419–426 (2003)
29. J Tumblin, H Rushmeier, Tone reproduction for realistic images. IEEE Comput. Graph. Appl. **13**(6), 42–48 (1993)

12

New inexact explicit thresholding/ shrinkage formulas for inverse problems with overlapping group sparsity

Gang Liu[1,2], Ting-Zhu Huang[2*], Xiao-Guang Lv[3] and Jun Liu[2]

Abstract

The least-square regression problems or inverse problems have been widely studied in many fields such as compressive sensing, signal processing, and image processing. To solve this kind of ill-posed problems, a regularization term (i.e., regularizer) should be introduced, under the assumption that the solutions have some specific properties, such as sparsity and group sparsity. Widely used regularizers include the ℓ_1 norm, total variation (TV) semi-norm, and so on. Recently, a new regularization term with overlapping group sparsity has been considered. Majorization minimization iteration method or variable duplication methods are often applied to solve them. However, there have been no direct methods for solving the relevant problems due to the difficulty of overlapping. In this paper, we proposed new inexact explicit shrinkage formulas for one class of these relevant problems, whose regularization terms have translation invariant overlapping groups. Moreover, we apply our results to TV deblurring and denoising problems with overlapping group sparsity. We use alternating direction method of multipliers to iteratively solve them. Numerical results verify the validity and effectiveness of our new inexact explicit shrinkage formulas.

Keywords: Overlapping group sparsity, Total variation, ADMM, Image deblurring, Regularization, Explicit shrinkage formula

1 Introduction

The least-square regression problems or inverse problems have been widely studied in many fields such as compressive sensing, signal processing, image processing, statistics, and machine learning. Regularization terms with sparse representations (for instance, the ℓ_1 norm regularizer) have been developed into an important tool in these applications, and a list of methods have been proposed [1–4]. These methods are based on the assumption that signals or images have a sparse representation, that is, only containing a few nonzero entries. The corresponding task is to solve the following problem

$$\min_{\mathbf{z}} \quad \|\mathbf{z}\|_1 + \frac{\beta}{2}\|\mathbf{z}-\mathbf{x}\|_2^2, \tag{1}$$

where $\mathbf{x} \in \mathbb{R}^n$ is a given vector, $\mathbf{z} \in \mathbb{R}^n$, $\|\mathbf{z}\|_p = \left(\sum_{i=1}^n |z_i|^p\right)^{\frac{1}{p}}$ with $(p = 1, 2)$ represents the ℓ_p norm of vector z. The first term of (1) is called the regularization term, the second term is called the fidelity term, and $\beta > 0$ is the regularization parameter.

To further improve the solutions, recent studies suggested to go beyond sparsity and took into account additional information about the underlying structure of the solutions [1, 2, 5]. In particular, a wide class of solutions which have specific "group sparsity" structure are considered. In this case, a group sparse vector can be divided into groups of components satisfying (a) only a few of groups contain nonzero values and (b) these groups are not needed to be sparse. If we insert such group vectors into a matrix as row vectors of the matrix, this matrix will only have few nonzero rows and these rows may be not sparse. This property is called "group sparsity" or "joint sparsity," and many literature have considered these new sparse problems [1–6]. These problems can be formulated as the following expression.

*Correspondence: tingzhuhuang@126.com
[2]Research Center for Image and Vision Computing/School of Mathematical Sciences, University of Electronic Science and Technology of China, Xiyuan Ave, 611731 Chengdu, China
Full list of author information is available at the end of the article

$$\min_{\mathbf{z}} \sum_{i=1}^{r} \|\mathbf{z}[\,i]\,\|_2 + \frac{\beta}{2}\|\mathbf{z}-\mathbf{x}\|_2^2, \tag{2}$$

where $|z_i|$ is the absolute value of z_i, and $\mathbf{z}[\,i]$ is the ith group of z with $\mathbf{z}[\,i] \cap \mathbf{z}[\,j] = \emptyset$ and $\bigcup_{i=1}^{r} \mathbf{z}[\,i] = \mathbf{z}$. "Group sparsity" solutions from (2) have better representation and have been widely studied both for convex and nonconvex cases [1, 3, 6–9].

More recently, overlapping group sparsity (OGS) have been considered [1, 3, 10–19]. These methods are based on the assumption that signals or images have a special sparse representation with OGS. The task is to solve the following problem

$$\min_{\mathbf{z}} \|\mathbf{z}\|_{2,1} + \frac{\beta}{2}\|\mathbf{z}-\mathbf{x}\|_2^2, \tag{3}$$

where $\|\mathbf{z}\|_{2,1} = \sum_{i=1}^{n} \|(z_i)_g\|_2$ is the generalized $\ell_{2,1}$ norm. Here, each $(z_i)_g$ is a group vector containing s (called group size) elements that surrounding the ith entry of $\mathbf{z}$. For example, $(z_i)_g = (z_{i-1}, z_i, z_{i+1})$ with $s = 3$. In this case, $(z_i)_g$, $(z_{i+1})_g$, and $(z_{i+2})_g$ contain the $(i+1)$th entry of $\mathbf{z}$, z_{i+1}, which means "overlapping" different from the form of group sparsity in (2). In particular, if $s = 1$, the generalized $\ell_{2,1}$ norm degenerates into the original ℓ_1 norm, and the relevant regularization problem (3) degenerates to (1).

To be more general, researchers considered the weighted generalized $\ell_{2,1}$ norm $\|z\|_{w,2,1} = \sum_{i=1}^{n} \|w_g \circ (z_i)_g\|_2$ (we only consider that each group has the same weight, which means translation invariant) instead of the former generalized $\ell_{2,1}$ norm. The task can be extended to

$$\min_{\mathbf{z}} \|\mathbf{z}\|_{w,2,1} + \frac{\beta}{2}\|\mathbf{z}-\mathbf{x}\|_2^2, \tag{4}$$

where w_g is a nonnegative real vector with the same size as $(z_i)_g$ and "$\circ$" is the point-wise product or Hadamard product. For instance, $w_g \circ (z_i)_g = ((w_g)_1 z_i, (w_g)_2 z_{i+1}, (w_g)_3 z_{i+2})$ with $s = 3$ as the former example. In particular, the weighted generalized $\ell_{2,1}$ norm degenerates into the generalized $\ell_{2,1}$ norm if each entry of w_g equals to 1.

The problems (3) and (4) have been considered in [1, 14–19]. They solve the relevant problems by using variable duplication methods (variable splitting, latent/auxilliary variables, etc.). For example, Deng et al. in [1] introduced a diagonal matrix G in their method. However, matrix G is not easy to find and would break the structure of the coefficient matrix, which makes the difficulty of solving solutions under high-dimensional vector cases. Moreover, it is difficult to extend this method to the matrix case.

Considering the matrix case of the problem (4), we can get

$$\min_{A} \|A\|_{W,2,1} + \frac{\beta}{2}\|A-X\|_F^2, \tag{5}$$

where $X, A \in \mathbb{R}^{m\times n}$, $\|A\|_{W,2,1} = \sum_{i=1}^{m}\sum_{j=1}^{n} \|W_g \circ (A_{i,j})_g\|_F$. Here, each $(A_{i,j})_g$ is a group matrix containing $K_1 \times K_2$ (called group size) elements that surround the (i,j)th entry of A. It is defined by

$$W_g \circ (A_{i,j})_g = \begin{bmatrix} (W_g)_{1,1}A_{i-l_1,j-l_2} & (W_g)_{1,2}A_{i-l_1,j-l_2+1} & \cdots & (W_g)_{1,K_2}A_{i-l_1,j+r_2} \\ (W_g)_{2,1}A_{i-l_1+1,j-l_2} & (W_g)_{2,2}A_{i-l_1+1,j-l_2+1} & \cdots & (W_g)_{2,K_2}A_{i-l_1+1,j+r_2} \\ \vdots & \vdots & \ddots & \vdots \\ (W_g)_{K_1,1}A_{i+r_1,j-l_2} & (W_g)_{K_1,2}A_{i+r_1,j-l_1+1} & \cdots & (W_g)_{K_1,K_2}A_{i+r_1,j+r_2} \end{bmatrix} \in \mathbb{R}^{K_1\times K_2},$$

where $l_1 = \lfloor \frac{K_1-1}{2} \rfloor$, $l_2 = \lfloor \frac{K_2-1}{2} \rfloor$, $r_1 = \lfloor \frac{K_1}{2} \rfloor$, $r_2 = \lfloor \frac{K_2}{2} \rfloor$ (with $l_1+r_1+1 = K_1$ and $l_2+r_2+1 = K_2$) and $\lfloor x \rfloor$ denotes the largest integer less than or equal to x. In particular, if $K_2 = 1$, this problem degenerates to the former vector case (4). If $K_1 = K_2 = 1$, this problem degenerates to the original ℓ_1 regularization problem (1) for the matrix case.

Much more recently, Chen et al. [10] considered the problem (5) when $(W_g)_{i,j} \equiv 1$ for $i = 1, \cdots, K_1$, $j = 1, \cdots, K_2$. They used an iterative algorithm to solve this problem based on the principle of majorization minimization (MM). Their experiments showed that their method on solving the problem (3) is more efficient than other variable duplication methods (variable splitting, latent/auxilliary variables, etc.) [1, 14–19]. Although their method is efficient for computing the solution, it may cost more time since the solution is obtained after many iterations rather than a direct step computation.

To our knowledge, there have been no direct methods for solving the relevant problems (3), (4), and (5). In this paper, we propose new inexact explicit shrinkage formulas for solving them directly, which are direct methods, and faster or more convenient than other methods, for instance, the most recently MM iteration method [10]. This new direct method can save more time than the MM method in [10] while getting very similar solutions. Numerical results are given to show the effectiveness of our new explicit shrinkage formulas. Moreover, the new method can be used in application for solving the subproblem in many other OGS problems, such as the wavelet regularizer with OGS in compressive sensing and the total variation (TV) regularizer with OGS in image restoration [11, 12]. For example, we will apply our results to image restoration using TV with OGS in this work similarly as [11, 12]. Moveover, we expend the works, only considering ATV in [11, 12] to both ATV and ITV (more detail can be referred to Section 4). Numerical results will show that our method can save more time with getting very similar results.

The outline of the paper is as follows. In Section 2, we deduce our explicit shrinkage formulas in detail for the OGS problems (3), (4), and (5). In Section 3, we propose some extensions for these shrinkage formulas. In Section 4, we apply our results to image deblurring and denoising problems with OGS TV. Numerical results are given in Section 5. Finally, we conclude this paper in Section 6.

2 OGS shrinkage

2.1 Classic shrinkage

In this subsection, we will review the original shrinkage formulas and their principles, since our new explicit OGS shrinkage formulas are based on them. Firstly, we give the following definition.

Definition 1. Define shrinkage mappings Sh_1 and Sh_2 from $\mathbb{R}^N \times R^+$ to $\mathbb{R}^N$ by

$$\mathrm{Sh}_1(\mathbf{x},\beta)_i = \mathrm{sgn}(x_i)\max\left\{|x_i| - \frac{1}{\beta}, 0\right\}, \tag{6}$$

$$\mathrm{Sh}_2(\mathbf{x},\beta) = \frac{\mathbf{x}}{\|\mathbf{x}\|_2}\max\left\{\|\mathbf{x}\|_2 - \frac{1}{\beta}, 0\right\}, \tag{7}$$

where the expression is taken to be zero when the second factor is zero, and "sgn" represents the signum function indicating the sign of a number. More precisely, $\mathrm{sgn}(x) = 0$ if $x = 0$, $\mathrm{sgn}(x) = -1$ if $x < 0$, and $\mathrm{sgn}(x) = 1$ if $x > 0$.

The shrinkage (6) is known as soft thresholding and occurs in many algorithms related to sparsity since it is the proximal mapping for the ℓ_1 norm. The shrinkage (7) is known as high-dimensional shrinkage formula.

Now, we consider solving the following problems

$$\min_{\mathbf{z}} \quad \|\mathbf{z}\|_p + \frac{\beta}{2}\|\mathbf{z}-\mathbf{x}\|_2^2, \; p = 1, 2. \tag{8}$$

The minimizer of (8) with $p = 1$ is the following equation.

$$\arg\min_{\mathbf{z}} \quad \|\mathbf{z}\|_1 + \frac{\beta}{2}\|\mathbf{z}-\mathbf{x}\|_2^2 = \mathrm{Sh}_1(\mathbf{x},\beta). \tag{9}$$

Due to the additivity and separability of both the ℓ_1 norm and the square of the ℓ_2 norm, Eq. (9) can be deduced easily by the following formula:

$$\min_{\mathbf{z}} \quad \|\mathbf{z}\|_1 + \frac{\beta}{2}\|\mathbf{z}-\mathbf{x}\|_2^2 = \sum_{i=1}^{n} \min_{z_i} |z_i| + \frac{\beta}{2}|z_i - x_i|^2. \tag{10}$$

The minimizer of (8) with $p = 2$ is the following equation.

$$\arg\min_{\mathbf{z}} \quad \|\mathbf{z}\|_2 + \frac{\beta}{2}\|\mathbf{z}-\mathbf{x}\|_2^2 = \mathrm{Sh}_2(\mathbf{x},\beta). \tag{11}$$

This formula is deduced by the Euler equation of (8) with $p = 2$. In particular, without considering $\mathbf{x} = \mathbf{0}$ or $\mathbf{z} = \mathbf{0}$, we obtain the following Euler equation.

$$\beta(\mathbf{z}-\mathbf{x}) + \frac{\mathbf{z}}{\|\mathbf{z}\|_2} \ni \mathbf{0}, \tag{12}$$

$$\left(1 + \frac{1}{\beta}\frac{1}{\|\mathbf{z}\|_2}\right)\mathbf{z} - \mathbf{x} \ni \mathbf{0}. \tag{13}$$

We can easily get that the necessary condition is that the vector $\mathbf{z}$ is parallel to the vector $\mathbf{x}$. That is, $\frac{\mathbf{z}}{\|\mathbf{z}\|_2} = \frac{\mathbf{x}}{\|\mathbf{x}\|_2}$. Substituting into (13) and considering $\mathbf{x} = \mathbf{0}$ or $\mathbf{z} = \mathbf{0}$, we obtain the formula (7). More details can be referred to [20, 21].

Our new explicit OGS shrinkage formulas are based on these observations, especially the properties of additivity, separability, and parallelity.

Remarks 1. The problem (2) is easy to be solved by a simple shrinkage formula, which is not used in this work. More details can be referred to [1, 22, 23].

2.2 The OGS shrinkage formulas

Now, we focus on the problem (3) firstly. The difficulty of this problem is "overlapping." Therefore, we must take some special techniques to avoid "overlapping." That is the point of our new explicit OGS shrinkage formulas.

It is obvious that the first term of the problem (3) is additive and separable. So if we find some relative rules such that the second term of the problem (3) has the same properties with the same variable as the first term, one approximate solution of (3) can be easily found similar as classic shrinkage.

Assuming periodic boundary condition (PBC) (boundary condition (BC) is necessary in OGS functionals) is used here, we observe that each entry z_i of the vector $\mathbf{z}$ would appear exactly s times in the first term. Therefore, to hold on the uniformity of vectors $\mathbf{z}$ and $\mathbf{x}$, we need to multiply the second term by s. To maintain the invariability of the problem (3), after some manipulations, we have

$$\begin{aligned} f_m(\mathbf{z}) &= \min_{\mathbf{z}} \|\mathbf{z}\|_{2,1} + \frac{\beta}{2}\|\mathbf{z}-\mathbf{x}\|_2^2 \\ &= \min_{\mathbf{z}} \sum_{i=1}^{n} \|(z_i)_g\|_2 + \frac{\beta}{2s} s\|\mathbf{z}-\mathbf{x}\|_2^2 \\ &= \min_{\mathbf{z}} \sum_{i=1}^{n} \|(z_i)_g\|_2 + \frac{\beta}{2s}\sum_{i=1}^{n} \|(z_i)_g - (x_i)_g\|_2^2 \\ &= \min_{\mathbf{z}} \sum_{i=1}^{n} \left(\|(z_i)_g\|_2 + \frac{\beta}{2s}\|(z_i)_g - (x_i)_g\|_2^2\right), \end{aligned} \tag{14}$$

where $(x_i)_g$ is similarly defined as $(z_i)_g$ in Section 1.

For example, we set $s = 3$ and define $(z_i)_g = (z_i, z_{i+1}, z_{i+2})$. The generalized $\ell_{2,1}$ norm $\|\mathbf{z}\|_{2,1}$ can be treated as the generalized ℓ_1 norm of generalized points, whose entry $(z_i)_g$ is also a vector, and the absolute value of each entry is treated as the ℓ_2 norm of $(z_i)_g$. See Fig. 1a intuitively, where the top line is the vector $\mathbf{z}$, the rectangles with dashed line are original $(z_i)_g$, and the rectangles with solid line are the generalized points. In the case of PBC, we know that each line of Fig. 1a is translated equal. We operate the vector $\mathbf{x}$ similarly as the vector $\mathbf{z}$. Putting these generalized points (rectangles with solid line in the figure) as the columns of a matrix, we can regard $s\|\mathbf{z}-\mathbf{x}\|_2^2$ as the matrix Frobenius norm $\| \left[(z_i)_g\right] - \left[(x_i)_g\right] \|_F^2$, where $\left[(z_i)_g\right]$ is a matrix as in Fig. 1a with every line being its row ($\left[(x_i)_g\right]$ is similar). This is why the second equality in (14) holds.

Therefore, generally, for each i of the last line of (14), from the Eqs. (7) and (11), we can obtain

$$\begin{aligned}
&\arg\min_{(z_i)_g} \|(z_i)_g\|_2 + \frac{\beta}{2s}\|(z_i)_g - (x_i)_g\|_2^2 = \mathrm{Sh}_2\left((x_i)_g, \frac{\beta}{s}\right), \\
&(z_i)_g = \max\left\{\|(x_i)_g\|_2 - \frac{s}{\beta}, 0\right\}\frac{(x_i)_g}{\|(x_i)_g\|_2}, \\
&(z_i)_g = \left(\left[(z_i)_g\right]_1, \left[(z_i)_g\right]_2, \ldots, \left[(z_i)_g\right]_s\right).
\end{aligned} \tag{15}$$

Similarly as Fig. 1a, for each i, the ith entry x_i (or z_i) of the vector $\mathbf{x}$ (or $\mathbf{z}$) may appear s times, so we need to compute each z_i for s times in s different groups.

However, the results from (15) are not able to be satisfied simultaneously, because the results z_i in s different groups are different from (15). Moreover, for each i of the last line in (14), the result (15) is given by that the vector $(z_i)_g$ is parallel to the vector $(x_i)_g$ due to Section 2.1.

Notice this point and ignore that $(z_i)_g = \mathbf{0}$ or $(x_i)_g = \mathbf{0}$, particularly for $s = 4$ and $(z_i)_g = (z_{i-1}, z_i, z_{i+1}, z_{i+2})$, the vector $\mathbf{z}$ can be split as follows,

$$\begin{aligned}
\mathbf{z} = \;& (z_1,\ z_2,\ z_3,\ z_4,\ z_5,\ \cdots,\ z_{n-2},\ z_{n-1},\ z_n\ ) \\
= & +\tfrac{1}{4}\ (z_1,\ z_2,\ z_3,\ 0,\ 0,\ \cdots,\ 0,\ 0,\ z_n\ ) \\
& +\tfrac{1}{4}\ (z_1,\ z_2,\ z_3,\ z_4,\ 0,\ \cdots,\ 0,\ 0,\ 0\ ) \\
& +\tfrac{1}{4}\ (0,\ z_2,\ z_3,\ z_4,\ z_5,\ \cdots,\ 0,\ 0,\ 0\ ) \\
& +\ \cdots \\
& +\tfrac{1}{4}\ (z_1,\ 0,\ 0,\ 0,\ 0,\ \cdots,\ z_{n-2},\ z_{n-1},\ z_n\ ) \\
& +\tfrac{1}{4}\ (z_1,\ z_2,\ 0,\ 0,\ 0,\ \cdots,\ 0,\ z_{n-1},\ z_n\ ).
\end{aligned} \tag{16}$$

Let $(z_i)'_g = (0, \cdots, 0, z_{i-1}, z_i, z_{i+1}, z_{i+2}, 0, \cdots, 0)$ be the expansion of $(z_i)_g$, with $(z_1)'_g = (z_1, z_2, z_3, 0, \cdots, 0, z_n)$, $(z_{n-1})'_g = (z_1, 0, 0, \cdots, 0, z_{n-2}, z_{n-1}, z_n)$, $(z_n)'_g = (z_1, z_2, 0, \cdots, 0, z_{n-1}, z_n)$. Let $(x_i)'_g = (0, \cdots, 0, x_{i-1}, x_i, x_{i+1}, x_{i+2}, 0, \cdots, 0)$ be the expansion of $(x_i)_g$ similarly as $(z_i)'_g$. Then, we have $\mathbf{z} = \frac{1}{4}\sum_{i=1}^{n}(z_i)'_g$, and $\mathbf{x} = \frac{1}{4}\sum_{i=1}^{n}(x_i)'_g$. Moreover, we can easily obtain that $\|(z_i)'_g\|_2 = \|(z_i)_g\|_2$ and $\|(x_i)'_g\|_2 = \|(x_i)_g\|_2$ for every i.

On one hand, the Euler equation of $f_m(\mathbf{z})$ (with $s = 4$) is given by

$$\beta(\mathbf{z}-\mathbf{x}) + \frac{(z_1)'_g}{\|(z_1)'_g\|_2} + \cdots + \frac{(z'_n)_g}{\|(z'_n)_g\|_2} \ni \mathbf{0}, \tag{17}$$

$$\begin{aligned}
\frac{\beta}{4}\sum_{i=1}^{n}\left((z_i)'_g - (x_i)'_g\right) + \frac{(z_1)'_g}{\|(z_1)'_g\|_2} + \cdots + \frac{(z_i)'_g}{\|(z_i)'_g\|_2} + \cdots \\
+ \frac{(z_n)'_g}{\|(z_n)'_g\|_2} \ni \mathbf{0}.
\end{aligned} \tag{18}$$

From the deduction of the two-dimensional shrinkage formula (7) in Section 2.1, we know that the necessary condition of minimizing the ith term of the last line in (14)

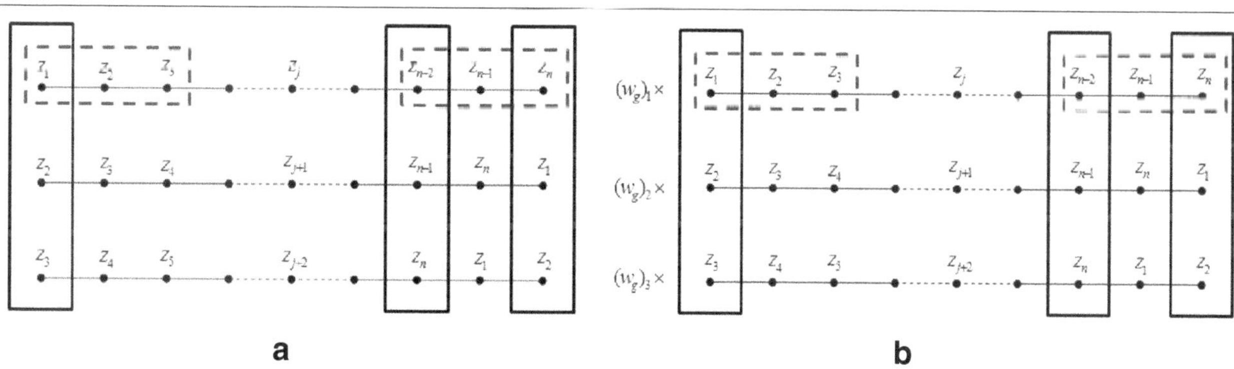

Fig. 1 Vector case. The *top line* is the vector **z**, the *rectangles with dashed line* are original group vector $(z_i)_g$, and the *rectangles with solid line* are the generalized points, which changes the overlapping to non-overlapping. **a** Vector. **b** Weighted vector

is that $(z_i)_g$ is parallel to $(x_i)_g$. That is, $(z_i)'_g$ is parallel to $(x_i)'_g$ for every i. For example,

$$(z_2)'_g = (z_1, z_2, z_3, z_4, 0, \cdots, 0, 0) // (x_2)'_g = (x_1, x_2, x_3, x_4, 0, \cdots, 0, 0). \tag{19}$$

Then, we obtain $\frac{(z_i)'_g}{\|(z_i)'_g\|_2} = \frac{(x_i)'_g}{\|(x_i)'_g\|_2}$. Therefore, (18) can be changed to

$$\frac{\beta}{4}\sum_{i=1}^{n}\left((z_i)'_g - (x_i)'_g\right) + \frac{(x_1)'_g}{\|(x_1)'_g\|_2} + \cdots + \frac{(x_i)'_g}{\|(x_i)'_g\|_2} + \cdots + \frac{(x_n)'_g}{\|(x_n)'_g\|_2} \ni \mathbf{0}, \tag{20}$$

$$\beta(\mathbf{z} - \mathbf{x}) + \frac{(x_1)'_g}{\|(x_1)'_g\|_2} + \cdots + \frac{(x_i)'_g}{\|(x_i)'_g\|_2} + \cdots + \frac{(x_n)'_g}{\|(x_n)'_g\|_2} \ni \mathbf{0}, \tag{21}$$

$$\mathbf{z} \ni \mathbf{x} - \frac{1}{\beta}\left(\frac{(x_1)'_g}{\|(x_1)'_g\|_2} + \cdots + \frac{(x_i)'_g}{\|(x_i)'_g\|_2} + \cdots + \frac{(x_n)'_g}{\|(x_n)'_g\|_2}\right), \tag{22}$$

for each component, we obtain

$$z_i \ni x_i - \frac{1}{\beta}\left(\frac{x_i}{\|(x_{i-2})'_g\|_2} + \frac{x_i}{\|(x_{i-1})'_g\|_2} + \frac{x_i}{\|(x_i)'_g\|_2} + \frac{x_i}{\|(x_{i+1})'_g\|_2}\right). \tag{23}$$

Therefore, when $(z_i)_g \neq \mathbf{0}$ and $(x_i)_g \neq \mathbf{0}$, we find a minimizer of (3) on the direction that all the vectors $(z_i)_g$ are parallel to the vectors $(x_i)_g$. In addition, when $\frac{\beta}{4}\left((z_i)'_g - (x_i)'_g\right) + \frac{(z_i)'_g}{\|(z_i)'_g\|_2} = 0$, (18) holds, then $\left(\frac{\beta}{4} + \frac{1}{\|(z_i)'_g\|_2}\right)(z_i)'_g = (x_i)'_g$; therefore, $\frac{(z_i)'_g}{\|(z_i)'_g\|_2} = \frac{(x_i)'_g}{\|(x_i)'_g\|_2}$ holds. Moreover, because of the strict convexity of $f_m(\mathbf{z})$, we know that the minimizer is unique.

On the other hand, when $(z_i)_g = \mathbf{0}$ or $(x_i)_g = \mathbf{0}$, our method may not obtain good results. When $(x_i)_g = \mathbf{0}$, we know that the minimizer of the subproblem $\min_{(z_i)_g} \|(z_i)_g\|_2 + \frac{\beta}{2s}\|(z_i)_g - (x_i)_g\|_2^2$ is exactly that $(z_i)_g = \mathbf{0}$. When $(x_i)_g \neq \mathbf{0}$ and the minimizer of the subproblem $\min_{(z_i)_g} \|(z_i)_g\|_2 + \frac{\beta}{2s}\|(z_i)_g - (x_i)_g\|_2^2$ is that $(z_i)_g = \mathbf{0}$ (since the parameter β/s is two small), our method is not able to obtain good results. For example, that $(z_i)_g = \mathbf{0}$ while $(z_{i+1})_g \neq \mathbf{0}$ makes the element z_i in $\mathbf{z}$ take different values in different subproblems. However, we can obtain an approximate minimizer in this case, which is that the element z_i is a simple summation of corresponding subproblems containing z_i. We will show that in the experiments of Section 5, the approximate minimizer is also good. Moreover, when we take this problem as a subproblem of the image processing problem, we can set the parameter β to be large enough to avoid this drawback.

In addition, from (23), we can know that the element z_i of the minimizer can be treated as in s subproblems independently and then we combine them. In conclusion, after some manipulations, we can get the following two general formulas.

(I)

$$\arg\min_{\mathbf{z}} \quad \|\mathbf{z}\|_{2,1} + \frac{\beta}{2}\|\mathbf{z} - \mathbf{x}\|_2^2 = \mathrm{Sh}_{\mathrm{OGS}}(\mathbf{x}, \beta), \tag{24}$$

with

$$\mathrm{Sh}_{\mathrm{OGS}}(\mathbf{x}, \beta)_i = z_i = \max\left\{1 - \frac{1}{\beta}F(x_i), 0\right\} x_i. \tag{25}$$

Here, for instance, when group size $s = 4$, $(z_i)_g = (z_{i-1}, z_i, z_{i+1}, z_{i+2})$ in $\|\mathbf{z}\|_{2,1}$ and $(x_i)_g$ is defined similarly as $(z_i)_g$, we have $F(x_i) = \left(\frac{1}{\|(x_{i-2})_g\|_2} + \frac{1}{\|(x_{i-1})_g\|_2} + \frac{1}{\|(x_i)_g\|_2} + \frac{1}{\|(x_{i+1})_g\|_2}\right)$. The $\|(x_j)_g\|_2$ is contained in $F(x_i)$ if and only if $\|(x_j)_g\|_2$ has the component x_i, and we follow the convention $(1/0) = 1$ in $F(x_i)$ because $\|(x_i)_g\|_2 = 0$ implies $x_i = 0$ and the value of $F(x_i)$ is insignificant in (25).

(II)

$$\arg\min_{\mathbf{z}} \quad \|\mathbf{z}\|_{2,1} + \frac{\beta}{2}\|\mathbf{z} - \mathbf{x}\|_2^2 = \mathrm{Sh}_{\mathrm{OGS}}(\mathbf{x}, \beta), \tag{26}$$

with

$$\mathrm{Sh}_{\mathrm{OGS}}(\mathbf{x}, \beta)_i = z_i = G(x_i) \cdot x_i. \tag{27}$$

Here, symbols are the same as (I), and $G(x_i) = \max\left(\frac{1}{s} - \frac{1}{\beta\|(x_{i-2})_g\|_2}, 0\right) + \max\left(\frac{1}{s} - \frac{1}{\beta\|(x_{i-1})_g\|_2}, 0\right) + \max\left(\frac{1}{s} - \frac{1}{\beta\|(x_i)_g\|_2}, 0\right) + \max\left(\frac{1}{s} - \frac{1}{\beta\|(x_{i+1})_g\|_2}, 0\right)$.

We call them Formula (I) and Formula (II) respectively throughout this paper. When β is sufficiently large or sufficiently small, the former two formulas are the same. For the other values of β, from the experiments, we find that Formula (II) is better approximate to the MM iteration method than Formula (I), so we list the algorithm for Formula (II) as follows for finding the minimizer of (3). If we would use Formula (I), we can only change the final three steps, which does not change the computation cost.

We can see that Algorithm 1 only needs 2 times of convolution computations with time complexity $n*s$, which is just the same time complexity as one-step iteration in the MM method of [10]. Therefore, our method is much more efficient than the MM method or other variable duplication methods. In Section 5, we will give the numerical experiments for comparison between our method and the MM method. Moreover, if $s = 1$, our Algorithm 1 degenerates to the classic soft thresholding as our formula (27) degenerates to (6). Moreover, when β is sufficiently large or sufficiently small, the results of our new explicit algorithm are almost the same as the MM method (after 20 or more iteration steps) in [10].

Algorithm 1 Direct shrinkage algorithm for the minimization problem (3)

Input:

Given vector $\mathbf{x} = (x_1, x_2, \cdots, x_n)$, group size s, parameter β.

Compute:

1. Definition of $(x_i)_g$, for example $(x_i)_g = \left(x_{i-s_l}, \cdots, x_i, \cdots, x_{i+s_r}\right)$, with $s_l = [\frac{s-1}{2}]$ and $s_r = [\frac{s}{2}]$ ($s_l + s_r + 1 = s$).
   $\mathbf{w}$ = ones(1,s) = [1,1,$\cdots$,1].
   If s is even, then $\mathbf{w} =[\mathbf{w}, 0]$ and $s = s+1$.
   Let $\mathbf{w}_r$ = fliplr($\mathbf{w}$) be fliping $\mathbf{w}$ over 180 degrees.
2. Compute $X_n = \left(\|(x_1)_g\|_2, \cdots, \|(x_i)_g\|_2, \cdots, \|(x_n)_g\|_2\right)$, by convolution of $\mathbf{w}$ and $\mathbf{x}$.
3. Compute $X_n' = \max\left(\frac{1}{s} - 1./(X_n \cdot \beta), 0\right)$ point-wise.
4. Compute $G(x_l)$ by correlation of $\mathbf{w}$ and X_n', or by convolution of $\mathbf{w}_r$ and X_n'.
5. Compute $\mathbf{z}$ by multiplying $\mathbf{x}$ with $G(x_i)$ point-wise.

Remarks 2. Our new explicit algorithm can be treated as an average estimation algorithm for solving all the overlapping group subproblem independently. In Section 5, our numerical experiments show that the results of our two formulas are almost the same as the MM method in [10] when β is sufficiently large or sufficiently small and is approximate to the MM method for other β. Moreover, this point illuminates that the OGS regularizers coincide to maintain the property of the classical sparse regularizers and smoothen the local regions. For instance, TV with OGS can preserve edges and simultaneously overcome staircases in our experiments and [11, 12].

For the problems (4) (See Fig. 1b intuitively) and (5), we can get similar algorithms. We omit the evolution details here and give them as in Appendix. Here, we directly give the algorithm for solving the problem (5) as follows under Formula (II).

We can also see that Algorithm 2 only needs 2 times of convolution computations with time complexity $n * s$, which is just the same time complexity as one-step iteration in the MM method of [10]. Therefore, our method is much more efficient than the MM method.

Algorithm 2 Direct shrinkage algorithm for the minimization problem (4)

Input:

Given matrix A, group size $s = K_1 \times K_2$, parameter β, weight vector W_g.

Compute:

1. Pad W_g to be an odd-by-odd matrix.
   If K_1 is even, then $W_g =$[zeros$(1, K_2)$; W_g] and $K_1 = K_1 + 1$.
   If K_2 is even, then $W_g =$[zeros$(K_1, 1)$, W_g] and $K_2 = K_2 + 1$.
   Let W_g^r = fliplr(W_g) be fliping W_g over 180 degrees.
2. Compute
   $A_{temp} = \left(\|(A_{1,1})_g\|_2, \cdots, \|(A_{1,K_2})_g\|_2; \cdots; \|(A_{K_1,1})_g\|_2, \cdots, \|(A_{K_1,K_2})_g\|_2\right)$, by correlation of $W_g.^2$ (pointwise square) and A, or by convolution of $W_g^r.^2$ and A.
3. Compute $A_{temp}' = \max\left(\frac{1}{\|W_g\|_F^2} - 1./(A_{temp} \cdot \beta), 0\right)$ pointwise.
4. Compute $G(A_{i,j})$ by convolution of $W_g.^2$ and A_{temp}', or by correlation of $W_g^r.^2$ and A_{temp}'.
5. Compute X by multiplying A with $G(A_{i,j})$ point-wise.

3 Several extensions

3.1 Other boundary conditions

In Section 2, we gave the explicit shrinkage formulas for one class of OGS problems (3), (4), and (5). In order to achieve a simply deduction, we assume that PBC is used. One may confuse that whether PBC is always good for regularization problems such as signal processing or image processing, since natural signals or images are often asymmetric. However, in these problems, assuming a kind of boundary condition is necessary for simplifying the problem and making the computation possible [24]. There are kinds of BCs, such as zero BC (ZBC) and reflective BC. PBC is often used in optimization because it can be computed fast as above or other methods, for example, computation of matrix of block circulant with circulant blocks (BCCB) by fast Fourier transforms [20, 21, 24].

In this section, we consider other BC such as ZBC and reflective BC. For simplification, we only consider the vector case, while the results can be easily expanded to the matrix case as that in Section 2. When ZBC is used, we can expand the original vectors (or signal) $\mathbf{z}$ $(= (z_i)_{i=1}^n)$ and $\mathbf{x}$ by two s-length vectors on the both hands of the original vectors, which are $\tilde{\mathbf{z}}$ $(= \left[\tilde{z}_{-s}, \cdots, \tilde{z}_{-1}, (\tilde{z}_i)_{i=1}^n, \tilde{z}_{n+1}, \cdots, \tilde{z}_{n+s}\right] = \left[\mathbf{0}_s, (z_i)_{i=1}^n, \mathbf{0}_s\right])$ and $\tilde{\mathbf{x}}$, respectively. Then, the results and algorithms in Section 2 are similar as the case of PBC on $\tilde{\mathbf{z}}$ and $\tilde{\mathbf{x}}$. See Fig. 2a intuitively.

Moreover, our numerical results will show that the results from a different BC are almost the same in practice (see Section 5.1); and according to the definition of weighted generalized $\ell_{2,1}$ norm, ZBC seems better than PBC. Therefore, we will choose the ZBC to solve the problems (3), (4), and (5) in the following sections.

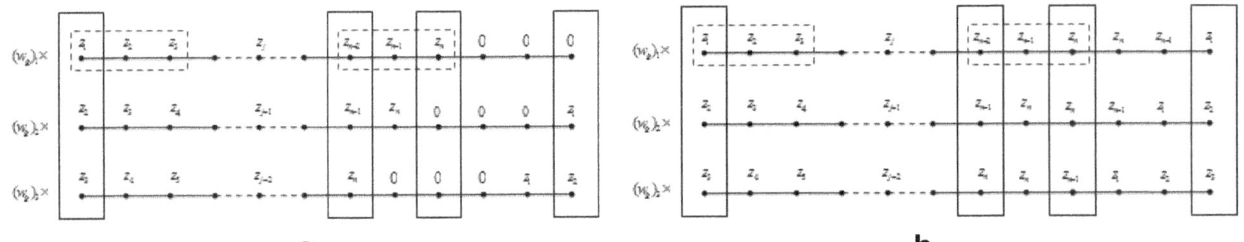

Fig. 2 The case of expanding vectors with other BCs to use the formulas under PBC. **a** ZBC. **b** Reflective BC

When reflective BC is assumed, the results are also the same. We only need to extend the original vector $\mathbf{x}$ to $\hat{\mathbf{x}}$ with reflective BC. See Fig. 2b intuitively.

3.2 Nonpositive weights and different weights in groups

In this section, we show that the weight vector w_g and matrix W_g in the former sections can contain arbitrary s entries with arbitrary real numbers. On one hand, the zero value can be the arbitrary entries of the weight vector or matrix. For example, the original ℓ_1 regularization problem (1) can be seen as a special form of weighted generalized $\ell_{2,1}$ norm regularization problems (4) with $s = 1$ or $s = 3$ and $w_g = (1, 0, 0)$ for the example of Fig. 1b.

On the other hand, the norm is with the property of positive homogeneity, $\|(-1) \cdot w_g\|_{any} = \|w_g\|_{any}$, where "$\|\cdot\|_{any}$" can be arbitrary norm such as $\|\cdot\|_1$ and $\|\cdot\|_2$. Therefore, for the regularization problems (4) (or (5)), the weight vector w_g (or matrix W_g) is the same as $|w_g|$ (or $|W_g|$), where $|\cdot|$ the point-wise absolute value. Therefore, in general, our results are true, whatever the number of entries included in the weight vector w_g and matrix W_g and whatever the real number value of these entries.

However, if the weight w_g is dependent on the group index i, for example, $w_{i,g} \circ (z_i)_g = \left((w_{i,g})_1 z_i, (w_{i,g})_2 z_{i+1}, (w_{i,g})_3 z_{i+2}\right)$, we cannot solve the relevant problems easily since our method fails. As we mentioned above, we only focus on that the weight w_g is independent on the group index i, that is, $w_{i,g} = w_g$ for all i, which means that it is with translation invariant overlapping groups.

4 Applications in TV regularization problems with OGS

The TV regularizer was firstly introduced by Rudin et al. [25] (ROF), which is widely used in many fields, i.e., denoising and deblurring problems [20, 21, 26–31]. Several fast algorithms such as Chambelle [26] and fast TV deconvolution (FTVd) [20, 21] have been proposed. Its corresponding minimization task is

$$\min_f \|f\|_{\mathrm{TV}} + \frac{\mu}{2}\|Hf - g\|_2^2, \tag{28}$$

where $\|f\|_{\mathrm{TV}} := \sum_{1\le i,j\le n} \|(\nabla f)_{i,j}\|_2 = \sum_{1\le i,j\le n} \sqrt{|(\nabla_x f)_{i,j}|^2 + |(\nabla_y f)_{i,j}|^2}$ (called isotropic TV, ITV), or $\|f\|_{TV} := \sum_{1\le i,j\le n} \|(\nabla f)_{i,j}\|_1 = \sum_{1\le i,j\le n} |(\nabla_x f)_{i,j}| + |(\nabla_y f)_{i,j}|$ (called anisotropic TV, ATV), H denotes the blur matrix, and g denotes the given observed image with blur and noise. Operator $\nabla : \mathbb{R}^{n^2} \to \mathbb{R}^{2\times n^2}$ denotes the discrete gradient operator (under PBC) which is defined by $(\nabla f)_{i,j} = ((\nabla_x f)_{i,j}, (\nabla_y f)_{i,j})$, with

$$(\nabla_x f)_{i,j} = \begin{cases} f_{i+1,j} - f_{i,j} & \text{if } i < n, \\ f_{1,j} - f_{n,j} & \text{if } i = n, \end{cases} \quad (\nabla_y f)_{i,j} = \begin{cases} f_{i,j+1} - f_{i,j} & \text{if } j < n, \\ f_{i,1} - f_{i,n} & \text{if } j = n, \end{cases}$$

for $i,j = 1, 2, \cdots, n$, where $f_{i,j}$ refers to the $((j-1)n + i)$th entry of the vector f (it is the (i,j)th pixel location of the $n \times n$ image, and this notation remains valid throughout the paper unless otherwise specified). Notice that H is a matrix of BCCB structure when PBC are applied [24].

Recently, Selesnick et al. [13] proposed an OGS TV regularizer to one-dimensional signal denoising. They applied the MM method to solve their model. Their numerical experiments showed that their method can overcome staircase effects effectively and get better results. However, their method has the disadvantages of the low speed of computation and the difficulty to be directly extended to the two-dimensional image case. More recently, Liu et al. [11] proposed an OGS TV regularizer for two-dimensional image denoising and deblurring under Gaussian noise, and Liu et al. [12] proposed an OGS TV regularizer for image deblurring under impulse noise. Both of them used a variable substitution method and the ADMM framelet with an inner MM iteration for solving the subproblems similar as (3). Therefore, their methods may spend more time than our methods. Moreover, when the MM method is used in the inner iterations in [11, 12], they can only solve the ATV case but not the ITV case with OGS while our methods can solve both ATV and ITV cases.

Firstly, we define the ATV case with OGS under Gaussian noise and impulse noise respectively which is similar as [11, 12]. For the Gaussian noise case,

$$\min_f \|(\nabla_x f)\|_{W,2,1} + \|(\nabla_y f)\|_{W,2,1} + \frac{\mu}{2}\|Hf - g\|_2^2. \quad (29)$$

For the impulse case,

$$\min_f \|(\nabla_x f)\|_{W,2,1} + \|(\nabla_y f)\|_{W,2,1} + \mu\|Hf - g\|_1. \quad (30)$$

We call the former model as TV OGS L_2 model and the latter model as TV OGS L_1 model, respectively.

Then, we defined the ITV case with OGS. For Gaussian noise and impulse noise, we only change the former two terms of (29) and (30) respectively by $\|A\|_{W,2,1}$. Here, A is a high-dimensional matrix (or tensor) with each entry $A_{i,j} = ((\nabla_x f)_{i,j}; (\nabla_y f)_{i,j})$.

Remarks 3. In particular in this work, A can be treated as $((\nabla_x f); (\nabla_y f))$ for simplicity in vector and matrix computation, although the computation of $\|A\|_{W,2,1}$ is should be treated as pointwise with $A_{i,j} = ((\nabla_x f)_{i,j}; (\nabla_y f)_{i,j})$ since the computation on A is almost all pointwise.

Moreover, we consider constrained model as listing in [11, 12, 32]. For any true digital image, its pixel value can attain only a finite number of values. Hence, it is natural to require all pixel values of the restored image to lie in a certain interval $[a, b]$, see [32] for more details. In general, with the easy computation and the certified results in [32], we only consider all the images located on the standard range $[0, 1]$. Therefore,we define a projection operator $\mathcal{P}_\Omega$ on the set $\Omega = \{f \in \mathbb{R}^{n\times n} | 0 \le f \le 1\}$,

$$\mathcal{P}_\Omega(f)_{i,j} = \begin{cases} 0, & f_{i,j} < 0, \\ f_{i,j}, & f_{i,j} \in [0,1], \\ 1, & f_{i,j} > 1. \end{cases} \quad (31)$$

4.1 Constrained TV OGS L_2 model

For the constrained model (the ATV case) (called CATVOGSL2), we have

$$\min_{u\in\Omega, v_x, v_y, f} \Big\{\|(v_x)\|_{W,2,1} + \|(v_y)\|_{W,2,1} + \frac{\mu}{2}\|Hf - g\|_2^2 : \\ v_x = \nabla_x f, v_y = \nabla_y f, u = f\Big\}. \quad (32)$$

The augmented Lagrangian function [33] of (32) is

$$\begin{aligned}\mathcal{L}(v_x, v_y, u, f; \lambda_1, \lambda_2, \lambda_3) = {} & \|v_x\|_{W,2,1} - \lambda_1^T(v_x - \nabla_x f) + \frac{\beta_1}{2}\|v_x - \nabla_x f\|_2^2 \\ & + \|v_y\|_{W,2,1} - \lambda_2^T(v_y - \nabla_y f) + \frac{\beta_1}{2}\|v_y - \nabla_y f\|_2^2 \\ & - \lambda_3^T(u - f) + \frac{\beta_2}{2}\|u - f\|_2^2 + \frac{\mu}{2}\|Hf - g\|_2^2,\end{aligned} \quad (33)$$

where $\beta_1, \beta_2 > 0$ are penalty parameters and $\lambda_1, \lambda_2, \lambda_3 \in \mathbb{R}^{n^2}$ are the Lagrange multipliers.

The solving algorithm is according to the scheme of ADMM in Gabay [34], and we refer to some applications in image processing which can be solved by ADMM, e.g., [35–43]. For a given $\left(v_x^k, v_y^k, u^k, f^k; \lambda_1^k, \lambda_2^k, \lambda_3^k\right)$, the next iteration $(v_x^{k+1}, v_y^{k+1}, u^{k+1}, f^{k+1}; \lambda_1^{k+1}, \lambda_2^{k+1}, \lambda_3^{k+1})$ is generated as follows:

1. Fix $f = f^k$, $\lambda_1 = \lambda_1^k, \lambda_2 = \lambda_2^k, \lambda_3 = \lambda_3^k$ and minimize (33) with respect to v_x, v_y, and u. With respect to v_x and v_y,

$$\begin{aligned} v_x^{k+1} &= \arg\min \|v_x\|_{W,2,1} - {\lambda_1^k}^T(v_x - \nabla_x f^k) + \tfrac{\beta_1}{2}\|v_x - \nabla_x f^k\|_2^2 \\ &= \arg\min \|v_x\|_{W,2,1} + \tfrac{\beta_1}{2}\|v_x - \nabla_x f^k - \tfrac{\lambda_1^k}{\beta_1}\|_2^2, \end{aligned} \quad (34)$$

$$\begin{aligned} v_y^{k+1} &= \arg\min \|v_y\|_{W,2,1} - {\lambda_2^k}^T(v_y - \nabla_y f^k) + \tfrac{\beta_1}{2}\|v_y - \nabla_y f^k\|_2^2 \\ &= \arg\min \|v_y\|_{W,2,1} + \tfrac{\beta_1}{2}\|v_y - \nabla_y f^k - \tfrac{\lambda_2^k}{\beta_1}\|_2^2. \end{aligned} \quad (35)$$

It is obvious that problems (34) and (35) match the framework of the problem (5); thus, the minimizers of (34) and (35) can be obtained by using the formulas in Section 2.2.

With respect to u,

$$\begin{aligned} u^{k+1} &= \arg\min -{\lambda_3^k}^T(u - f^k) + \tfrac{\beta_2}{2}\|u - f^k\|_2^2 \\ &= \arg\min \tfrac{\beta_2}{2}\|u - f^k - \tfrac{\lambda_3^k}{\beta_2}\|_2^2. \end{aligned}$$

The minimizer is given explicitly by

$$u^{k+1} = \mathcal{P}_\Omega\left[f^k + \frac{\lambda_3^k}{\beta_2}\right]. \quad (36)$$

2. Compute f^{k+1} by solving the normal equation

$$\begin{aligned} &(\beta_1(\nabla_x^*\nabla_x + \nabla_y^*\nabla_y) + \mu H^*H + \beta_2 I)f^{k+1} \\ &- \nabla_x^*(\beta_1 v_x^{k+1} - \lambda_1^k) + \nabla_y^*(\beta_1 v_y^{k+1} - \lambda_2^k) + \mu H^* g \\ &+ \beta_2(u^{k+1} - \tfrac{\lambda_3^k}{\beta_2}), \end{aligned} \quad (37)$$

where "*" denotes the conjugate transpose, see [44] for more details. Since all the parameters are positive, the coefficient matrix in (37) are always invertible and symmetric positive-definite. In addition, H, ∇_x, ∇_y and their conjugate transpose have BCCB structure under PBC. We know that the computations with BCCB matrix can be very efficient by using fast Fourier transforms.

3. Update the multipliers via

$$\begin{cases} \lambda_1^{k+1} = \lambda_1^k - \gamma\beta_1(v_x^{k+1} - \nabla_x f^{k+1}), \\ \lambda_2^{k+1} = \lambda_2^k - \gamma\beta_1(v_x^{k+1} - \nabla_x f^{k+1}), \\ \lambda_3^{k+1} = \lambda_3^k - \gamma\beta_2(u^{k+1} - f^{k+1}). \end{cases} \quad (38)$$

Based on the discussions above, we present the ADMM algorithm for solving the convex CATVOGSL2 model (32), which is shown as Algorithm 3.

Algorithm 3 CATVOGSL2 for the minimization problem (32)

---

initialization:
Starting point $f^0 = g$, $k = 0$, $\beta_1, \beta_2, \gamma, \mu$, group size $K_1 \times K_2$, weighted matrix W_g, $\lambda_i^0 = 0, i = 1, 2, 3$.
iteration:
1. Compute v_x^{k+1} and v_y^{k+1} according to (34) and (35).
2. Compute u^{k+1} according to (36).
3. Compute f^{k+1} by solving (37).
4. update $\lambda_i^0 = 0, i = 1, 2, 3$ according to (38).
5. $k = k + 1$.

until a stopping criterion is satisfied.

Algorithm 3 is an application of ADMM for the case with two blocks of variables (v_x, v_y, u) and f. Thus, if the step (1) of Algorithm 3 can be solved exactly, its convergence is guaranteed by the theory of ADMM [34, 35, 37, 43]. Although step (1) of Algorithm 3 cannot be solved exactly, we can find a convergent series to ensure the convergence as [35]. Besides, our numerical experiments verify the convergence of Algorithm 3.

Remarks 4. For the image f, we use PBC for the fast computation. However, for v_x, v_y, we use ZBC. Because v_x and v_y denote the gradient of the image, ZBC seems better for the definition of the generalized norm $\ell_{2,1}$ on v_x and v_y as mentioned in Section 3.1. Therefore, the two kinds of BCs for the image and its gradient are different and independent. These remain valid throughout the paper unless otherwise specified.

For the ITV case, the constrained model (CITVOGSL2) is

$$\min_{u\in\Omega,A,f} \left\{ \|A\|_{W,2,1} + \frac{\mu}{2}\|Hf-g\|_2^2 : \ A = (\nabla_x f; \nabla_y f), u = f \right\}. \tag{39}$$

The detail will be presented in the next section for the constrained TV OGS L_1 model. We call this relevant algorithm CITVOGSL2.

4.2 Constrained TV OGS L_1 model

For the constrained model (the ITV case) (called CITVOGSL1), we have

$$\min_{u\in\Omega,A,f} \{\|A\|_{W,2,1} + \mu\|z\|_1 : z = Hf - g, A = (\nabla_x f; \nabla_y f), u = f\}. \tag{40}$$

The augmented Lagrangian function of (40) is

$$\begin{aligned}\mathcal{L}(v_x, v_y, z, w, f; \lambda_1, \lambda_2, \lambda_3, \lambda_4) = \|A\|_{W,2,1} - \left(\lambda_1^T, \lambda_2^T\right)\left(A - \begin{bmatrix}\nabla_x f \\ \nabla_y f\end{bmatrix}\right) \\ + \frac{\beta_1}{2}\left\|A - \begin{bmatrix}\nabla_x f \\ \nabla_y f\end{bmatrix}\right\|_2^2 \\ + \mu\|z\|_1 - \lambda_3^T(z - (Hf - g)) \\ + \frac{\beta_2}{2}\|z - (Hf - g)\|_2^2 \\ - \lambda_4^T(w - f) + \frac{\beta_3}{2}\|w - f\|_2^2,\end{aligned} \tag{41}$$

where $\beta_1, \beta_2, \beta_3 > 0$ are penalty parameters and $\lambda_1, \lambda_2, \lambda_3, \lambda_4 \in \mathbb{R}^{n^2}$ are the Lagrange multipliers.

For a given $(A^k, z^k, u^k, f^k; \lambda_1^k, \lambda_2^k, \lambda_3^k, \lambda_4^k)$, the next iteration $(A^{k+1}, z^{k+1}, u^{k+1}, f^{k+1}; \lambda_1^{k+1}, \lambda_2^{k+1}, \lambda_3^{k+1}, \lambda_4^{k+1})$ is generated as follows:

1. Fix $f = f^k$, $\lambda_1 = \lambda_1^k, \lambda_2 = \lambda_2^k, \lambda_3 = \lambda_3^k, \lambda_4 = \lambda_4^k$ and minimize (41) with respect to A, z, and u. With respect to A,

$$\begin{aligned}A^{k+1} &= \arg\min \|A\|_{W,2,1} - (\lambda_1^T, \lambda_2^T)(A - \begin{bmatrix}\nabla_x f^k \\ \nabla_y f^k\end{bmatrix}) + \tfrac{\beta_1}{2}\left\|A - \begin{bmatrix}\nabla_x f^k \\ \nabla_y f^k\end{bmatrix}\right\|_2^2, \\ \begin{bmatrix}A_1^{k+1} \\ A_2^{k+1}\end{bmatrix} &= \arg\min \|A\|_{W,2,1} + \tfrac{\beta_1}{2}\left\|\begin{bmatrix}A_1 \\ A_2\end{bmatrix} - \begin{bmatrix}\nabla_x f^k \\ \nabla_y f^k\end{bmatrix} - \begin{bmatrix}\lambda_1^k/\beta_1 \\ \lambda_2^k/\beta_1\end{bmatrix}\right\|_2^2.\end{aligned} \tag{42}$$

It is obvious that problem (42) match the framework of the problem (5), thus the solution of (42) can be obtained by using the formulas in Section 2.2.

With respect to z,

$$\begin{aligned}z^{k+1} &= \arg\min \mu\|z\|_1 - {\lambda_3^k}^T\left(z - (Hf^k - g)\right) + \tfrac{\beta_2}{2}\|z - (Hf^k - g)\|_2^2 \\ &= \arg\min \mu\|z\|_1 + \tfrac{\beta_2}{2}\|z - (Hf^k - g) - \tfrac{\lambda_3^k}{\beta_2}\|_2^2.\end{aligned}$$

The minimization with respect to z can be given by (6) and (9) explicitly, that is,

$$z^{k+1} = \mathrm{sgn}\left\{Hf^k - g + \frac{\lambda_3^k}{\beta_2}\right\} \circ \max\left\{|Hf^k - g + \frac{\lambda_3^k}{\beta_2}| - \frac{\mu}{\beta_2}, 0\right\}. \tag{43}$$

With respect to u,

$$\begin{aligned}u^{k+1} &= \arg\min -{\lambda_4^k}^T\left(u - f^k\right) + \tfrac{\beta_3}{2}\|u - f^k\|_2^2 \\ &= \arg\min \tfrac{\beta_3}{2}\|u - f^k - \tfrac{\lambda_4^k}{\beta_3}\|_2^2.\end{aligned}$$

The minimizer is given explicitly by

$$u^{k+1} = \mathcal{P}_\Omega\left[f^k + \frac{\lambda_4^k}{\beta_3}\right]. \tag{44}$$

2. Compute f^{k+1} by solving the following normal equation similarly as the last section.

$$\begin{aligned}&(\beta_1(\nabla_x^*\nabla_x+\nabla_y^*\nabla_y)+\beta_2H^*H+\beta_3 I)f^{k+1}\\&=\nabla_x^*(\beta_1A_1^{k+1}-\lambda_1^k)+\nabla_y{}^*(\beta_1A_2^{k+1}-\lambda_2^k)+\\&H^*(\beta_2z^{k+1}-\lambda_3^k)+\beta_2H^*g+\beta_3(u^{k+1}-\tfrac{\lambda_4^k}{\beta_3}).\end{aligned}\tag{45}$$

3. Update the multipliers via

$$\begin{cases}\begin{bmatrix}\lambda_1^{k+1}\\\lambda_2^{k+1}\end{bmatrix}=\begin{bmatrix}\lambda_1^{k}\\\lambda_2^{k}\end{bmatrix}-\gamma\beta_1\left(\begin{bmatrix}A_1^{k+1}\\A_2^{k+1}\end{bmatrix}-\begin{bmatrix}\nabla_xf^{k+1}\\\nabla_yf^{k+1}\end{bmatrix}\right),\\ \lambda_3^{k+1}=\lambda_3^k-\gamma\beta_2(z^{k+1}-(Hf^{k+1}-g)),\\ \lambda_4^{k+1}=\lambda_4^k-\gamma\beta_3(u^{k+1}-f^{k+1}).\end{cases}\tag{46}$$

Based on the discussions above, we present the ADMM algorithm for solving the convex CITVOGSL1 model (40), which is shown as Algorithm 4.

Algorithm 4 CITVOGSL1 for the minimization problem (40)

---

initialization:
Starting point $f^0=g$, $k=0$, $\beta_1,\beta_2,\beta_3,\gamma,\mu$, group size $K_1\times K_2$, weighted matrix W_g, $\lambda_i^0=0$, $i=1,2,3,4$.
iteration:
1. Compute A^{k+1} according to (42), and compute z^{k+1} according to (43).
2. Compute u^{k+1} according to (44).
3. Compute f^{k+1} by solving (45).
4. update $\lambda_i^0=0$, $i=1,2,3,4$ according to (46).
5. $k=k+1$.

until a stopping criterion is satisfied.

Algorithm 4 is an application of ADMM for the case with two blocks of variables (A,z,u) and f. Thus, if step (1) of Algorithm 4 can be solved exactly, its convergence is guaranteed by the theory of ADMM [34, 35, 37, 43]. Although step (1) of Algorithm 4 cannot be solved exactly, we can find a convergent series to ensure the convergence as [35]. Besides, our numerical experiments verify the convergence of Algorithm 4.

For the ATV case, the constrained model (CATVOGSL1) is

$$\min_{u\in\Omega,v_x,v_y,f}\left\{\begin{array}{l}\|(v_x)\|_{W,2,1}+\|(v_y)\|_{W,2,1}+\mu\|z\|_1:\\ z=Hf-g,v_x=\nabla_xf,v_y=\nabla_yf,u=f\end{array}\right\}.\tag{47}$$

The detail has been presented in the last section for the constrained TV OGS L_2 model. We call this relevant algorithm CATVOGSL1.

5 Numerical results

In this section, we present several numerical results to illustrate the performance of the proposed method. All experiments are carried out on a desktop computer using MATLAB 2010a. Our computer is equipped with an Intel Core i3-2130 CPU (3.4 GHz), 3.4 GB of RAM, and a 32-bit Windows 7 operation system.

5.1 Comparison with the MM method for one-dimensional signal denoising

As an illustrative example, we apply our direct formulas in one-dimensional signal denoising. We only compare the results of our explicit shrinkage formulas with the most recent MM iteration method proposed in [10] as a simple example. The one-dimensional group sparse signal $\mathbf{z}$ is shown in the top left of Fig. 3. The noisy signal $\mathbf{x}$ in the top right of Fig. 3 is obtained by adding independent white Gaussian noise with standard deviation $\sigma=0.5$ same as [10]. The denoising model is as follows

$$\arg\min_{\mathbf{z}}\quad \mathrm{Fun}(\mathbf{z})=\|\mathbf{z}\|_{2,1}+\frac{\beta}{2}\|\mathbf{z}-\mathbf{x}\|_2^2.\tag{48}$$

We test several different parameters β (any integer from 1 to 50) for the comparison. The results of our formulas are almost the same as the MM method for almost these βs. Due to the limit space, we only illustrate parts of the results ($\beta=3,10,15$) in Fig. 3. We take both our Formulas (I) and (II) for the comparison with the MM method. From the figure, we can visually see that our results by our two kinds of formulas are almost the same as the MM method for different β. Moreover, for the practical problems, $\beta=10$, can be better than others from the figure. This shows that our formulas are feasible, useful, and effective, because our method can only need the same computation cost as one-step iteration in the MM method while the MM method needs 25 iterations as [10]. That means our method is 25 times faster than the MM method. From the bottom line of Fig. 3, we can see that the MM method may take 5 iterations to be convergent, which is the reason why Liu et al. in [11] and Liu et al. in [12] only choose 5 inner iterations. However, in this case, our method is still 5 times faster than the MM method.

Remarks 5. Although this model is not the best model for signal denoising, it shows the superiority of our method. Moreover, the authors in [10] also take similar experiments.

5.2 Comparison with the MM method on problem (5)

In this section, we give another example by solving the matrix case problem (5) for comparison with the MM method. Without loss of generality, we let $A=\mathrm{rand}(100,100)$ and $A(45:55,45:55)=0$ in the

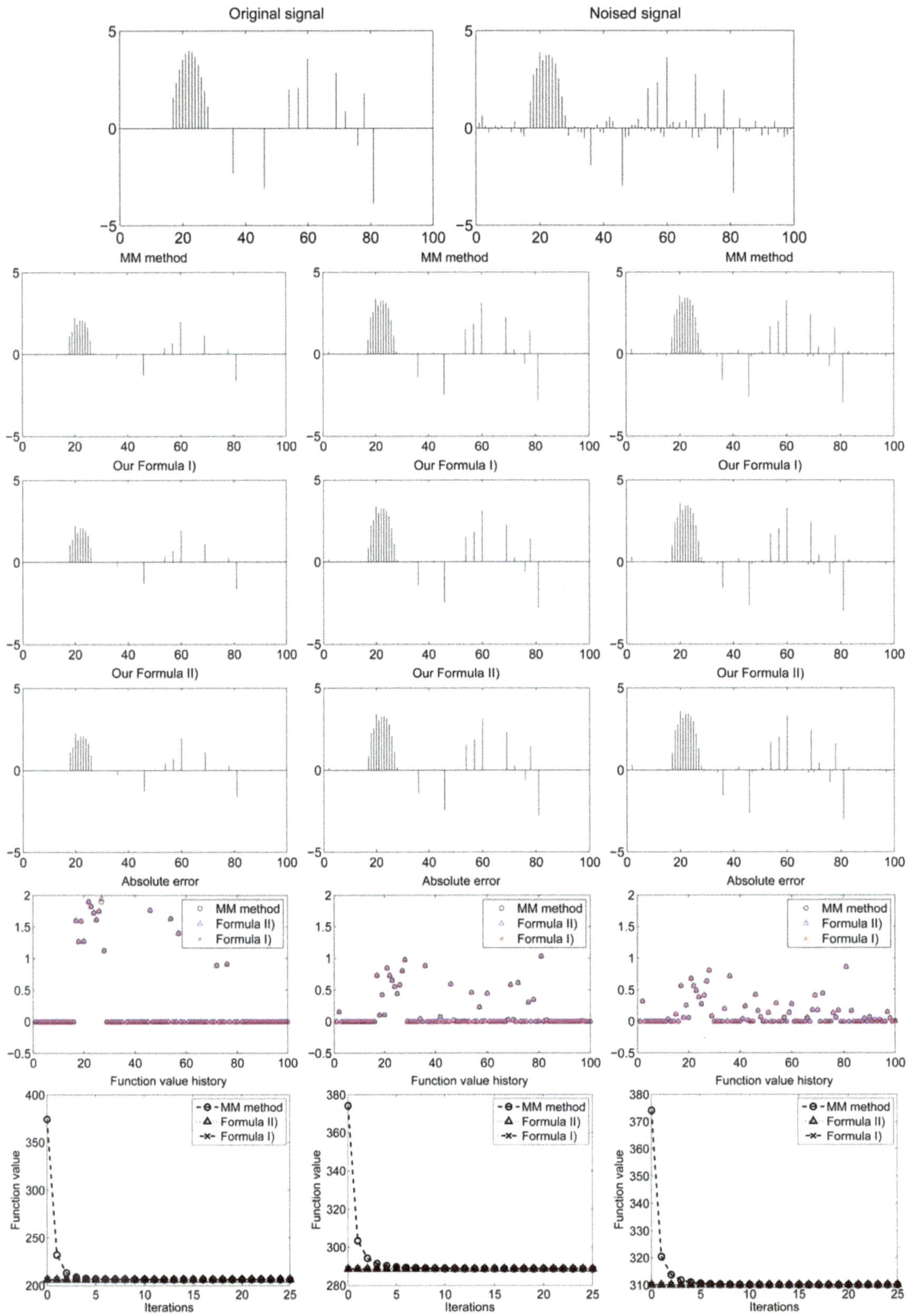

Fig. 3 Comparison between our method and the MM method for one-dimensional signal denoising. *Second row* to *bottom row*, from *top* to *bottom*, results of the MM method, our Formula (I), our Formula (II) respectively, absolute error, and function value history. From *left* to *right*, $\beta = 3, 10, 15$ respectively

example. In particular, we set weighted matrix $W_g \in \mathbb{R}^{3\times 3}((W_g)_{i,j} \equiv 1)$ and list (5) again

$$f_A(X) = \min_X \text{Fun}(X) = \|X\|_{2,1} + \frac{\beta}{2}\|X - A\|_F^2. \quad (49)$$

We use different parameters β and use the number of MM iteration by 20 steps. For comparison, we expand our result by explicit shrinkage formulas to length 20 (the same as the MM iteration steps). The values of the object function Fun(X) against iterations are illustrated in the top line in Fig. 4. The cross-sectional elements of the minimizer A are shown in the bottom line in Fig. 4. We choose both ZBC and PBC for our method and the MM method. It can be seen that there is only a little difference between the two kinds of BCs.

It is obvious that our method is much more efficient than the MM method since our time complexity is just the same as one-step iteration in the MM method due to the explicit shrinkage formula. From the top line in Fig. 4, we observe that the MM method is also fast because it only needs less than 20 steps (sometimes 5) to converge. The related error of the function value of two methods is much less than 1 % for different parameters β. From the bottom line in Fig. 4, we can see that when $\beta \leq 1$ which is sufficiently small, and when $\beta \geq 30$ which is sufficiently large, our result is almost the same as the final results by the MM iteration method. This shows that the results by our method are almost the same as the MM method. When $1 < \beta < 30$, the minimizer computed by our method is approximate to the MM method both on the error of the function value and on the minimizer X.

In Table 1, we show the numerical comparison of our method and the MM method on three parts, the related error of function value (ReE$_{f_A}$), related error of minimizer X (ReE$_X$), and the mean absolute error of X (MAE$_X$). The three terms are defined by ReE$_{f_A} = \frac{|f_{A\text{MM}} - f_{A\text{ours}}|}{f_{A\text{ours}}}$, ReE$_X = \frac{\|X_{\text{MM}} - X_{\text{ours}}\|_F}{\|X_{\text{ours}}\|_F}$, and MAE$_X$ = mean absolute error of $(X_{\text{MM}} - X_{\text{ours}})$, where X_{MM} and X_{ours} are the final solution X of the MM method and our method, respectively. $f_{A\text{MM}}$ and $f_{A\text{ours}}$ are the final function value of MM and our method, respectively.

From the table and the figure, we can see that our formula can almost get the same results as the MM method when β is sufficiently large and approximate results similar as the MM method for other β. This is another proof for the feasibility of our formula.

Similarly as weighted matrix $W_g \in \mathbb{R}^{3\times 3}((W_g)_{i,j} \equiv 1)$, we also tested other weighted matrix for more than 1000 times. For example, when $A = \text{rand}(100, 100)$ and $A(A >= 0.5) = 0$ (or $A(A <= 0.5) = 1$) (every element is in $[0, 1]$), we find that, generally, $\beta \leq \frac{\|w_g\|_2}{\sqrt{s}}$ is sufficiently small and $\beta \geq 30 \cdot \frac{\|W_g\|_2}{\sqrt{s}}$ is sufficiently large. These results illustrate that the former when $W_g \in \mathbb{R}^{3\times 3}((W_g)_{i,j} \equiv 1)$ $\beta \geq 30$ which is sufficiently large again. However, in practice, if β is too large, then $A = X$, and the minimization problem (49) is meaningless. In our experiments after more than 1000 tests, we find that, when every element of A is in $[0, 1]$, $30 \cdot \frac{\|W_g\|_2}{\sqrt{s}} \leq \beta \leq 300 \cdot \frac{\|W_g\|_2}{\sqrt{s}}$ is sufficiently large but not too large to make the minimization problem meaningless. In this case, that means we may tune the parameter to be better in practice. On the other hand, when some elements of A are not in $[0, 1]$, we can first project or stretch A into the region $[0, 1]$ and then choose the better parameter β. Moreover, although the parameter is not chosen to be in this region, the result can also be treated as an approximate solution with small error. That means our formulas are feasible, useful, and effective.

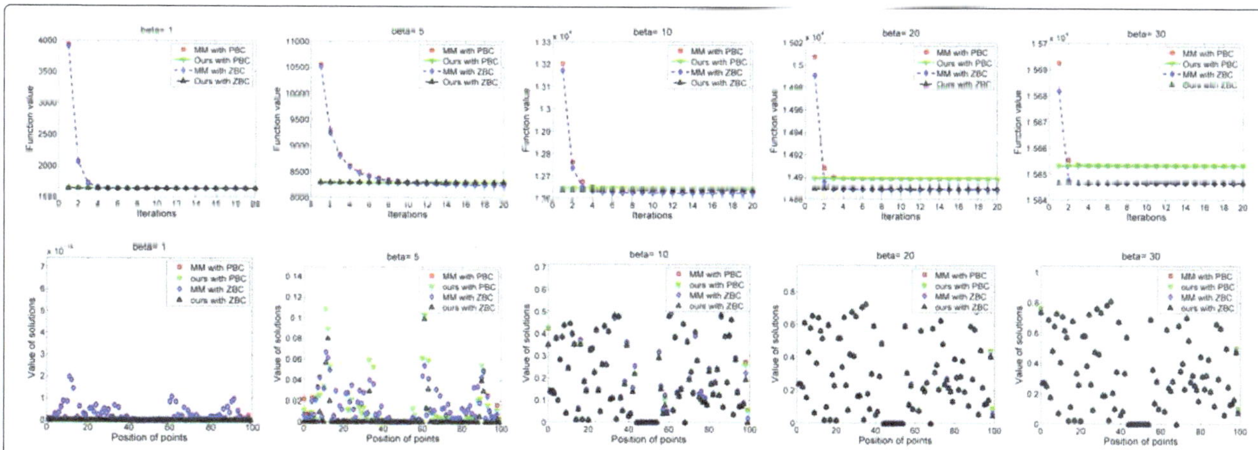

Fig. 4 Comparison between our method and the MM method on two kinds of BCs, ZPC, and PBC. *Top row*, the function value history. *Bottom row*, the cross line of the final minimizer—matrix X. From *left* to *right*, $\beta = 1, 5, 10, 20, 30$

Table 1 Numerical comparison of our method and the MM method for two kinds of BCs, ZBC, and PBC

BC	$\beta =$	1	5	7	10	15	20	30	50
ZBC	ReE_{f_A}	5.9e-14	0.0055	0.0028	6.5e-4	1.4e-4	5.4e-5	1.4e-5	2.6e-6
	ReE_X	—	—	0.0207	0.0017	1.9e-4	4.7e-5	7.5e-6	7.8e-7
	MAE_X	3.4e-15	0.0094	0.0130	0.0067	0.0029	0.0016	6.9e-4	2.4e-4
PBC	ReE_{f_A}	6.5e-14	0.0053	0.0025	4.1e-4	5.5e-5	1.1e-5	6.3e-7	1.3e-6
	ReE_X	—	—	0.0193	0.0014	1.3e-4	2.9e-5	4.3e-6	4.5e-7
	MAE_X	3.8e-15	0.0097	0.0130	0.0063	0.0026	0.0014	5.9e-4	2.1e-4

ReE of $f_A(X)$ denotes the related error of final function value fun(X). ReE of X denotes related error of minimizer X. MAE of X denotes the mean absolute error of X

5.3 Comparison with TV methods and TV with OGS with inner iteration MM methods for image deblurring and denoising

In this section, we compare all our algorithms with other methods. All the test images are shown in Fig. 5, one 1024×1024 image as (a) Man and three 512×512 images as (b) Car, (c) Parlor, and (d) Housepole.

The quality of the restoration results is measured quantitatively by using the peak signal-to-noise ratio (PSNR) in decibel (dB) and the relative error (ReE):

$$\text{PSNR} = 10 \log_{10} \frac{n^2 \text{Max}_I^2}{\|f - \bar{f}\|_2^2}, \quad \text{ReE} = \frac{\|f - \bar{f}\|_2}{\|\bar{f}\|_2},$$

where $\bar{f}$ and f denote the original and restored images, respectively, and Max_I represents the maximum possible pixel value of the image. In our experiments, $\text{Max}_I = 1$.

The stopping criterion used in our work is set to be as other methods

$$\frac{|\mathcal{F}^{k+1} - \mathcal{F}^k|}{|\mathcal{F}^k|} < 10^{-5}, \tag{50}$$

where $\mathcal{F}^k$ is the objective function value of the respective model in the kth iteration.

We compare our methods with some other methods, such as Chan et al.'s TV method proposed in [32], Liu et al.'s method proposed in [11], and Liu et al.'s method proposed in [12]. Both the latter two methods [11, 12] are with inner iteration MM methods for OGS TV problems, where the number of the inner iterations is set 5 by them. In particular, for a fair comparison, we set weighted matrix $W_g \in \mathbb{R}^{3\times3}((W_g)_{i,j} \equiv 1)$ in all the experiments of our methods as in [11, 12].

5.3.1 Experiments for the constrained TV OGS L_2 model

In this section, we compare our methods (CATVOGSL2 and CITVOGSL2) with some other methods, such as Chan et al.'s method proposed in [32] (Algorithm 1 in [32] for the constrained TV-L2 model) and Liu et al.'s method proposed in [11].

We set the penalty parameters $\beta_1 = 35$, $\beta_2 = 20$, for the ATV case, $\beta_1 = 100$, $\beta_2 = 20$ for the ITV case and relax parameter $\gamma = 1.618$. The blur kernels are generated by MATLAB built-in function (i)`fspecial('average',9)` for 9×9 average blur. We generate all blurring effects using the MATLAB built-in function `imfilter(I,psf,'circular','conv')` under PBC with "`I`" as the original image and "`psf`" as the blur kernel. We first generated the blurred images operating on images (a)–(c) by the former Gaussian blurs

Fig. 5 Original images. The versions of the images in this paper are special formats which are converted by Photoshop from the sources downloaded from `http://decsai.ugr.es/cvg/dbimagenes/`. **a** Man. **b** Car. **c** Parlor. **d** Housepole

Table 2 Numerical comparison of Chan [32], Liu [11], CATVOGSL2, and CITVOGSL2 for images (a)–(c) in Fig. 5

Images		(a) Man				(b) Car				(c) Parlor			
Method	μ	Itr	PSNR	Time	ReE	Itr	PSNR	Time	ReE	Itr	PSNR	Time	ReE
Chan [32]	0.5	15	30.34	6.02	0.0730	11	31.13	2.00	0.0426	12	31.70	2.01	0.0511
Liu [11]	1	12	30.60	9.06	0.0708	12	31.68	3.93	0.0386	13	32.46	4.21	0.0464
CAOL2	1	7	30.62	*3.37*	0.0711	9	31.68	*1.76*	0.0399	7	32.40	*1.34*	0.0472
CIOL2	1	8	30.59	*3.70*	0.0716	11	31.60	*2.04*	0.0403	8	32.03	*1.61*	0.0492

PSNR dB, *Time* s, *Itr* iterations, μ regularization parameter ($\times 10^5$), *CAOL2* CATVOGSL2, *CIOL2* CITVOGSL2

and further corrupted by zero mean Gaussian noise with BSNR = 40. The BSNR is given by

$$\text{BSNR} = 10\log_{10}\frac{\text{variance of}g}{\text{variance of }\eta}.$$

where g and η are the observed image and the noise, respectively.

The numerical results of the three methods are shown in Table 2. We have tuned the parameters for all the methods as in Table 2. From Table 2, we can see that PSNR and ReE of our methods (both ATV and ITV cases) are almost same as [11], which applied MM inner iterations to solve the subproblems (34) and (35) (only for the ATV case). However, each outer iteration of our methods is nearly twice faster than [11] from the experiments. The time of each outer iteration of our methods is almost the same as the traditional TV method in [32]. In Fig. 6, we display the restored "Parlor" images from different algorithms. We can see that OGS TV regularizers can get clearer edges on the "desk" of the image than TV regularizer.

Now, we compute the complexity of each step of our methods and Liu et al.'s method [11]. Firstly, we know that

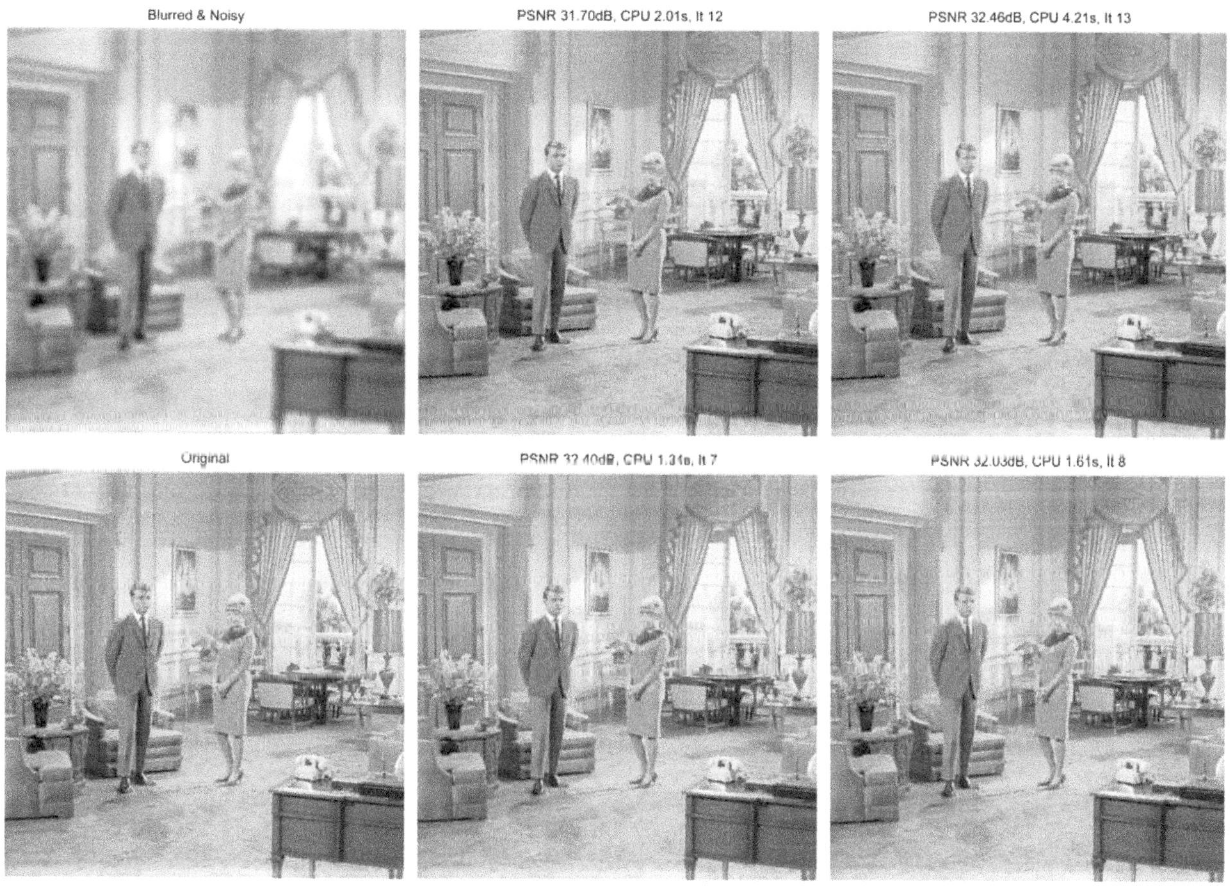

Fig. 6 *Top row*: blurred and noisy image (*left*), restoration images of Chan [32] (*middle*), Liu [11] (*right*). *Bottom row*: original image (*left*), restoration images of CATVOGSL2 (*middle*), CITVOGSL2 (*right*)

the complexity of all the methods is $512 \times 512 \times 18 \times 4$ (4 times $n \log_2 n$) except the OGS subproblems. Then, for the OGS subproblems, the MM method in [11] with five-step inner iteration, the complexity is $512 \times 512 \times 90 \times 2$ (2 subproblems). The complexity of our methods is $512 \times 512 \times 18 \times 2$ in CATVOGSL2 and $512 \times 512 \times 18$ in CITVOGSL2. Therefore, the total complexity of [11] is $512^2 \times 252$, the total complexity of CATVOGSL2 is $512^2 \times 108$, and the total complexity of CITVOGSL2 is $512^2 \times 90$. That means each step of our methods is more than double faster than the inner iteration method in [11]. In the next section, the common computation parts of our methods and the inner iteration method are much more, and then our methods are only nearly double faster.

Remarks 6. We do not list the results of image (d) because they are almost the same as images (b) and (c). Moreover, when $\beta_1 < 30$ which is not sufficiently large, the numerical results are also good while we did not list them. This shows that the approximate part of our shrinkage formula is also good, and that when the inner step in [11, 12] is chosen to be 5, the numerical experiments are convergent although they did not find a convergence control sequence. In addition, in our experiments, we find that the results of the ATV case are better than the ITV case, which is a little different from the classic TV methods. Moreover, we find that if the parameters of [11] are chosen to be the same as ours, the solutions of our method and their method are always the same on PSNR and visual presentation while our method may save more time (it remains valid for [12]).

5.3.2 Experiments for the constrained TV OGS L_1 model

In this section, we compare our methods (CATVOGSL1 and CITVOGSL1) with some other methods, such as Chan et al.'s method proposed in [32] (Algorithm 2 in [32] for the constrained TV-L1 model) and Liu et al.'s method proposed in [12].

Similarly as the last section, we set the penalty parameters $\beta_1 = 80$, $\beta_2 = 2000$, $\beta_3 = 1$, for the ATV case, $\beta_1 = 80$, $\beta_2 = 2000$, $\beta_3 = 1$, for the ITV case and relax parameter $\gamma = 1.618$. The blur kernel is generated by MATLAB built-in function `fspecial('gaussian',7,5)` for 7×7 Gaussian blur with standard deviation 5. We first generated the blurred images operating on images (b)–(d) by the former Gaussian blur and further corrupted them by salt-and-pepper noise from 30 to 40 %. We generate all noise effects by MATLAB built-in function `imnoise(Bl,'salt & pepper',level)` with "`Bl`" the blurred image and fix the same random matrix for different methods.

The numerical results by the three methods are shown in Table 3. We have tuned the parameters manually to give the best PSNR improvement for Chan [32] as in Table 3 for

Table 3 Numerical comparison of Chan [32], Liu [12], CATVOGSL1, and CITVOGSL1 for images (a)–(d) in Fig. 5

Is	N	Chan [32]				Liu [12]			
	L	μ/Itr	PSNR	Time	ReE	Itr	PSNR	Time	ReE
(a)	30	25/130	29.59	61.57	0.0796	35	30.92	30.23	0.0683
	40	18/ 99	28.85	46.41	0.0866	37	30.11	31.54	0.0749
(b)	30	25/128	29.71	22.90	0.0501	35	31.81	12.93	0.0393
	40	20/ 98	28.59	17.80	0.0570	40	30.70	14.99	0.0447
(c)	30	26/138	30.21	25.65	0.0607	35	32.15	12.84	0.0485
	40	22/105	29.10	19.27	0.0689	37	31.04	13.67	0.0551
(d)	30	26/127	30.41	23.17	0.0975	36	32.47	13.15	0.0769
	40	20/ 95	29.44	17.43	0.1091	39	31.43	14.27	0.0867
		CATVOGSL1				CITVOGSL1			
(a)	30	32	31.17	*17.97*	0.0663	41	31.34	*21.56*	0.0651
	40	32	30.10	*18.03*	0.0751	37	30.06	*19.95*	0.0754
(b)	30	26	31.77	*5.97*	0.0395	29	31.75	*6.26*	0.0396
	40	26	30.47	*6.12*	0.0459	29	30.22	*6.38*	0.0472
(c)	30	25	32.30	*5.82*	0.0477	30	32.08	*6.75*	0.0489
	40	26	31.01	*5.76*	0.0553	27	30.50	*6.19*	0.0587
(d)	30	25	32.54	*5.68*	0.0764	29	32.51	*6.75*	0.0766
	40	28	31.35	*6.40*	0.0875	32	31.06	*7.33*	0.0905

PSNR dB, *Time* s, *Is* images, *NL* noise level (%)

different images. For Liu [12], we choose the given parameters μ default as 100, 80 for 30 and 40 %, respectively. For our method CATVOGSL1, we set μ as 180, 140 for 30 and 40 %, respectively. For our method CITVOGSL1, we set μ as 140, 100 for 30 and 40 %, respectively. In the experiments, we find that the parameters of our methods are robust and have wide range to choose. Therefore, we set the same μ for different images.

From Table 3, we can also see that PSNR and ReE of our methods (both ATV and ITV cases) are almost the same as Liu [12], which applied MM inner iterations to solve the subproblems (34) and (35) (only for the ATV case). However, each outer iteration of our methods is nearly twice faster than Liu [12] from the experiments. The time of each outer iteration of our methods is almost the same as the traditional TV method in Chan [32]. Moreover, we can also see that sometimes ATV is better than ITV and sometimes on the contrary for OGS TV. Finally, in Fig. 7, we display the degraded image, the original image, and the restored images for 30 % level of noise on image (d) by the four methods. From the figure, we can see that both our methods and Liu [12] can get better edges (handrail and window) than Chan [32].

6 Conclusions

In this paper, we propose the explicit shrinkage formulas for one class of OGS regularization problems with translation invariant overlapping groups. These formulas can be extended to several other regularization OGS problems as a subproblem in many fields. In this work, we apply our results in OGS TV regularization problems—deblurring and denoising problems. Furthermore, we also extend the image deblurring problems with OGS ATV in [11, 12] to both ATV and ITV cases. Since the formulas are very simple, these results can be easily extended to many other applications such as multichannel deconvolution and compress sensing, which we will consider in the future. In addition, in the experiments, we only choose all the entries of the weight matrix W_g equal to 1. We will test for other weights in the future on more experiments in order to choose the better or best weights for some other applications.

Fig. 7 *Top row*: blurred and noisy image (*left*), restoration images of Chan [32] (*middle*), Liu [12] (*right*). *Bottom row*: original image (*left*), restoration images of CATVOGSL1 (*middle*), CITVOGSL1 (*right*)

Appendix

For problem (4), similar to (14), we can get

$$\begin{aligned}
f_w(\mathbf{z}) &= \min_{\mathbf{z}} \|\mathbf{z}\|_{w,2,1} + \frac{\beta}{2}\|\mathbf{z}-\mathbf{x}\|_2^2 \\
&= \min_{\mathbf{z}} \sum_{i=1}^{n} \|w_g \circ (z_i)_g\|_2 + \frac{\beta}{2\sum_{k=1}^{s}(w_g)_k^2} \sum_{k=1}^{s} (w_g)_k^2 \|\mathbf{z}-\mathbf{x}\|_2^2 \\
&= \min_{\mathbf{z}} \sum_{i=1}^{n} \|w_g \circ (z_i)_g\|_2 + \frac{\beta}{2\|w_g\|_2^2} \sum_{i=1}^{n} \|w_g \circ ((z_i)_g - (x_i)_g)\|_2^2 \\
&= \min_{\mathbf{z}} \sum_{i=1}^{n} \|w_g \circ (z_i)_g\|_2 + \frac{\beta}{2\|w_g\|_2^2} \sum_{i=1}^{n} \|w_g \circ (z_i)_g - w_g \circ (x_i)_g\|_2^2 \\
&= \min_{\mathbf{z}} \sum_{i=1}^{n} \left(\|w_g \circ (z_i)_g\|_2 + \frac{\beta}{2\|w_g\|_2^2} \|w_g \circ (z_i)_g - w_g \circ (x_i)_g\|_2^2 \right).
\end{aligned} \tag{51}$$

All the symbols are the same as before. Similarly as before, we know that the necessary condition of minimizing the ith term of the last line in (51) is that $(w_g \circ (z_i)_g)$ is parallel to $(w_g \circ (x_i)_g)$. That is, $\frac{w_g \circ (z_i)_g}{\|w_g \circ (z_i)_g\|_2} = \frac{w_g \circ (x_i)_g}{\|w_g \circ (x_i)_g\|_2}$ for every i. On the other hand, Let $W = \mathrm{diag}(w_g)$ be a diagonal matrix with diagonal being the vector w_g, then $w_g \circ (x_i)_g = W(x_i)_g = W^T(x_i)_g$. If the vector $\mathbf{x}'$ (the same size as $(x_i)_g$) is parallel to the vector $\mathbf{z}'$, we have $\mathbf{x}' = \alpha \mathbf{z}'$. Then, $W\mathbf{x}' = W\alpha\mathbf{z}' = \alpha W \mathbf{z}'$. We obtain that the vector $W\mathbf{x}'$ is also parallel to the vector $W\mathbf{z}'$. Therefore, $\frac{W(z_i)_g}{\|W(z_i)_g\|_2} = \frac{W(x_i)_g}{\|W(x_i)_g\|_2}$, and each of them is a unit vector. Then, $\frac{W^T W(z_i)_g}{\|W(z_i)_g\|_2} = \frac{W^T W(x_i)_g}{\|W(x_i)_g\|_2}$. That is,

$$\frac{w_g \circ w_g \circ (z_i)_g}{\|w_g \circ (z_i)_g\|_2} = \frac{w_g \circ w_g \circ (x_i)_g}{\|w_g \circ (x_i)_g\|_2}.$$

In particular, we first consider that $s = 4$, $(z_i)_g = (z_{i-1}, z_i, z_{i+1}, z_{i+2})$ and $w_g = (w_1, w_2, w_3, w_4)$. We mark $(w_g \circ (z_i)_g)'$ as the expansion of $(w_g \circ (z_i)_g)$ similarly as $(z_i)_g'$, and we can get $\mathbf{x} = \frac{1}{\sum_{i=1}^{4} w_i} \sum_{i=1}^{n} (w_g \circ (z_i)_g)'$. Then, the Euler equation of $f_w(\mathbf{z})$ is given by

$$\beta(\mathbf{z}-\mathbf{x}) + \frac{(w_g \circ w_g \circ (z_1)_g)'}{\|(w_g \circ (z_1)_g)'\|_2} + \cdots + \frac{(w_g \circ w_g \circ (z_n)_g)'}{\|(w_g \circ (z_n)_g)'\|_2} \ni \mathbf{0}, \tag{52}$$

$$\frac{\beta}{\sum_{i=1}^{4} w_i^2} \sum_{i=1}^{n} \left(w_g \circ (z_i)_g - w_g \circ (x_i)_g\right) + \frac{\left(w_g \circ w_g \circ (z_1)_g\right)'}{\|\left(w_g \circ (z_1)_g\right)'\|_2} + \cdots + \frac{\left(w_g \circ w_g \circ (z_n)_g\right)'}{\|\left(w_g \circ (z_n)_g\right)'\|_2} \ni \mathbf{0}, \tag{53}$$

$$\beta(\mathbf{z}-\mathbf{x}) + \frac{(w_g \circ w_g \circ (x_1)_g)'}{\|(w_g \circ (x_1)_g)'\|_2} + \cdots + \frac{(w_g \circ w_g \circ (x_n)_g)'}{\|(w_g \circ (x_n)_g)'\|_2} \ni \mathbf{0}, \tag{54}$$

$$\mathbf{z} \ni \mathbf{x} - \frac{1}{\beta}\left(\frac{(w_g \circ w_g \circ (x_1)_g)'}{\|(w_g \circ (x_1)_g)'\|_2} + \cdots + \frac{(w_g \circ w_g \circ (x_n)_g)'}{\|(w_g \circ (x_n)_g)'\|_2}\right). \tag{55}$$

For each component, we obtained

$$z_i \ni x_i - \frac{1}{\beta}\left(\frac{w_4^2 x_i}{\|w_g \circ (x_{i-2})_g\|_2} + \frac{w_3^2 x_i}{\|w_g \circ (x_{i-1})_g\|_2} + \frac{w_2^2 x_i}{\|w_g \circ (x_i)_g\|_2} + \frac{w_1^2 x_i}{\|w_g \circ (x_{i+1})_g\|_2}\right). \tag{56}$$

Another expression is as follows.

$$z_i \ni \left(\frac{w_4^2 x_i}{\|w_g\|_2^2} - \frac{1}{\beta}\frac{w_4^2 x_i}{\|w_g \circ (x_{i-2})_g\|_2}\right) + \left(\frac{w_3^2 x_i}{\|w_g\|_2^2} - \frac{1}{\beta}\frac{w_3^2 x_i}{\|w_g \circ (x_{i-1})_g\|_2}\right) + \cdots + \left(\frac{w_1^2 x_i}{\|w_g\|_2^2} - \frac{1}{\beta}\frac{w_1^2 x_i}{\|w_g \circ (x_{i+1})_g\|_2}\right). \tag{57}$$

Similarly as Algorithm 1, we obtain the following algorithm for finding the minimizer of (4).

Algorithm 5 Direct shrinkage algorithm for the minimization problem (4)

Input:

Given vector $\mathbf{x}$, group size s, parameter β, weight vector $w_g = (w_1, w_2, \cdots, w_s)$.

Compute:

Definition of $(x_i)_g$, for example $(x_i)_g = (x_{i-s_l}, \cdots, x_i, \cdots, x_{i+s_r})$, with $s_l = [\frac{s-1}{2}]$ and $s_r = [\frac{s}{2}]$.

If s is even, then $w_g = [0, w_g]$ and $s = s+1$.

Let w_g^r = fliplr(w_g) be fliping w_g over 180 degrees.

Compute $X_n = \left(\|(x_1)_g\|_2, \cdots, \|(x_i)_g\|_2, \cdots, \|(x_n)_g\|_2\right)$, by correlation of $w_g.^2$ (pointwise square) and $\mathbf{x}$, or by convolution of $w_g^r.^2$ and $\mathbf{x}$.

Compute $X_n' = \max\left(\frac{1}{\|w_g\|_2^2} - 1./(X_n \cdot \beta), 0\right)$ pointwise.

Compute $F(x_i)$ by convolution of $w_g.^2$ and X_n', or by correlation of $w_g^r.^2$ and X_n'.

We can also see that Algorithm 2 only needs 2 times of convolution computations with time complexity $n * s$, which is just the same time complexity as one-step iteration in the MM method in [10]. Therefore, our method is much more efficient than the MM method.

Here, due to the properties of inequalities, without loss of generality, let $\mathbf{x} \in [0,1]^n$, then we obtain that if $\beta \le \frac{\|w_g\|_2}{\sqrt{s}} \le \frac{\|w_g\|_2}{\|x_g\|_2} = \frac{\|w_g\|_2^2}{\|w_g\|\|x_g\|_2} \le \frac{\|w_g\|_2^2}{\|w_g \cdot x_g\|_2}$, β is sufficiently small. However, we do not directly know why β is sufficiently large. In Section 5, after more than 1000 tests, we find that $\beta \ge 30 \cdot \frac{\|w_g\|_2}{\sqrt{s}}$ is sufficiently large generally.

For problem (5), similar to (51), we can obtain

$$\begin{aligned} f_W(A) &= \min_A \|A\|_{W,2,1} + \tfrac{\beta}{2}\|A-X\|_F^2 \\ &= \min_A \textstyle\sum_{i=1}^m \sum_{j=1}^n \|W_g \circ (A_{i,j})_g\|_F + \\ &\quad \tfrac{\beta}{2\sum_{k_1=1}^{K_1}\sum_{k_2=1}^{K_2}(W_g)_{k_1,k_2}^2} \textstyle\sum_{k_1=1}^{K_1}\sum_{k_2=1}^{K_2}(W_g)_{k_1,k_2}^2 \|A-X\|_F^2 \\ &= \min_A \textstyle\sum_{i=1}^m \sum_{j=1}^n \|W_g \circ (A_{i,j})_g\|_F + \\ &\quad \tfrac{\beta}{2\|W_g\|_F^2} \textstyle\sum_{i=1}^m \sum_{j=1}^n \|W_g \circ ((A_{i,j})_g - (X_{i,j})_g)\|_F^2 \\ &= \min_A \textstyle\sum_{i=1}^m \sum_{j=1}^n \|W_g \circ (A_{i,j})_g\|_F + \\ &\quad \tfrac{\beta}{2\|W_g\|_F^2} \textstyle\sum_{i=1}^m \sum_{j=1}^n \|W_g \circ (A_{i,j})_g - W_g \circ (X_{i,j})_g\|_F^2 \\ &= \min_A \textstyle\sum_{i=1}^m \sum_{j=1}^n \left(\|W_g \circ (A_{i,j})_g\|_F + \tfrac{\beta}{2\|W_g\|_F^2}\|W_g \circ (A_{i,j})_g - W_g \circ (X_{i,j})_g\|_F^2\right). \end{aligned} \tag{58}$$

For example, we set $K_1 = 2, K_2 = 2$ and define $(A_{i,j})_g = (A_{i,j}, A_{i,j+1}; A_{i+1,j}, A_{i+1,j+1})$ and $W_g = (W_{1,1}, W_{1,2}; W_{2,1}, W_{2,2})$. Similar as the vector case, each $(A_{i,j})_g$ is a matrix with $\|(A_{i,j})_g\|_F = \sqrt{\sum_{k_1=1}^{K_1}\sum_{k_2=1}^{K_2}((A_{i,j})_g)_{k_1,k_2}^2}$. Notice that the Frobenius norm of a matrix is equal to the ℓ_2 norm of a vector reshaped by the matrix. Then, the Euler equation of $f_w(\mathbf{z})$ is given by

$$\begin{aligned} \beta(A-F) &+ \frac{(W_g \circ W_g \circ (A_{1,1})_g)'}{\|(W_g \circ (A_{1,1})_g)'\|_2} + \cdots \\ &+ \frac{(W_g \circ W_g \circ (A_{n,n})_g)'}{\|(W_g \circ (A_{n,n})_g)'\|_2} \ni \mathbf{0}, \end{aligned} \tag{59}$$

where $((A_{i,j})_g)'$ is defined similarly as $((z_i)_g)'$, which is an expansion of $(A_{i,j})_g$. These symbols remain consistent as default through this paper.

$$\begin{aligned} \beta(A-X) &+ \frac{(W_g \circ W_g \circ (X_{1,1})_g)'}{\|(W_g \circ (X_{1,1})_g)'\|_2} + \cdots \\ &+ \frac{(W_g \circ W_g \circ (X_{n,n})_g)'}{\|(W_g \circ (X_{n,n})_g)'\|_2} \ni \mathbf{0}, \end{aligned} \tag{60}$$

$$\begin{aligned} A \ni X - \frac{1}{\beta}\Bigg(&\frac{(W_g \circ W_g \circ (X_{1,1})_g)'}{\|(W_g \circ (X_{1,1})_g)'\|_2} + \cdots \\ &+ \frac{(W_g \circ W_g \circ (X_{n,n})_g)'}{\|(W_g \circ (X_{n,n})_g)'\|_2}\Bigg). \end{aligned} \tag{61}$$

For each component, we obtained

$$\begin{aligned} A_{i,j} \ni X_{i,j} - \frac{1}{\beta}\Bigg(&\frac{W_{2,2}^2 X_{i,j}}{\|W_g \circ (X_{i-1,j-1})_g\|_2} + \frac{W_{2,1}^2 X_{i,j}}{\|W_g \circ (X_{i-1,j})_g\|_2} + \\ &\frac{W_{1,2}^2 X_{i,j}}{\|W_g \circ (X_{i,j-1})_g\|_2} + \frac{W_{1,1}^2 X_{i,j}}{\|W_g \circ (X_{i,j})_g\|_2}\Bigg). \end{aligned} \tag{62}$$

Therefore, we can obtain a similar algorithm on the former formula (62) for finding the minimizer of (5).

Competing interests

The authors declare that they have no competing interests.

Authors' contributions

GL carried out the main part of this manuscript. TH participated in the discussion and carried out the numerical results. XL and JL participated in the discussion and corrected the English errors. All authors read and approved the final manuscript.

Acknowledgements

The work of Gang Liu, Ting-Zhu Huang, and Jun Liu was supported by the 973 Program (2013CB329404), NSFC (61370147), and Fundamental Research Funds for the Central Universities (ZYGX2013Z005). The work of Xiao-Guang Lv is supported by NSFC (61401172).

Author details

[1]College of Applied Mathematics, Chengdu University of Information Technology, Xuefu Road, 610225 Chengdu, China. [2]Research Center for Image and Vision Computing/School of Mathematical Sciences, University of Electronic Science and Technology of China, Xiyuan Ave, 611731 Chengdu, China. [3]School of Science, Huaihai Institute of Technology, Cangwu Road, 222005 Lianyungang, China.

References

1. W Deng, W Yin, Y Zhang, in *Proc. SPIE*. Group sparse optimization by alternating direction method (SPIE, San Diego, US, 2013), pp. 88580R–15
2. J Huang, T Zhang, The benefit of group sparsity. Ann. Stat. **38**(4), 1978–2004 (2010). doi:10.1214/09-AOS778
3. R Chartrand, B Wohlberg, in *IEEE International Conference on Acoustics, Speech and Signal Processing (ICASSP)*. A nonconvex ADMM algorithm for group sparsity with sparse groups (IEEE, Vancouver, BC, 2013), pp. 6009–6013
4. M Stojnic, F Parvaresh, B Hassibi, On the reconstruction of block-sparse signals with an optimal number of measurements. IEEE Trans. Signal Process. **57**(8), 3075–3085 (2009)
5. L Jacob, G Obozinski, J-P Vert, in *Proceedings of the 26th Annual International Conference on Machine Learning. ICML '09*. Group lasso with overlap and graph lasso (ACM, New York, 2009), pp. 433–440
6. A Majumdar, RK Ward, in *IEEE International Conference on Acoustics, Speech and Signal Processing (ICASSP)*. Classification via group sparsity promoting regularization (IEEE, Taipei, 2009), pp. 861–864
7. E Elhamifar, R Vidal, in *IEEE Conference on Computer Vision and Pattern Recognition (CVPR)*. Robust Classification using structured sparse representation (IEEE, Providence, 2011), pp. 1873–1879
8. J Gao, Q Shi, TS Caetano, Dimensionality reduction via compressive sensing. Pattern Recognit. Lett. **33**(9), 1163–1170 (2012)
9. YN Liu, F Wu, YT Zhuang, Group sparse representation for image categorization and semantic video retrieval. Sci. China Inform. Sci. **54**(10), 2051–2063 (2011)
10. P-Y Chen, IW Selesnick, Translation-invariant shrinkage/thresholding of group sparse signals. Signal Process. **94**(0), 476–489 (2014)
11. J Liu, T-Z Huang, IW Selesnick, X-G Lv, P-Y Chen, Image restoration using total variation with overlapping group sparsity. Inform. Sci. **295**(20), 232–246 (2015)
12. G Liu, T-Z Huang, J Liu, X-G Lv, Total variation with overlapping group sparsity for image deblurring under impulse noise. Plos One. **10**(4), 0122562 (2015)
13. IW Selesnick, P-Y Chen, in *IEEE International Conference on Acoustics, Speech and Signal Processing (ICASSP)*. Total variation denoising with overlapping group sparsity (IEEE, Vancouver, BC, 2013), pp. 5696–5700
14. M Figueiredo, J Bioucas-Dias, in *Signal Processing with Adaptive Sparse Structured Representations - SPARS11*. An alternating direction algorithm for (overlapping) group regularization, (Edinburgh, Scotland, UK, 2011)
15. R Jenatton, J-Y Audibert, F Bach, Structured variable selection with sparsity-inducing norms. J. Mach. Learn. Res. **12**, 2777–2824 (2011)
16. M Kowalski, Sparse regression using mixed norms. Appl. Comput. Harmonic Anal. **27**(3), 303–324 (2009)
17. G Peyré, JM Fadili, in *EUSIPCO*. Group sparsity with overlapping partition functions, (Barcelona, Spain, 2011), pp. 303–307
18. F Bach, R Jenatton, J Mairal, G Obozinski, Structured sparsity through convex optimization. Stat. Sci. **27**(4), 450–468 (2012)

19. I Bayram, in *IEEE International Conference on Acoustics, Speech and Signal Processing (ICASSP)*. Mixed norms with overlapping groups as signal priors (IEEE, Prague, 2011), pp. 4036–4039
20. Y Wang, J Yang, W Yin, Y Zhang, A new alternating minimization algorithm for total variation image reconstruction. SIAM J. Imaging Sci. **1**(3), 248–272 (2008)
21. J Yang, W Yin, Y Zhang, Y Wang, A fast algorithm for edge-preserving variational multichannel image restoration. SIAM J. Imaging Sci. **2**(2), 569–592 (2009). doi:10.1137/080730421
22. B Wohlberg, R Chartrand, J Theiler, in *IEEE International Conference on Acoustics, Speech and Signal Processing (ICASSP)*. Local principal component pursuit for nonlinear datasets (IEEE, Kyoto, 2012), pp. 3925–3928
23. P Sprechmann, I Ramirez, G Sapiro, YC Eldar, C-HiLasso: a collaborative hierarchical sparse modeling framework. IEEE Trans. Signal Process. **59**(9), 4183–4198 (2011)
24. PC Hansen, JG Nagy, DP O'Leary, *Deblurring images: matrices, spectra, and filtering*, 1st edn. (Society for Industrial and Applied Mathematics, Philadelphia, United States, 2006)
25. LI Rudin, S Osher, E Fatemi, Nonlinear total variation based noise removal algorithms. Physica D: Nonlinear Phenomena. **60**(1–4), 259–268 (1992)
26. A Chambolle, An algorithm for total variation minimization and applications. J. Math. Imaging Vis. **20**(1–2), 89–97 (2004)
27. G Liu, T-Z Huang, J Liu, High-order TVL1-based images restoration and spatially adapted regularization parameter selection. Comput. Math. Appl. **67**(10), 2015–2026 (2014)
28. X-L Zhao, W Wang, T-Y Zeng, T-Z Huang, MK Ng, Total variation structured total least squares method for image restoration. SIAM J. Sci. Comput. **35**(6), 1304–1320 (2013)
29. X-L Zhao, F Wang, MK Ng, A new convex optimization model for multiplicative noise and blur removal. SIAM J. Imaging Sci. **7**(1), 456–475 (2014)
30. C Chen, MK Ng, X-L Zhao, Alternating direction method of multipliers for nonlinear image restoration problems. IEEE Trans. Image Process. **24**(1), 33–43 (2015)
31. L-J Deng, H Guo, T-Z Huang, A fast image recovery algorithm based on splitting deblurring and denoising. J. Comput. Appl. Math. **287**, 88–97 (2015)
32. R Chan, M Tao, X Yuan, Constrained total variation deblurring models and fast algorithms based on alternating direction method of multipliers. SIAM J. Imaging Sci. **6**(1), 680–697 (2013)
33. J Nocedal, SJ Wright, *Numerical optimization*. 1431–8598. (Springer, New York, 2006), p. 664
34. D Gabay, B Mercier, A dual algorithm for the solution of nonlinear variational problems via finite element approximation. Comput. Math. Appl. **2**(1), 17–40 (1976)
35. J Eckstein, D Bertsekas, On the Douglas-Rachford splitting method and the proximal point algorithm for maximal monotone operators. Math. Programm. **55**(1–3), 293–318 (1992). doi:10.1007/BF01581204
36. R Glowinski, *Numerical methods for nonlinear variational problems*. 1434–8322. (Springer, Berlin Heidelberg, 1984), p. 493
37. B He, H Yang, Some convergence properties of a method of multipliers for linearly constrained monotone variational inequalities. Oper. Res. Lett. **23**(3–5), 151–161 (1998)
38. E Esser, Applications of Lagrangian-based alternating direction methods and connections to split Bregman. Los Angeles: UCLA CAM report, 9–31 (2009). ftp://arachne.math.ucla.edu/pub/camreport/cam09-31.pdf
39. M Ng, P Weiss, X Yuan, Solving constrained total-variation image restoration and reconstruction problems via alternating direction methods. SIAM J. Sci. Comput. **32**(5), 2710–2736 (2010)
40. X Zhang, M Burger, X Bresson, S Osher, Bregmanized nonlocal regularization for deconvolution and sparse reconstruction. SIAM J. Imaging Sci. **3**(3), 253–276 (2010)
41. J Yang, Y Zhang, W Yin, A fast alternating direction method for TVL1-L2 signal reconstruction from partial fourier data. IEEE J. Selected Topics Signal Process. **4**(2), 288–297 (2010)
42. X Zhang, M Burger, S Osher, A unified primal-dual algorithm framework based on Bregman iteration. J. Sci. Comput. **46**(1), 20–46 (2011)
43. S Boyd, N Parikh, E Chu, B Peleato, J Eckstein, Distributed optimization and statistical learning via the alternating direction method of multipliers. Foundations Trends Mach. Learn. **3**(1), 1–122 (2011)
44. C Wu, X Tai, Augmented Lagrangian method, dual methods, and split Bregman iteration for ROF, vectorial TV, and high order models. SIAM J. Imaging Sci. **3**(3), 300–339 (2010)

13

Multiview video plus depth transmission via virtual-view-assisted complementary down/upsampling

Zhi Jin[1,2*], Tammam Tillo[1], Jimin Xiao[1] and Yao Zhao[3]

Abstract

Multiview video plus depth is a popular 3D video format which can provide viewers a vivid 3D feeling. However, its requirements in terms of computational complexity and transmission bandwidth are more than that of conventional 2D video. To mitigate these limitations, some works have proposed to reduce the amount of transmitted data by adopting different resolutions for different views, and consequently, the transmitted video is called mixed resolution video. In order to further reduce the transmitted data and maintain good quality at the decoder side; in this paper, we propose a down/upsampling algorithm for 3D multiview video which systematically takes into account the video encoder and decoder. At the encoder side, the rows of the two adjacent views are downsampled following an interlacing and complementary fashion, whereas, at the decoder side, the discarded pixels are recovered by fusing the virtual view pixels with the directional interpolated pixels from the complementary downsampled views. Moreover, the patterns of the texture surrounding the discarded pixels are used to aid the data fusion, so as to enhance edges recovery. Meanwhile, with the assistance of virtual views, at the decoder side, the proposed approach can effectively recover the discarded high-frequency details. The experimental results demonstrate the superior performance of the proposed framework.

Keywords: Low bit rate video transmission, Video coding, Multiview video plus depth (MVD), Depth map, DIBR, Virtual view

1 Introduction

The development of 3D technologies and communication networks makes 3D video applications increasingly popular. An example of this technology is 3D multiview television which allows to cover a wide view angle of the scene. However, delivering a large number of high-quality views to end users is a challenging task due to the limitations of data transmission and storage capacity [1]. Therefore, some advanced video coding standards have been proposed to compress video data, such as H.264/MVC [2] and HEVC [3–6]. Besides that, some data representation formats for the 3D multiview data have been adopted for efficient representation. One widely accepted format is the multiview video plus depth (MVD) format [7], which consists of textures and the associated per-pixel depth data (the latter describes the geometric relationship between objects in the scene and the capturing cameras [8]). Since this format allows any intermediate view within a certain range to be generated, with the assistance of the depth-image-based rendering (DIBR) technique [9]. Therefore, it can, to a large extent, reduce the number of transmitted views. However, the required data of 3D multiview video is still very large.

Because of the above reasons, many works have focused on reducing the amount of transmitted data at the encoder side and recovering it at the decoder side for low bit rate transmission [10–12]. Additionally, given that depth maps consist of large homogeneous areas, they require less transmission bit rate compared with texture [13]. Therefore, the reduction and recovery of texture data have drawn more attention than that of the depth data. In order to speed up the encoding process and reduce the overall

*Correspondence: zhi.jin10@student.xjtlu.edu.cn
[1] Department of Electrical and Electronic Engineering, Xi'an Jiaotong-Liverpool University, Ren Ai Road 111, 215123 Suzhou, People's Republic of China
[2] Department of Electrical Engineering and Electronics, University of Liverpool, Merseyside L69 3BX, Liverpool, UK
Full list of author information is available at the end of the article

bit rate, in [14], Garcia et. al adopted a low-resolution (LR) and full-resolution (FR) frames mixed video sequence. Based on this, a mixed resolution coding approach was proposed where the first M frames in the sequence were encoded at FR and the rest frames were coded at LR. Mixed resolution (MR) view frameworks have been proposed in [15, 16] for multiview video coding, where at least one view is coded at LR, while the others are coded at FR. These frameworks can reduce the amount of transmitted and stored data in comparison with the full FR framework. In [17, 18], a MR-MVD framework was adopted at the encoder side and 3D warping-generated virtual views were utilized to recover FR frames from the LR frames at the decoder side. With proper reconstruction algorithms at the decoder side, MR-MVD framework can well enhance the overall transmission efficiency.

In terms of reducing the transmission and storage data size, for stereoscopic video, the MR format can save 3/8 transmitted texture data compared with FR format, while downsampling both views by half can reduce the amount of transmitted texture data to 1/2 of the original format. Furthermore, the neighboring views have the same frame size which is suitable for the MVC coding approach. Two frame-compatible coding frameworks for stereoscopic video were proposed in [19, 20]. In [19], two views were decimated to half of their original size by downsampling filters, which were designed based on frame contents and the targeted interpolation coefficients. At the decoder side, the reconstructed frames were demultiplexed and interpolated into the full resolution. Since the frame downsampling pattern at the encoder side was evaluated as a function of the interpolation method, it can, to a large extent, reduce the interpolation errors. Unfortunately, this paradigm is not suitable for video applications because the downsampling pattern is frame dependent which means that the temporal redundancy cannot be efficiently removed by the video encoder. Thus, for video applications, there is a need to use temporally static downsampling patterns. In [20], an opposite parity packing arrangement for stereoscopic video was proposed to multiplex the two views. At the decoder side, the optimal disparity vector for each block was obtained from calculating the smallest differences between the matched blocks in coarsely interpolated FR left view and right view. However, in this work, some theoretical supports for the proposed view packing arrangement were missing. Moreover, although both [19, 20] had shown superior performance than other state-of-the-art frame-compatible coding frameworks, they mainly focus on stereoscopic video which may not be optimal for multiview or MVD video.

In this paper, a *systematical* down/upsampling framework for MVD video is proposed to enhance the coding performance at low bit rate (Fig. 1). In the proposed downsampling approach, the rows of two adjacent texture views are downsampled following an interlacing and complementary pattern, before compression. The aim of this downsampling approach is to facilitate the upsampling at the decoder side while exploiting the feature of 1D DIBR. The LR views will be upsampled by fusing the virtual view pixels with directional interpolated pixels with the aid of the pattern of the texture surrounding the discarded pixels. This approach has two benefits. Firstly, the high frequency information contained in one LR view can be properly utilized to upsample the other LR view through the generated virtual views. Secondly, the proposed directional interpolation approach can overcome the limitation of the virtual-view-based upsampling mechanism, which suffers in the areas corresponding to depth map discontinuity. Hence, by taking advantage of these two strategies, the discarded pixels can be recovered effectively. Experimental results have shown that the proposed algorithm achieves superior performance with respect to the filter-based interpolation algorithms and other state-of-the-art algorithms. The proposed upsampling approach will be named directional data fusion upsampling (DDFU) throughout this paper.

The rest of this paper is organized as follows. Section 2 describes the details of the proposed downsampling algorithm, and the upsampling algorithm is introduced in Section 3. The generalization of the proposed method is presented in Section 4, and experimental results are presented in Section 5. Finally, the conclusions are in Section 6.

2 Proposed interlacing-and-complementary-row-downsampling

A proper downsampling approach for multiview video needs to take into account the fact that different views cover almost the same scene. Hence, between neighboring views, there is considerable amount of inter-view redundancy. In this work, an interlacing-and-complementary-row-downsampling method is proposed by taking the features of proposed upsampling and multiview video into account, as shown in Fig. 2.

In the view multiplexing approaches, the generated sequence is from the mapping of two downsampled views. The process of multiplexing is carried out before the video encoding stage. There are variety of options for both the downsampling and view combination [21]. Due to inter-view redundancy, interlacing-and-complementary downsampling approaches could maintain more information than the non interlacing and complementary ones. In the following parts and aided with a graphical example, three downsampling approaches will be compared. In these three scenarios, we assume that two calibrated cameras in a parallel configuration setting and the same image plane (the most common camera configuration) are used to record an uneven bar structure

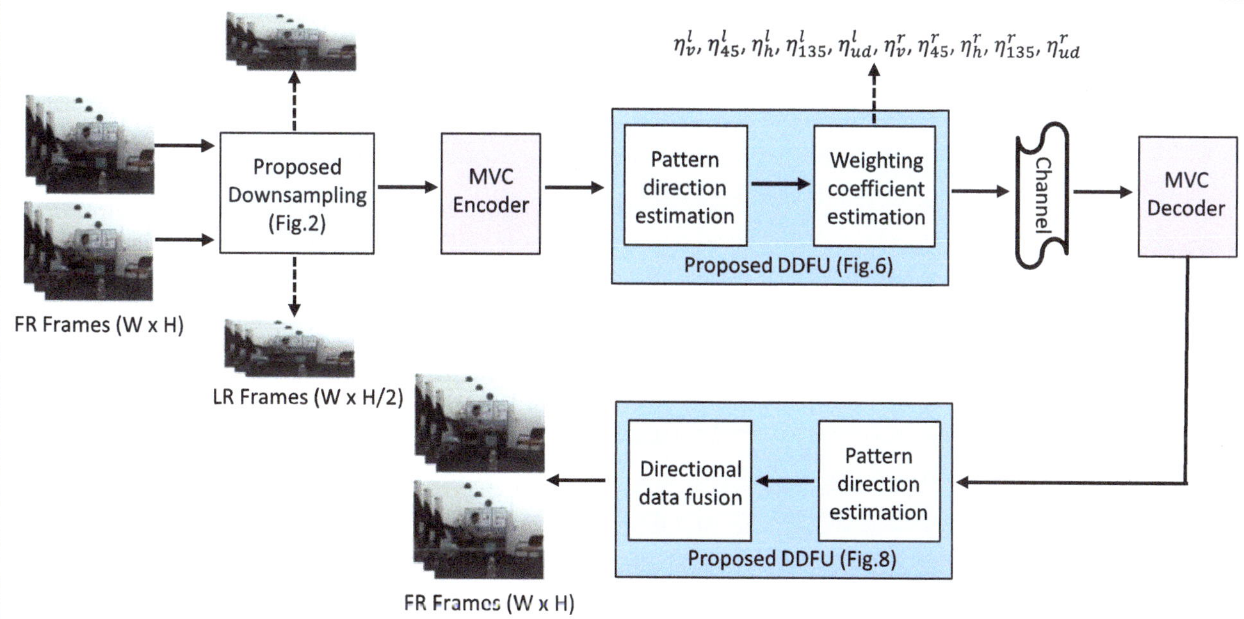

Fig. 1 Framework of the proposed down/upsampling method. Framework of the proposed down/upsampling method for a stereo video

(similar to the artistic gymnastics apparatus), as shown in Fig. 3d. Figure 3a–c shows the front, side, and top view of the stereoscopic orthographic projection of the scene, respectively. The viewed scene of the first and second cameras is shown in Fig. 4a. The output of the vertical interlacing and complementary downsampling approach (i.e., column-wise downsampling) is shown in Fig. 4b, where the gray areas indicate the "discarded areas" during downsampling process. It is possible to see that the left black bar of the uneven bar structure is missing in both views. Hence, neither intra-view or inter-view interpolation can help to recover this part. This is because the column-wise downsampling approach causes some "blind areas," where objects cannot be seen in any of the two views. The "blind areas" could be easily seen in Fig. 5, which shows the top view of the prospective projection of a scene with two pinhole cameras. The area enclosed by red lines can be viewed by both cameras, whereas, the yellow and blue bands indicate discarded areas in view 1 and view 2, respectively, due to the column-wise downsampling. Some areas (indicated by black) inevitably end up being discarded in both views; thus, any object falling in any of these black areas cannot be recovered by inter-view interpolation, and consequently, these areas are called "blind areas".

Compared with column-wise downsampling, the output of row-wise one is shown in Fig. 4c. It indicates that the proposed interlacing-and-complementary-row-downsampling will almost always guarantee that the object can be seen in the rows of one of the two views, except for some small objects with a one-pixel-width projection size in the camera plane. Nevertheless, the probability of this situation is low, and it also happens for the column-wise downsampling approach. Consequently, this row-wise downsampling can better exploit the warping feature of the DIBR technique and, as a result, can enhance the upsampling performance.

The chessboard downsampling approach can be regarded as the combination of row- and column-wise ones. It is able to achieve highest intra-view upsampling

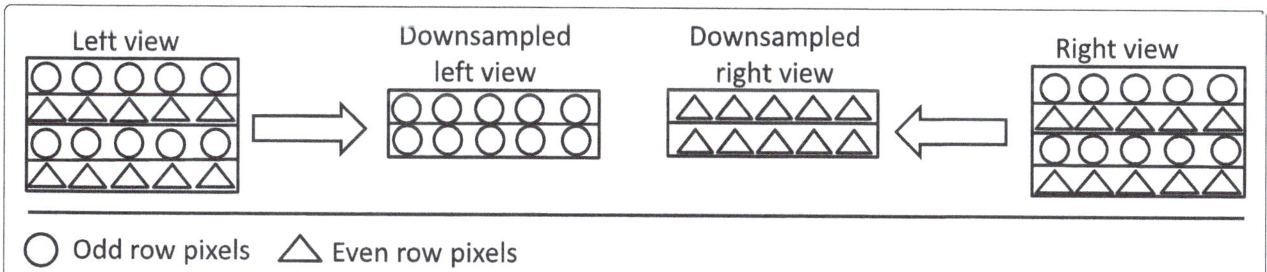

Fig. 2 The proposed downsampling method. The proposed interlacing-and-complementary-row-downsampling process for a stereo video

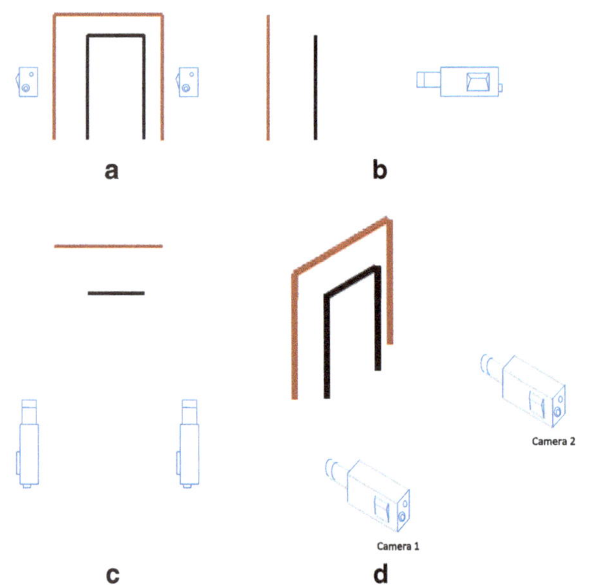

Fig. 3 Captured scene description. **a**, **b**, and **c** show the front, side, and top view of the stereoscopic orthographic projection of uneven bar structure viewed by two cameras in a parallel configuration setting as depicted in (**d**)

performance, since each to-be-filled pixel has four adjacent pixels in both horizontal and vertical directions which provide more information during interpolation. However, the chessboard pattern usually requires a comparatively higher bit rate due to low spatial and temporal correlations [19]. Furthermore, for each row of the chessboard downsampled views, it is possible to notice that the top view of the prospective projection of a scene is similar to the one shown in Fig. 5. Therefore, it could be conjectured that the chessboard approach also suffers from some "blind areas"; thus, its performance is better than the column-wise approach while being worse than the row-wise approach.

3 Virtual view-assisted directional data fusion upsampling

In order to reduce the required resources and the amount of transmitted data, downsampling of the texture sequences is performed before the compression stage. In this paper, motivated by the findings in Section 2, the downsampled texture frames are generated by discarding the even rows in the left view and the odd rows in the right view of the stereo video, respectively.

Let the left and right FR frames be defined as $\mathbf{V}_f^l$ and $\mathbf{V}_f^r$, respectively, with size $W \times H$, and the downsampled left and right LR frames as $\mathbf{V}_l^l$ and $\mathbf{V}_l^r$, respectively, with size $W \times H/2$. Figure 6 shows the main stages of the proposed FR recovery mechanism. The downsampled views are expanded to their original size with the positions of the discarded pixels left empty (this stage is indicated by ① in Fig. 6). The expanded left view is represented by $\mathbf{V}_e^l$ where $V_e^l(2n, m) = 0, 1 \leq n \leq H/2, 1 \leq m \leq W$, whereas, the expanded right view is represented by $\mathbf{V}_e^r$ where $V_e^r(2n-1, m) = 0, 1 \leq n \leq H/2, 1 \leq m \leq W$. Then, in the second stage indicated by ② based on the direction estimation results, a directional interpolation method is used to generate the corresponding interpolated frames, and these are denoted by $\mathbf{V}_i^l$ and $\mathbf{V}_i^r$ for the left and right views, respectively. Meanwhile, in the third stage ③, the DIBR technique is applied on the expanded frames using the corresponding depth maps in order to generate the virtual views at the counterpart viewpoints, i.e., the left side virtual view $\mathbf{V}_v^l$ is generated by the right side expanded view. As a consequence, all the even rows in the left virtual view are warped from the even rows in the right view. Similarly, for the right virtual view, all the odd rows are warped from the odd rows in the left view. Therefore, based on the above design which aims to make the recovery of discarded pixels work in synergy with the downsampling stage, the virtual view becomes a potential source of information to efficiently recover the discarded pixels. Virtual views and directional interpolated views as the outputs of the two parallel stages,② and ③, are fused to generate the final FR frames at stage ④. This fusion process is driven by the pattern direction of the texture around each of the discarded pixels, so as to exploit the potential of stages ② and ③.

3.1 PCA-based pattern direction estimation

Knowing the dominant direction of the texture surrounding each discarded pixel allows better exploitation of the

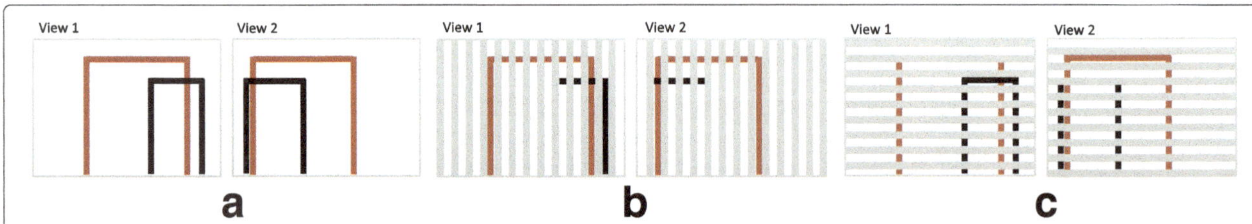

Fig. 4 The outputs of different downsampling methods. **a** the left side and right side of each frame shown the captured scene by the corresponding cameras without downsampling, respectively; **b** the output of the vertical downsampling method (i.e., column-wise downsampling); **c** the output of the interlacing and complementary row-wise downsampling method

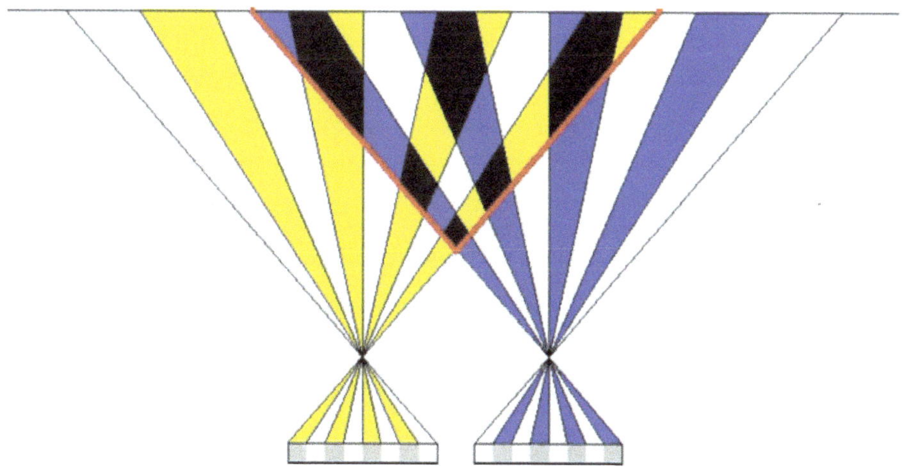

Fig. 5 The *top view* of the prospective projection. The *top view* of the prospective projection of a scene using a pinhole camera model for the column-wise downsampling approach; area which could be viewed by both cameras (before the downsampling process) is enclosed in *red*; *yellow* bands indicate areas that cannot be seen in view 1 due to the column-wise downsampling approach; unviewed areas in view 2 due to the downsampling are indicated by *blue*. Areas depicted in black cannot be viewed in both views, those are called "blind areas"

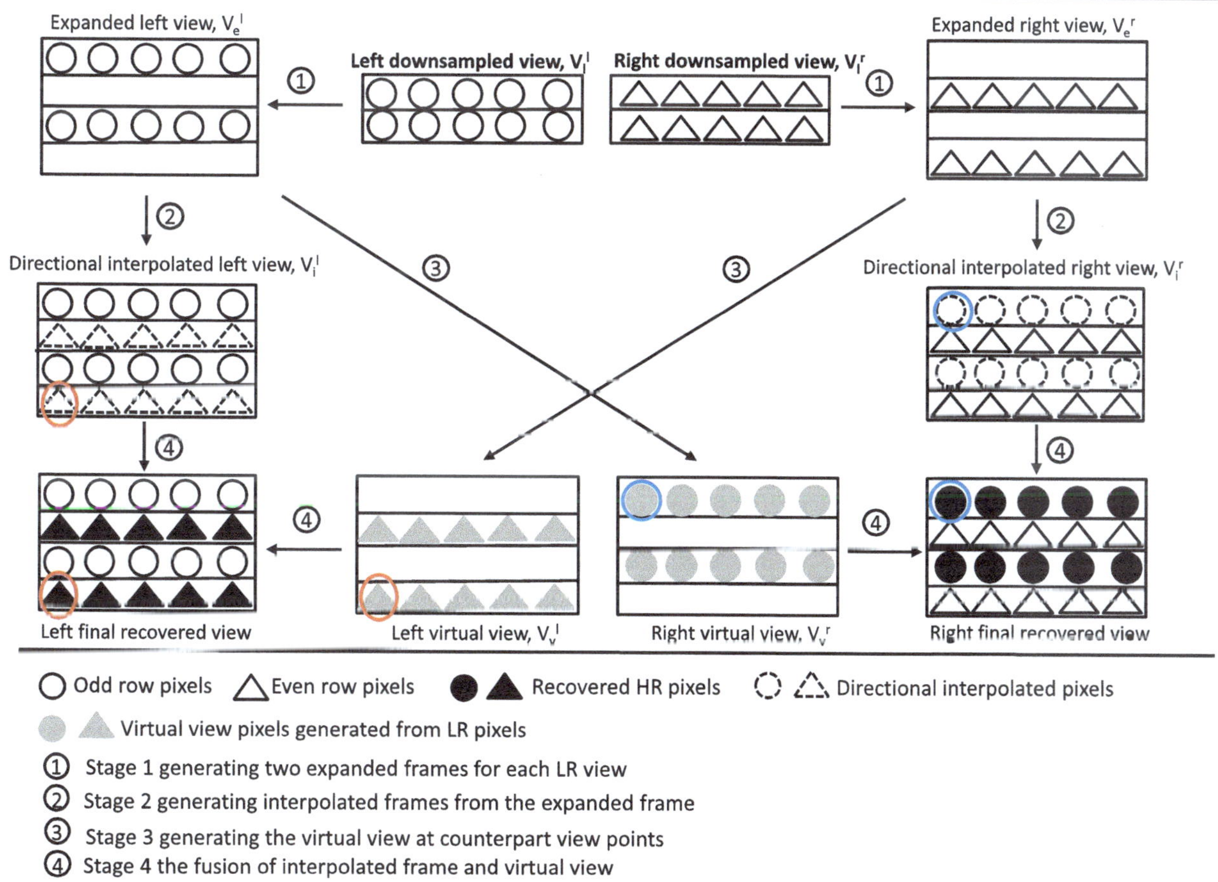

Fig. 6 The proposed discarded pixels recovery process

virtual and interpolated frames in recovering the discarded pixels. For example, texture patterns with horizontal edges usually cannot be accurately estimated from their upper and lower neighbors. Hence, exploiting the virtual view pixels can greatly help to recover such pixels.

To get the pattern direction, in this work, a principal components analysis (PCA)[22]-based method is used. This approach evaluates the gradients of the surrounding pixels for each discarded pixel, and then, the dominant direction is determined [23], where PCA can be implemented by evaluating the singular value decomposition (SVD) [24] of the data.

In general, the gradient at $V(x,y)$ can be obtained by $\nabla V(x,y) = [\partial V(x,y)/\partial x, \partial V(x,y)/\partial y]^T$, and this could be approximated for discrete applications as

$$\nabla V(x,y) \approx \begin{pmatrix} \frac{1}{2}(V(x+\Delta,y) - V(x-\Delta,y)) \\ \frac{1}{2}(V(x,y+\Delta) - V(x,y-\Delta)) \end{pmatrix} \tag{1}$$

$\Delta = 1$ offers the best approximation, however, taking into account that half of the rows are discarded, then Δ needs to be 2 while evaluating the gradients of the surrounding pixels of a discarded pixel. This ensures that $V(x+\Delta,y)$, $V(x-\Delta,y)$, $V(x,y+\Delta)$, and $V(x,y-\Delta)$ are available[1].

It is worth noticing that the horizontal neighbors (i.e., left and right neighbors) of the discarded pixels are unavailable; therefore, the dominant direction for each discarded pixel will be inferred from the four corner pixels of a 3×3 overlapping window centered at the discarded pixel. For example, the discarded pixel p_5, in Fig. 7, has two discarded neighbors, namely p_4 and p_6, so in order to maintain an equivalent number of neighbors and symmetric structure around p_5, the two pixels p_2 and p_8 will not be taken into account while evaluating the dominant pattern direction. In other words, only the gradients of the corner pixels p_1, p_3, p_7, and p_9 will be evaluated[2]. The gradients of the surrounding pixels of the discarded pixel at position (x,y) will be then arranged into a 4×2 matrix $\mathbf{G}$, as follows:

$$\mathbf{G} = \begin{bmatrix} \nabla V(x-1,y-1)^T \\ \nabla V(x-1,y+1)^T \\ \nabla V(x+1,y-1)^T \\ \nabla V(x+1,y+1)^T \end{bmatrix} \tag{2}$$

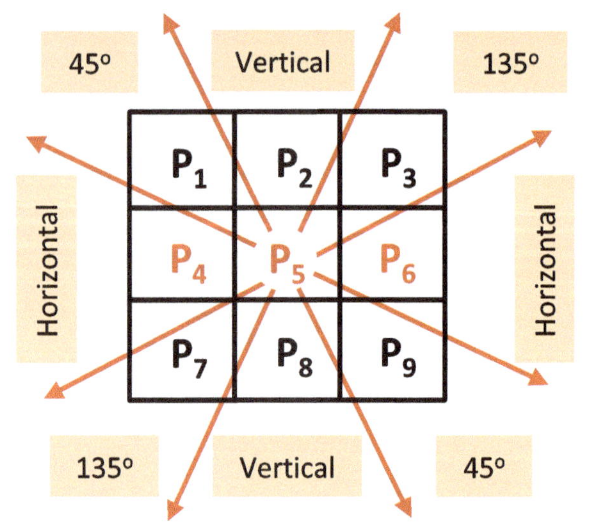

Fig. 7 Missing pixels classification based on texture pattern directions. The overlapping window centered at the discarded pixel p_5. The dominant pattern direction will be categorized into five groups. In this figure, only the remarkably dominant patterns are shown which are horizontal, 45° diagonal, vertical, and 135° diagonal directions

The SVD of the matrix $\mathbf{G}$ will be computed as $\mathbf{G} = \mathbf{USV^T}$, where $\mathbf{S}$ is a 4×2 diagonal matrix and the ratio of the diagonal elements in $\mathbf{S}$ (i.e., S_{11}/S_{22}) represents the energy of the dominant gradient. $\mathbf{U}$ and $\mathbf{V}$ are orthogonal matrices with size 4×4 and 2×2, respectively, and the angle of dominant gradient is $\theta = \arctan(v_{21}/v_{11})$. For the remarkably dominant gradient (i.e., $S_{11}/S_{22} \geq Th$ where Th is a threshold to define the remarkably dominant gradient), this angle will be used to determine the pattern directions of the discarded pixel, which are horizontal, 45° diagonal, vertical, and 135° diagonal directions as shown in Fig. 7. For the pixels from texture uniform areas whose energy in all four directions is almost equal, there is no remarkably dominant directional pattern (i.e., $S_{11}/S_{22} < Th$), will be classified into the "undefined" direction category. This process will be carried forward at both the encoder and decoder sides.

3.2 Weighting coefficient estimation of directional data fusion

Since all of the discarded pixels are classified into five categories: *horizontal*, 45° *diagonal*, *vertical*, 135° *diagonal*, and *undefined direction*, the directional interpolated frames $\mathbf{V}_i^l$ and $\mathbf{V}_i^r$ are generated based on this classification. The *horizontal pixels* are the average of corresponding four nearest corner pixels, and the *undefined directional pixels* are recovered by vertical interpolation, since the vertical neighbors are the closest to the discarded pixels. For the other three directions' pixels, they are recovered by linearly interpolation along the pattern direction. In the fourth stage of the proposed upsampling algorithm, the discarded pixels are recovered by fusing the interpolated pixels with the virtual view pixels in order to exploit the advantages of both types of approach and to compensate the compression distortion.

To recover the discarded information, at the fusion stage, each discarded pixel is filled by a weighted average of the counterpart pixels in V_v and V_i as shown:

$$\hat{V}^l(2n,m) = \eta^l V_i^l(2n,m) + (1-\eta^l)V_v^l(2n,m) \tag{3}$$

The value of the weighting coefficient, η^l, is in the range [0, 1]. This value, in theory, should be evaluated for each missing pixel, and it determines the relative contribution of the directional interpolated pixel with respect to the virtual view pixel. The fusing coefficients could be obtained by minimizing the L2 distance between the recovered pixels and their counterpart original pixels, as follows.

$$\sum_{m=1}^{W}\sum_{n=1}^{H/2}\left(\hat{V}^l(2n,m) - V_f^l(2n,m)\right)^2 \tag{4}$$

Holes and disoccluded areas in the virtual views are excluded during the fusion process and in these areas; the discarded pixels are directly recovered by directional interpolation. Since the original FR frame is only available at the encoder side, this means that all the fusing coefficients need to be transmitted for each frame to the decoder side. Obviously, this makes the pixel-by-pixel estimation of the weighting coefficient impractical. Hence, in this paper, a direction-mask-based weighting coefficient estimation is proposed, which can hugely reduce the transmitted side information. At the encoder and decoder sides, all the discarded pixels will be classified into five categories and respectively represented by five binary masks $\mathbf{M_h}$, $\mathbf{M_{45}}$, $\mathbf{M_v}$, $\mathbf{M_{135}}$, and $\mathbf{M_{ud}}$. In this way, the binary value "1" in $\mathbf{M_h}$ indicates that the discarded pixel in that position has a horizontal texture pattern; in this case, the same position in $\mathbf{M_v}$, $\mathbf{M_{45}}$, $\mathbf{M_{135}}$, and $\mathbf{M_{ud}}$ will have "0" binary value. For each direction, one weighting coefficient will be estimated by (4). Therefore, Eq. (3) could be rewritten in matrix format, while taking into account the five pattern categories, as follows:

$$\begin{aligned}\hat{\mathbf{V}}^l = {} & \eta_h^l \mathbf{M_h^l}.*\mathbf{V}_i^l + (1-\eta_h^l)\mathbf{M_h^l}.*\mathbf{V}_v^l \\ & +\eta_{45}^l \mathbf{M_{45}^l}.*\mathbf{V}_i^l + (1-\eta_{45}^l)\mathbf{M_{45}^l}.*\mathbf{V}_v^l \\ & +\eta_v^l \mathbf{M_v^l}.*\mathbf{V}_i^l + (1-\eta_v^l)\mathbf{M_v^l}.*\mathbf{V}_v^l \\ & +\eta_{135}^l \mathbf{M_{135}^l}.*\mathbf{V}_i^l + (1-\eta_{135}^l)\mathbf{M_{135}^l}.*\mathbf{V}_v^l \\ & +\eta_{ud}^l \mathbf{M_{ud}^l}.*\mathbf{V}_i^l + (1-\eta_{ud}^l)\mathbf{M_{ud}^l}.*\mathbf{V}_v^l\end{aligned} \tag{5}$$

where $\hat{\mathbf{V}}^l$ denotes the recovered frame. The operation .* represents the element-by-element multiplication of two matrixes. The optimal weighting coefficient for each direction can be obtained applying (4) on each direction.

Given that the encoder and decoder work on the same set of data to estimate the pattern direction, there is no need to transmit the five direction masks and only the directional weighting coefficients for the left view (i.e., η_h^l, η_{45}^l, η_v^l, η_{135}^l, and η_{ud}^l) and the right view, need to be estimated at the encoder side and transmitted to the decoder side. Obviously, the overhead bit cost of transmitting the weighting coefficients is negligible in comparison to the bit cost of texture and depth map. In the experimental results section, the term "DDFU" will be used to refer to this proposed full version scheme. In addition, DDFU can be simplified to only transmit the weighting coefficients of the first frame, which will be used for the fusion of all the remaining frames as well. This simplification is possible as the content of each frame does not change significantly, especially for the sequences with slow motion. Based on this observation, the simplified approach can further reduce the amount of transmitted side information with little quality degradation. In the experimental section, the term "DDFU (first frame η)" will be used to refer to this simplified scheme.

Although directional data fusion process happens at both encoder and decoder sides, it has different targets. At the encoder side, it involves the original FR frames to estimate the optimal weighting coefficients for all the directions. Then, as the outputs, these coefficients are sent to the decoder side and used in the fusion process to reconstruct the FR frames. A graphic representation of the proposed directional data fusion process at the decoder side is shown in Fig. 8. By receiving the weighting coefficients and using the same pattern direction estimation process, virtual view pixels and directional interpolated pixels can be fused to generate the recovered frame using Eq. (5).

4 Generalization of proposed down/upsampling method

The proposed down/upsampling with virtual view-assisted data fusion algorithm can also be applied to MVD video in addition to stereo video. Since in this kind of video more neighboring views and the corresponding depth maps are available, at a given viewpoint, more virtual view versions can be utilized. With the aid of these virtual views, the quality of the final recovered FR views can be considerably improved. As depicted in Fig. 9 in a N views multiview video, V_e^1 is the expanded view at viewpoint 1 and similarly for V_e^2. V_i^1 is the directional interpolated view at viewpoint 1, and V_v^{21}, V_v^{31}, and V_v^{N1} are the virtual views generated from the adjacent views 2, 3, N at viewpoint 1. In this case, the discarded pixels are recovered by fusing interpolated pixels with one of the available virtual views that gives the minimum differences when compared with original FR pixels. Subsequently, the fusion coefficients are transmitted. In this way, the proposed algorithm can also effectively recover the FR frames.

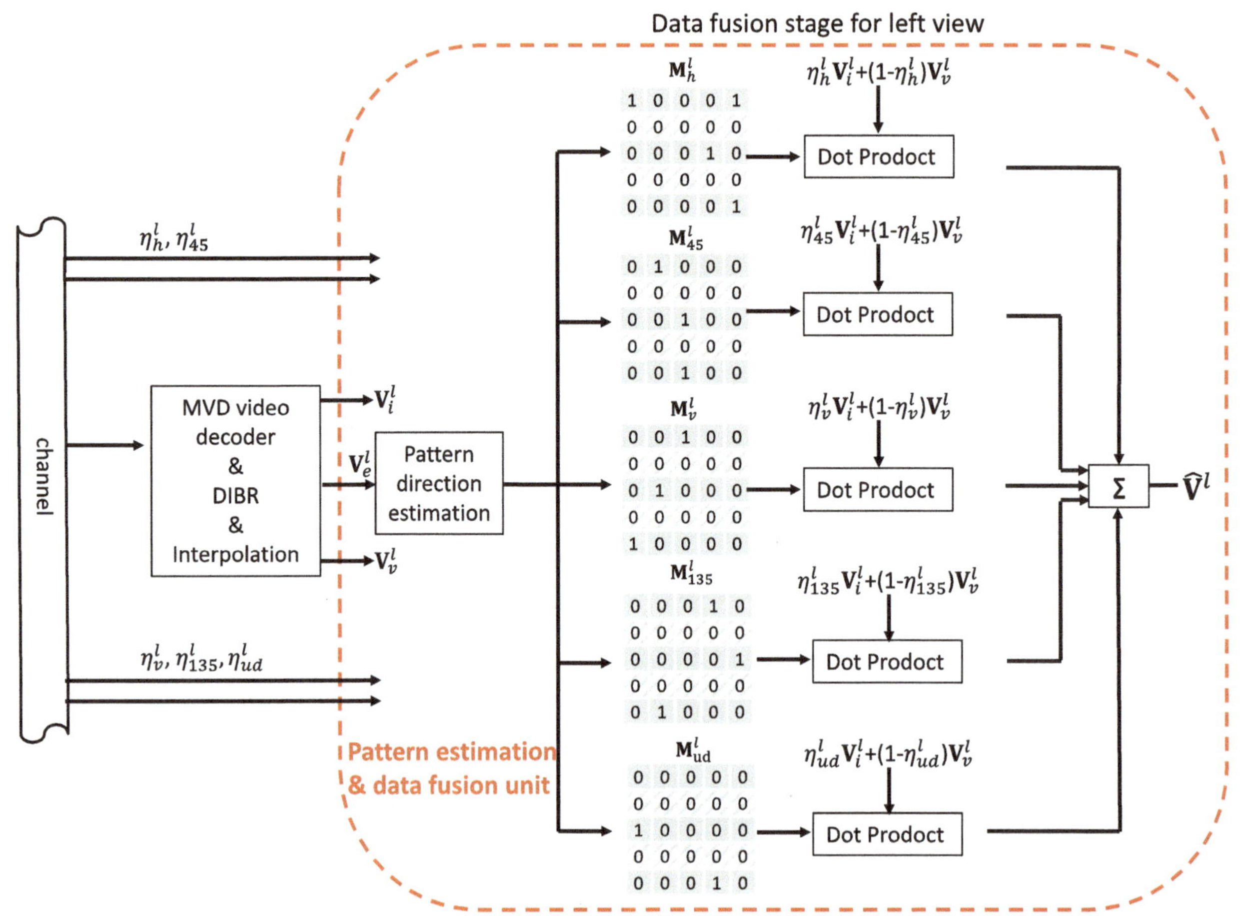

Fig. 8 Directional data fusion process. The process of data fusion by directional weighting coefficients and corresponding directional binary masks

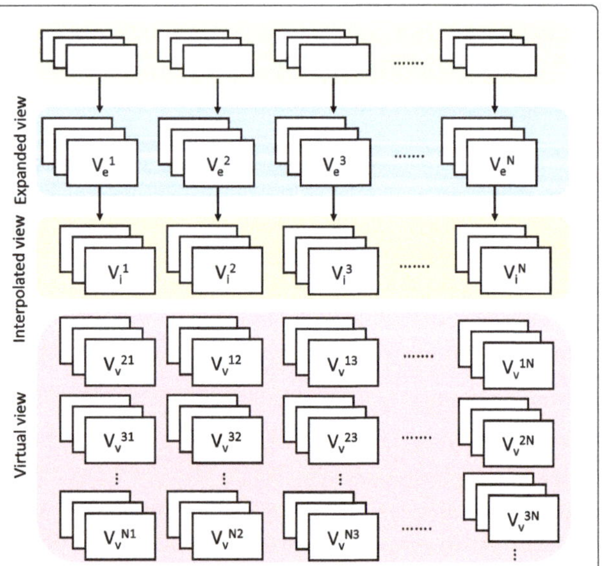

Fig. 9 The proposed algorithm for multiview video

5 Experimental results

To objectively evaluate the performance of the proposed method, several experiments were conducted on public 3D video datasets [25, 26] "Doorflower," "Kendo," "Dog," "Balloons," "Newspaper," and "Undo-Dancer". Some parameters and content characteristics of the testing sequences are listed in Table 1 for reference. For each sequence, both the left and right views had been interlacing and complementary row downsampled with a factor 2 before encoding. JMVC 5.0 [27] was used for compression, and six different QPs, namely 34, 37, 40, 43, 46, and 49, were used to code the texture and depth map sequences. The temporal GOP size and the total number of encoded frames was 8 and 80, respectively, while the delta QP and the differential QP between the base layer and sublayer in hierarchical-B picture structure was set to zero in all layers. The virtual views at the decoder side were rendered using a 1D DIBR technique from one reference view to another view without any postprocessing (i.e., no hole filling).

Table 1 The parameters and characteristics of each used sequence

	Size	Camera	Left	Right	Content's motion
Doorflower	1024 × 768	Fixed	View10	View08	Moderate
Kendo	1024 × 768	Moving	View03	View05	Complex
Dog	1280 × 960	Fixed	View38	View39	Medium
Balloons	1024 × 768	Moving	View03	View05	Complex
Newspaper	1024 × 768	Fixed	View04	View06	Simple
Undo-Dancer	1920 × 1088	Moving	View02	View05	Complex

5.1 Performance evaluation on stereoscopic video

The first set of simulations aims to evaluate the effectiveness of the proposed approach by comparing the rate distortion performance with FR video coding approach, filter-based approach, and the state-of-the-art approaches [19, 28]. Figure 10 shows the coding performance of these methods on all of the testing sequences. In the comparison, a 6-tap Lanczos filter has been used at encoder and decoder sides for down/upsampling, respectively, for the filter-based approach. From Fig. 10, the

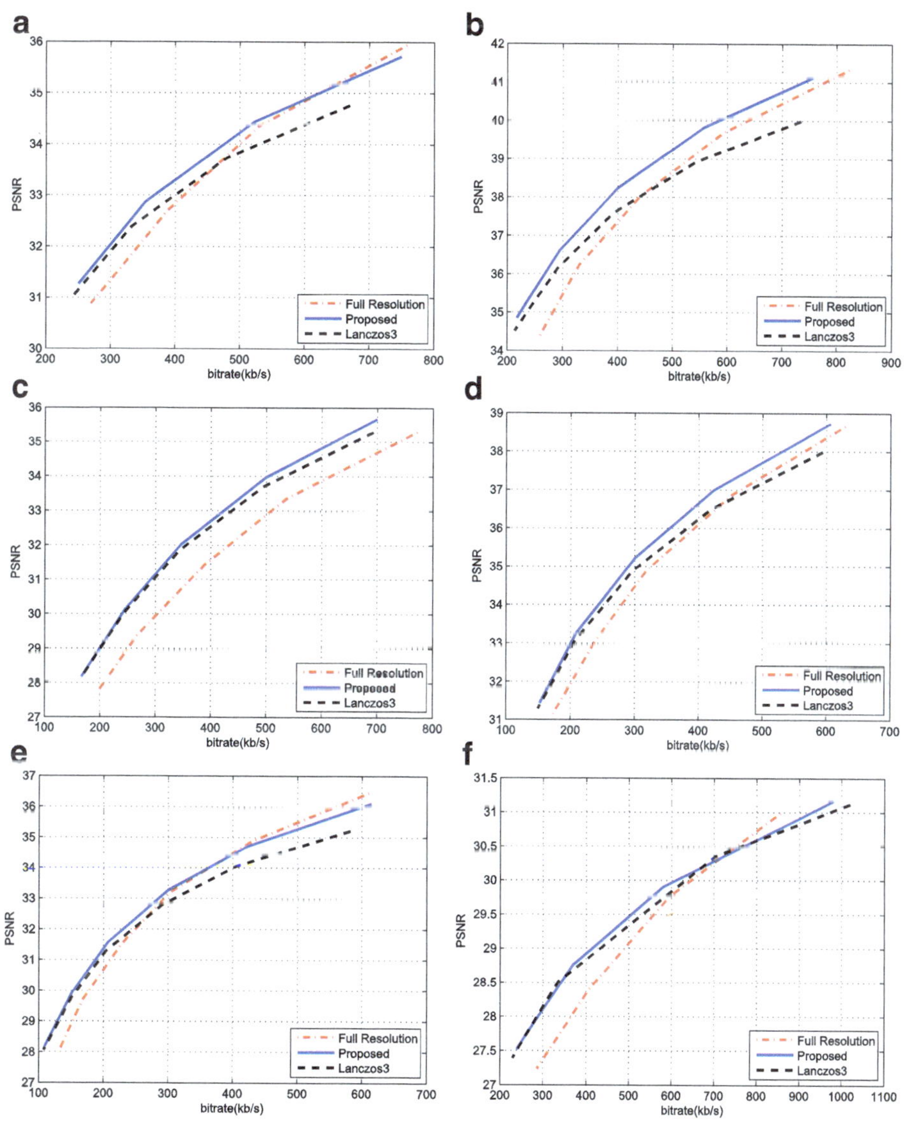

Fig. 10 The rate distortion curves. The rate distortion curves for the testing sequences. **a** Doorflower. **b** Kendo. **c** Dog. **d** Balloons. **e** Newspaper. **f** Dancer

effectiveness of the proposed approach over the matched-filter approach and FR coding at low bit rate can be appreciated for all the testing sequences. The proposed method has higher PSNR results than the matched-filter method, and the maximum PSNR gains can be 0.81 and 0.76 dB on sequences "Undo-Dancer" and "Doorflower", respectively. This is due to the proposed fusion mechanism that can well preserve the edges. In addition, the better quality of the depth maps is the more contributions the virtual view pixels can make. Generally, the matched-filter approach has good coding performance on the smooth areas; therefore, for the sequences (e.g. "Dog") containing more smooth areas, it is comparable to the proposed method. However, if more texture is contained in the sequences, the proposed method can be more advantageous over the matched filter method. In summary, the average PSNR gains across different bit rates for all the sequences range from 0.18 to 0.45 dB.

To further evaluate the effectiveness of the proposed method, two fair comparisons with the frame-compatible coding methods in [19, 28][3] are carried out by adopting the same testing sequences with the same resolution used in their work, the same coding standard, and the same coding parameters as [19, 28]. The comparison results of [19] are shown in Fig. 11, where LF1 represents the direct downsampling (i.e., even rows in both left and right views are discarded without low pass filtering), CAIS represents the proposed method in [19], and IF1 and IF2 are two interpolation filters with coefficients $\{1, -5, 20, 20, -5, 1\}/32$ and $\{-3, 28, 8, -1\}/32$, respectively, as proposed in [19]. The depth sequences used in the proposed DDFU are generated by the method presented in [29], and depth map bit rates have been included in the total bit rate. Indicated by these results, the gain of the proposed method is larger than that of [19]. Figure 12 shows the comparison results of the proposed approach with respect to [28]. In [28], two views are asymmetrically downsampled in frame-compatible coding and the left view has higher quality. Separately comparing the left view and right view, it is obvious that even the left view in the proposed work contains less information (downsampled by factor of 2) than the left view in [28] (downsampled by factor of 8/5); the left view recovery performance of the proposed work is comparable to [28]. Meanwhile, the recovery performance of the right view of the proposed approach is much higher than that in [28]. From the comparison with frame-compatible coding methods, the superiority of the proposed method is due to jointly take into account the features of down/upsampling, the inter-view redundancy, and virtual views.

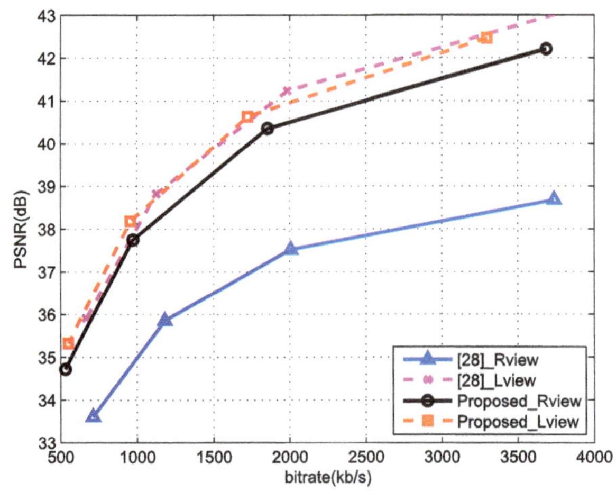

Fig. 12 Comparison with [28]. The rate distortion curves for the testing sequence balloons for the proposed approach and [28]

The visual results of zoomed-in parts of sequences "Doorflower" and "Undo-dancer" are shown in Fig. 13. It is possible to note that the edges recovered by DDFU are sharper than those recovered by the benchmark method. Although the proposed DDFU recovered frame also has some blurred areas, nevertheless, it still achieves a better visual quality than matched-filter-interpolated frame. Figure 13d shows a portion of the original left view of "Undo-Dancer", and its recovered versions using

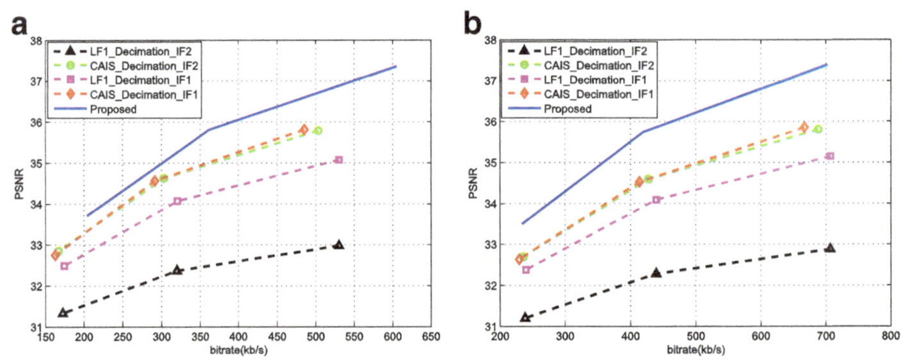

Fig. 11 Comparison with [19]. The rate distortion curves for the testing sequences. **a** Doorflower. **b** Laptop, for the proposed approach and [19]

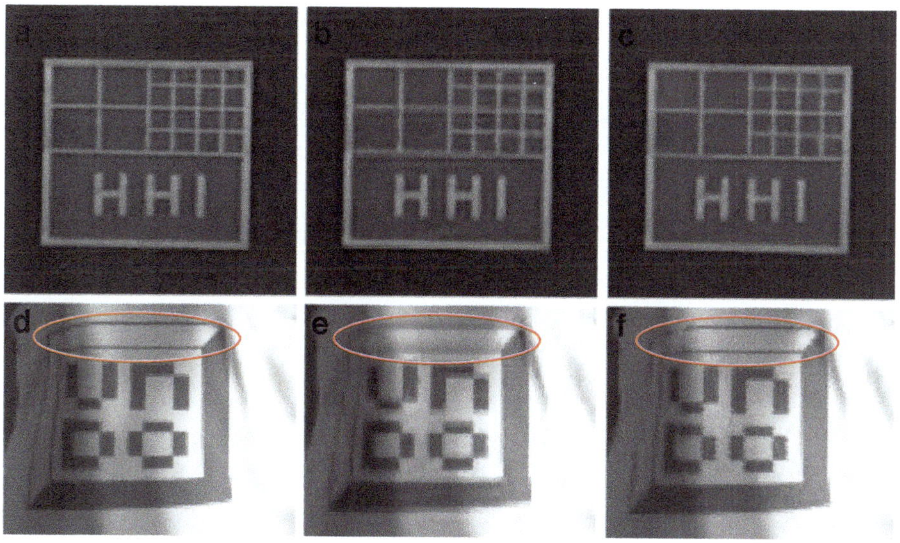

Fig. 13 The visual results. Comparison between proposed DDFU method and benchmark method. **a**, **b**, and **c** are the results of Original, Benchmark, and DDFU, respectively, on zoomed-in part of the sequence Doorflower; **d**, **e**, and **f** are the results of Original, Benchmark, and DDFU, respectively, on zoomed-in part of the sequence Undo-Dancer

the matched-filter-based approach and the proposed approach in Fig. 13e, f, respectively. Since the one-pixel-wide edge is difficult to recover properly by using only the surrounding pixels, the advantage of the DDFU method is more obvious in the highlighted areas by a red ellipse in Fig. 13e, f. From this comparison, it can be seen that the proposed approach can recover the one-pixel-wide edge without blurring.

5.2 Performance of each stage of the proposed method

In this subsection, several experiments have been conducted to validate the necessity and effectiveness of

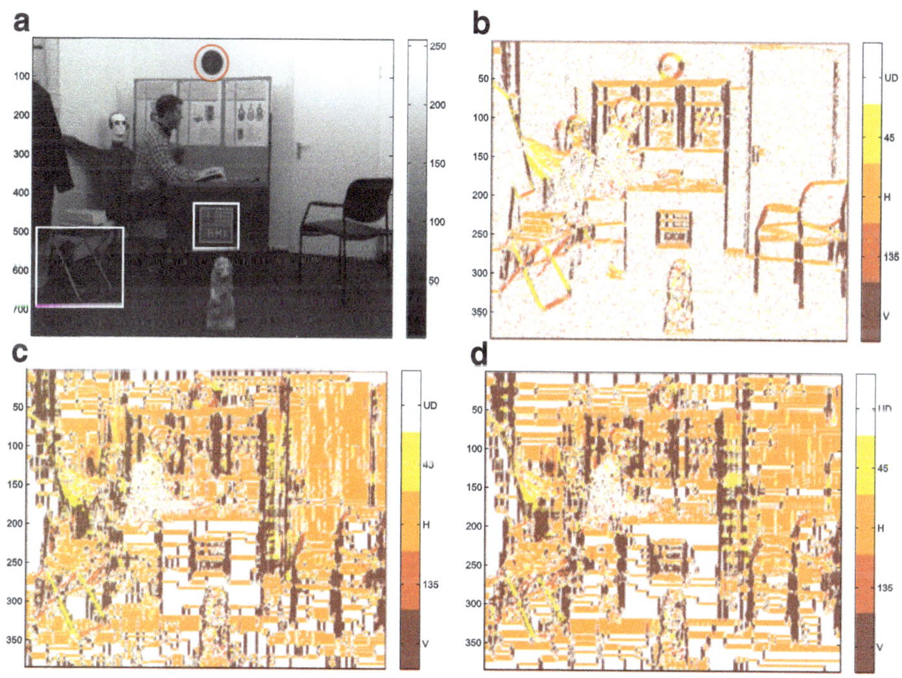

Fig. 14 The pattern direction estimation results. **a** Original texture; the pattern direction estimation results on (**b**) original uncompressed texture, **c** compressed texture with QP = 34, and (**d**) compressed texture with QP = 40; the colors *dark red*, *red*, *orange*, *yellow*, and *white* represent vertical, 135° diagonal, horizontal, 45° diagonal, and undefined direction pixels, respectively. (For clearness, the directional estimation results on the discarded pixels are scaled up to the same size as the original texture; their real size is shown on the *y*-axis of each figure)

Table 2 The two-views-all-frames average values of η_h, η_v, η_{45}, η_{135}, and η_{ud} for each sequence and for different QPs

				Doorflower			
	QP		34	37	40	43	46	49
Average η		η_h	0.23	0.24	0.30	0.37	0.47	0.55
		η_v	0.97	0.95	0.91	0.89	0.86	0.82
		η_{45}	0.77	0.76	0.78	0.77	0.75	0.75
		η_{135}	0.80	0.80	0.79	0.80	0.82	0.79
		η_{ud}	0.58	0.61	0.62	0.65	0.64	0.63
				Undo-Dancer			
	QP		34	37	40	43	46	49
Average η		η_h	0.11	0.16	0.23	0.36	0.50	0.58
		η_v	0.97	0.98	0.98	0.96	0.95	0.93
		η_{45}	0.63	0.72	0.79	0.85	0.83	0.80
		η_{135}	0.62	0.72	0.79	0.82	0.81	0.78
		η_{ud}	0.19	0.28	0.40	0.53	0.62	0.66
				Kendo			
	QP		34	37	40	43	46	49
Average η		η_h	0.83	0.81	0.77	0.73	0.68	0.64
		η_v	0.95	0.89	0.86	0.81	0.78	0.74
		η_{45}	0.94	0.91	0.89	0.84	0.77	0.71
		η_{135}	0.93	0.90	0.87	0.83	0.80	0.75
		η_{ud}	0.86	0.83	0.79	0.75	0.69	0.66
				Balloons			
	QP		34	37	40	43	46	49
Average η		η_h	0.86	0.85	0.82	0.77	0.72	0.65
		η_v	0.95	0.95	0.95	0.94	0.91	0.87
		η_{45}	0.95	0.95	0.93	0.90	0.87	0.82
		η_{135}	0.95	0.95	0.94	0.91	0.89	0.87
		η_{ud}	0.93	0.93	0.91	0.88	0.84	0.78
				Dog			
	QP		34	37	40	43	46	49
Average η		η_h	0.95	0.95	0.92	0.89	0.85	0.77
		η_v	1.00	0.98	0.95	0.94	0.89	0.80
		η_{45}	0.99	0.96	0.94	0.90	0.84	0.78
		η_{135}	0.99	0.97	0.95	0.93	0.86	0.78
		η_{ud}	0.99	0.97	0.95	0.91	0.90	0.82
				Newspaper			
	QP		34	37	40	43	46	49
Average η		η_h	0.89	0.89	0.86	0.85	0.81	0.77
		η_v	1.00	1.00	1.00	1.00	0.99	0.96
		η_{45}	1.00	1.00	1.00	0.98	0.94	0.93
		η_{135}	1.00	1.00	1.00	0.98	0.96	0.94
		η_{ud}	0.98	0.99	0.98	0.93	0.92	0.89

Table 3 The upsampling performance comparison on multiview video

			Kendo				
	QP		34	37	40	43	46	49
	Bit rate (kb/s)		1073	808	608	476	361	283
PSNR		Lanc	37.12	35.88	34.38	32.79	30.94	29.10
		Pro	38.31	36.71	34.95	33.19	31.26	29.39
	ΔPSNR		1.19	0.83	0.57	0.40	0.32	0.29
			Balloons				
	QP		34	37	40	43	46	49
	Bit rate(kb/s)		1150	824	577	426	309	245
PSNR		Lanc	36.54	34.97	33.12	31.37	29.54	27.86
		Pro	37.07	35.32	33.36	31.54	29.67	28.00
	ΔPSNR		0.53	0.36	0.23	0.17	0.14	0.14
			Newspaper				
	QP		34	37	40	43	46	49
	Bit rate(kb/s)		1134	809	572	426	317	253
PSNR		Lanc	34.12	32.90	31.44	29.94	28.23	26.55
		Pro	34.85	33.41	31.77	30.16	28.39	26.67
	Δ PSNR		0.73	0.51	0.33	0.22	0.15	0.12

direction estimation and data fusion step in the proposed algorithm. Since the category of the to-be-filled pixels is determined by the estimated texture pattern, accurate pattern direction estimation plays an important role in the fusion process. Therefore, to verify its effectiveness, Fig. 14b shows the pattern estimation result on the uncompressed frame, whereas, Fig. 14c, d shows the estimation results on the compressed frame with $QP = 34$ and $QP = 40$, respectively. For reference, Fig. 14a shows the original uncompressed texture frame from the "Doorflower" sequence with three highlighted parts containing clear patterns. Different colors are used to distinguish the

Table 4 The two-view all-frame average values of η_h, η_v, η_{45}, η_{135}, and η_{ud} for three multiview sequences and different QPs

			Kendo				
	QP	34	37	40	43	46	49
	η_h	0.91	0.89	0.85	0.79	0.71	0.62
	η_v	0.99	0.96	0.91	0.87	0.82	0.75
Average η	η_{45}	0.98	0.94	0.91	0.86	0.81	0.71
	η_{135}	0.96	0.91	0.89	0.84	0.77	0.72
	η_{ud}	0.95	0.91	0.85	0.77	0.70	0.62
			Balloons				
	QP	34	37	40	43	46	49
	η_h	0.94	0.91	0.86	0.82	0.74	0.63
	η_v	1.00	0.99	0.95	0.91	0.87	0.80
Average η	η_{45}	0.97	0.96	0.94	0.89	0.84	0.78
	η_{135}	1.00	0.98	0.93	0.90	0.84	0.78
	η_{ud}	0.99	0.95	0.92	0.88	0.82	0.74
			Newspaper				
	QP	34	37	40	43	46	49
	η_h	0.87	0.86	0.84	0.83	0.78	0.73
	η_v	1.00	1.00	1.00	1.00	0.99	0.95
Average η	η_{45}	1.00	1.00	1.00	0.99	0.96	0.92
	η_{135}	1.00	1.00	1.00	0.99	0.95	0.92
	η_{ud}	0.99	0.99	0.98	0.96	0.93	0.87

five directions, so the colors dark red, red, orange, yellow, and white are used to represent vertical, 135° diagonal, horizontal, and 45° diagonal edges and the undefined pattern areas, respectively. In this paper, areas are regarded as undefined pattern areas when $S_{11}/S_{22} \leq Th$ where $Th = 4$. The accuracy of the adopted pattern detection algorithm could be appreciated from Fig. 14b, c. By comparing these two figures, the direction estimation results of the three highlighted parts are almost the same. It demonstrates that the accuracy of the pattern estimation is barely affected by the compression distortion.

To show the level of contribution of the virtual views in the fusion stage and how the texture pattern direction influences the fusing process, the average fusion coefficients η_h, η_v, η_{45}, η_{135}, and η_{ud} of both views are reported in Table 2 for the six testing sequences and different QPs. The smaller the value of η is, the more important the virtual view pixels are for the recovery of discarded pixels. In the fusion stage, the contribution of virtual view depends on several factors, such as the adopted DIBR technique and depth map quality. It is worth noticing that even with advanced rendering techniques, the generated virtual view may still face a problem in generating high-quality and aligned texture around depth discontinuous areas, where the adopted directional interpolation can compensate well. From this table, it can be seen that virtual view þixels are more important to recover the pixels with horizontal pattern than other directions. On the other hand, the directional interpolated frame is more important to recover the pixels with vertical pattern. For example, the average η_h for the "Undo-Dancer" sequence at $QP = 34$ is 0.11 versus $\eta_v = 0.97$. The η_{45}, η_{135}, and η_{ud} values for the two diagonal patterns and undefined pattern lay in between the horizontal and vertical cases, which are $0.63, 0.62, 0.19$ for $QP = 34$, respectively. Moreover, it should be noted that for the sequence "Undo-Dancer" which is a computer graphic sequence and consequently has an accurate depth map, the virtual view pixels provide a greater contribution to the final recovered FR frames, in all five directions, with respect to other sequences. As expected, this contribution is remarkably higher for the horizontal pattern.

5.3 Performance evaluation on multiview video

When testing on multiview video, the View1, View3, and View5 of sequences "Kendo" and "Balloons" and the View2, View4, and View6 of sequence "Newspaper" are adopted. For multiview testing, the same codec setting up is used as the two-view testing and after proposed downsampling method, each LR view is in half size of its corresponding FR view. The performance comparison at the decoder side is shown in Table 3. From the

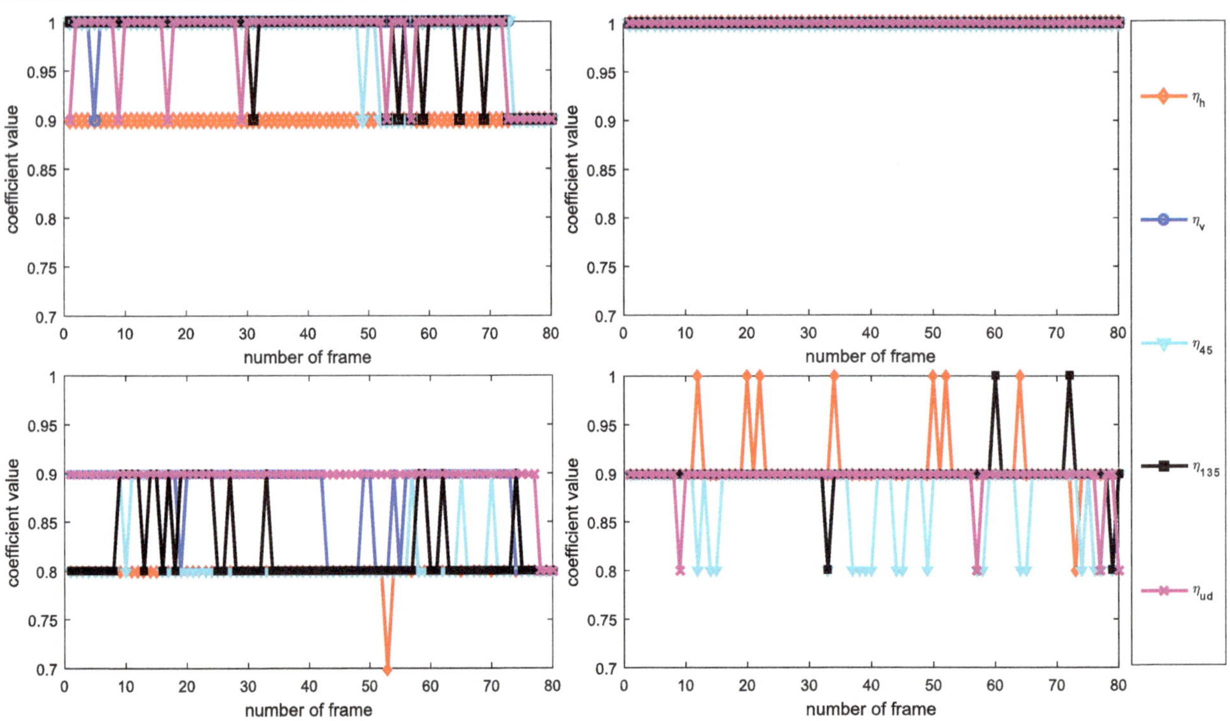

Fig. 15 The variance of five η coefficients. The pane shows the five coefficients for the sequence "Dog". The *top left* and *right* figures of the pane are the weighting coefficients of the left and right views, respectively, when QP = 34; the *bottom left* and *right* figures of the pane are the weighting coefficients of the left and right views, respectively, when QP = s46

reported results, it is possible to note that the proposed down/upsampling method can also work properly in multiview video system. In this case, the highest PSNR gain can be up to 1.19dB for the "Kendo" sequence. Compared with the two-view video case, the PSNR gains of multiview video become higher and the average gain for all the sequences and all QPs is 0.4 db. These gains are obtained due to the availability of multiple virtual view candidates. This ensures that the more suitable virtual view pixels are merged with the interpolated view. Moreover, compared with the average η value for two-view testing, in multiview testing, the virtual view has more impacts at low bit rate (Table 4).

5.4 Performance of simplified method

In the basic implementation of the DDFU algorithm, the encoder needs to transmit the five weighting coefficients, η, for each frame and each view. Clearly, it needs to evaluate them by minimizing (4). However, given that in most cases, there are no major changes in the scene content, it is reasonable to assume that those coefficients do not change very much from frame to frame; hence it is not necessary to evaluate them for each frame. This assumption could be verified by Fig. 15 which shows the trend of the weighting coefficients versus frame number. Thus, one way to reduce the overhead transmission of the proposed approach is using directional

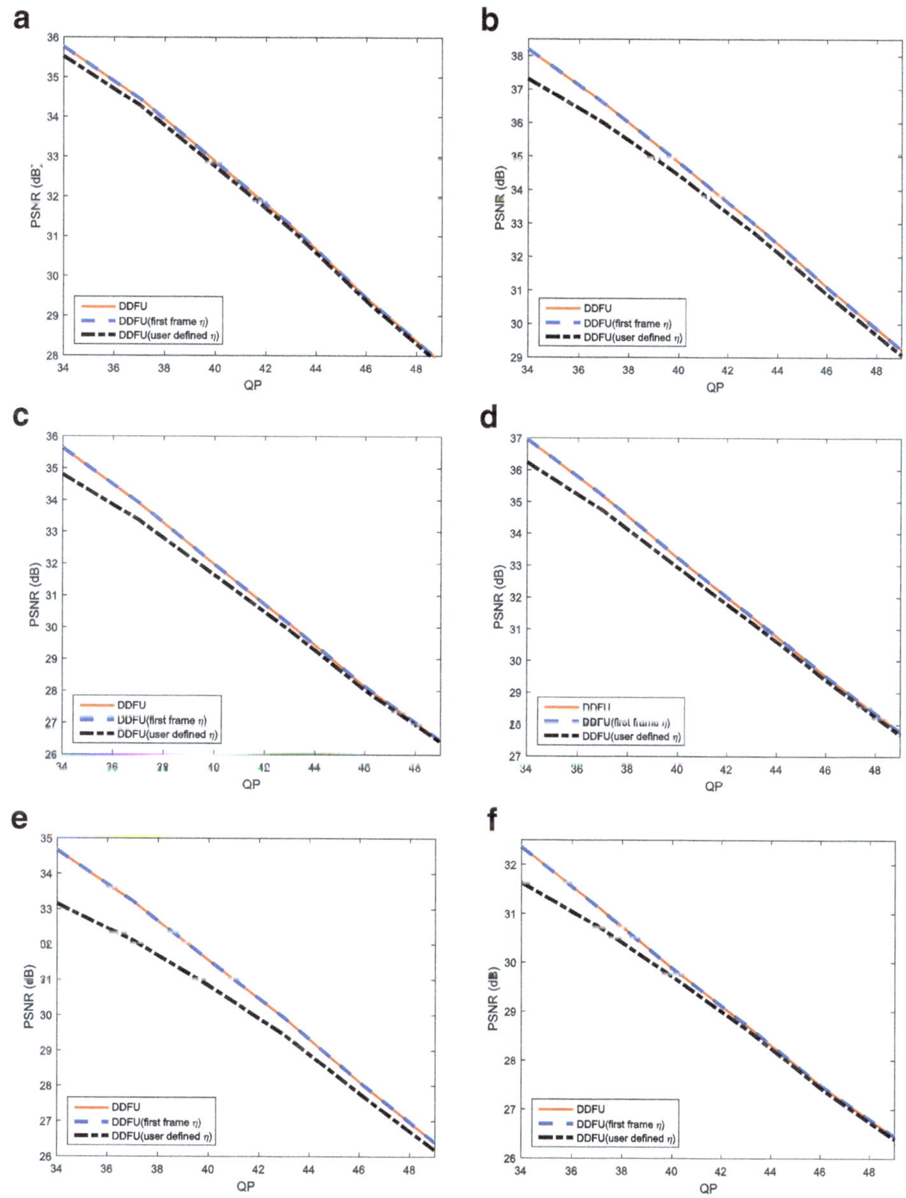

Fig. 16 The comparisons of proposed simplified approach. The comparisons of proposed simplified approach with full version approach and user-defined coefficient approach on the testing sequences (**a**) Doorflower, **b** Kendo, **c** Dog, **d** Balloons, **e** Newspaper, and **f** Dancer

weighting coefficients of the first frame for the whole sequence (called DDFU (first frame η)). In this approach, the weighting coefficients are only estimated for the first frame and then used for the whole sequence. To verify the effectiveness of this simplified approach, its performance has been compared with the proposed DDFU approach and another simplified approach called in the following DDFU (user defined η), the latter adopts user defined coefficients at the decoder side for the whole sequence. The pre-set values for the DDFU (user defined η) used are $\eta_v = 1$, $\eta_h = 0$, $\eta_{45} = 0.5$, $\eta_{135} = 0.5$, and $\eta_{ud} = 1$ for the left and right views which means that all vertical edges and undefined pattern areas are recovered by the directional interpolation algorithm. All recovered horizontal edges are obtained from the virtual view pixels, and the two diagonal direction pixels are obtained by equally fusing the directional interpolated pixels with the virtual view pixels. The results of this comparison are listed in Fig. 16.

From Fig. 16, it can be seen that DDFU and DDFU (first frame η) have almost the same performance for all sequences, which demonstrates the validity and effectiveness of the simplified approach. By comparing the results of DDFU and DDFU (user defined η), the importance of adapting the coefficients to the scene content can be appreciated. The results in Fig. 16 show that the performance of DDFU (first frame η) are better than that of DDFU (user defined η). This is due to the η values for the DDFU (first frame η) are based on the content of the testing sequence, if the content of the sequence does not vary hugely frame by frame, neither does the value of η, while the values of the predetermined η are user defined values, which means they do not take the content of the sequence into account. The performance of DDFU (user defined η) highly depends on how close the predetermined values are to the frame-by-frame evaluated coefficients.

6 Conclusions

In this paper, an interlacing-and-complementary-row-downsampling method is employed on the two adjacent views of a multiview video at the encoder side to reduce the transmitted data. This downsampling method allows the proposed directional data fusion upsampling (DDFU) algorithm to recover the discarded pixels by exploiting the information of the downsampled views and the corresponding virtual views. In the proposed upsampling approach, edge directions around the discarded pixels are estimated by principal components analysis. This information is subsequently used to steer the fusion of the virtual view with the directional interpolated pixels. The aim behind this is to exploit the inter-view redundancy to minimize the overall system distortion, which is a combination of the compression distortion and the distortion introduced by the downsampling process. Therefore, different from filter-based interpolation algorithms, the advantages of virtual views have been exploited by the proposed method. Moreover, it has been shown that the proposed algorithm achieves superior performance in comparison with filter-based interpolation algorithms and the state-of-the-art algorithms. The future work will be devoted to exploiting the temporal correlation in video sequences to control the fusion process.

Endnotes

[1] The pixels on the boarder of the frame will be filled by filter-based interpolation without estimating their pattern directions.

[2] Although using p_2 and p_8 may seem beneficial, the lack of p_4 and p_6 will negatively affect the direction estimation due to the non-symmetric set of pixels. Nevertheless, p_2 and p_8 will be used in pattern estimation of the following discarded pixel, p_6.

[3] All the results of [19, 28] have been obtained from the author and the paper, respectively.

Competing interests

The authors declare that they have no competing interests.

Authors' contributions

ZJ proposed the framework of this work, carried out the whole experiments, and drafted the manuscript. TT supervised the whole work, participated in its design, offered useful suggestions, and helped to modify the manuscript . JX and YZ participated in the discussion of this work and helped to polish the manuscript. All authors read and approved the final manuscript.

Acknowledgements

This work was supported by the National Natural Science Foundation of China (NO. 61210006 and NO. 61501379), Jiangsu Science and Technology Programme (BK20150375), the 973 program (No.2012CB316400), and the Program for Changjiang Scholars and Innovative Research Team in University (No. IRT201206).

Author details

[1] Department of Electrical and Electronic Engineering, Xi'an Jiaotong-Liverpool University, Ren Ai Road 111, 215123 Suzhou, People's Republic of China. [2] Department of Electrical Engineering and Electronics, University of Liverpool, Merseyside L69 3BX, Liverpool, UK. [3] Institute of Information Science, Beijing Jiaotong University, No.3 Shangyuancun Haidian District, Beijing, People's Republic of China.

References

1. A Vetro, T Wiegand, G Sullivan, Overview of the stereo and multiview video coding extensions of the H.264/MPEG-4 AVC standard. Proc. IEEE. **99**, 626–642 (2011)
2. M Flierl, A Mavlankar, B Girod, Motion and disparity compensated coding for multiview video. IEEE Trans. Circ. Syst. Video Technol. **17**, 1474–1484 (2007)
3. C Yan, Y Zhang, J Xu, F Dai, J Zhang, Q Dai, F Wu, Efficient parallel framework for HEVC motion estimation on many-core processors. IEEE Trans. Circ. Syst. Video Technol. **24**, 2077–2089 (2014)
4. C Yan, Y Zhang, J Xu, F Dai, L Li, Q Dai, F Wu, A highly parallel framework for HEVC coding unit partitioning tree decision on many-core processors. IEEE Signal Process. Lett. **21**, 573–576 (2014)
5. C Yan, Y Zhang, F Dai, X Wang, L Li, Q Dai, Parallel deblocking filter for HEVC on many-core processor.Electron. Lett. **50**, 367–368 (2014)

6. C Yan, Y Zhang, F Dai, J Zhang, L Li, Q Dai, Efficient parallel HEVC intra-prediction on many-core processor. Electron. Lett. **50**, 805–806 (2014)
7. P Merkle, A Smolic, K Muller, T Wiegand, Efficient prediction structures for multiview video coding. IEEE Trans. Circ. Syst. Video Technol. **17**, 1461–1473 (2007)
8. C Fehn, in *Conference Record of the Thirty-Seventh Asilomar Conference on Signals, Systems and Computers*. A 3D-TV system based on video plus depth information, vol. 2 (IEEE, 2003), pp. 1529–1533
9. C Fehn, in *Electronic Imaging, International Society for Optics and Photonics*. Depth-image-based-rendering (DIBR), compression, and transmission for a new approach on 3d-tv (SPIE, USA, 2004), pp. 93–104
10. J Wu, Y Xing, G Shi, L Jiao, in *16th IEEE International Conference on Image Processing (ICIP)*. Image compression with downsampling and overlapped transform at low bit rates (IEEE, Cairo, Egypt, 2009), pp. 29–32
11. V Nguyen, Y Tan, W Lin, in *IEEE International Symposium on Circuits and Systems (ISCAS)*. Adaptive downsampling/upsampling for better video compression at low bit rate (IEEE, Seattle, USA, 2008), pp. 1624–1627
12. X Wu, X Zhang, X Wang, Low bit-rate image compression via adaptive down-sampling and constrained least squares upconversion. IEEE Trans. Image Process. **18**, 552–561 (2009)
13. C Fehn, *A 3D-TV approach using depth-image-based rendering (DIBR), VIIP 03*. (ACTA, Calgary, Canada, 2003)
14. D Garcia, T Fonseca, R Queiroz, in *18th IEEE International Conference on Image Processing (ICIP)*. Video compression complexity reduction with adaptive down-sampling (IEEE, Brussels, Belguim, 2011), pp. 745–748
15. H Sawhney, Y Guo, K Hanna, R Kumar, in *SIGGRAPH*. Hybrid stereo camera: an IBR approach for synthesis of very high resolution stereoscopic image sequences (ACM, NY, USA, 2001), pp. 451–460
16. P Aflaki, M Hannuksela, J Hakkinen, P Lindroos, M Gabbouj, in *17th IEEE International Conference on Image Processing (ICIP)*. Subjective study on compressed asymmetric stereoscopic video (IEEE, Hong Kong, 2010), pp. 4021–4024
17. D Garcia, C Dorea, R Queiroz, Super resolution for multiview images using depth information. IEEE Trans. Circ. Syst. Video Technol. **22**, 1249–1256 (2012)
18. Z Jin, T Tillo, C Yao, J Xiao, Y Zhao, Virtual view assisted video super-resolution and enhancement. IEEE Trans. Circ. Syst. Video Technol. **26**(3), 467–478 (2016)
19. Y Zhang, X Ji, H Wang, Q Dai, Stereo interleaving video coding with content adaptive image subsampling. IEEE Trans.Circ. Syst. Video Technol. **23**, 1097–1108 (2013)
20. X Ji, Y Zhang, Q Dai, Opposite parity packing arrangement for stereoscopic video coding.Electron. Lett. **47**, 1180–1181 (2011)
21. A Vetro, in *17th IEEE International Conference on Image Processing (ICIP)*. Frame compatible formats for 3D video distribution (IEEE, Hong Kong, 2010), pp. 2405–2408
22. I Jolliffe, *Principal Component Analysis*. (Springer, New York, 2002)
23. B Yang, Z Gao, X Zhang, in *IEEE International Conference on Multimedia and Expo (ICME)*. Principal components analysis-based edge-directed image interpolation (IEEE, Melbourne, Australia, 2012), pp. 580–585
24. G Stewart, On the early history of the singular value decomposition. SIAM Rev. **35**, 551–566 (1993)
25. Fraunhofer Heinrich Hertz Institute, MPEG 3D video Sequences. http://ftp.hhi.de/HHIMPEG3DV/. Accessed 17 June 2012
26. Nagoya University, FTV Test Sequences (2008). http://www.tanimoto.nuee.nagoya-u.ac.jp/mpeg/mpeg_ftv.html. Accessed 17 June 2012
27. Y Chen, P Pandit, S Yea, C Lim, in *Joint Video Team (JVT) of ISO/IEC/MPEG and ITU-T/VCEG, Doc. JVT-AE207*. Draft reference software for MVC (Joint Video Team, London, UK)
28. J Chiang, S Chen, in *2014 IEEE Global Conference on Signal and Information Processing (GlobalSIP)*. Frame-compatible asymmetric stereo video coding considering human perception (IEEE, Atlanta, USA, 2014), pp. 956–959
29. Z Wang, Z Zheng, in *IEEE Conference on Computer Vision and Pattern Recognition (CVPR)*. A region based stereo matching algorithm using cooperative optimization (IEEE, Anchorage, USA), pp. 1–8

14

Making texture descriptors invariant to blur

Michael Gadermayr[1*] and Andreas Uhl[2]

Abstract

Besides a high distinctiveness, robustness (or invariance) to image degradations is very desirable for texture feature extraction methods in real-world applications. In this paper, focus is on making arbitrary texture descriptors invariant to blur which is often prevalent in real image data. From previous work, we know that most state-of-the-art texture feature extraction methods are unable to cope even with minor blur degradations if the classifier's training stage is based on idealistic data. However, if the training set suffers similarly from the degradations, the obtained accuracies are significantly higher. Exploiting that knowledge, in this approach the level of blur of each image is increased to a certain threshold, based on the estimation of a blur measure. Experiments with synthetically degraded data show that the method is able to generate a high degree of blur invariance without loosing too much distinctiveness. Finally, we show that our method is not limited to ideal Gaussian blur.

Keywords: Texture recognition, Feature extraction, Robustness, Invariance

1 Introduction

For some decades, texture classification [1–16] has been a fundamental challenge in image processing. On the one hand, texture descriptors have to capture all intrinsic image properties. These are properties that contain distinctive information (for discrimination) and do not depend on the image acquisition conditions. On the other hand, extrinsic properties (i.e., properties that vary with different acquisition conditions) should not be captured, in order to maintain invariance to specific properties.

In the following, we focus on blur which is usually caused by defocus, motion, or chromatic aberrations. Although in case of good image acquisition conditions blur can mostly be prevented effectively, in many real-world scenarios, this degradation still features a present problem. One specific medical application prone to non-idealistic images is endoscopy. Firstly, it is quite difficult to adjust the distance between the lens and the surface, which is a source for defocus aberrations. Furthermore, the permanent activity of the bowel in combination with a difficult handling of the endoscope is a source for motion blur. There is significant literature on texture classification from endoscopic images such as celiac disease diagnosis [17, 18], small bowel tumor detection [19], and colon cancer detection [20].

Currently, there is only limited literature on blur-invariant texture feature extraction. Most common approaches exploit either the blur-invariant Fourier phase information of the image [4, 5] or are based on the image moment method [3]. Furthermore, previous work [21] showed that highly distinctive state-of-the-art texture features [6–9] are in general extremely sensitive to blur.

We have learned from previous work [21] that a systematic degradation, prevalent in the evaluation set, in general affects the classification accuracy by far less if the images in the training set similarly suffer from the degradation. In the referenced paper, this knowledge has been exploited by dividing the image data sets into smaller sets that are similar with respect to the level of a specific image degradation. This method is referred to as degradation adaptive texture classification. The main restriction of degradation adaptive classification is that the distribution of the degradations must be similar in the training and the evaluation set. For example, if the training set contains only idealistic images and the evaluation set contains majorly strongly degraded images, the framework does not work as there

*Correspondence: Michael.Gadermayr@lfb.rwth-aachen.de
[1] Institute of Imaging and Computer Vision, RWTH Aachen University, Kopernikusstr. 16, 52074 Aachen, Germany
Full list of author information is available at the end of the article

are no training images which are similar to the strongly degraded evaluation set images.

1.1 Contribution and related work

In this work, we propose a general methodology to make image descriptors invariant to blur. Compared to degradation adaptive classification [21], such invariant texture features can be utilized more generally, as no assumptions on the distribution of the degradations in the training and the evaluation set must be made. Instead of focusing on a specific descriptor, our approach can be understood as a pre-processing technique and thereby can be applied to arbitrary texture feature extraction methods. This makes the approach highly generic as for a certain problem definition; the most appropriate state-of-the-art machine learning stage can be applied. A completely different concept is identified in the case of blur-invariant methods from literature [3–5]. These approaches apply specific concepts during feature extraction to ignore information which changes between differing blur levels.

From previous work [21], we know that blurring does not delete much distinctive information but leads to a different image representation (in the case of non-invariant feature extraction). The method proposed in this paper exploits this knowledge by equalizing the level of blur, prior to the actual feature extraction. This is done by means of an iterative algorithm based on specific blur metrics from literature.

In a comprehensive experimental evaluation, we focus on classification tasks with idealistic (blur-free) training data and distorted (blurred) evaluation data. This specific scenario is investigated because previous work [21] showed that non-invariant methods in that case are generally unable to deliver acceptable outcome. In contrast, considering classification tasks with similar degradations in the training and the evaluation set, non-invariant image descriptors have proven to be mostly quite effective [21]. Furthermore, the accuracies can be increased even more using adaptive classification. It has been shown [22] that the classification model can be adjusted effectively to partly corrupted data (during training) without noticing a strong impact on the final overall classification rates. Focusing on a scenario with idealistic training data and distorted evaluation data, these effects during classification can be eliminated to emphasize on the feature extraction stage only. Similar strategies are conducted for evaluation of scale-invariant texture descriptors [23, 24]. In real-world applications, the investigated scenario is applicable for instance if having idealistic (e.g., manually selected) training data in combination with partly degraded evaluation set data. This can be the case in specific medical applications as already mentioned above [17–20].

1.2 Outline

This paper is structured as follows. In Section 2, the methodology of making texture image descriptors blur-invariant is introduced. In Section 3, the classification improvements are presented and discussed. Section 4 finally concludes this paper.

2 Blur-invariance framework

In the following, we assume that a blurred image can be modeled by convolution of an ideal image with a Gaussian kernel. In this case, blur is referred to as Gaussian blur. The major idea of this work is to adaptively add Gaussian blur to an image to reach a specific blur level which is previously specified and is the same for all images (in the training and the evaluation set). This is done recursively by the blur-equalizing function E, as shown in Eq. (1):

$$E(I) = \begin{cases} I & , \text{if } B(I) \geq \Theta \,, \\ E(I * G) & , \text{if } B(I) < \Theta \end{cases} . \tag{1}$$

The original image I is recursively convolved with a Gaussian kernel G until the blur measure B based on the image reaches the threshold Θ. The procedure is outlined graphically in Fig. 1.

Having perfect Gaussian blur in the images and disregarding that the desired blur strength Θ is not achieved exactly, the method leads to theoretically perfect invariance. For that, we exploit that a convolution of two Gaussian kernels is another Gaussian kernel and that convolution is associative. In the discrete case, besides that Θ is not achieved exactly, an error is furthermore obtained due to the discrete image sampling.

Figure 2 shows how the blur level of an example texture image is increased during blur equalization.

2.1 Choosing the threshold

Although the main part of the algorithm is highly simple, there remain some open issues: One question is, how to choose the threshold Θ. From theoretical point of view, this threshold adjusts the degree of invariance. If Θ is chosen to be smaller than the blur measure of some images in a database, a perfect equalization cannot be realized. Contrarily, if this value is chosen too large, the removal of high frequency information might affect the distinctiveness of the final feature vectors. To put it into a nutshell, Θ is the regulating parameter between high distinctiveness and a

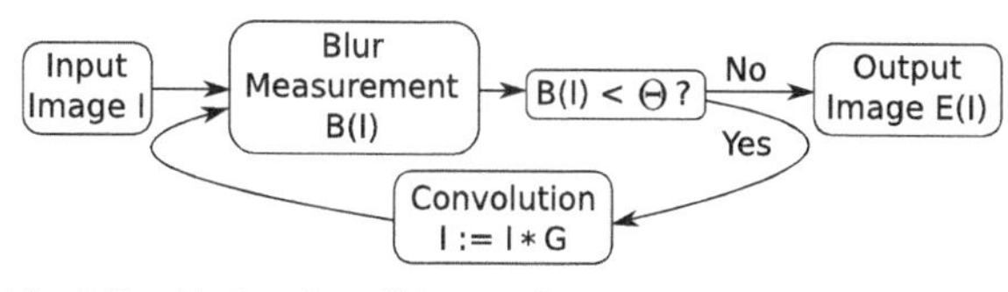

Fig. 1 Graphical outline of blur equalization

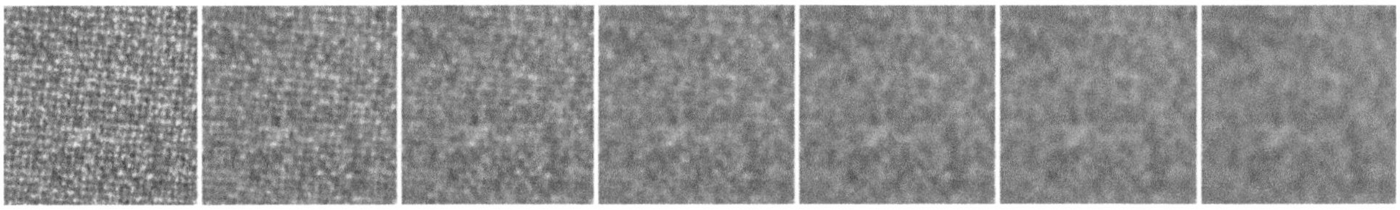

Fig. 2 From left to right, the increased blur level during blur equalization is shown using an example texture image [27]. Between two neighboring images, exactly one convolution cycle (Fig. 1) is applied

high level of invariance. In our experiments, we choose different Θ to investigate this correlation.

2.2 Blur measurement

Another question is how to estimate the blur level (measured by a function B) in an image. For this purpose, we rely on previous work on non-reference blur measurement. We investigate two common methods [25, 26] as well as one method which does not directly estimate blur, but a related property.

- In the case of the first blur measure (B_{Ma}) which has been introduced by Marziliano et al. [25], an edge detector is applied to find the vertical edges. Then, the local extrema are detected which correspond to the start and end points of edges. Finally, the blur metric is achieved by computing the ratio between the average edge length (i.e. the distance between start and end pixel) and the average edge magnitude (i.e. the gray value difference). A high average edge length (and/or a low edge magnitude) indicates that the edges are blurred. In opposite, short edges (and/or high edge magnitudes) means that the edges are sharp.
- Another blur measure (B_{Cr}) has been introduced by Crete et al. [26]. In this case, the intensity variations between neighboring pixels of the original image are compared with the intensity variations of a low-pass filtered version of the image. A high variation indicates that the original image is sharp, whereas a low variation means that the original image is already blurred. Low-pass filtering is done (as proposed [26]) using an averaging filter with a size of 3×3 pixels.
- Finally, we will compare the two elaborated blur measurement techniques with the simple contrast property B_{Co} which is computed by summing up the squared differences of neighboring gray values [13] (based on horizontal and vertical neighbors). This at first sight inappropriate method is used in order to get more insight and to investigate the importance of a good blur measurement in our approach.

We investigate the behavior with these three different measures which are not built for our problem definition. Firstly, blur measures are usually not built to measure blur in textured images but rather in natural scenes. Furthermore, these metrics are constructed to measure the perceptual image degeneration and not the Gaussian σ.

As it is not clear, which properties of a blur measure are important in case of our scenario, in a first step, we investigate them with respect to two prediction rates. The first one (called "intra-class prediction") measures the ability to decide which of two textures of the same class is stronger blurred with respect to the Gaussian σ. The second one (called "inter-class prediction") measures the ability to decide which of two textures of different classes is stronger blurred, which is supposed to be the more difficult task. Finally, we compare these two rates with the final classification performances in order to detect correlations.

3 Experiments

3.1 Setup

The experiments are based on the Kylberg texture database [27], consisting of 28 classes with 160 unique texture patches per class, captured at a single scale. Each unique database (four of them are available) contains 40 patches per class (i.e., the total number of images per set is 1120). Blurred images for the main experiments (Figs. 3, 4, and 5) are achieved by simulation, using a Gaussian kernel with varying variances σ. We generate images on nine different Gaussian blur levels, leading from (theoretically) $\sigma = 0$, indicating the original image, to $\sigma = 4$. For additional experiments (Figs. 6 and 7), images are filtered with averaging as well as median filters with sizes between 1×1 (unfiltered) and 9×9 pixels. For training, one Kylberg database ("A") is used without any simulated blur. For evaluation, we use a separate database ("B") with simulated blur of variable strengths (and types). For final classification, we utilize the linear support vector classifier [28]. In the experiments, the behavior with ten different blur measure thresholds (Θ) are investigated. They are defined by the first to the tenth ten-quantile of the blur measures in the original evaluation set (e.g., in the case of the first threshold, 10 % of the images are less and 90 % are more blurry).

For blur equalization E (Eq. (1)), we use a Gaussian kernel with $\sigma = 1$ and a size of 3×3 pixels which turned out to be appropriate. Using a kernel with a larger

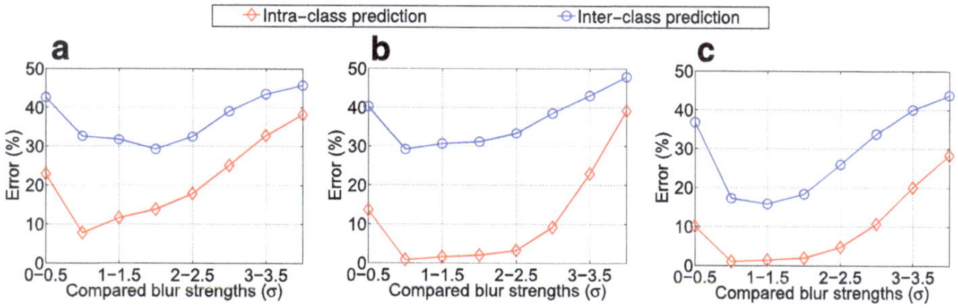

Fig. 3 Blur measurement performances achieved with the three blur metrics

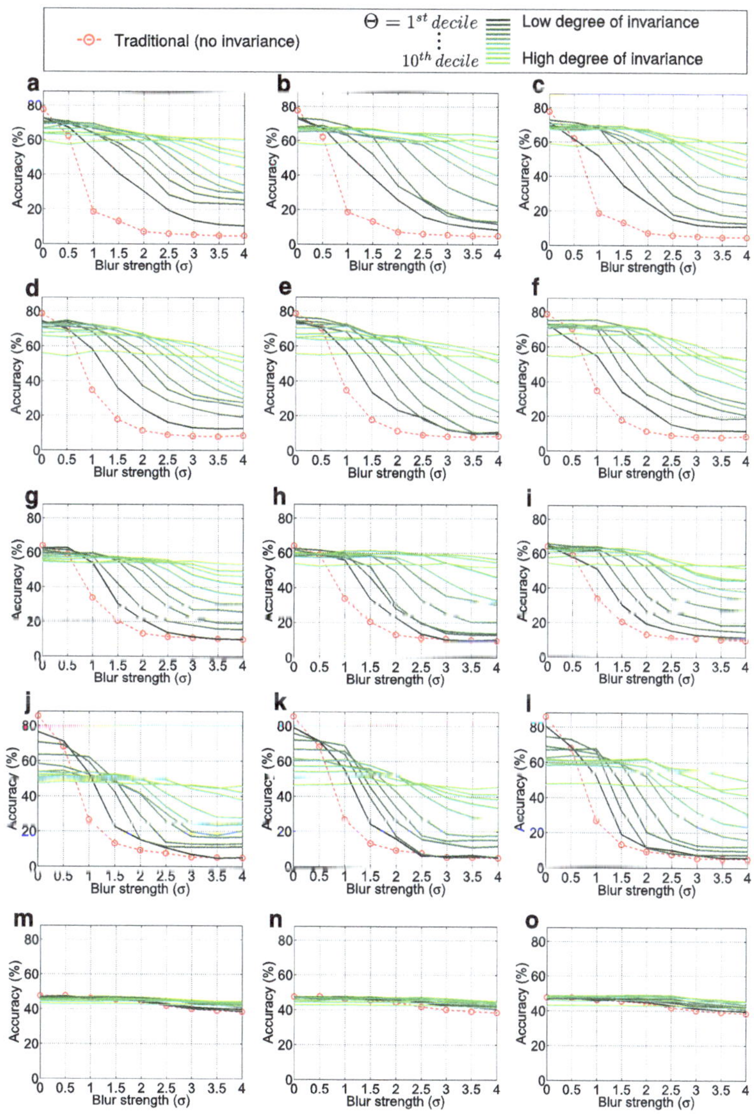

Fig. 4 Classification accuracies achieved with different configurations

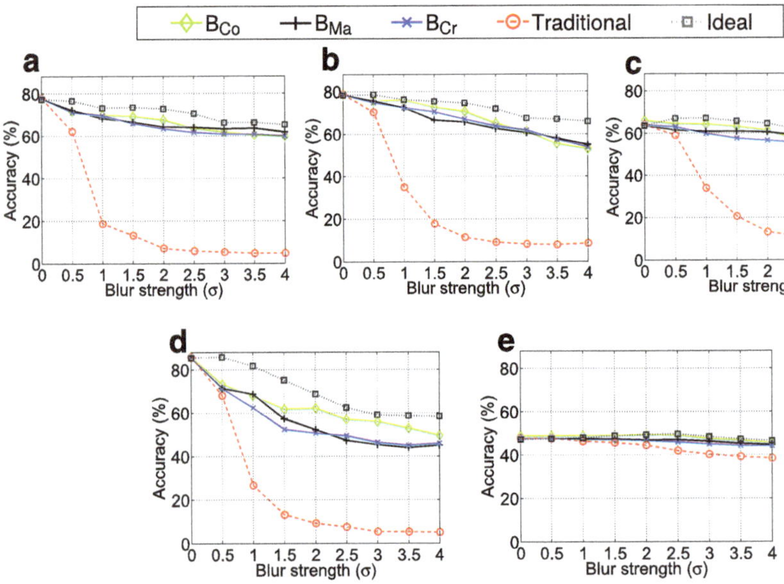

Fig. 5 Classification accuracies if optimizing the threshold Θ, separately for the respective blur strength. The lines correspond to the top most points in the respective subplots in Fig. 4

variance, execution time can be reduced, but the accuracy might decrease slightly. A too small variance would lead to unnecessarily high computational expenses.

We do not apply any preprocessing before blur estimation. Experimentation (applying noise removal methods) did not lead to improved outcomes.

For feature extraction, five different well known techniques are investigated:

- Local binary patterns [16] (MRLBP):
  describes a texture by utilizing the joint distribution of pixel intensity differences represented by binary patterns. We deploy a multi-resolution version of the uniform patterns (capturing only patterns with at most 2 bitwise transitions) using the standard eight-neighborhood with a radius of one and two pixels. Multi-resolution in this case means that the

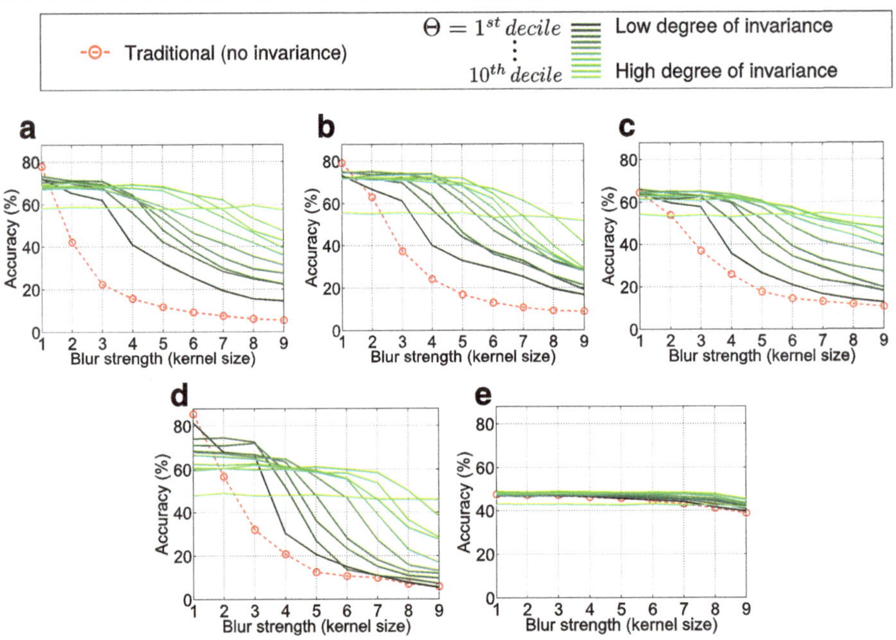

Fig. 6 Classification accuracies with B_{Co}, different thresholds (Θ) and different averaging filters (*horizontal axis*) reaching from 1 × 1 to 9 × 9 pixels

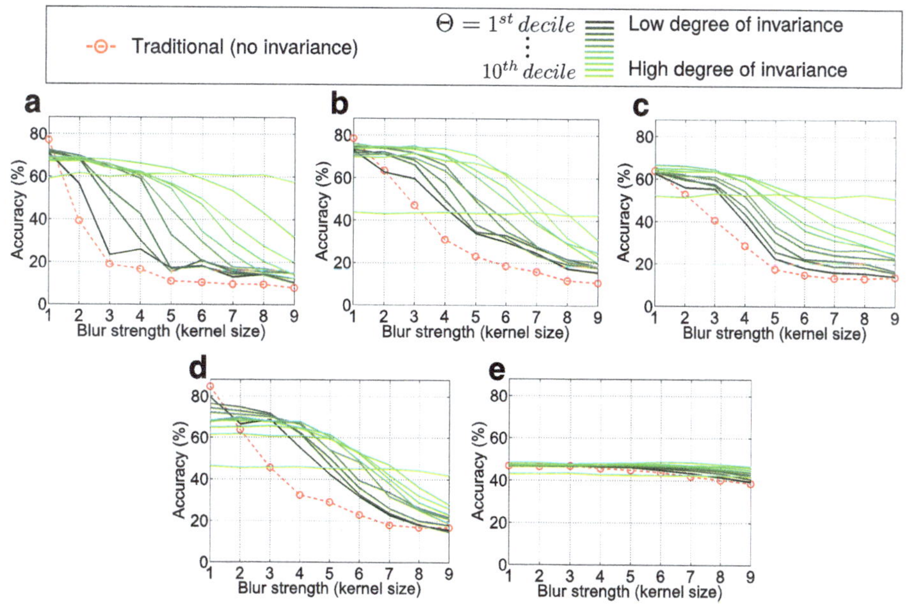

Fig. 7 Classification accuracies with B_{Co}, different thresholds (Θ) and different median filters (*horizontal axis*) reaching from 1 × 1 to 9 × 9 pixels

feature vector for LBP with a radius of one and LBP with a radius of two pixels are concatenated. Due to its high distinctiveness, this feature is commonly used, although it is highly sensitive to blur [29].

- Extended local binary patterns (ELBP) [30]:
ELBP is an edge-based derivative of local binary patterns. This descriptor is utilized with eight neighbors and a radius of one pixel. This feature is known to be similarly distinctive compared to LBP but slightly more robust to blur [29].
- Multi-fractal spectrum [7] (FRA):
The local fractal dimension is computed for each pixel using three different types of measures for computing the local density. The feature vector is built by concatenation of these fractal dimensions. This feature is investigated because of its high discriminative power.
- Edge co-occurrence matrix [9] (ECM):
After applying eight differently orientated directional filters, the orientation is determined for each pixel, followed by masking out pixels with a gradient magnitude below some threshold t. Finally, the ECM is achieved by computing the gray-level co-occurrence matrix of these data and a specified displacement v. For the experiments, t is set to 25 % of the maximum response and the displacement vector $v = (1, 1)$ is used. This feature is investigated as it could be, in opposite to the others, interpreted as a shape feature.
- Local phase quantization (LPQ) [5]:
LPQ is based on the short-term Fourier transform, computed over a square local neighborhood. As it extracts (low frequency) phase information, this feature is declared to be robust to image blur. We choose a local neighborhood of 15 × 15 pixels, which turned out to be appropriate for our problem definition.

3.2 Results

In a first experiment, focus is on the blur measures if being applied to the image textures. We would like to find out how effective Gaussian blur can be determined in an intra-class as well as in an inter-class sense (as defined in Section 2). The blur step between two compared textures in both cases is fixed to $\sigma = 0.5$. In Fig. 3, the computed prediction errors for these two cases are presented, separately for the compared blur strengths (shown on the horizontal axes) and for all blur measures.

We notice that especially in combination with strongly blurred images, the dedicated blur measures B_{Cr} and B_{Ma} are unable (50 % error is achieved with guessing) to effectively measure the applied blur in an inter class sense. Interestingly, the rather simple contrast measure B_{Co} seems to be even slightly more accurate in case of our problem definition. Intra-class prediction is (as supposed) easier and the measures B_{Ma} and B_{Co} are similarly competitive. B_{Cr} exhibits the highest error rates. Comparing images with stronger blur (e.g., images with $\sigma = 3$ and $\sigma = 3.5$) in general is more difficult, which is obvious as these images are more similar as far as perception is concerned. We suppose that an accurate intra-class prediction necessarily is required for our problem definition, because we assume that images of the same class should be similarly blurred in order to achieve a small feature distance. However, we do not know if an accurate inter-class prediction is necessary. Therefore, now we will focus on the final classification accuracies, achieved with the new approach, applying Eq. (1) to all images, and the different metrics.

In Fig. 4, the main results are presented. Training is done using original (non-blurred) images and evaluation is done with different Gaussian blur levels. For varying blur strengths in the evaluation set (horizontal axes), one subplot shows the traditional classification accuracies (dotted line) as well as the rates achieved with our approach. The solid lines indicate the rates obtained with our approach in combination with varying thresholds where a dark line indicates a small threshold and vice versa. First, we notice that all features, apart from LPQ, are highly sensitive to blur changes considering the traditional classification case without our pre-processing stage. Even with a small blur strength (e.g., $\sigma = 1$), the rates drop significantly. Considering the different solid lines, it can be seen that a certain degree of invariance definitely can be achieved with our new pre-processing technique. The invariance is (as supposed) more distinct with larger thresholds Θ. But (also as supposed) the high degree of invariance in case of large thresholds faces lower accuracies in case of optimal images without blur. Consequently, in the case of our pre-processing method, obviously, distinctiveness has to be sacrificed for the gained invariance. However, this is not a big surprise as we know that adding significant blur slightly decreases the distinctiveness of features [21]. It is much more interesting that the distinctiveness obviously does not strongly decrease in most cases. Especially, if considering MRLBP, ELBP, or ECM, the accuracies remain quite stable with increasing thresholds, especially if compared with the blur-invariant LPQ method. However, we notice that Θ which regulates the invariance should be carefully chosen depending on the expected blur strengths as mostly distinctiveness must be sacrificed for a higher degree of invariance.

In Fig. 5, for each feature extraction method, each blur metric and each blur strength, the accuracies achieved with the optimal Θ for the specific blur strengths are plotted and compared with the classification rates achieved with traditional classification. Thereby, the blur measures can be effectively compared with each other. Additionally, we show the ("ideal") accuracies that could be achieved if the blur measure would be able to exactly determine the Gaussian σ. These rates cannot be achieved in a real-world scenario (as the real Gaussian σ is unknown); however, they inform us about the effectiveness of the utilized blur metrics. Quite interestingly, we notice that B_{Co} works very effectively in case of moderate degradations (i.e., σ between 1 and 2) and is even able to outperform the dedicated blur metrics B_{Ma} and B_{Cr}. Especially in combination with the MFS feature, the contrast measure seems to be highly competitive. This is quite astonishing, as the B_{Co} measure definitely is not built for blur measurement. However, our analysis (Fig. 3) already showed that it might work for our problem definition considering intra-class and inter-class blur prediction performance. Obviously, the sophisticated perceptual blur metrics are not optimally suited. Considering stronger blur, apart from the MFS feature, B_{Ma} is the best or at least highly competitive. Considering the plots in Fig. 3, it could be deduced that a good intra-class blur prediction is decisive (as B_{Cr} in general seems to be less appropriate). Moreover, a good inter-class prediction seems to be also important if considering B_{Co} which works effectively (and corresponds to a good inter-class prediction rate) in the lower blur range. We suppose that a good inter-class prediction is important to limit the required threshold Θ. If some textures correspond to outlying blur estimations, the threshold must be set high in order to achieve a certain degree of invariance. Regarding the "ideal" lines, it can be seen that the best blur measures are mostly quite competitive. Especially in case of MRLBP, ECM, and LPQ, the gap is mostly lower than 5 %. However, considering the MFS subplot, we notice that the new approach could profit even more from a more appropriate blur metric.

Finally, we investigate the effect of our pre-processing approach in case of "non-Gaussian" blur. As our method (see Eq. (1)) is based on the assumption that a blurred image is obtained by convolution of an ideal image with a Gaussian kernel, it could be supposed that the prevalence of a different kind of blur corrupts the invariance. Therefore, in Figs. 6 and 7, the same experiment as in Fig. 4 with averaging and median filtered images (the filter size is given on the horizontal axis) and the probably most appropriate blur metric B_{Co} are presented. We notice that the impact of averaging filtered images (Fig. 6) instead of Gaussian-filtered images is quite small. With all features, similarly good compromises between a high invariance and a high distinctiveness can be obtained. The performances of traditional classification can be outperformed significantly in each case. Especially until a size of 4×4 pixels, the loss of accuracy compared to idealistic images (1×1) is in general negligible. Is has to be mentioned that the curves cannot be directly compared to the curves in Fig. 4, because the degradation is different and thereby the traditional classification rates (dashed lines) are also different.

Considering the classification rates in case of median filtered images (see Fig. 7), it can be seen that the significantly different kind of blur (compared to Gaussian blur) actually has an impact on the invariance. Especially in the range of large kernel sizes (above 5×5 pixels) combined with lower thresholds, accuracy increases are lower compared to Gaussian or averaging filtered images. This is supposed to be due to the different properties between

Gaussian blur and blur due to median filtering. However, if considering lower blur levels (up to a kernel size of 5×5 pixels), which are probably more relevant in practice, the level of invariance is still worthwhile. From these results, it can be concluded that our method is definitely not limited to image databases suffering from Gaussian blur.

4 Conclusions

We have proposed a generic approach to make texture features invariant to blur. By equalizing the blur level, a high degree of invariance can be achieved without losing too much distinctiveness, which is of high relevance for practical usage. Even the robustness of a dedicated blur-robust descriptor can be improved furthermore. With all of the three tested blur measures, competitive results can be obtained. However, depending on the respective feature extraction method and the blur strength, either the contrast-based B_{Co} mostly in case of lower blur strengths or the more elaborated technique B_{Ma} leads to the best compromise between accuracy and invariance. Furthermore, we showed that the performance could be improved again if more appropriate blur measures would be available. Finally, it has been proven that although it assumes Gaussian-blurred images, our method can be successfully applied even in case of other kinds of blur.

Competing interests

The authors declare that they have no competing interests.

Authors' contributions

MG and AU designed the study. MG developed the methodology, performed the analysis, and wrote the manuscript. AU obtained the funding and revised the manuscript. Both authors read and approved the final manuscript.

Acknowledgments

This work is partially funded by the Austrian Science Fund (FWF) under Project No. 24366.

Author details

[1]Institute of Imaging and Computer Vision, RWTH Aachen University, Kopernikusstr. 16, 52074 Aachen, Germany. [2]Department of Computer Sciences, University of Salzburg, Jakob-Haringer-Str. 2, 5020 Salzburg, Austria.

References

1. L Liu, P Fieguth, G Kuang, H Zha, in *Proceedings of the IEEE International Conference on Computer Vision (ICCV'11)*. Sorted random projections for robust texture classification (IEEE, Barcelona, Spain, 2011), pp. 391–398
2. L Sifre, S Mallat, in *Proceedings of the IEEE Conference on Computer Vision and Pattern Recognition (CVPR'13)*. Rotation, scaling and deformation invariant scattering for texture discrimination (IEEE, Columbus, USA, 2013), pp. 1233–1240
3. J Liu, T Zhang, Recognition of the blurred image by complex moment invariants. Pattern Recogn. Lett. **26**(8), 1128–1138 (2005)
4. KM Saipullah, DH Kim, A robust texture feature extraction using the localized angular phase. Multimed. Tools Appl. **59**(3), 717–747 (2012)
5. V Ojansivu, J Heikkilä, in *Proceedings of the International Conference on Image and Signal Processing (ICISP'08)*. Blur insensitive texture classification using local phase quantization, vol. 5099 (Springer LNCS, Boston, USA, 2008), pp. 236–243
6. T Ojala, M Pietikäinen, T Mäenpää, Multiresolution gray-scale and rotation invariant texture classification with local binary patterns. IEEE Trans. Pattern. Anal. Mach. Intell. **24**(7), 971–987 (2002)
7. Y Xu, H Ji, C Fermüller, Viewpoint invariant texture description using fractal analysis. Int. J. Comput. Vis. **83**(1), 85–100 (2009)
8. N Dalal, B Triggs, in *Proceedings of the IEEE Conference on Computer Vision and Pattern Recognition, (CVPR'05)*. Histograms of oriented gradients for human detection, vol. 1 (IEEE, San Diego, USA, 2005), pp. 886–893
9. R Rautkorpi, J Iivarinen, in *Proceedings of the International Conference on Image Analysis and Recognition (ICIAR'04)*. A novel shape feature for image classification and retrieval, vol. 3211 (Springer LNCS, Porto, Portugal, 2004), pp. 753–760
10. H Jegou, M Douze, C Schmid, P Perez, in *Proceedings of the IEEE Conference on Computer Vision and Pattern Recognition (CVPR'10)*. Aggregating local descriptors into a compact image representation (IEEE, San Francisco, USA, 2010), pp. 3304–3311
11. F Perronnin, Y Liu, J Sanchez, H Poirier, in *Proceedings of the IEEE Conference on Computer Vision and Pattern Recognition (CVPR'10)*. Large-scale image retrieval with compressed Fisher vectors (IEEE, San Francisco, USA, 2010), pp. 3384–3391
12. J Sánchez, F Perronnin, T Mensink, JJ Verbeek, Image classification with the Fisher vector: Theory and practice. Int. J. Comput. Vis. (IJCV). **105**(3), 222–245 (2013)
13. RM Haralick, I Dinstein, K Shanmugam, Textural features for image classification. IEEE Trans. Syst. Man Cybern. **3**, 610–621 (1973)
14. M Varma, A Zisserman, A statistical approach to material classification using image patch exemplars. IEEE Trans. Pattern. Anal. Mach. Intell. (TPAMI). **31**(11), 2032–2047 (2009)
15. M Cimpoi, S Maji, I Kokkinos, S Mohamed, A Vedaldi, in *Proceedings of the IEEE Conference on Computer Vision and Pattern Recognition (CVPR'14)*. Describing textures in the wild (IEEE, Columbus, USA, 2014), pp. 3606–3613
16. T Ojala, M Pietikäinen, D Harwood, A comparative study of texture measures with classification based on feature distributions. Pattern Recogn. **29**(1), 51–59 (1996)
17. A Vécsei, T Fuhrmann, M Liedlgruber, L Brunauer, H Payer, A Uhl, Automated classification of duodenal imagery in celiac disease using evolved fourier feature vectors. Comput. Methods Prog. Biomed. **95**, 68–78 (2009)
18. EJ Ciaccio, CA Tennyson, SK Lewis, S Krishnareddy, G Bhagat, P Green, Distinguishing patients with celiac disease by quantitative analysis of videocapsule endoscopy images. Comput. Methods Programs Biomed. **100**(1), 39–48 (2010)
19. DJC Barbosa, J Ramos, CS Lima, in *Proceedings of the 30th Annual International Conference of the IEEE Engineering in Medicine and Biology Society (EMBS'08)*. Detection of small bowel tumors in capsule endoscopy frames using texture analysis based on the discrete wavelet transform (IEEE, Vancouver, Canada, 2008), pp. 3012–3015
20. M Häfner, M Liedlgruber, A Uhl, A Vécsei, F Wrba, Delaunay triangulation-based pit density estimation for the classification of polyps in high-magnification chromo-colonoscopy. Comput. Methods Prog. Biomed. **107**(3), 565–581 (2012)
21. M Gadermayr, A Uhl, in *Proceedings of the IEEE International Conference on Image Processing 2014 (ICIP'14)*. Degradation adaptive texture classification (IEEE, Paris, France, 2014)
22. M Gadermayr, S Hegenbart, A Uhl, in *Proceedings of 22nd IEEE International Conference on Pattern Recognition (ICPR'14)*. Scale-adaptive texture classification (IEEE, Stockholm, Sweden, 2014)
23. S Hegenbart, A Uhl, A Vécsei, G Wimmer, Scale invariant texture descriptors for classifying celiac disease. Med. Image Anal. **17**(4), 458–474 (2013)
24. J Zhang, T Tan, Brief review of invariant texture analysis methods. Pattern Recogn. **35**(3), 735–747 (2002)
25. P Marziliano, F Dufaux, S Winkler, T Ebrahimi, G Sa, in *Proceedings of the IEEE International Conference on Image Processing (ICIP'02)*. A no-reference perceptual blur metric (IEEE, Rochester, USA, 2002), pp. 57–60
26. F Crete, T Dolmiere, P Ladret, M Nicolas, in *Proceedings of SPIE, Electronic Imaging Symposium Conf Human Vision and Electronic Imaging*. The blur effect: perception and estimation with a new no-reference perceptual blur metric, vol. 6492 (SPIE, San Jose, USA, 2007), pp. 64920–6492011

27. G Kylberg, The kylberg texture dataset v. 1.0. External report (Blue series) 35, Centre for Image Analysis, Swedish University of Agricultural Sciences and Uppsala University, Uppsala, Sweden (2011). http://www.cb.uu.se/~gustaf/texture/. Accessed 01 June 2014
28. RE Fan, KW Chang, CJ Hsieh, XR Wang, CJ Lin, LIBLINEAR: a library for large linear classification. J. Mach. Learn. Res. **9**, 1871–1874 (2008)
29. S Hegenbart, A Uhl, A Vécsei, in *Proceedings of the 7th International Symposium on Image and Signal Processing and Analysis (ISPA'11)*. Impact of endoscopic image degradations on LBP based features using one-class SVM for classification of celiac disease (IEEE, Dubrovnik, Croatia, 2011), pp. 715–720
30. S Liao, X Zhu, Z Lei, L Zhang, S Li, in *Advances in Biometrics*. Learning multi-scale block local binary patterns for face recognition (Springer, Berlin, Heidelberg, 2007), pp. 828–837

15

Semi-supervised feature selection for audio classification based on constraint compensated Laplacian score

Xu-Kui Yang[1], Liang He[2], Dan Qu[1*], Wei-Qiang Zhang[2] and Michael T. Johnson[3]

Abstract

Audio classification, classifying audio segments into broad categories such as speech, non-speech, and silence, is an important front-end problem in speech signal processing. Dozens of features have been proposed for audio classification. Unfortunately, these features are not directly complementary and combining them does not improve classification performance. Feature selection provides an effective mechanism for choosing the most relevant and least redundant features for classification. In this paper, we present a semi-supervised feature selection algorithm named Constraint Compensated Laplacian score (CCLS), which takes advantage of the local geometrical structure of unlabeled data as well as constraint information from labeled data. We apply this method to the audio classification task and compare it with other known feature selection methods. Experimental results demonstrate that CCLS gives substantial improvement.

Keywords: Audio classification, Semi-supervised feature selection, Locality preserving, Constraint information

1 Introduction

Initial classification of audio segments into broad categories such as speech, non-speech, and silence provides useful information for audio content understanding and analysis [1], and it has been used in a variety of commercial, forensic, and military applications [2]. Most audio classification systems involve two processing stages: feature extraction and classification. There is a considerable amount of literature on audio classification regarding different features [3] or classification methods [4]. Many features [5] have been developed to improve classification accuracy. Nevertheless, using all of these features in a classification system may not enhance but instead degrade the performance. The underlying reason is that there can be irrelevant, redundant, and even contradictory information among these features. Choosing the most relevant features to improve the classification accuracy is a challenging problem [6].

Feature selection methods can be divided into three categories: supervised, unsupervised, and semi-supervised. Supervised approaches require a large quantity of labeled data, and they are apt to ignore the internal structure of data by focusing too much on label information. Unsupervised feature selection fails to extract more discriminative features which may yield worse performance. Semi-supervised feature selection focuses on maximizing data effectiveness by using labeled and unlabeled data together [7]. In this case, the amount of unlabeled data is much larger than that of labeled data. Semi-supervised algorithms have attracted attention for their ability to model the intrinsic structure of data.

Approaches to feature selection are generally categorized into filter, wrapper, and embedded techniques. Filter methods use scores or confidences to evaluate the importance of features in the learning tasks and include algorithms such as Laplacian score (LS) [8], constraint score (CS) [9], and constrained Laplacian score (CLS) [10, 11]. Wrapper approaches evaluate different subsets of features and select the one with the best performance. The embedded model techniques search for the most relevant and effective features for models. The most common embedded methods are regularization-based [12], including LASSO, elastic net, or ridge regression. Since filter approaches can be applied to a broad

* Correspondence: qudanqudan@sina.com
[1]Zhengzhou Information Science and Technology Institute, Zhengzhou, China
Full list of author information is available at the end of the article

range of classification and learning methods, they have been widely used for their better generalization properties.

For audio classification, it is computationally challenging to evaluate the features' properties by testing them individually [13] or analyzing their characteristics individually [14]. Although some recent work on feature selection algorithms has focused on improving these weaknesses [15, 16], an efficient and effective method has yet to be developed. This is primarily because most approaches rely on labeled data. It is hard to get sufficient labeled data for the evaluation of features' scores in practical applications. Thus, semi-supervised feature selection can play an important role.

In this paper, we propose a novel semi-supervised filter method called constraint compensated Laplacian score (CCLS), which is similar to Laplacian score. The difference is that CCLS uses constraint information generated from a small amount of labeled data to compensate for the construction of local structure and global structure instead of unsupervised construction. Hence, CCLS has better locality discrimination ability than LS.

The outline of this paper is as follows: the background and motivation of this paper are given in Section 2. Section 3 enumerates several main methods used in feature selection. The CCLS method is presented in Section 4. Section 5 depicts the experimental setup and analyzes the results. Finally, conclusions are given in Section 6.

2 Semi-supervised feature selection for audio classification

Audio segmentation is the task of splitting an audio stream into segments of homogeneous content. Given a predefined set of audio classes, the process of segmentation involves joint boundary detection and classification, resulting in identification of segment regions as well as classification of those regions. Assuming that an audio signal has been divided into a sequence of audio segments using fixed window segmentation, our works focus on categorizing these audio segments into a set of predefined audio classes. Although there may be some differences between the traditional definition of audio classification and that in our work, the essential issues are the same.

Figure 1 illustrates the process of audio classification. In an audio classification system, every audio signal is first divided into mid-length segments which range in duration from 0.5 to 10 s. After this, the selected features are extracted for each segment using short-term overlapping frames. The sequence of short-term features in each segment is used to compute feature statistics, which are used as inputs to the classifier. In the final classification stage, the classifier determines a segment-by-segment decision.

In audio analysis and classification, there are dozens of features which can be used. A number of novel feature extraction methods have been proposed in recent years [17–19]. In this paper, some classical and widely used acoustic features are selected for feature selection sources. Widely used time-domain features [5] include short-term energy [20], zero-crossing rate [21], and entropy of energy [22]. Common frequency-domain features include spectral centroid, spectral spread, spectral entropy [23], spectral flux, spectral roll-off, Mel-frequency cepstral coefficients (MFCCs), and chroma vector [24].

There is a lot of complementary information among these features which can improve classification accuracy when used together; however, there is also a lot of redundant and even contradictory information which can degrade performance. It is hard to judge which combination of features is most likely to have a positive effect on classification. Furthermore, it is computationally infeasible to select the optimal feature subset by

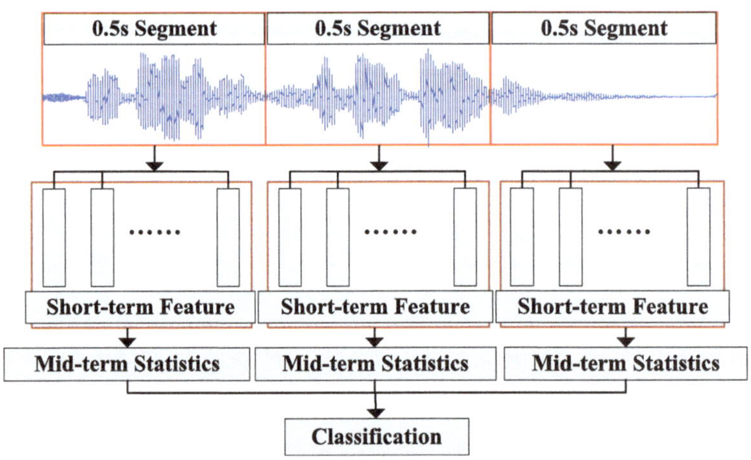

Fig. 1 The audio classification framework

exhaustive search. Thus, it is important to implement an effective feature selection method for this task.

Most supervised feature selection methods are dependent on labeled data. Unfortunately, it is difficult to obtain sufficient labeled data for audio classification, while unlabeled data is readily available. Semi-supervised feature selection methods can take good use of both labeled and unlabeled data; thus, this approach is more practical.

3 Related work

Let the training dataset with N instances be $X = \{\mathbf{x}_i \in \mathbb{R}^M\}$, where $i = 1, 2, \cdots, N$. Let $\mathbf{f}_1, \mathbf{f}_2, \cdots, \mathbf{f}_M$ denote the corresponding feature vectors, where f_{ri} denotes the rth feature of $\mathbf{x}_i$, where $r = 1, 2, \cdots, M$. In semi-supervised learning, the training dataset X can be divided into two subsets. The first contains data $X^l = \{\mathbf{x}_1, \mathbf{x}_2, \cdots, \mathbf{x}_L\}$ with labels $Y^l = \{y_1, y_2, \cdots, y_L\}$, where $y_i = 1, 2, \cdots, C$ and C is the number of classes. The second set has only the unlabeled data $X^u = \{\mathbf{x}_{L+1}, \mathbf{x}_{L+2}, \cdots, \mathbf{x}_N\}$.

3.1 Laplacian score

Laplacian score is a recently proposed unsupervised feature selection method [8]. The basic idea is to evaluate features according to their locality preserving ability. If two data points are close to each other, they belong to the same class with high probability, so local structure is more important than global structure. The Laplacian score of the rth feature is a measure of local compactness computed as follows:

$$L_r = \frac{\sum_{i,j}\left(f_{ri} - f_{rj}\right)^2 S_{ij}}{\sum_i (f_{ri} - u_r)^2 D_{ii}}, \tag{1}$$

where $u_r = \sum_{i=1}^{N} f_{ri}/N$ denotes the mean of the rth feature of the whole data set. $\mathbf{D}$ is a diagonal matrix with $D_{ii} = \Sigma_j S_{ij}$, and $\mathbf{S}$ denotes the similarity matrix whose elements are defined as follows:

$$S_{ij} = \begin{cases} w_{ij} & \text{if } \mathbf{x}_i \text{ and } \mathbf{x}_j \text{ are neighbors} \\ 0 & \text{otherwise.} \end{cases} \tag{2}$$

The similarity w_{ij} between $\mathbf{x}_i$ and $\mathbf{x}_j$ is defined by:

$$w_{ij} = e^{-\frac{\left\|\mathbf{x}_i - \mathbf{x}_j\right\|^2}{2\sigma^2}}, \tag{3}$$

where σ is a constant. $\mathbf{x}_i$ and $\mathbf{x}_j$ are considered to be neighbors if $\mathbf{x}_i$ is among the k nearest neighbors of $\mathbf{x}_j$ or $\mathbf{x}_j$ is among the k nearest neighbors of $\mathbf{x}_i$ in terms of Euclidean distance.

In the score function in Eq. 1, the numerator indicates the locality preserving the power of f_r, with smaller values indicating more local compactness in the feature space. The denominator is the weighted global variance of f_r. Thus, the criterion of the Laplacian score approach is to minimize the relative local compactness given by Eq. 1.

3.2 Constraint score

Constraint score is a supervised feature selection algorithm [9] which requires a relatively small amount of labeled data. For any pair of instances $(\mathbf{x}_i, \mathbf{x}_j)$ in the labeled data set X^l, there is a constraint assigned, either must-link (ML) or cannot-link (CL). The ML constraint is constructed if $\mathbf{x}_i$ and $\mathbf{x}_j$ have the same label, and the CL constraint is formed when $\mathbf{x}_i$ and $\mathbf{x}_j$ belong to different classes. Then, ML and CL constraints are grouped into two sets Ω_{ML} and Ω_{CL}, respectively.

In the constraint score approach, the pairwise constraints between all pairs of data points are generated using the data labels, and a score function is computed as the following:

$$C_r = \frac{\sum_{(\mathbf{x}_i, \mathbf{x}_j) \in \Omega_{\mathrm{ML}}} \left(f_{ri} - f_{rj}\right)^2}{\sum_{(\mathbf{x}_i, \mathbf{x}_j) \in \Omega_{\mathrm{CL}}} \left(f_{ri} \; f_{rj}\right)^2}. \tag{4}$$

This score represents a ratio of pairwise distances between same-class pairs and different-class pairs. Features are selected through minimizing this constraint score, with maximizes class separability.

3.3 Constrained Laplacian score

3.3.1 The score function

Constrained Laplacian score [10, 11] combines the above methods. The objective function of CLS is as follows:

$$\phi_r = \frac{\sum_{i,j}\left(f_{ri} - f_{rj}\right)^2 \mathcal{S}_{ij}}{\sum_{i,j}\left(f_{ri} - \alpha_{rj}^i\right)^2 \mathcal{D}_{ii}}, \tag{5}$$

where $\mathcal{S}_{ij} = S_{ij} + N_{ij}$, with S_{ij} computed as in Eq. 2 from both labeled and unlabeled data and N_{ij} is given as follows:

$$\mathcal{N}_{ij} = \begin{cases} -w_{ij} & \text{if } \mathbf{x}_i \text{ and } \mathbf{x}_j \text{ are neighbors and } (\mathbf{x}_i, \mathbf{x}_j) \in \Omega_{\mathrm{ML}} \\ w_{ij}^2 & \text{if } \left[\mathbf{x}_i \text{ and } \mathbf{x}_j \text{ are neighbors and } (\mathbf{x}_i, \mathbf{x}_j) \in \Omega_{\mathrm{CL}}\right] \text{ or } \left[\mathbf{x}_i \text{ and } \mathbf{x}_j \text{ are not neighbors and } (\mathbf{x}_i, \mathbf{x}_j) \in \Omega_{\mathrm{ML}}\right] \\ 0 & \text{otherwise.} \end{cases} \tag{6}$$

In addition, $D_{ii} = \Sigma_j Sij$, and α_{rj}^i is defined as follows:

$$\alpha_{rj}^i = \begin{cases} f_{rj} & \text{if } (\mathbf{x}_i, \mathbf{x}_j) \in \Omega_{\mathrm{CL}} \\ u_r & \text{if } i = j \text{ and } x_i \in \mathrm{X}^u \\ f_{ri} & \text{otherwise.} \end{cases} \tag{7}$$

CLS combines Laplacian score to represent the internal structure characteristics of the entire data space and constraint score to incorporate class separability of the labeled data. However, this algorithm may be not suitable for some scenarios, as discussed in the next section.

3.3.2 The shortcomings of CLS

CLS uses constraint information from labeled data to help construct the local structure, represented by a matrix with elements N_{ij}. The elements of the similarity matrix used for local structure construction are $S_{ij} = S_{ij} + N_{ij}$, where S_{ij} is defined as follows:

$$S_{ij} = \begin{cases} 0 & \text{if } \mathbf{x}_i \text{ and } \mathbf{x}_j \text{ are neighbors and } (\mathbf{x}_i, \mathbf{x}_j) \in \Omega_{\mathrm{ML}}' \\ w_{ij} + w_{ij}^2 & \text{if } [\mathbf{x}_i \text{ and } \mathbf{x}_j \text{ are neighbors and } (\mathbf{x}_i, \mathbf{x}_j) \in \Omega_{\mathrm{CL}}'] \text{ or} \\ & [\mathbf{x}_i \text{ and } \mathbf{x}_j \text{ are not neighbors and } (\mathbf{x}_i, \mathbf{x}_j) \in \Omega_{\mathrm{ML}}'] \\ w_{ij} & \text{if } [\mathbf{x}_i \text{ and } \mathbf{x}_j \text{ are neighbors}] \text{ and } [\mathbf{x}_i \in \mathrm{X}^u \text{ or } \mathbf{x}_j \in \mathrm{X}^u] \\ 0 & \text{otherwise.} \end{cases} \tag{8}$$

As shown in Eq. 8, when two samples with the same labels are close to each other, the similarity between them is set to 0. In other words, CLS does not use close neighbors to construct the local structure. However, these example pairs are of high importance, because the local structures of neighbors are the most reliable. The preservation of such structure is an important measure of feature quality.

Moreover, when the constraint information from labeled data has conflicts with the local structure, CLS adds an additional item w_{ij}^2 to the similarity. This is problematic for several reasons, including because S_{ij} may be greater than 1 (for example, $w_{ij} = 0.9$, $S_{ij} = 1.71$). This conflict may appear in two cases, when two samples are close to each other but have different labels, or when two samples are far from each other but have the same label. In the first case, we would like to decrease the similarity because of the label differences, but the CLS formula instead increases the similarity with the added term. In the second case, we would like to increase the similarity, as the CLS formula does, but only to a limited degree because w_{ij} is close to 0 and thus w_{ij}^2 is very close to 0.

4 Constraint compensated Laplacian score

4.1 Score function

The main advantage of the Laplacian score approach is its locality-preserving ability. However, due to the lack of prior supervised information, the accuracy of this method is not high. The constraint score approach selects features based on a small amount of labeled data but ignores unlabeled data. CLS combines these approaches, but it neglects some important factors in estimations of local structures and supervised information.

To address these problems, we propose a new feature selection algorithm called constraint compensated Laplacian score (CCLS). The score function to be minimized is defined as follows:

$$\eta_r = \frac{\sum_{i,j} \left(f_{ri} - f_{rj}\right)^2 (S_{ij} + \bar{N}_{ij})}{\Sigma_r + \Sigma_r^b - \Sigma_r^w}, \tag{9}$$

where

$$\bar{N}_{ij} = \begin{cases} 1 - w_{ij} & \mathbf{x}_i \text{ and } \mathbf{x}_j \text{ are neighbors and } (\mathbf{x}_i, \mathbf{x}_j) \in \Omega_{\mathrm{ML}}. \\ -\gamma w_{ij} & \mathbf{x}_i \text{ and } \mathbf{x}_j \text{ are neighbors and } (\mathbf{x}_i, \mathbf{x}_j) \in \Omega_{\mathrm{CL}}. \\ \lambda & \mathbf{x}_i \text{ and } \mathbf{x}_j \text{ are not neighbors and } (\mathbf{x}_i, \mathbf{x}_j) \in \Omega_{\mathrm{ML}}. \\ 0 & \text{otherwise,} \end{cases} \tag{10}$$

where γ and λ are required parameters to be determined. S_{ij} is computed as in Eq. 2 using both labeled and unlabeled data, and

$$\Sigma_r = \sum_i (f_{ri} - \mu_r)^2 D_{ii} \tag{11}$$

$$\Sigma_r^b = \sum_c n_c \left(u_r^{(c)} - \mu_r^l\right)^2 \tag{12}$$

$$\Sigma_r^w = \sum_c n_c \left(\sigma_r^{(c)}\right)^2, \tag{13}$$

where n_c is the number of instances of the cth class, $\mu_r^l = \sum_{i|\mathbf{x}_i \in \mathrm{X}^l} f_{ri}/L$ is the mean of the rth feature of the labeled dataset, $\mu_r^{(c)} = \sum_{i|y_i = c} f_{ri}/L_c$ and $\left(\sigma_r^{(c)}\right)^2$ denote the mean and variance of the rth feature of the cth class, and L_c is the number of instances which belong to the cth class in the labeled dataset X_l.

4.2 Benefits of the new approach

The proposed CCLS approach integrates the LS and CS techniques under a unified semi-supervised framework, with two additional improvements: more accurate estimation of local structure and variance.

Table 1 Statistics of the UCI data sets

Dataset	Size	*M*	*C*	*N*	*L*
Ionoshpere	351	34	2	176	20
Segment	2310	19	7	1155	350
Soybean	47	35	4	24	12
Vehicle	846	18	4	400	100

M number of potential features, *C* number of classes, *N* number of instances, *L* number of labeled instances

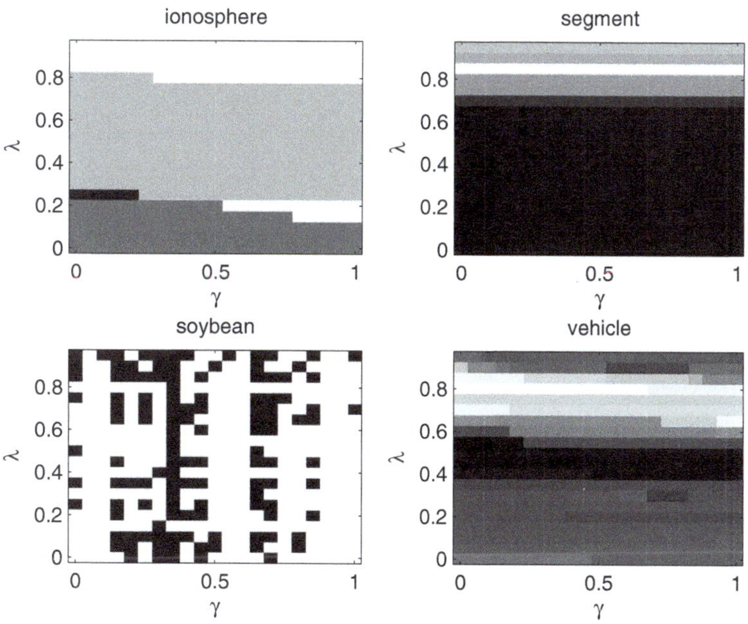

Fig. 2 Performance as a function of γ and λ. *Whiter shades* indicate better performance, *darker shades* indicate worse performance

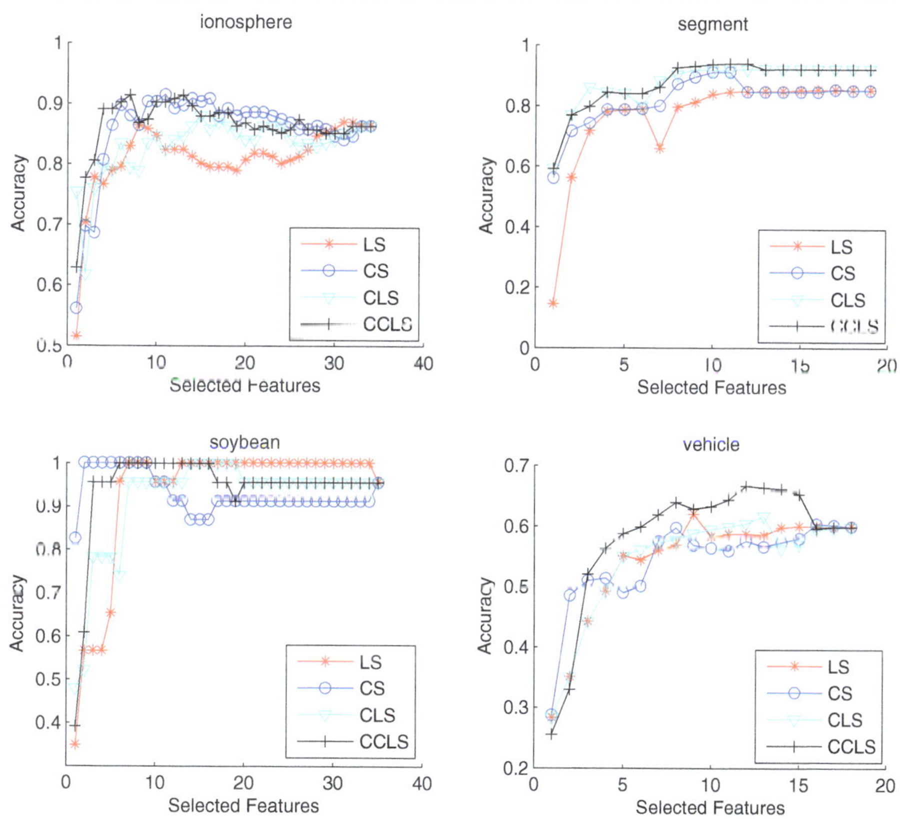

Fig. 3 Accuracy as a function of the number of selected features on four UCI data sets

4.2.1 The estimation of local structure

With respect to the calculation of within-class variance represented by the numerator of Eq. 9, the new CCLS method improves over CLS in the following aspects:

- When $\mathbf{x}_i$ and $\mathbf{x}_j$ are neighbors and also in the same labeled class, it is more certain that $\mathbf{x}_i$ is similar to $\mathbf{x}_j$. It is more intuitive to increase the similarity between them, as represented in Eq. 10, rather than set it to zero as in Eq. 6 of CLS.
- When $\mathbf{x}_i$ and $\mathbf{x}_j$ are neighbors and in two different labeled classes, any local structure between $\mathbf{x}_i$ and $\mathbf{x}_j$ may mislead feature selection. Thus, it is appropriate to decrease the similarity instead of increasing them, as now represented in the second case of Eq. 9.
- When $\mathbf{x}_i$ and $\mathbf{x}_j$ are not neighbors but are in the same labeled class, they can be still considered as neighbors. In such most cases, however, the value of w_{ij} is very close to 0 because the distance between the points is large, so rather than using this distance as a weight, the new approach uses a controllable constant λ.

4.2.2 The estimation of variance

The new CCLS approach improves the accuracy of the variance of estimation. In the CLS approach, the variance estimation ignores the inner-class covariance of labeled data which has good discriminative ability. Moreover, CLS directly sums the variances of unlabeled data with that of labeled pairs from different classes. Specifically, in CLS, the variance of the rth feature vector $\mathbf{f}_r$ is given as follows:

$$\sum_{i,j}\left(f_{ri}-\alpha^{i}_{rj}\right)^2\mathcal{D}_{ii} = \sum_{i|\mathbf{x}_i\in X^u}(f_{ri}-\mu_r)^2\mathcal{D}_{ii} + \sum_{i,j|(\mathbf{x}_i,\mathbf{x}_j)\in\Omega_{CL}}\left(f_{ri}-f_{rj}\right)^2\mathcal{D}_{ii}. \qquad (14)$$

In the proposed CCLS approach, the denominator of Eq. 9 shows that both inter-class covariance and inner-class covariance are used to estimate variance. This approach is motivated by the discrimination of these two types of covariance given by linear discriminative analysis

Table 2 Average accuracy of four different algorithms on UCI data sets

Algorithms	LS	CS	CLS	CCLS
Ionoshpere	80.87 ± 6.26	85.60 ± 7.22	83.55 ± 5.02	**86.35 ± 4.66**
Segment	76.26 ± 16.9	81.75 ± 8.10	75.31 ± 8.48	**82.29 ± 8.48**
Soybean	89.32 ± 15.7	93.04 ± 4.50	83.11 ± 19.2	**94.29 ± 11.5**
Vehicle	54.02 ± 9.27	54.07 ± 7.45	55.80 ± 8.82	**58.07 ± 11.2**

Best results of each experiment are set in bold

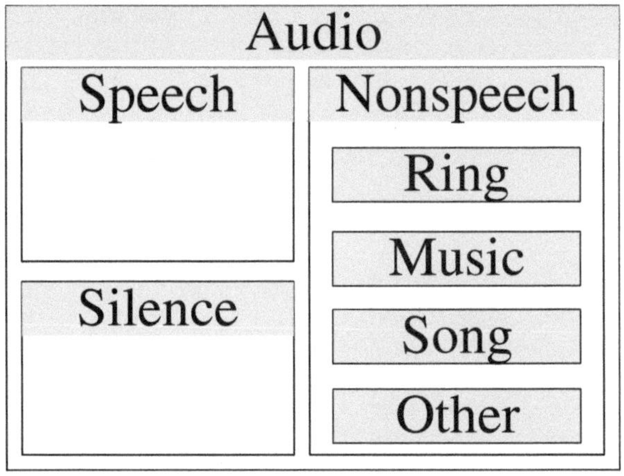

Fig. 4 The audio classes in telephone channel

[25]. Thus, a relevant feature should be correlated not only with larger variance of unlabeled data but also with larger inter-class covariance and smaller inner-class covariance.

4.3 Comparison of LS, CS, and CCLS

To illustrate the performance of LS, CS, CLS, and the proposed CCLS algorithm, we compare these four algorithms on several high-dimensional machine learning databases [26] including Ionoshpere, Image Segmentation, Soybean, and Vehicle datasets. The data set information and the sizes of whole training data set and labeled data set are detailed in Table 1. A nearest neighbor (1-NN) classifier with Euclidean distance is employed for classification.

To determine the parameters γ and λ, which govern how the rules affect feature selection performance, we experimentally vary parameter pairs from 0 to 1 with 0.05 intervals. The results are shown in Fig. 2, with lighter shades indicating better performance. Although the pattern is inconsistent, performance is generally

Table 3 Individual classification accuracy of different features

Feature	Dimensional	Accuracy		
		Mean	STD	Mean and STD
Zero-crossing rate	1	73.73	74.86	75.10
Short-term energy	1	45.81	46.03	69.41
Energy entropy	1	71.86	69.10	74.99
Spectral centroid	2	79.19	74.49	84.79
Spectral entropy	1	69.33	74.27	76.86
Spectral flux	1	79.21	69.09	77.86
Spectral roll-off	1	71.80	74.20	74.13
MFCCs	13	**84.26**	**86.44**	**87.66**
Harmonic	2	69.99	82.90	83.13
Chroma vector	12	83.49	83.87	83.73
All	35	81.07	85.97	86.04

Best results of each experiment are set in bold

Table 4 Averaged accuracy of different algorithms (400 labeled segments)

Alg.	Spec	ReliefF	LS	CS	CLS	CCLS
Ave.	85.26 ± 4.66	86.41 ± 2.79	85.32 ± 3.07	84.62 ± 2.92	83.40 ± 4.56	**88.46 ± 3.67**
Opt.	89.97	90.90	89.08	88.95	88.27	**91.14**
Num.	23	**19**	33	39	47	26

Best results of each experiment are set in bold

better when γ is high and λ is near 0.5. Thus, the parameters γ and λ are set to be 0.9 and 0.5, respectively, in all of our experiments followed.

The experimental results on UCI data sets are shown in Fig. 3 and Table 2. Figure 3 shows the plots for accuracy versus the number of selected features, and Table 2 compares the averaged accuracy across these cases. It can be seen that the performance of CS and CCLS is better than that of LS in all cases. This illustrates that constraint information from labeled data and local geometrical structure from unlabeled data are complementary, and using them in conjunction can be useful for feature selection.

5 Experiments and results

To further illustrate the effectiveness of CCLS, it is compared to several established feature selection methods. These include spectral feature selection (Spec) [27], ReliefF [28], Laplacian score, constraint score, and constrained Laplacian score.

5.1 Data and experimental setup

Experiments were performed using audio signals under telephone channel. Thus, each audio segment may contain speech, non-speech, or silence, with more detailed classes as shown in Fig. 4. "Speech" indicates direct dialogues between the calling and called users, when the call is connected, while "silence" implies the segment with comfort noise. "Non-speech" can be sub-classified into four types: ring, music, song, and other. "Ring" contains the single-tone, dual-tone, or multi-tone used for dialing or waiting warning, "music" and "song" are the waiting music before the call is connected or the environmental noise when the phone is in call. "Others" includes special sounds, such as laugh, barking, coughing, or other isolated sounds. Mixed segments, such as speech over music, are excluded from the dataset.

The database used here has been collected and manually labeled by Tsinghua University. It contains about 7 h of audio from 837 real telephone recordings. The speaker in each recording is different, as is the background music. The corpus consists of 3.4 h of speech data, 0.2 h of ring data, 0.1 h of music data, 0.1 h of song data, and 0.02 h of other data.

According to the label, an audio signal, which contains speech or non-speech, is divided into several 0.5-s segments. For each segment, all features mentioned in Section 2 are extracted based on the short-term analysis,

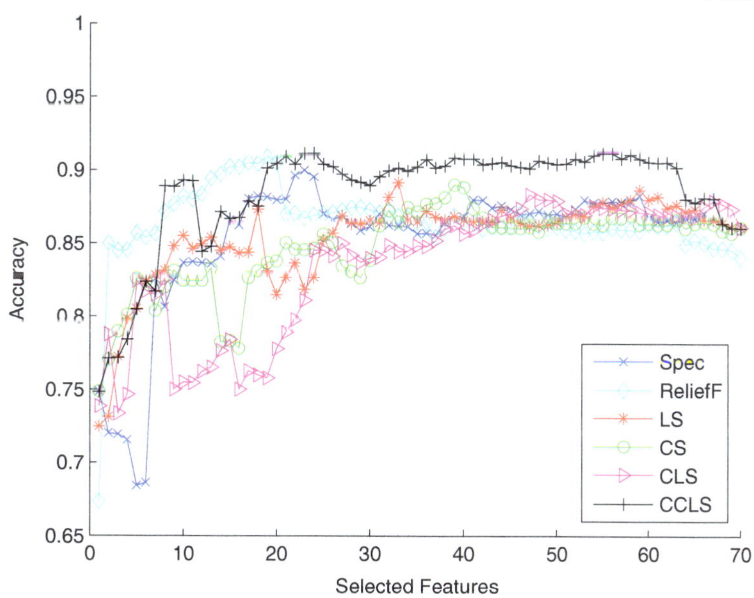

Fig. 5 Accuracy as a function of the number of selected features

and the dimension of short-term feature is 35. The frame length and frame step size are 32 and 10 ms, respectively. Then, the two mid-term statistics, mean value, and standard deviation are drawn per feature, so that the dimension of mid-term statistics vector is 70.

For feature selection, we choose 2000 speech segments and 2000 non-speech segments, with only 400 randomly chosen labeled segments. The γ value is set to 0.9 and $\lambda = 0.5$. We compare CCLS with unsupervised Laplacian score, as well as supervised constraint score, constrained Laplacian score, Spec, and ReliefF. We use a development dataset containing 200 speech segments and 200 non-speech segments to choose the optimal feature subset. The test dataset includes 500 speech segments and 500 non-speech segments.

In all experiments, the k-nearest neighborhood (KNN) classifier with Euclidean distance and $k = 5$ is utilized for classification after feature selection. To avoid the influence of the classifier, the training datasets of the classifier for all experiments are kept the same.

5.2 Experimental results

Ten types of short-term features extracted are listed in Table 3. Two statistics, mean and standard deviation are used as the mid-term representation of the audio segments. Table 3 shows the classification accuracy of different features for audio classification. The top three best features are MFCCs, chroma vector, and spectral centroid, and the worst feature is short-term energy. Moreover, using all of these features does not improve but rather decreases the accuracy, as seen by comparing MFCC accuracy to that using all features, indicating that there is redundant and even contradictory information among the features. Thus, it is valuable to use feature selection as a preprocessing module.

Table 4 compares the averaged accuracy (Ave.), optimized accuracy (Opt.), and the optimized number (Num.) of features among all evaluated methods, and the value after the symbol "±" denotes the standard deviation. Results indicate that the performance is significantly improved by using the first d features selected from the ranking list of features generated by feature selection algorithms. This supports the hypothesis that there is redundant and even contradictory information among the original feature space and that a feature selection algorithm can remove irrelevant and redundant features effectively.

CCLS is superior to other evaluated methods not only in terms of averaged accuracy but also in terms of optimized accuracy. In contrast, CLS has the lowest averaged accuracy and optimized accuracy. This is because the estimations of local structure and variance are not accurate for CLS method as described in Section 4.2.

Figure 5 shows accuracy vs. the number of selected features. It can be seen that the performance of CCLS is significantly better than that of Spec, Laplacian score, constraint score, and constrained Laplacian score. This supports that combining supervised information with data structures to evaluate the relevance of features is useful in feature selection.

Table 5 Performance of supervised and semi-supervised methods with 200 labeled segments

Algorithms	Spec	ReliefF	CS	CCLS
Ave.	83.33 ± 3.94	85.53 ± 5.92	81.06 ± 3.34	**87.62 ± 3.50**
Opt.	87.20	89.24	87.68	**91.64**
Num.	57	**35**	55	40

Best results of each experiment are set in bold

To explore the influence of the numbers of labeled segments on the performance of the algorithm, different numbers of labeled data are used. The averaged accuracy, optimized accuracy, and the optimized number of features on the condition of 200 and 800 labeled segments are summarized in Tables 5 and 6, respectively. Comparing Table 4 with Tables 5 and 6, it is easy to conclude that the performance improves as the number of labeled data segments increases from 200 to 800. The CCLS is best in terms of averaged accuracy and optimized accuracy regardless of the number of labeled segments. The performance of ReliefF is always better than others in terms of optimized number of features.

Figure 6 shows the plots of accuracy vs. the number of selected features and the amount of labeled data. However, it should also be noticed that the performances of CCLS and ReliefF do not drop rapidly when decreasing the amount of labeled data to 200, while the CS and Spec algorithms are unable to select relevant features.

In all cases, there are many irrelevant features, almost two-thirds, which can be removed to achieve the best performance. This not only improves classification accuracy but also reduces the time complexity of classification.

Figure 7 shows the plots of average accuracy vs. the number of labeled data segments. The average accuracy increases with the addition of labeled data, asymptoting between 500 and 700 segments. The CCLS algorithm outperforms the other algorithms significantly.

After the optimal feature subset has been selected, classification is done on the test data set. The results are listed in Table 7. The optimal feature subset selected

Table 6 Performances of supervised and semi-supervised methods with 800 labeled segments

Algorithms	Spec	ReliefF	CS	CCLS
Ave.	86.76 ± 2.82	88.47 ± 4.69	87.24 ± 3.48	**89.72 ± 3.71**
Opt.	89.49	92.27	91.45	**92.71**
Num.	55	**18**	23	25

Best results of each experiment are set in bold

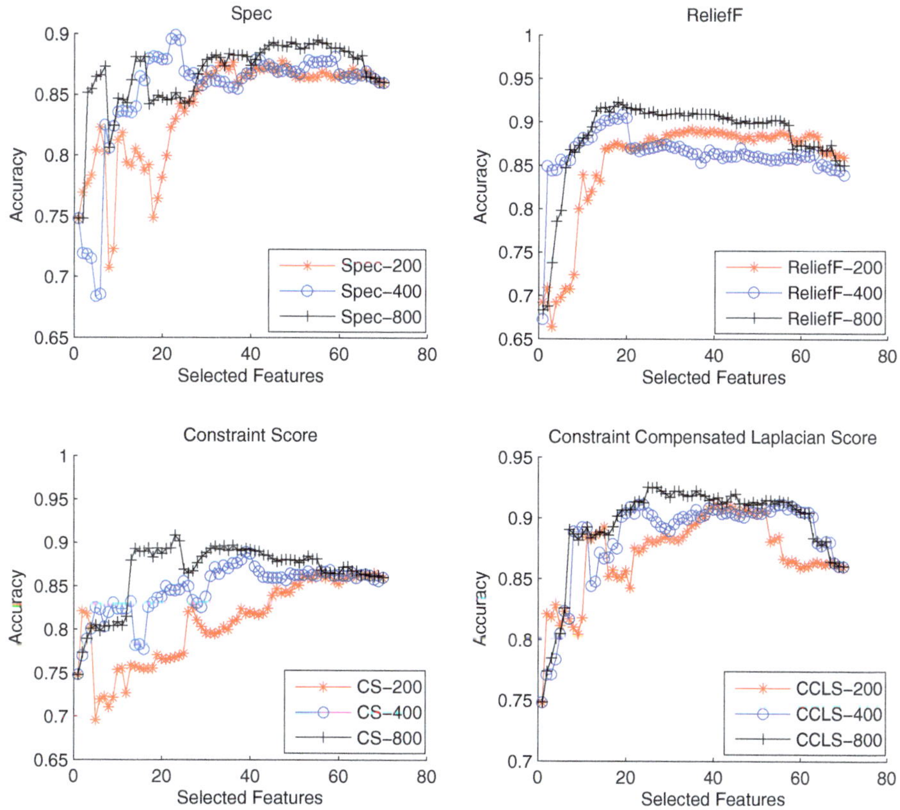

Fig. 6 Accuracy as a function of number of selected features and number of labeled data segments

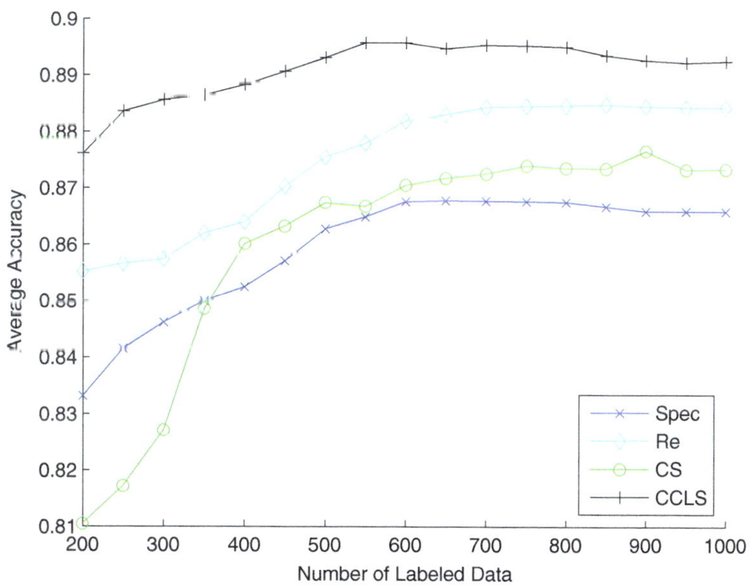

Fig. 7 Average accuracy as a function of the number of labeled data segments

Table 7 Accuracy of different algorithms on test dataset, using feature subsets tuned over the development dataset (400 labeled segments)

Algorithms	Spec	ReliefF	LS	CS	CLS	CCLS
Accuracy	88.87	90.24	89.77	89.02	87.35	**91.06**

Best results of each experiment are set in bold

from development dataset improves the performance on the test dataset. Though CCLS still outperforms other algorithms, the accuracy differences between algorithms are relatively small. However, the average accuracy of CCLS across all feature subsets is much high than the other algorithms, which indicates that the algorithm is more robust to feature subset selection than the comparative methods.

6 Conclusions

In this paper, we have presented a semi-supervised filter-based feature selection method. The new CCLS method integrates locality preservation across unlabeled data and label consistency within labeled data. Experimental results show that the proposed algorithm outperforms Spec, ReliefF, LS, and CS for audio classification.

As mentioned in Section 5.2, the performance of CCLS was not as good as that of ReliffF in terms of optimized number of features. This may indicate that there are some redundancy features in the optimum feature set selected by CCLS method. Several studies have addressed influences of such redundancy [11, 29, 30]. To improve the generalization quality of CCLS, there are primarily two areas for the future work: (1) Improvement of discrimination ability across audio classes, for example, more accurate estimation of local structure and variance, and (2) redundancy should be further removed from CCLS optimal feature sets.

Competing interests

The authors declare that they have no competing interests.

Acknowledgements

This work was supported in part by the National Natural Science Foundation of China (No. 61175017, No. 61403415, No. 61370034, and No. 61403224).

Author details

[1]Zhengzhou Information Science and Technology Institute, Zhengzhou, China. [2]Department of Electronic Engineering, Tsinghua University, Beijing, China. [3]Department of Electrical and Computer Engineering, Marquette University, Milwaukee, WI, USA.

References

1. S Zahid, F Hussain, M Rashid, MH Yousaf, HA Habib, Optimized audio classification and segmentation algorithm by using ensemble methods. Math. Probl. Eng. **2015**, 11 (2015). doi:10.1155/2015/209814. Article ID 209814
2. T Hirvonen, *Speech/Music Classification of Short Audio Segments* (Proc. 2014 IEEE International Symposium on Multimedia, Taichung, 2014). Dec. 10–12
3. Y Vaizman, B McFee, G Lanckriet, Codebook-based audio feature representation for music information retrieval. IEEE/ACM Transactions on Audio, Speech, and Language Processing **22**(10), 1483–1493 (2014)
4. P Mahana, G Singh, Comparative analysis of machine learning algorithms for audio signals classification. International Journal of Computer Science and Network Security **15**(6), 49–55 (2015)
5. Giannakopoulos, T, & Pikrakis, A (2014). *Introduction to Audio Analysis: A MATLAB Approach*. Elsevier Academic Press.
6. Z Zhao, H Liu, *Spectral Feature Selection for Data Mining (Data Mining and Knowledge Discovery Series)*(Chapman and Hall-CRC, Boca Raton, FL, USA, 2012)
7. H Yan, J Yang, Locality preserving score for joint feature weights learning. Neural Netw. **69**, 126–134 (2015)
8. X He, D Cai, P Niyogi, *Laplacian Score for Feature Selection* (in Proc NIPS, Vancouver, BC, Canada, 2005)
9. D Zhang, S Chen, Z Zhou, Constraint score: a new filter method for feature selection with pairwise constraints. Pattern Recogn. **41**(5), 1440–1451 (2008)
10. K Benabdeslem, M Hindawi, *Constrained Laplacian Score for Semi-Supervised Feature Selection* (in Proc ECML-PKDD, Athens, Greece, 2011), pp. 204–218
11. K Benabdeslem, M Hindawi, Efficient semi-supervised feature selection: constraint, relevance and redundancy. IEEE Trans. Knowl. Data Eng. **26**(5), 1131–1143 (2014)
12. B Efron, T Hastie, I Johnstone, R Tibshirani, Least angle regression. Ann. Stat. **25**, 407–449 (2004)
13. Mckinney, M, & Breebaart, J (2003). Features for Audio and Music Classification. *Proc. the International Symposium on Music Information Retrieval* (pp. 151-158). Baltimore, USA: Library of Congress.
14. Hao Jiang, Lie Lu, and HongJiang Zhang, A robust audio classification and segmentation method, Microsoft Research, No. MSR-TR-2001-79.
15. R Fiebrink, I Fujinaga, *Feature Selection Pitfalls and Music Classification* (in proc International Conference on Music Information Retrieval, Victoria, Canada, 2006)
16. Maria Markaki (2011). *Selection of Relevant Features for Audio Classification tasks*. Doctoral Dissertation, Department of Computer Science, University of Crete.
17. JT Geiger, B Schuller, G Rigoll, *Large-Scale Audio Feature Extraction and SVM for Acoustic Scene Classification* (in Proc Applications of Signal Processing to Audio and Acoustics, New Paltz, 2013), pp. 1–4
18. T Ramalingam, P Dhanalakshmi, Speech/music classification using wavelet based feature extraction techniques. J. Comput. Sci. **10**(1), 34–44 (2014)
19. S Zubair, F Yan, W Wang, Dictionary learning based sparse coefficients for audio classification with max and average pooling. Digital Signal Processing **23**(5), 960–970 (2013)
20. C Panagiotakis, G Tziritas, A speech/music discriminator based on RMS and zero-crossings. IEEE Trans. Multimedia **7**(1), 155–166 (2005)
21. E Scheirer, M Slaney, *Construction and Evaluation of a Robust Multifeature Speech/Music Discriminator* (in Proc. ICASSP, Munich, Germany, 1997)
22. Giannakopoulos, T, Pikrakis, A, Theodoridis, S (2008). Gunshot detection in audio streams from movies by means of dynamic programming and Bayesian networks. *Proc. of ICASSP* (pp. 21-24). Las Vegas, USA: IEEE
23. H Misra, S Ikbal, H Bourlard, and H HERMANSKY, "Spectral entropy based feature for robust ASR, in Proc. ICASSP, 2004.
24. MA Bartsch, GH Wakefield, Audio thumbnailing of popular music using chroma-based representations. IEEE Trans. Multimedia **7**(1), 96–104 (2005)
25. K Fikunaga, *Introduction to Statistical Pattern Recognition*, 2nd edn. (Academic, San Diego, 1990)
26. C Blake, E Keogh, CJ Merz, *UCI Repository of Machine Learning Databases* (Department of Information and Computer Science, University of California, Irvine, 1998). http://www.ics.uci.edu/~mlearn/MLRepository.html
27. Zhao, Z, & Liu, Hin (2007). Spectral Feature Selection for Supervised and Unsupervised Learning. *Proc. ICML* (pp. 1151-1157). New York, USA: ACM
28. M Robnik-Šikonja, I Kononenko, Theoretical and empirical analysis of ReliefF and RReliefF. Mach. Learn. **53**, 23–69 (2003)
29. C Ding, HC Peng, *Minimum Redundancy Feature Selection from Microarray Gene Expression Data* (in Proc. IEEE CSB, Stanford, CA, 2003), pp. 523–528
30. L Yu, H Liu, Efficient feature selection via analysis of relevance and redundancy. J. Mach. Learn. **5**, 1205–1224 (2004)

Permissions

All chapters in this book were first published by Springer; hereby published with permission under the Creative Commons Attribution License or equivalent. Every chapter published in this book has been scrutinized by our experts. Their significance has been extensively debated. The topics covered herein carry significant findings which will fuel the growth of the discipline. They may even be implemented as practical applications or may be referred to as a beginning point for another development.

The contributors of this book come from diverse backgrounds, making this book a truly international effort. This book will bring forth new frontiers with its revolutionizing research information and detailed analysis of the nascent developments around the world.

We would like to thank all the contributing authors for lending their expertise to make the book truly unique. They have played a crucial role in the development of this book. Without their invaluable contributions this book wouldn't have been possible. They have made vital efforts to compile up to date information on the varied aspects of this subject to make this book a valuable addition to the collection of many professionals and students.

This book was conceptualized with the vision of imparting up-to-date information and advanced data in this field. To ensure the same, a matchless editorial board was set up. Every individual on the board went through rigorous rounds of assessment to prove their worth. After which they invested a large part of their time researching and compiling the most relevant data for our readers.

The editorial board has been involved in producing this book since its inception. They have spent rigorous hours researching and exploring the diverse topics which have resulted in the successful publishing of this book. They have passed on their knowledge of decades through this book. To expedite this challenging task, the publisher supported the team at every step. A small team of assistant editors was also appointed to further simplify the editing procedure and attain best results for the readers.

Apart from the editorial board, the designing team has also invested a significant amount of their time in understanding the subject and creating the most relevant covers. They scrutinized every image to scout for the most suitable representation of the subject and create an appropriate cover for the book.

The publishing team has been an ardent support to the editorial, designing and production team. Their endless efforts to recruit the best for this project, has resulted in the accomplishment of this book. They are a veteran in the field of academics and their pool of knowledge is as vast as their experience in printing. Their expertise and guidance has proved useful at every step. Their uncompromising quality standards have made this book an exceptional effort. Their encouragement from time to time has been an inspiration for everyone.

The publisher and the editorial board hope that this book will prove to be a valuable piece of knowledge for researchers, students, practitioners and scholars across the globe.

List of Contributors

Yang Yu and Peng Han
School of Marine Science and Technology, Northwestern Polytechnical University, Xi'an, China

WenwuWang
Centre for Vision Speech and Signal Processing, University of Surrey, Guildford, UK

Li Gao, Ruimin Hu, Xiaochen Wang, Gang Li, Yuhong Yang and Weiping Tu
State Key Laboratory of Software Engineering, Wuhan University, Wuhan, China
National Engineering Research Center for Multimedia Software, the Key Laboratory of Multimedia and Network Communication Engineering, Computer School, Wuhan University, Wuhan, China
Collaborative Innovation Center of Geospatial Technology, Wuhan, China

Pedro Gil-Jiménez, Hilario Gómez-Moreno, Roberto López-Sastre and Saturnino Maldonado-Bascón
Signal Theory and Communication Department, Universidad de Alcala, Crta Madrid-Barcelona km 33.600, Alcala de Henares, Spain

Xin Liu and Chang-Chun Bao
Speech and Audio Signal Processing Lab, School of Electronic Information and Control Engineering, Beijing University of Technology, 100124 Beijing, China

Igor Djurović
Electrical Engineering Department, University of Montenegro, Podgorica, Montenegro

Giuseppe Baruffa and Fabrizio Frescura
Department of Engineering, University of Perugia, via G. Duranti 93, I-06125 Perugia, Italy

C. Gottschlich
Institute for Mathematical Stochastics, University of Goettingen, Goldschmidtstr. 7, 37077 Goettingen, Germany

D. H. Thai
Institute for Mathematical Stochastics, University of Goettingen, Goldschmidtstr. 7, 37077 Goettingen, Germany
Statistical and Applied Mathematical Science Institute (SAMSI), 19 T. W. Alexander Drive, 27709-4006 NC, USA

Suryanarayana M. Muddala, Roger Olsson and Mårten Sjöström
Department of Information and Communication Systems, Mid Sweden University, Holmgatan 10, 85170 Sundsvall, Sweden

S. Jafar Hosseini and Helder Araujo
Institute of Systems and Robotics, Department of Electrical and Computer Engineering, University of Coimbra, Coimbra, Portugal

Hyun-Seung Yang
Korea Advanced Institute of Science and Technology (KAIST), School of Computing, 291 Daehak-ro, Yuseong-gu, Daejeon 34141, Republic of Korea

Jin-Woo Koh
Korea Advanced Institute of Science and Technology (KAIST), School of Computing, 291 Daehak-ro, Yuseong-gu, Daejeon 34141, Republic of Korea
Agency for Defense Development (ADD), P.O. Box 35, Yuseong, Daejeon 34188, Republic of Korea

Emanuele Maiorana and Patrizio Campisi
Roma Tre University, Department of Engineering, Section of Applied Electronics, Via Vito Volterra 62, 00146 Rome, Italy

Gang Liu
College of Applied Mathematics, Chengdu University of Information Technology, Xuefu Road, 610225 Chengdu, China
Research Center for Image and Vision Computing/ School of Mathematical Sciences, University of Electronic Science and Technology of China, Xiyuan Ave, 611731 Chengdu,
China

Ting-Zhu Huang and Jun Liu
Research Center for Image and Vision Computing/ School of Mathematical Sciences, University of Electronic Science and Technology of China, Xiyuan Ave, 611731 Chengdu, China

Xiao-Guang Lv
School of Science, Huaihai Institute of Technology, Cangwu Road, 222005 Lianyungang, China

Tammam Tillo and Jimin Xiao
Department of Electrical and Electronic Engineering, Xi'an Jiaotong-Liverpool University, Ren Ai Road 111, 215123 Suzhou, People's Republic of China

Zhi Jin
Department of Electrical and Electronic Engineering, Xi'an Jiaotong-Liverpool University, Ren Ai Road 111, 215123 Suzhou, People's Republic of China
Department of Electrical Engineering and Electronics, University of Liverpool, Merseyside L69 3BX, Liverpool, UK

Yao Zhao
Institute of Information Science, Beijing Jiaotong University, No.3 Shangyuancun Haidian District, Beijing, People's Republic of China

Michael Gadermayr
Institute of Imaging and Computer Vision, RWTH Aachen University, Kopernikusstr. 16, 52074 Aachen, Germany

Andreas Uhl
Department of Computer Sciences, University of Salzburg, Jakob-Haringer-Str. 2, 5020 Salzburg, Austria

Xu-Kui Yang and Dan Qu
Zhengzhou Information Science and Technology Institute, Zhengzhou, China

Liang He and Wei-Qiang Zhang
Department of Electronic Engineering, Tsinghua University, Beijing, China

Michael T. Johnson
Department of Electrical and Computer Engineering, Marquette University, Milwaukee, WI, USA

Index